T0314028

Design for Maintainability

Wiley Series in Quality & Reliability Engineering

Dr. Andre Kleyner
Series Editor

The Wiley Series in Quality & Reliability Engineering aims to provide a solid educational foundation for both practitioners and researchers in the Q&R field and to expand the reader's knowledge base to include the latest developments in this field. The series will provide a lasting and positive contribution to the teaching and practice of engineering. The series coverage will contain, but is not exclusive to:

- Statistical methods
- Physics of failure
- Reliability modeling
- Functional safety
- Six-sigma methods
- Lead-free electronics
- Warranty analysis/management
- Risk and safety analysis

Wiley Series in Quality & Reliability Engineering

Reliability Culture: How Leaders can Create Organizations that Create Reliable Products
by Adam P. Bahret
March 2021

Design for Maintainability
edited by Louis J. Gullo and Jack Dixon
February 2021

Design for Excellence in Electronics Manufacturing
by Cheryl Tulkoff and Greg Caswell
January 2021

Lead-free Soldering Process Development and Reliability
edited by Jasbir Bath
September 2020

Automotive System Safety: Critical Considerations for Engineering and Effective Management
by Joseph D. Miller
February 2020

Prognostics and Health Management: A Practical Approach to Improving System Reliability Using Condition-Based Data
by Douglas Goodman, James P. Hofmeister, and Ferenc Szidarovszky
April 2019

Improving Product Reliability and Software Quality: Strategies, Tools, Process and Implementation, 2nd Edition
by Mark A. Levin, Ted T. Kalal, and Jonathan Rodin
April 2019

Practical Applications of Bayesian Reliability
by Yan Liu and Athula I. Abeyratne
April 2019

Dynamic System Reliability: Modeling and Analysis of Dynamic and Dependent Behaviors
by Liudong Xing, Gregory Levitin, and Chaonan Wang
March 2019

Reliability Engineering and Services
by Tongdan Jin
March 2019

Design for Safety
by Louis J. Gullo and Jack Dixon
February 2018

Thermodynamic Degradation Science: Physics of Failure, Accelerated Testing, Fatigue and Reliability
by Alec Feinberg
October 2016

Next Generation HALT and HASS: Robust Design of Electronics and Systems
by Kirk A. Gray and John J. Paschkewitz
May 2016

Reliability and Risk Models: Setting Reliability Requirements, 2nd Edition
by Michael Todinov November 2015

Applied Reliability Engineering and Risk Analysis: Probabilistic Models and Statistical Inference
by Ilia B. Frenkel, Alex Karagrigoriou, Anatoly Lisnianski, and Andre V. Kleyner
September 2013

Design for Reliability
edited by Dev G. Raheja and Louis J. Gullo
July 2012

Effective FMEAs: Achieving Safe, Reliable, and Economical Products and Processes Using Failure Modes and Effects Analysis
by Carl Carlson
April 2012

Failure Analysis: A Practical Guide for Manufacturers of Electronic Components and Systems
by Marius Bazu and Titu Bajenescu
April 2011

Reliability Technology: Principles and Practice of Failure Prevention in Electronic Systems
by Norman Pascoe
April 2011

Improving Product Reliability: Strategies and Implementation
by Mark A. Levin and Ted T. Kalal
March 2003

Test Engineering: A Concise Guide to Cost-Effective Design, Development and Manufacture
by Patrick O'Connor
April 2001

Integrated Circuit Failure Analysis: A Guide to Preparation Techniques
by Friedrich Beck
January 1998

Measurement and Calibration Requirements for Quality Assurance to ISO 9000
by Alan S. Morris
October 1997

Electronic Component Reliability: Fundamentals, Modelling, Evaluation, and Assurance
by Finn Jensen
1995

Design for Maintainability

Edited by

Louis J. Gullo
Northrop Grumman Corporation (NGC), Roy, Utah, USA

Jack Dixon
JAMAR International, Inc., Florida, USA

This edition first published 2021

Registered Offices
John Wiley & Sons, Inc., 111 River Street, Hoboken, NJ 07030, USA
John Wiley & Sons Ltd, The Atrium, Southern Gate, Chichester, West Sussex, PO19 8SQ, UK

Editorial Office
The Atrium, Southern Gate, Chichester, West Sussex, PO19 8SQ, UK

For details of our global editorial offices, customer services, and more information about Wiley products visit us at www.wiley.com.

Wiley also publishes its books in a variety of electronic formats and by print-on-demand. Some content that appears in standard print versions of this book may not be available in other formats.

Library of Congress Cataloging-in-Publication Data

Names: Gullo, Louis J., editor. | Dixon, Jack, 1948- editor.
Title: Design for maintainability / edited Louis J. Gullo, Raytheon Missile Systems, Arizona, USA, Jack Dixon, JAMAR International, Inc., Florida, USA.
Description: Hoboken : Wiley, 2021. | Series: Wiley series in quality & reliability engineering | Includes bibliographical references and index.
Identifiers: LCCN 2020030480 (print) | LCCN 2020030481 (ebook) | ISBN 9781119578512 (cloth) | ISBN 9781119578413 (adobe pdf) | ISBN 9781119578505 (epub)
Subjects: LCSH: Maintainability (Engineering)
Classification: LCC TS174 .D47 2021 (print) | LCC TS174 (ebook) | DDC 620/.0045–dc23
LC record available at https://lccn.loc.gov/2020030480
LC ebook record available at https://lccn.loc.gov/2020030481

Cover Design: Wiley
Cover Image: © videnko / Shutterstock, © aapsky / Shutterstock, © Dim Dimich / Shutterstock, © Jacques Tarnero / Shutterstock

Set in 9.5/12.5pt STIXTwoText by SPi Global, Chennai, India

10 9 8 7 6 5 4 3 2 1

Printed and bound by CPI Group (UK) Ltd, Croydon, CR0 4YY

From Louis J. Gullo

To my wife Diane for her love and support through the many days I spent researching, writing, and editing for this project. To my children, Louis J. Gullo, Jr, Stephanie M. Geer, Catherine F. Gullo, Christina T. Gullo, and Nicholas A. Gullo for their love and understanding as my work on the project took time away from them and my grandchildren.

From Jack Dixon

To my wife, Margo, who has supported me during the long process of creating this book.

Contents

Series Editor's Foreword by Dr Andre Kleyner

The Wiley Series in Quality & Reliability Engineering aims to provide a solid educational foundation for researchers and practitioners in the field of quality and reliability engineering, and to expand the knowledge base by including the latest developments in these disciplines.

The importance of quality and reliability to a system can hardly be disputed. Product failures in the field inevitably lead to losses in the form of repair costs, warranty claims, customer dissatisfaction, and in extreme cases, loss of life.

With each year engineering systems are becoming more and more complex with added functions and capabilities, and longer expected life-cycles. However, the reliability and availability requirements remain the same or grow even more stringent due to the rising expectations of the product end-user.

As the complexity of electronic systems grows, so does the scheduled maintenance time due to the complexity of repair. However, it doesn't have to be so. A system designed with a focus on future repairs will inevitably be easier and cheaper to maintain. Design for Maintainability (DfMn) should be part of a design process of any repairable system or product to ensure this focus occurs.

The book you are about to read is written by experts in the field of reliability and maintainability. It provides the methodologies, guidance, and suggestions on how to deliver more effective and efficient maintenance of systems and products.

The DfMn methods and tools described in this book, such as a data-driven maintenance process, testability, Condition-Based Maintenance (CBM), Prognostics and Health Management (PHM), machine learning, Reliability Centered Maintenance (RCM), maintainability predictions, and many others should become part of the design process of any repairable system, with the ultimate goal of optimizing system availability to the highest level possible.

However, despite its obvious importance, quality and reliability education is paradoxically lacking in today's engineering curriculum. Very few engineering schools offer degree programs or even a sufficient variety of courses in reliability or maintainability methods. The reason for this is difficult to explain. Perhaps engineering schools prefer educating their students on how the systems work, but not how they fail. However, in order to design a truly reliable system, it is important to understand the mechanisms of failure and the ways to prevent them.

The topics of maintainability and reliability data analysis, warranty management, CBM, RCM, PHM, and other related topics receive, as mentioned before, little coverage in today's engineering student curriculum, despite being essential to delivering failure-free systems. Therefore, the majority of maintainability, reliability, and asset management practitioners receive their professional training from colleagues at organizations they belong to, professional seminars, and professional publications. The book you are about to read is another step in closing this educational gap. It is intended to provide additional learning opportunities for a wide range of readers from graduate-level students to seasoned engineering professionals.

We are confident that this book, as well as this entire book series, will continue Wiley's tradition of excellence in technical publishing, and provide a lasting and positive contribution to the teaching and practice of reliability and quality engineering.

Preface

The idea for this book came from the fruits of our collaboration on our previous book, entitled *Design for Safety*, which was published by John Wiley and Sons in 2018. In the course of writing the *Design for Safety* book over a five-year period, we realized that Design for Maintainability was also an important aspect of engineering design for excellence. The editors' combined 80 years of professional hands-on and managerial-level experience, involved with reliability, maintainability, systems engineering, logistics, and system safety engineering, has provided the foundation upon which this book was built. Further enhancing the substance of the book's material is an additional 80 years of varied experience possessed by the contributing authors. While the focus of *Design for Maintainability* is obviously on maintainability, we have endeavored to capture maintainability's relationship to the larger world of systems engineering and emphasize the connections it has to numerous engineering disciplines.

Even though there are combined Design for Maintainability and Design for Safety experiences that impacted engineering design achievements on an equal level, the mental attitude that a person must have to respond to Design for Maintainability situations is in many aspects the opposite of a person's approach to Design for Safety. In Design for Maintainability (DfMn), the goal is to plan for the least amount of maintenance as possible. Design for Safety (DfSa) provides an understanding that we can never have enough safety measures and capabilities. When more system safety design practices, processes, and design features are developed, the safer the system or product is, which provides a major benefit to the customers or users of the system or product. In contrast, in designing for maintainability, the mantra should be "The less, the better." More maintainability design practices, processes, and design features that reduce maintenance makes the system or product better. Design for Maintainability teaches engineering practitioners that they need maintainability design practices to achieve the least amount of maintenance to realize cost and time objectives for the system or product operation and sustainment over its lifetime. The ultimate preferred situation is no maintenance. If the system or product does not fail, most maintenance is not needed. The challenge in maintainability is how to design a system or product where maintenance is minimized or avoided. The more automated streamlined processes and robust design features that reduce maintenance, the better the system or product is from the customer's perspective.

This book is intended for practicing engineers in multiple fields of discipline and for engineering managers of all types. It is written for those people who know how to perform maintenance on a system or product, but don't know how to improve the maintenance processes so the system or product will cost less and be more efficient in the future.

There are many real-world situations that are discussed in this book, where maintainability engineering processes, practices, and design features reduce or eliminate the need for maintenance. While the reliance on maintenance may diminish, the need for maintenance is prevalent for today's complex systems and products. As engineering practitioners build an understanding of designing for maintainability, they should, at the same time, implement design improvements that improve design reliability and reduce the potential for hazards by assessing the safety risks of a design.

Acknowledgments

We would like to express our sincere thanks and gratitude to all the authors and to the Wiley editors who contributed to this book and supported us to achieve the vision we had for this project. We would like to thank David Franck, James Kovacevic, and Anne Meixner for their major contributions in writing several chapters of this book as shown:

David Franck

Chapter 3: Maintainability Program Planning and Management
Chapter 4: Maintenance Concept
Chapter 12: Maintainability Testing and Demonstration
Appendix A: System Maintainability Design Verification Checklist

James Kovacevic

Chapter 6: Maintainability Analysis and Modeling
Chapter 7: Maintainability Predictions and Task Analysis
Chapter 9: Condition-Based Maintenance and Design for Reduced Staffing
Chapter 15: Design for Availability
Chapter 16: Design for Supportability
Chapter 17: Special Topics

Anne Meixner

Chapter 3: Maintainability Program Planning and Management
Chapter 13: Design for Test and Testability

Their expertise in their respective fields of engineering greatly contributed to these important chapters of the book. Working with all of these authors made our job as editors very enjoyable.

We would additionally like to acknowledge David Franck for his contributions to Chapter 5. Also, a further acknowledgement goes to James Kovacevic for his substantial contribution to the Maintainability Requirements section of Chapter 5.

We would like to acknowledge Fred Schenkelberg for his contributions in Chapter 1 related to the 8 Factors of Design for Maintainability, copyrighted by Accendo Reliability.

We would like to acknowledge J.P. "Hoffy" Hofmeister for his contributions to Chapter 9 using reference material from his book entitled *Prognostics and Health Management: Practical Approach to Improving System Reliability Using Condition-Based Data*, published by John Wiley and Sons in 2019.

We would like to acknowledge Alex Gnusin for his contribution of multiple figures to Chapter 13, Design for Test and Testability, from his work *Introduction to Design for Test* found at website: http://sparkeda.com/technology/

We would like to thank John Cooper from Reliability Express, LLC, for his help in editing several chapters in the book. He greatly supported our editorial efforts at a time when we were in a time crunch and attempting to deliver the manuscript to the publisher by the deadline date.

We would like to thank the following folks for their support of Chapter 8, Design for Machine Learning: Michael Jensen from 4ATMOS Technologies, LLC, and Robert Stoddard from Carnegie Mellon University (CMU), and Software Engineering Institute (SEI).

Louis J. Gullo
Jack Dixon

Introduction: What You Will Learn

Chapter 1: Design for Maintainability Paradigms (Gullo and Dixon)

This chapter describes what it means to Design for Maintainability (DfMn) and why DfMn is important. This chapter includes maintainability factors for design consideration and 9 DfMn paradigms. The paradigms as shown in this chapter and throughout this book provide the reader with the means to achieve a maintainable design. Using this book as a guide, the reader will be equipped with the basic elements of maintainable design that can be applied to any system or product, given an adequate amount of preparation and forethought, so that the resulting system or product design can be easily and economically maintained to obtain customer satisfaction.

Chapter 2: History of Maintainability (Gullo)

This chapter explores the history, origins, and evolution of maintainability and maintenance engineering. It covers their definitions and the key standards that launched maintainability as an important engineering discipline. This chapter also explains the difference between maintainability and maintenance engineering. It also describes in detail the original maintainability program roadmap that was developed at the advent of DfMn. The various important maintainability-related tasks are described. The chapter concludes with a brief description of MIL-HDBK-470, which provides a culmination of the beneficial standards, requirements, and lessons learned over the last 60 years.

Chapter 3: Maintainability Program Planning and Management (Franck and Meixner)

This chapter describes the tasks involved in designing for maintainability, the need for maintainability program management, the basic elements of a maintainability program plan, and the relationship of maintainability engineering to other engineering disciplines. This chapter provides an introduction to the system/product life-cycle and stresses the importance of early considerations of maintainability to ensure proper design. The earlier that the design characteristics are measured against requirements, the less time and funding are required to make changes to the design. This chapter also emphasizes that thorough and thoughtful maintainability program planning is the most effective tool in the maintainability manager's toolkit. It is important to successful design that maintainability engineering be integrated into the overall design team throughout the development program.

Chapter 4: Maintenance Concept (Franck)

This chapter describes the maintenance concept as one of the most basic guiding principles for comprehensive system, subsystem, and component design. This chapter explains how the maintenance concept is central to defining and influencing sustainment plans and capabilities to allow for full life-cycle operability. The chapter includes an introduction to maintainability requirements, the categories of maintenance, the levels of maintenance, and an introduction to Integrated Logistics Support (ILS), along with the 12 elements of Integrated Product Support (IPS).

Chapter 5: Maintainability Requirements and Design Criteria (Gullo and Dixon)

A full explanation of maintainability requirements, guidelines, and design criteria is provided in this chapter, along with a tie-in to Appendix A, System Maintainability Design Verification Checklist. This chapter describes how requirements are developed and evolve through the systems engineering process. This chapter has focused on developing, implementing, and using maintainability requirements, guidelines, criteria, and checklists. It explains the use of design criteria in the evaluation process and as input to the design process and equipment selection. Application of the design criteria by the design team is emphasized because it will ensure compliance with maintainability requirements and will support the optimization of the system's sustainability. The development of good requirements is strongly emphasized throughout this chapter since they are the key to developing a system that successfully meets the needs of the customer, and they will allow for the efficient and cost-effective maintenance of the system or product.

Chapter 6: Maintainability Analysis and Modeling (Kovacevic)

The chapter describes the various types of maintainability modeling and analysis techniques, including functionality analysis, functional block diagrams, maintainability allocations, maintainability design trade studies, maintainability design evaluations, maintainability task analysis, maintainability modeling, and Level of Repair Analyses (LORAs) as well as numerous additional types of maintainability analyses. The reader will be encouraged to conduct or participate in multiple design iterations in an effort to balance the maintainability, reliability, and cost of the system in the given operating environments.

Chapter 7: Maintainability Predictions and Task Analysis (Gullo and Kovacevic)

This chapter describes various methodologies for performing maintainability predictions based on a popular standard, and describes the method for performing a Maintenance Task Analysis (MTA), which is used to support the maintainability prediction process. Maintainability predictions and a detailed MTA are crucial steps to ensuring that the asset will be able to achieve the desired performance objectives over its life-cycle. Using these techniques iteratively, the designer and asset owners can ensure that the best possible design is put forth within the assigned or estimated financial constraints for maintenance and logistics support over the asset's life-cycle.

Chapter 8: Design for Machine Learning (Gullo)

This chapter discusses the meaning of Machine Learning and Deep Learning, and the differences between Machine Learning (ML), Artificial Intelligence (AI), and Deep Learning (DL). This chapter explains what Machine Learning is and how it supports Design for Maintainability activities that facilitate Preventative Maintenance Checks and Services (PMCS), Digital Prescriptive Maintenance (DPM), Prognostics and Health Management (PHM), Condition-based Maintenance (CBM), Reliability-Centered Maintenance (RCM), Remote Maintenance Monitoring (RMM), Long Distance Support (LDS), and Spares Provisioning (SP).

Chapter 9: Condition-based Maintenance and Design for Reduced Staffing (Gullo and Kovacevic)

This chapter explains how Condition-based Maintenance (CBM) is a predictive maintenance technique to enable smart maintenance decisions by monitoring the conditions of an asset and interpreting these data to detect off-nominal behaviors or anomalies. This chapter further explains how CBM is able to dramatically reduce the man-hours needed to maintain an asset by performing maintenance only when it is needed on the specific components of the asset. Several methods are described, which will help the reader identify what maintenance should be performed on the asset. Two methods are discussed in detail in this chapter. One is Failure Modes and Effects Analysis (FMEA), which identifies the failure modes of an asset and is used to develop risk mitigation actions. A second topic is Reliability-centered Maintenance (RCM), which utilizes the FMEA and a logic tree to decide what maintenance or design actions to take.

Chapter 10: Safety and Human Factors Considerations in Maintainable Design (Dixon)

This chapter is divided into two sections: one section for safety considerations, and a second section for human factors considerations. This chapter provides an understanding of how maintainability, safety, and human factors are intimately related to each other and how they interact to help achieve a successful maintainable design. As with other areas of concern discussed in this book, it is imperative that the human user be considered early and throughout the entire design and development process. This chapter provides a sampling of the considerations needed to ensure that systems are designed to be safe and ergonomically effective to use. This chapter also presents a sampling of the analysis techniques that can be used by the reader to achieve cost-effective maintenance while protecting the maintenance personnel and making their job easier.

Chapter 11: Design for Software Maintainability (Gullo)

This chapter describes what software maintainability is and how maintainability impacts the software design. It also presents three basic types of software maintenance methods and the five primary life-cycle processes that may be performed during the software life-cycle. The chapter also highlights some of the most relevant IEEE standards. The information in this chapter will help the reader in their effort to design for software maintainability so that maintenance will be performed economically, effectively, and efficiently over the system's life-cycle.

Chapter 12: Maintainability Testing and Demonstration (Franck)

This chapter explains the analysis, test, and demonstration processes that are used to verify the DfMn "goodness" of a system or product to determine whether or not the design satisfies specified maintainability requirements. This chapter emphasizes a maintainability test and demonstration philosophy that embraces the totality of work products from the design and development spectrum of activities. The detailed procedure for performing a Maintainability Demonstration (M-Demo) using an established standard, and an example of an M-Demo, are provided in this chapter.

Chapter 13: Design for Test and Testability (Meixner and Gullo)

This chapter defines the engineering role of designing for maintainability concurrent with Design for Test and Design for Testability. This chapter explains how Design for Test considers designing test capability as applied during development and production, while Design for Testability refers to the analysis of the test capability applied during development, production, and sustainment. Design for Testability refers to changes in the circuit design to increase test coverage, probability of fault detection, and probability of fault isolation that benefits maintainability throughout the system or product life-cycle. In a sense, Design for Testability is the gift to maintainability engineering that keeps on giving.

Chapter 14: Reliability Analyses (Dixon)

This chapter focuses on several of the more common types of reliability analyses and models. This chapter explains how reliability analysis efforts support other system engineering design characteristics, such as testability, maintainability, and availability. An engineering practitioner can focus on a reliability strategy or a maintainability strategy to achieve the desired level of availability. This chapter explains that when reliability is not under control, more complicated issues may arise, like manpower shortages, spare part availability, logistic delays, lack of repair facilities, and complex configuration management costs that negatively impact maintainability and availability. Several types of reliability analyses and modeling are covered, including Reliability Block Diagrams (RBDs), reliability allocations, and reliability mathematical models. Also, the techniques of reliability prediction, Fault Tree Analysis (FTA), and Failure Modes, Effects, and Criticality Analysis (FMECA) are covered in detail.

Chapter 15: Design for Availability (Kovacevic)

This chapter describes how availability is a vital metric needed by organizations that own expensive electronic system assets requiring continuous operation. Availability is a powerful metric that allows asset owners to understand how ready their assets are to perform intended missions while their assets are deployed for continuous 24 hour/7 day (24/7) operations. Availability allows asset owners to understand the impact to their probability of mission success as a result of poor maintainability, reliability, or logistical delays. This chapter describes what availability is, the types of availability, and how availability is calculated. By leveraging the various types of availability, organizations can identify reliability, maintainability, or supportability issues. Availability provides the most cost-effective solutions for asset designers as they perform a delicate balance between designing for reliability and designing for maintainability during design trade study analyses.

Chapter 16: Design for Supportability (Kovacevic)

This chapter describes design for supportability including detailed analyses and plans. This chapter explains how designing an asset for supportability is both a product of the design of the asset, as well as the service support and logistics systems that are put in place to ensure the asset can be operated and maintained efficiently in its working environment. This chapter provides several examples to explain why it is important to consider design for supportability during the system design and development phase of a program. The engineering practitioner will find examples of planning elements with applications of design for maintainability that also benefit design for supportability. The 12 elements of IPS that were initially mentioned in Chapter 4 are discussed in detail here as necessary steps to consider in ILS planning for a comprehensive design for supportability.

Chapter 17: Special Topics (Dixon)

This chapter includes three special topics and their applications that challenge the status quo of maintenance practices and offer insight on how to design for maintainability for the future. The topics included are:

- Reducing Active Maintenance Time with Single Minute Exchange of Dies (SMED) by James Kovacevic.
- How to Use Big Data to Enable Predictive Maintenance by Louis J. Gullo.
- Self-correcting Circuits and Self-healing Materials for Improved Maintainability, Reliability, and Safety by Louis J. Gullo.

Appendix A: System Maintainability Design Verification Checklist (Franck)

The checklist in Appendix A provides a wide-ranging collection of suggested guidelines, design criteria, practices, and check points that draw from numerous knowledgeable sources and references. Among these sources are a plethora of standards, various military guidance documents, industry training and guidance manuals, white papers and reports, and many years of experience of the authors.

This checklist is intended to provide these guidelines, design criteria, practices, and check points for thought-provoking discussions, design trade studies, and considerations by a design team to instigate design requirements generation for specification of maintainability design aspects of the product, its use, and its maintenance, and to verify the acceptance of the implementation of these design requirements in the product design. This maintainability design verification checklist is not intended to be a complete, all-encompassing list of all possible maintainability design considerations for any and all products for any customer. The reader may want to tailor this checklist for the items that are applicable to the particular product, customer, use application, and service support situation.

1

Design for Maintainability Paradigms

Louis J. Gullo and Jack Dixon

1.1 Why Design for Maintainability?

Maintenance for most systems and products is too costly, and it takes a long time to repair them. If the system or product owner is not covered by an extensive warranty from the Original Equipment Manufacturer (OEM), the owner is burdened by the cost to keep the system or product in service. By designing a system or product for maintainability, it will be serviceable and supportable without causing a cost and schedule overburden on the owner. Serviceable means the system or product is easy to repair. Supportable means the system or product is cost-effectively maintained and able to be restored to usable condition.

Only through knowledge of a specific system's design, performance, and construction can a person understand how to Design for Maintainability (DfMn) for any particular system to minimize the cost of maintaining that system. Anyone designing for maintainability should realize that there is no substitute for first-hand knowledge of a system's operating characteristics, architecture, and design topology. The most important parts of this knowledge are understanding how the system is assembled and disassembled, learning how it performs when functioning as intended by the designer, verifying how the system performs when applied under nominal-case and worst-case environmental stress conditions, and anticipating faulty conditions that need to be corrected through maintenance actions and servicing.

1.1.1 What is a System?

A system is defined as a network or group of interdependent components and operational processes that work together to accomplish the objectives and requirements of the system. The design process of any system should ensure that everybody involved in using the system or developing the system gains something they need, avoiding the allure of sacrificing one critical part of the system design in favor of another critical part of the system. This context includes customers, system operators, system owners, maintenance personnel, suppliers, system developers, the community, and the environment [1].

NOTE: For the purposes of simplicity and consistency, the terms "system" and "product" are used interchangeably in this book. When one or the other term is used, the intent is to apply the thought to both terms.

1.1.2 What is Maintainability?

Maintainability is a very important aim of a system while executing its functions in accordance with its system requirements to accomplish its objectives. Maintainability can be different things to different people. Maintainability can be expressed as a process to develop the methods and techniques to perform maintenance of systems and equipment in field applications. Maintainability is composed of both qualitative and quantitative design characteristics. Qualitatively speaking, maintainability is the

Design for Maintainability, First Edition. Edited by Louis J. Gullo and Jack Dixon.

engineering discipline that focuses on maintenance performance by the system being designed, integrated with maintenance personnel with specialized skills and training that use prescribed maintenance procedures and support equipment at each of the designated levels of repair. Quantitatively speaking, maintainability is a time-based function of the design that maximizes simplicity and gains economy of resources to restore system operation to a specified condition following the loss of critical system performance. Maintainability is a metric that drives the system design towards decreasing system downtime and minimizing the time to restore full operation of a complex system following a system failure. No matter how maintainability design metrics are defined, all types of system users measure the system performance (or lack thereof) in their own ways. Usually, the system user's top maintainability-related metric is the cost to maintain or sustain a system. The next top metric is the time it takes to restore a down system to full operation. As an example, an automobile owner may be mostly concerned with the cost of ownership and having few visits to the repair shop. An airline company may be mostly concerned with keeping flights on schedule. The measures to assess these customer concerns may or may not include factors totally determined by the design, therefore the way in which a customer measures the maintainability of a product in use may not be meaningful to that product's designer. This means that a translation from the user's concerns to measures more appropriate for design may be necessary to ensure customer success [2].

Maintainability is applied to systems that are repairable. There is no need to apply maintainability to a non-repairable system. There are two easy ways to know whether a system is non-repairable. First, if the system is discarded when it fails, this is a non-repairable system. If there are no parts in the system that need to be replaced to restore a system back to perfect working condition and able to perform at peak performance, then the system is non-repairable.

1.1.3 What is Testability?

Maintainability leverages the advantages established by testability. Testability, as an engineering discipline, may be considered an important subset of maintainability. Testability is a design characteristic that provides system status, an indication of a system fault condition, and a means to isolate a faulty condition in a timely and efficient manner. The system status indicators, which are provided through highly testable design architecture, will include determinations that the system is operable, inoperable, or degraded. Design for Testability (DfT) is just as important as DfMn in a design. Usually, a system designed with good testability features will be very maintainable. DfT not only benefits the maintainer in the field with effective and efficient test capabilities to ensure maintenance is performed quickly and correctly. DfT also benefits the system or product installer in the field when the system or product is deployed to the customer site and requires skilled technical personnel to initially set up the system or product and place the system or product into operation as the customer desires. The field installer will use the DfT capabilities to perform system or product checkout to ensure all functions are working as intended for the specific customer application.

The cost of DfT implementation has multiple paybacks. The system or product DfT capabilities not only positively affect customer user and maintainer performance in the field applications, but also positively affect production operations performance. DfT is a factor that improves Design for Manufacturing (DfMa). DfMa focuses on engineering a design for ease of assembly during a system or product's initial build processes. DfMa integrates the design of the system or product with the successful manufacturing concepts of Just-In-Time (JIT) and Lean Manufacturing to assemble and test new designs in a high volume manufacturing flow. "The ability to detect and isolate faults within a system, and to do so efficiently and cost effectively, is important not only in the field, but also during manufacturing. All products must be tested and verified prior to release to the customer. Paying attention to testability concerns up front will pay benefits during the testing phases of manufacturing. Therefore, a great deal of attention must be paid to ensuring that all designs incorporate features that allow testing to occur without a great deal of effort" [2].

1.2 Maintainability Factors for Design Consideration

Designers should consider several factors during the design and development process to balance the needs of the functional system design with the needs of the maintenance personnel in the field who attempt to restore faulty systems back in service. These factors provide benefits to the system operators and maintainers in terms of time, functionality, complexity, and cost during system service and support. The total cost of ownership and maintenance personnel well-being may be severely impacted if designers don't pay attention to these factors in the system or product designs [3]. The "8 Factors" for designers to consider when designing for maintainability are:

1. Part standardization
2. Structure modularization
3. Kit packaging
4. Part interchangeability
5. Human accessibility
6. Fault detection
7. Fault isolation
8. Part identification

1.2.1 Part Standardization

The designer should standardize common usage piece parts and components (such as screws, washers, nuts, and bolts) to minimize spare parts inventory for customer repair service in the various field applications. The designer should select from the smallest set of parts (e.g., one screw instead of 10 different types of screw) with as much compatibility as possible. Minimizing spare parts inventory is just one benefit. Keeping the design simple is difficult to do, but the payoff is huge in terms of fewer parts to keep in storage, less weight to transport parts to the field, smaller sets of tools, lower complexity of maintenance procedures and operations, shorter downtime, easier repairs, and overall a more efficient organization to conduct the maintenance. Imagine the efficiencies gained in an organization that no longer struggles with the difficulties of determining which screw goes where [3].

1.2.2 Structure Modularization

The designer should create a standard design template composed of standard dimensions, shapes, sizes, power, interfaces (mechanical and electrical), signal connections, and modular elements. Using standard structures with a minimal set of design options to choose from provides for design modularity and interchangeability. If the designer expects to develop different products with different features, using a standard structure from a baseline product model, then the hardware evolution will be simpler and easier to maintain in the field. Using a standard structure and interfaces allows the interchange of compatible parts to alter functionality without changing the majority of the product. Consider an example of a standard light bulb for a floor lamp. The lamp owner can select different types of light bulbs based on the desired power output (30 W vs. 60 W), color, filtering, bulb life, brightness, and cost to illuminate a room, or even the technology used (e.g., incandescent, compact fluorescent lamps (CFLs), or light-emitting diode (LED) lamps), while knowing that all these bulb types will fit into the same light bulb socket on the lamp [3].

1.2.3 Kit Packaging

The designer should plan to collocate and package all piece parts and components required for a particular maintenance task into a single kit. The kit may include screws, washers, o-rings, gaskets, and lubricants. The maintenance task will be performed much quicker if the required items needed to correctly perform maintenance of a product are placed together in a kit and contained in a single package. The kit package should also contain an inventory list of the names and quantities of the items in the kit [3].

1.2.4 Part Interchangeability

The designer should avoid developing custom sizes, shapes, connections, and fittings for a part or component. A custom size or fitting means a custom part is being designed, which reduces the chances of part interchangeability. Standardized parts mean the parts are interchangeable between multiple sources. A custom part means a single source for the part is highly probable, which reduces or eliminates the potential for multiple sources for the same part. A custom part means lack of compatibility with other similar types of parts. Custom parts result in an increasing amount of total part numbers for spare parts in inventory, since the custom parts will not be interchangeable with parts from other sources of similar type parts that are commonly used across a wide range of product applications. Also, custom parts cause limitations on future design changes if the designer wants to stay in that custom form factor as the design evolves, requiring extra design effort to manage and control the part drawing dimensions and tolerances [3].

1.2.5 Human Accessibility

The designer should provide adequate room or space between design structure dimensions for access to reach parts and components that need to be replaced. The designer that does not pay attention to this maintainability factor increases the risks of the maintenance personnel in the field being impaired or injured during performance of their maintenance tasks. If an item requires replacement or adjustment as part of the expected maintenance actions, then the designer should allow for access by the maintainer in the field for the different environments possible, such as cold harsh weather. The designer should consider the tools required by the maintainer to perform the maintenance actions when deciding how much access space is needed. The designer should provide for access panels that can easily be removed to allow for more room for maintainers to work. The panels should also be easy to replace to minimize downtime. Finally, the designer should consider the maintainer's viewing angles, lighting needs, and the skills and experience required [3].

1.2.6 Fault Detection

Fault detection is a key factor in performing maintenance. Fault detection could occur simply by performing a visual inspection or using other human senses such as hearing, smell, or touch. On complex systems, however, fault detection may not be so obvious. For complex systems, for example, involving the replacement of electronic circuit cards, a designer should provide the necessary information to the maintainer to know the cause of a system problem and to determine how to remedy the problem. This information could be derived from the system itself using embedded test diagnostics, or from external support test equipment connected to the system during maintenance actions. For example, a bicycle flat tire is obvious to the bicycle rider, and fault detection may occur by visual inspection or by hearing a change in the sound or feeling a vibration during the course of a ride. For complex systems, the system must inform the user that an event occurred that requires attention and to provide information on the presence of a fault that cannot be detected through human senses. By minimizing the need for inspection with tools and exhaustive disassembly to perform diagnostic tasks, the time and cost to perform maintenance is minimized [3].

1.2.7 Fault Isolation

Once a fault is detected, it should be easy to isolate. The designer should create the system diagnostics capability of providing the maintainer with as much information as possible about the type of fault detected and the location of that fault. The designer should consider the fact that the maintainer needs to fix a problem. The maintainer needs to know all evidence that a fault has occurred, any data collected by the system concerning the fault symptoms, as well as the possible failure mechanisms that could cause these symptoms. Once the system can determine the potential failure mechanisms, it may apply detailed diagnostics to triage the failure mechanisms with the other indicators of the system performance to isolate the cause, and give information to the maintenance personnel on what to replace and where to replace it. For example, the blown fuse fault event leads to the indication of which circuit lost power to enable the maintainer to replace the fuse, and then determine why the fuse was blown, such as a short circuit. Replacing the fuse is only a remedy to the immediate problem, but it may not resolve the root cause of the blown fuse. It could take significant time to troubleshoot the electronic circuits that could be overloaded by an intermittent short circuit and cause more blown fuses. A failure event, such as a blown fuse, could be simply a faulty or defective fuse, or it could be as complex as a short circuit in a part of the system that is unrelated to the circuit with the blown fuse. Also, the initial failure event can cause degraded performance and residual damage in other circuits and components, resulting in latent defects that will fail at some point later in time. This residual damage could be secondary and tertiary effects of a failure mode, which can have significant ramifications in the future, if not resolved immediately when the initial failure event occurred. For this reason, the designer should ensure that the effects of faults possible within the design are contained to minimize damage caused by the failure of one item in the design or another design within the system [3].

1.2.8 Part Identification

The designer should label or mark each part with a standard naming convention for all parts used in the design. Each part should be marked with a unique identifier (e.g., a part number) and date code. This use of part identification streamlines the maintenance task processes by connecting the maintainer with more valuable documentation such as maintenance manuals, work instructions, and teardown procedures. The designer should be consistent in the use of naming conventions. The part identifications should be meaningful and memorable to avoid confusion and aid in the maintenance process to ramp up the learning curve of new maintainers for the system [3].

1.3 Reflections on the Current State of the Art

Maintainability has been successfully integrated into the mainstream of systems engineering design and development processes in many organizations. It is vigorously supported by management as a discipline that adds value to the systems and product development processes. Many maintainability analysis techniques have been created and revised over the years to make them more effective and efficient. The application of maintainability in product design and development has proven valuable in reducing customer total cost of ownership and OEM warranty costs. However, there are still many challenges facing today's system and product maintainability engineers. The maintainability discipline is a small and somewhat obscure engineering discipline. Maintainability engineering skills are shared across an organization's technology infrastructure. It may be difficult to identify exactly where maintainability tasks are performed in any one organization. An organization's accountability for maintainability tasks may be split across multiple engineering departments, meaning the accountability is not clearly established. No one engineering function is responsible for starting and completing a maintainability task. Maintainability functions should be more visible in an organization, especially where field support personnel provide services to customers, generating a large source of aftermarket revenue. While many organizations successfully implement maintainability, many ignore the benefits and valuable contributions to the design, and therefore suffer the consequences of delivering inferior systems and products to their customers due to poor maintainable designs.

NOTE: The term "organization" will be used throughout the book to refer to all system developer and customer entities to include businesses, companies, suppliers, operators, maintainers, and users of systems.

Other challenges include the ever-increasing complexity of systems being developed today and planned for the future. Instead of an organization developing one system at a time, organizations are developing System-of-Systems (SOSs). More and more customers demand that multiple systems are networked and integrated into a single control platform. This type of system development is called System-of-Systems Integration (SOSI). The SOSI development introduces complexity and new challenges in how to monitor and maintain the performance of integrated large systems, considering all the interactions between the systems and subsystems of a SOSI program.

Inadequate system and product specifications often lead to inadequate requirements for the maintainability discipline. Many times, boilerplate and generic specifications were provided to the system and product designers, leading to faulty designs because specific requirements were omitted, ambiguous, vague, or incomplete.

Change management is a constant concern and cause for problems throughout the system or product development process. Change management may be successfully executed during the development process, but is often overlooked in aftermarket system and product support over the entire life-cycle. One challenge for organizations is to ensure that as design changes are made, maintainability engineers are involved in the design change approval process to ensure that the changes are adequate from a maintenance perspective, and that they do not cause unintended consequences that could lead to time-consuming repairs and high costs to perform necessary maintenance actions in the field.

Another design challenge is to consider the human during the required maintenance activities performed in the field. The human often causes problems by their lack of understanding of how the design works or fails, or by abusing the system or product, thus introducing failures that were not caused by the system when properly operated and maintained. Often, the user or maintainer can be confused by the complexity of a system, by the user interface provided by the hardware and software of the system itself, or by the external support test equipment. The man–machine interfaces along with human limitations should always be considered during the design process. It is paramount to the successful installation and deployment of the system or product.

The goals of this book are to give the readers the processes, tools, and knowledge they need to be successful as maintainability engineers, and to help the readers to avoid pitfalls and remedy some of the problems previously mentioned, while building upon the many years of success experienced by maintainability engineers thus far. Next, nine maintainability paradigms are presented that will lead the reader to a more maintainable system or product design. This chapter and the entire book provide both good and bad examples so the reader can identify with real-world cases from which to learn.

1.4 Paradigms for Design for Maintainability

As stated earlier, maintainability is a process. If the right process is used, then the right results will follow. Knowledge of the right things comes from experience and practice. The experiences could come from the reader or from others. These experiences are called lessons learned. Lessons-learned experiences are practiced by the student of DfMn to gain knowledge in the most efficient way. The student learns to arrive at correct decisions through the accumulated knowledge from oneself and others. "One must keep becoming better by practicing. Take the example of swimming. One cannot learn to swim from books alone; one must practice swimming. It is okay to fail as long as mistakes are the stepping-stones to failure prevention. Thomas Edison was reminded that he failed 2000 times before the success of the light bulb. His answer, 'I never failed. There were 2000 steps in the process.' " [4].

A successful technique to solidify knowledge and gain maximum benefit from experiences of one's self or others is to use lessons learned in the form of paradigms. Paradigms should be easy to remember and to recall during critical times in the design process, such as brainstorming sessions during design reviews with members of an integrated design team. An ideal systems approach to designing new systems involves developing paradigms, standards, policies, and design process models for developers to follow as a pattern for their future design efforts. Paradigms are often called "words of wisdom" or "rules of thumb." The word "paradigm," which originated from the Greek language, is used throughout the content of this book to describe a way of thinking, a framework, and a model to use to conduct yourself in your daily lives as a maintainability engineer. A paradigm becomes the way you view the world, perceiving, understanding, and interpreting your environment, and helping you formulate a response to what you see and understand [1].

This book focuses on nine paradigms for managing and designing systems and products for maintainability. These nine paradigms are the most important criteria to consider when designing for maintainability. These paradigms are listed next, and are explained in separate clauses following the list.

Paradigm 1: Maintainability is inversely proportional to reliability.

Paradigm 2: Maintainability is directly proportional to testability and Prognostics and Health Monitoring (PHM).

Paradigm 3: Strive for ambiguity groups no greater than 3.

Paradigm 4: Migrate from scheduled maintenance to Condition-based Maintenance (CBM).

Paradigm 5: Consider the human as the maintainer.

Paradigm 6: Modularity speeds repairs.

Paradigm 7: Maintainability predicts downtime during repairs.

Paradigm 8: Understand the maintenance requirements.
Paradigm 9: Support maintainability with data.

1.4.1 Maintainability is Inversely Proportional to Reliability

The system design should strike a good balance between the DfMn purposes and the design for reliability purposes. As reliability goes up, maintainability goes down. As more system or product failures occur, the calculated reliability of a system goes down and the quantity of maintenance actions goes up. The time spent by maintenance personnel doing maintenance actions goes up, increasing system downtime over the life of the system.

If a system's reliability is 100%, there is no need for maintenance since the system will never fail, so the time to perform maintenance is zero. Think about the Maytag Repairman commercial in the US from the 1970s. The Maytag appliance repairman was bored and never went on service calls. The repairman was bored because the Maytag appliances never failed, and therefore, they did not need maintenance. The commercial drew the analogy that a bored maintenance man meant a reliable appliance. As a result of this commercial, Maytag appliances were one of the most popular brands in the US in the 1970s.

1.4.2 Maintainability is Directly Proportional to Testability and Prognostics and Health Monitoring

The more test capability that is added to an equipment design with embedded test diagnostics, health monitoring, and prognostics functions, the quicker and easier the equipment is to maintain and restore to service following an equipment failure. As equipment circuit and functional test coverage increases with higher percentages of fault detection and fault isolation, the better the maintainability of the equipment.

PHM technology enhances system maintainability, reliability, availability, safety, efficiency, and effectiveness. PHM includes wireless embedded sensors, predictive analytics, data storage, and risk mitigation through a comprehensive logistics infrastructure. "Through the use of embedded sensors for health monitoring and predictive analytics within embedded processors, prognostic solutions for predictive maintenance is a possibility. PHM is an enabler for system reliability and safety. We need innovative tools for discovering hidden problems, which usually turn up in rare events, such as an airbag that does not deploy when needed in a crash. In the case of an airbag design, some brainstorming needs to be done on questions such as 'Will the air bag open when it is supposed to?' 'Will it open at the wrong time?' 'Will the system give a false warning?' or, 'Will the system behave failsafe in the event of an unknown component fault?' " [1].

1.4.3 Strive for Ambiguity Groups No Greater Than 3

"The expected percentage of faults that can be detected and isolated to a specified or desired level of ambiguity must be determined as an important input to the logistics analysis process" [2]. An ambiguity group is the number of possible subassemblies of a product or item identified by manual or automated test procedures that might contain the failed hardware or software component. A corrective maintenance action should involve removal and replacement tasks with an ambiguity group no greater than three assemblies or components. Troubleshooting to locate and isolate the cause of a failure mode does not always result in a single cause for the failure symptoms observed. Many times, troubleshooters cannot determine whether the cause of a failure is the result of one or more assemblies or components in a system or product. To speed up the troubleshooting process, troubleshooters will remove and replace more than one assembly or component at a time to quickly restore functionality of a system or product, placing it back in-service. This method of removing and replacing more than one assembly or component at a time is called "shotgunning." Each of the items replaced are potential causes of a failure. Shotgunning is performed due to the ambiguities of the failure symptoms, leading a troubleshooter to multiple reasons in the failure cause assessment. The number of assemblies or components that must be replaced determines the size of an ambiguity group. The more assemblies or components in an ambiguity group, the more wasted materials, since there is probably only one assembly or component causing the problem, but the maintainer does not want to take the time to remove and replace each one in a sequential manner, and instead replaces them all at once to save time.

1.4.4 Migrate from Scheduled Maintenance to Condition-based Maintenance

Preventive maintenance activities should migrate away from scheduled maintenance tasks to CBM tasks. Time-based maintenance is convenient for planning when maintainers are needed to perform certain tasks to prevent mission-critical failures, but it costs more and leads to more wasted materials compared to CBM. CBM is performed only when the equipment needs it, based on the physical wear-out and degradation of the materials as determined from aging and Physics of Failure (PoF) models, environmental stress conditions, and measurements of parametric design characteristics as determined from embedded sensors. CBM may also include PHM technology.

1.4.5 Consider the Human as the Maintainer

During the system or product design and development process, the designer should imagine how a person will interact with the system or product to operate it or to perform maintenance tasks. The designer should always consider the abilities and the limitations of the human who is tasked to repair or replace a piece of equipment. The designer should consider accessibility, weight of components, placement of connectors, operational environment, and technical complexity. The designer should also allow for access by the maintainer in the field for the different environments that may be encountered, such as when wearing bulky restrictive clothing to protect against cold harsh weather.

A comprehensive maintenance training program should be developed that includes all man–machine interfaces of the system for access to various functions by operators and maintainers. Development of a complete training program for certifying the operators and maintainers requires not only recognizing the components and subsystems, but also understanding the total system. “Many training programs are focused only on the subsystem training. When this occurs it means that the certification of the person operating or maintaining the equipment is limited. The operating and maintenance personnel may therefore not realize that the total system can be affected by hidden hazards. For example, having only one power source may negatively affect a maintainer’s work. For safety-critical work, a redundant source would be a good mitigation to implement. However, if all sources of power are lost (prime, secondary, and emergency), the total system will not work until a correction is made. Scenarios must be developed that provide instructors and students with a realistic understanding of the whole system and how to protect people from harm” [1].

Maintenance training represents a major risk mitigation method for accidents and human-caused failures as a result of maintenance performed on complex systems. It is imperative that complete training be provided for correctly operating and maintaining the system at all times and for all likely scenarios.

1.4.6 Modularity Speeds Repairs

Usually there are insufficient requirements for modularity in system design specifications. Specification requirements need to address interoperability functions in adequate detail, especially for internal system mechanical and electrical interfaces, external system interfaces, user–hardware interfaces, and user–software interfaces.

Modular design and construction of equipment simplifies maintenance and repairs. Modularity minimizes the quantity and duration of human–machine interactions to conduct maintenance on a system, which ultimately has a positive impact on the system reliability, safety, testability, serviceability, and logistics. Plug-in, easily removable circuit boards, component piece parts, and assemblies with reasonable costs will facilitate removal and replacement. Limiting costs of the various modular assemblies will help ensure cost-effectiveness.

1.4.7 Maintainability Predicts Downtime During Repairs

Just as reliability engineering predicts when a failure may occur, maintainability engineering predicts how long a system is down awaiting repair after the failure occurs. Maintainability performs prediction analysis of removal and repair activities considering all maintenance task times required to restore a system to full operation. The total time, starting with the detection of a failure event to final test and check-out of a system to verify functionality for full service, should be the main prediction task for maintainability engineering. The total time calculations result in downtime predictions. The accuracy of these predictions is paramount to the success of any maintenance concept and program.

Planned downtime for proactive maintenance is imperative for smooth system operation, but it must be balanced. Too much planned downtime for proactive or corrective maintenance takes away valuable operating time from the asset. The maintainability engineer must determine the best allocation of planned maintenance time to be scheduled by system operators, which is balanced with the conditions of the system and the needs of the customer to perform missions, with the biggest payback in terms of reducing the probability of mission-critical failures.

1.4.8 Understand the Maintenance Requirements

Designers should spend a significant amount of time in the early stages of system design and development on system requirements generation and analysis. This is to avoid the potential for system failures later in a development program. Most system failures originate from bad or missing functional requirements in specifications. When a system fails as a result of bad or missing requirements, it is called a requirements defect. The causes of most requirement defects are incomplete, ambiguous, and poorly defined specifications. These requirement defects result in expensive later engineering changes in a program, adding new requirements one at a time over a long period of development time, following the program's requirements development phase. This long, drawn-out requirements change is called "scope creep." Often robust design requirement changes cannot be made because the development program is already delayed in transitioning to production, or design resources are not available for implementing new features [1].

A Level of Repair Analysis (LORA) provides a logical means to understand and analyze maintenance requirements, and must be conducted early in the design process when system or product requirements are being generated and analyzed. Minimizing life-cycle costs involves more than just a design. It requires a holistic approach to plan for and prepare for an asset. Having the right skills sets, knowing which skills and activities to outsource, as well as having the right tools, are vital to minimizing planned downtime and maintenance costs. A LORA is paramount to understanding the maintenance requirements upfront and planning for them.

1.4.9 Support Maintainability with Data

In order to provide effective maintenance, it is imperative that the maintainers are provided with the correct data and information. The data are usually divided into two categories: (i) design data; and (ii) maintenance task time data. Design data are the technical data related to the operation of the system, or data may be the artifacts of the development process related to describing how the design fails, such as the case with a Failure Modes and Effects Analysis (FMEA). The maintenance task time data uses experience-based or historical evidence to support how much time is predicted to perform each individual maintenance task. The number of times a repeated maintenance action must be performed and the probability of performing that maintenance task are based on the probability of the occurrence of failures extracted from the reliability analyses.

All too often, OEMs do not receive the necessary design data from their suppliers. Technical data packages that include assembly drawings, parts lists, Bill of Materials (BOMs), as-built drawings, and operating manuals are required from the supplier, but for a multitude of reasons, the design data are not provided by the suppliers in order to support the maintainability analysis. Furthermore, organizations need to ensure that all technical data from suppliers are provided in an electronic format suitable for the maintainers to use. This may include complete parts lists and BOMs with component parts and materials described by an approved taxonomy.

NOTE: These paradigms, which are referenced here, are cited throughout the course of this book. Table 1.1 *provides a guide to locating the various paradigms within the book.*

Table 1.1 Paradigm locations by chapter.

Paradigm number	Paradigm	1	2	3	4	5	6	7	8	9	10	11	12	13	14	15	16	17
Paradigm 1	Maintainability is inversely proportional to reliability	×			×										×	×		
Paradigm 2	Maintainability is directly proportional to testability and PHM	×	×					×	×	×				×				
Paradigm 3	Strive for ambiguity groups no greater than 3	×							×				×	×				
Paradigm 4	Migrate from scheduled maintenance to Condition-based Maintenance	×							×	×							×	×
Paradigm 5	Consider the human as the maintainer	×		×	×	×	×				×		×					×
Paradigm 6	Modularity speeds repairs	×				×						×						×
Paradigm 7	Maintainability predicts downtime during repairs	×					×	×		×						×		×
Paradigm 8	Understand the maintenance requirements	×	×	×	×	×					×						×	
Paradigm 9	Support maintainability with data	×	×	×			×	×				×	×		×	×	×	×

1.5 Summary

In conclusion, these DfMn paradigms help the designer do the right things at the right times to avoid unnecessary maintenance actions, prevent causes of excessive downtime, benefit the user and maintainer in the field, and contribute to lower total cost of ownership over the system or product life-cycle. By implementing early DfMn actions including the assorted maintainability tasks and analyses described in this book, while using the nine paradigms and incorporating robust design processes early, the designer will create a successful design for the customer with a potentially huge Return on Investment (ROI) over the life of the design. The ROI of DfMn can be as high as 1000% compared to the cost of doing nothing.

The paradigms as shown in this book provide the reader with the means to achieve a maintainable design. When releasing a maintainable design to a customer, given an adequate amount of preparation and forethought using this book as a guide, the result will be a design that can be easily and economically maintained with guaranteed customer satisfaction.

References

1 Gullo, L.J. and Dixon, J. (2018). *Design for Safety*. Hoboken, NJ: Wiley.
2 US Department of Defense (1997). *Designing and Developing Maintainable Products and Systems MIL-HDBK-470A, 1*. Washington, DC: Department of Defense.
3 Schenkelberg, F. (n.d.). 8 Factors of Design for Maintainability. Accendo Reliability. https://accendoreliability.com/8-factors-of-design-for-maintainability (accessed 10 August 2020).
4 Raheja, D. and Gullo, L.J. (2012). *Design for Reliability*. Hoboken, NJ: Wiley.

2

History of Maintainability

Louis J. Gullo

2.1 Introduction

This chapter explores the history of maintainability and maintenance engineering, their definitions, and the key standards that launched maintainability as an important engineering discipline. Before delving into this topic, however, let us first consider the meaning of these words and how they were derived. Maintainability and maintenance are rooted in the verb "to maintain," which means to continue, or to retain, or to keep in existence, or to keep in proper or good condition. The word derives from the old French word, *maintenir*, which is derived from the Medieval Latin word *manutenere*, which means "to hold in hand." Maintainable, therefore, means someone has the ability to take one or more certain types of actions to ensure the continued use of something, or to keep something in good working order. Furthermore, maintenance is the action that is taken to ensure the continued use of something that was intended to be maintainable.

2.2 Ancient History

History has recorded many situations in the past where people of ancient civilizations took steps to repair the tools and equipment they used to do their daily tasks. Some of these steps are still performed today. For instance, metal knives, saws, or chisels were created and sharpened as far back as 3000 BCE, and possibly even earlier. Today, knives need to be sharpened, but electronic knife sharpeners make the chore of blade sharpening quick and simple. Imagine the difficulty of sharpening a knife thousands or even millions of years ago. There is archeological proof that ancient humans used stone tools over 2.5 million years ago. Stones were shaped and sharpened for cutting implements and arrowheads. Stones were selected for knives from the surroundings based on the ease of sharpening, but were brittle and prone to damage with many uses over time. People began working with metal over 9000 years ago. Metal tools were an improvement over stone tools in terms of durability and sharpness of edges.

In ancient Egypt, papyrus scrolls document the work performed by skilled stonemasons in the creation of the Great Pyramid of Giza. The ancient Egyptian stonemasons used copper chisels and wooden mallets to quarry and carve the blocks of stone used to build the pyramids and other beautiful structures. The stonemason's tools, such as metal chisels and wedges, would accumulate high stresses with the constant pounding forces from hammering stones, and their sharp edges would be worn down with continued use. This meant the tools required maintenance on a peri-

Design for Maintainability, First Edition. Edited by Louis J. Gullo and Jack Dixon.

odic basis (e.g., hourly or daily) to restore sharp edges for effective and efficient use of the tools. Specialized metal tool craftsmen and blacksmiths would receive the tools from the stonemasons, sharpen the edges, and return the tools to the stonemasons in "like-new" condition. Metal tools could be used over and over again, until they could no longer be repaired and were discarded.

With the invention of the wheel around 3500 BCE, carts, chariots, and wagons were used on a widespread basis through early Egyptian, Greek, and Roman times. The cartwrights of these civilizations realized that a chariot or wagon should not be destroyed when a wheel became wobbly or dislocated from its axle. Maintenance procedures and cart designs were developed by these cartwrights to easily replace the wheels on the axles when they broke down. Axle designs were modified over time to decrease the friction on the wheel hubs and allow for easy removal and replacement. Figure 2.1 shows a depiction of an onager-drawn cart on the Sumerian "Standard of Ur" (c. 2500 BCE).

"Although humans have felt the need to maintain their equipment since the beginning of time, the beginning of modern engineering maintenance may be regarded as the development of the steam engine by James Watt (1736–1819) in 1769 in Great Britain. In the United States, the magazine *Factory* first appeared in 1882 and has played a pivotal role in the development of the maintenance field. In 1886, a book on maintenance of railways was published [1]."

2.3 The Difference Between Maintainability and Maintenance Engineering

Maintainability as an engineering discipline first appeared in military standard documents in the 1960s. It was at this time period that maintainability was given credence and placed in the mainstream for system and product engineering development. Many years of engineering development process evolution passed to achieve this milestone. The evolution of maintainability engineering began about 60 years prior to the creation of the first maintainability standards.

Maintenance engineering was defined in Department of Defense (DoD) Instruction 4151.12 [2], entitled *Policies Governing Maintenance Engineering within the Department of Defense*, released in June 1968. As defined in DoD Instruction 4151.12, maintenance engineering is that activity of equipment maintenance which develops concepts, criteria and technical requirements during the conceptual and acquisition phases to be applied and maintained in a current status during the operational phase to assure timely, adequate and economic maintenance support of weapons and equipment.

Although the DoD Instruction 4151.12 was in effect as of 1968, the more recent DoD Directive 5000.40 [3] is more specific in defining the objective of maintenance engineering as well as selected measures of maintainability. It defines Reliability and Maintainability (R&M) engineering as "that set of design, development, and manufacturing tasks by which R&M are

Figure 2.1 Early Sumerian cart, from the Standard of Ur. Source: Anonymous, http://www.alexandriaarchive.org/opencontext/iraq_ghf/ur_standard/ur_standard_8.jpg. Creative Commons CC0 License.

achieved," and maintainability as "the ability of an item to be retained in or restored to specified condition when maintenance is performed by personnel having specified skill levels, using prescribed procedures and resources, at each prescribed level of maintenance and repair." Further, it expands on the overall objectives of R&M activities by stating that maintainability engineering "shall reduce maintenance and repair time, number of tasks required for each preventive and corrective maintenance action, and the need for special tools and test equipment" [4]. Maintainability engineering and maintenance engineering are closely tied, but they are very different.

2.4 Early Maintainability References

Maintainability traces back to 1901 with the Army Signal Corps contract for development of the Wright brothers' airplane. In this contract, it stated that the aircraft should be "simple to operate and maintain" [1]. Some people believe that maintainability as an engineering discipline began in the time period of World War II (c. 1939–1945), and others think it began in the 1950s, when various types of engineering efforts focused, either directly or indirectly, on maintainability. An example of these efforts was published in 1956 by the United States Air Force (USAF) as a 12-part series of articles that appeared in a periodical entitled *Machine Design*. The term "preventive maintenance" was clearly defined in the 1950s. The term "maintainability" was defined and documented in 1964 in MIL-STD-778 [5], which states: "Maintainability is a characteristic of design and installation which is expressed as the probability that an item will conform to specified conditions within a given period of time when maintenance action is performed in accordance with prescribed procedures and resources."

Early published articles on maintainability covered topics that were both general and specific to the subject. Some of these articles included topics such as [6]:

- Designing electronic equipment for maintainability
- Design recommendations for hardware modularity by separating electronic equipment into units, assemblies, and subassemblies
- Design of covers and cases
- Design of wiring harnesses, cables, and connectors
- Design recommendations for maintenance access to hardware components in electronic equipment
- Design recommendations for test points
- Design of maintenance controls
- Factors to consider in designing visual displays
- Designing for equipment installation
- Consideration of maintenance support equipment, such as test equipment, mock-ups, and tools
- A systematic approach to preparing maintenance procedures
- Methods of presenting maintenance instructions

2.4.1 The First Maintainability Standards

The USAF initiated a program for developing an effective systems approach to maintainability, which in 1959 resulted in the maintainability specification, MIL-M-26512, entitled *Maintainability Program Requirements for Aerospace* [1, 7]. This military specification cited maintainability requirements for military systems. This specification led to three military standards concerning related maintainability subjects to support the establishment of system-level maintainability requirements. These three military standard documents are MIL-STD-470, MIL-STD-471, and MIL-HDBK-472. MIL-STD-470 [8] is the military standard entitled *Maintainability Program Requirements*. MIL-STD-471 [9], entitled *Maintainability Demonstration*, provided a means to verify maintainability requirements as specified by MIL-M-26512, MIL-STD-470, and other related specifications. MIL-HDBK-472 [10] is the military handbook entitled *Maintainability Prediction*, which offered guidelines for maintainability analysis techniques to ensure maintainability requirements as specified by MIL-M-26512 [7] and other military maintainability specifications are designed correctly prior to the verification and testing of the particular system. These three documents [8–10] were published by the United States DOD in 1966.

The first commercially available book on maintenance engineering was published in 1957 – the *Maintenance Engineering Handbook* [11]. In 1960, the first commercially available book on maintainability was published [12], *Electronic Maintainability*.

2.4.2 Introduction to MIL-STD-470

Military Standard 470 (MIL-STD-470) [8] contains the requirements for conducting a maintainability program. MIL-STD-470 is a military tri-service coordinated document, meaning that representatives from the US Army, Navy, and Air Force worked together to create and release this standard. MIL-STD-470 replaced seven separate maintainability-related documents owned between these three military services. The purpose of MIL-STD-470, as stated in the Forward of this standard, was "to establish maintainability programs through standard program requirements for Department of Defense procurements." MIL-STD-470 went on to state that the degree of maintainability achieved is a function of two things: "(i) the contractual requirements imposed and (ii) management emphasis" [8]. MIL-STD-470 was intended for application to all military systems and equipment where a maintainability program is appropriate. If a maintainability program was deemed appropriate for a military system, based on the expectations for significant engineering development and operational system development as prescribed in DOD Instruction 3200.6, then relevant maintainability requirements for the size and type of system would be included in the acquisition contract's documentation, including the Request for Proposal (RFP), the Operational Requirements Document (ORD), the top-level system specification, and the Statement of Work (SOW).

With the creation of MIL-STD-470, all United States DoD military program procurements for repairable systems were expected to influence and assure: (i) the development and use of adequate government contractor and customer maintenance management procedures; (ii) the leveraging of the latest government and industry technology with analytical techniques to benefit maintenance procedures over the life-cycle of the system; and (iii) the integration of maintainability with other engineering disciplines and logistics support functions. If these three focus areas were properly executed, a military program's maintainability tasks would assure a high degree of confidence that all maintainability-related mission requirements would be achievable.

Table 2.1 Maintainability program task outline from MIL-STD-470 [8].

Task number	Maintainability program task title
1	Maintainability Program Plan
2	Maintainability analysis
3	Maintenance inputs
4	Maintainability design criteria
5	Maintainability trade studies
6	Maintainability predictions
7	Vendor controls
8	Integration
9	Maintainability design reviews
10	Maintainability data system
11	Maintainability demonstration
12	Maintainability status reports

Table 2.1 is the outline of the detailed maintainability program requirements in the order as provided in MIL-STD-470. This outline was the original roadmap for a maintainability program. Table 2.1 shows the discrete task requirements for a possible maintainability program that a customer may expect from their contractor. The customer may require their contractor to break down these maintainability task requirements into separate detailed explanations in the contractor's Maintainability Program Plan (MPP). The contractor's maintainability program should reference the customer's contractual documents and provide a schedule of tasks with a logical sequence of maintainability engineering activities that are planned during the course of the engineering development or operational system development program.

MIL-STD-470 required that a MPP include the following contents, as a minimum:

- Program tasks (using the task list from Table 2.1 as guidance)
- Responsibilities
- Significant maintainability program milestones
- Timing
- Communications
- Interfaces
- Techniques

2.5 Original Maintainability Program Roadmap

Each of the maintainability program tasks as outlined in Table 2.1 are described in detail in the next 12 sections.

2.5.1 Task 1: The Maintainability Program Plan

The MPP is the first task to be performed by a system or equipment contractor on any program where the customer specifies quantitative maintainability requirements along with the overall mission requirements for a system or equipment. This task is accomplished by the contractor, and usually a preliminary MPP is included in the contractor's proposal in response to a customer's RFP. The purpose of the MPP is for the contractor to describe, in as much detail as appropriate, how it plans to conduct a maintainability program to satisfy the quantitative maintainability requirements [13]. The contractor's MPP should identify and define all steps that the contractor intends to accomplish to address all the tasks of the program as shown in Table 2.1, and all content and pertinent information as required by MIL-STD-470. If the contractor provides a preliminary MPP in its proposal to the customer, the contractor would be expected to expand and modify the MPP during the contract requirements definition and development phase of the program to create an MPP that would guide the contractor throughout the life of the development program. Much more detail on the MPP is provided in Chapter 3.

2.5.2 Task 2: Maintainability Analysis

Maintainability analysis as a task may be interpreted in different ways. Depending on an industry or government perspective, maintainability analysis functions may be performed with an engineering focus or a logistics focus, or an integrated focus where both engineering and logistics have equal weighting in terms of their importance during the system or equipment development phase of a program. Problems have arisen in the past concerning a difference in semantics between industry and government perspectives regarding maintainability analysis vs. maintenance analysis, and the engineering vs. logistics functional aspects of the maintainability discipline. Some groups think the terms maintainability analysis and maintenance analysis are synonymous, and the functions are fully integrated into their organizations. Some groups think that one task is an extension of the other and may be performed as one function. To others, maintainability analysis is allied to the design engineering function, while maintenance analysis is allied to the logistics function which is concerned with all the resources involved in maintenance performance, not just those resources controlled by the contractor and the design authority [13]. It is important to recognize the contributions of both engineering and logistics functions in the performance of maintainability analyses for the successful achievement of maintainability requirements on a program. MIL-STD-470 requires the contractor to define the interface between the maintainability program, the design engineering organization, and the logistics support organization.

The maintainability analysis task involves the generation of quantitative maintainability metrics that are a function of estimated or measured maintenance task times. Maintainability analyses will determine each maintenance task that is required to be performed by the user or maintainer in the field. Each maintenance task will be broken down into discrete maintenance actions with a time value allocated to them. Each of these maintenance action times will be summed up to a total maintenance task time for any particular maintenance task that is needed. Quantitative maintainability metrics, such as Mean-Time-to-Repair (MTTR), will be determined based on the distribution of discrete maintenance task times. These maintainability metrics will be allocated to the maintainability task requirements to all significant functional levels of the system/equipment in a design specification. The MPP will define the maintainability allocation technique to be employed during the development program. The maintainability analysis task is an iterative process that results in a preliminary maintainability allocation, followed by subsequent allocations that use refined data that develop the metrics to the point where they are deemed mature and ready to be specified in a design requirements document. Further maintainability analyses are performed later in the development process for accomplishing maintainability trade studies and predictions. The design should evolve as the maintainability analysis process is conducted, up to the point when the maintenance concept is determined and baselined.

2.5.3 Task 3: Maintenance Inputs

As a result of performing iterative design analyses for maintainability consideration, data are collected from multiple engineering sources and interfaces. The data may be raw in the early stages of the data collection process, but over time the data are analyzed, filtered, and simplified to groupings of data that are immediately useable for maintainability purposes. The data are provided as input to the maintenance concept preparation tasking and the maintenance planning process. Some are these input data are [13]:

- Contractor and government conceptual phase studies
- Design engineering reports
- Customer resources and constraints in terms of personnel, skills, tools, and cost
- Operational and support concepts
- Projected facility and support equipment availability
- Quantitative maintainability requirements and other mission requirements that affect maintainability

These inputs are translated into the detailed quantitative and qualitative requirements, which become maintainability metrics for allocations, maintainability design criteria, and maintainability design characterizations, which are incorporated into the system and equipment design specifications. The data are provided primarily as input to the detailed maintenance concept and the detailed maintenance plan. The maintenance concept begins as a broad statement from a customer. It is defined in greater detail by the contractor during the course of data collection and maintainability analyses, until the maintenance concept is baselined and released in a controlling document. When the maintenance concept is baselined, a detailed maintenance plan is prepared.

The detailed maintenance plan evolves as hardware and software are designed and released to meet the functional requirements of the system/equipment. As the hardware and software are released and tested, more data become available for the maintainability analysis. These data initially allow laser-sharp focus on, and provides the justification for, the time parameter for each maintenance task to be performed in the field application, but later the data focus widens to include additional resources for maintenance, such as number of personnel needed to perform the maintenance tasks, maintainer skill levels based on training and experience, maintenance and storage facility capacity and capabilities, and availability of tools and support equipment. The data are not only used for the maintainability analysis task, but also for the completion of the maintenance plan. The maintainability data are provided for the interdependent efforts to prepare and complete the maintenance concept, the maintainability analysis, and the maintenance plan, as part of a closed-loop process. A dataset change that affects one of these tasks, either positively or negatively, will affect all three tasks the same way. For this reason, it is imperative that there is close cooperation and seamless flow of data between all parties affected by the data as the dataset changes.

2.5.4 Task 4: Maintainability Design Criteria

Maintainability design criteria are used to analyze different design options when there are options available, in order to make a design decision on selecting the best option. As a result of various iterations of maintainability analyses to constantly refine the quantitative maintainability allocations, the maintainability design criteria will be generated and evolve over the course of the development program. These maintainability design criteria are implemented by maintainability design principles that may start as generic maintainability design guidelines, but later become program-specific design criteria and guidelines that are tailored for the particular needs of the program. Some sources of early maintainability design criteria and guidelines, which were available to the military system and equipment development industry, are MIL-STD-803 (three volumes), *Human Engineering Design Criteria for Aerospace Systems and Equipment* [14]; ASD TR 61-381, *Guide to Design of Mechanical Equipment for Maintainability* [15]; ASD TR 61-424, *Guide to Integrated System Design for Maintainability* [16]; and NAVSHIPS 94324, *Maintainability Design Criteria Handbook* [17].

2.5.5 Task 5: Maintainability Trade Studies

Maintainability analysis may result in design trade studies when multiple options exist for solving an engineering design problem or engineering a new system/equipment design. Maintainability design criteria are used to analyze different design options for consideration in design trade studies. A maintainability trade study involves the creation of a decision matrix with design criteria compared to courses of action or design options. Design criteria may include quantitative maintainability metrics, such as a predicted MTTR, time to perform a maintenance action, or cost to perform the maintenance. Usually design criteria are weighted on the decision matrix using pair-wise comparisons to give more importance to the criteria that are weighted higher than others. These design criteria ultimately become maintainability design characterizations, which are incorporated into the system and equipment design specifications, when a design option or course of action has been selected.

2.5.6 Task 6: Maintainability Predictions

Maintainability predictions are necessary to give credence to the selected design option following a maintainability trade study. The prediction provides the customer assurance that the quantitative maintainability requirements will be satisfied with a certain level of confidence. If the prediction results conclude that the maintainability requirements will not be met, the prediction provides recommendations on what design changes are needed during the development program to correct the situation and guarantee program success. Maintainability predictions are assessments of the quantitative maintainability allocations similar to maintainability analyses to assure that maintainability requirements will be met when the rubber meets the road (slang for when the customer uses the system or equipment). MIL-HDBK-472 [10] is the standard method for conducting maintainability predictions. The results of maintainability predictions are compared to Maintainability Demonstration (M-Demo) results to further ensure the maintainability requirements are met when the design is mature and ready for deployment. "The standard allows the procuring activity to specify the use of a particular technique or offer alternatives of the techniques in MIL-HDBK-472 for contractor selection. Further provision is made for the contractor to propose the use of a technique, not contained in MIL-HDBK-472, when he deems such is needed. In any event, the procuring activity will either specify or exercise the right of approval over the technique to be used" [13].

2.5.7 Task 7: Vendor Controls

Once all maintainability analyses, predictions, design criteria, trade studies, maintenance concepts, and maintenance plans are completed, quantitative and qualitative maintainability requirements must be generated and flow down to all hardware and software item vendors, suppliers, and subcontractors. These requirements must be written so the vendors, suppliers, and subcontractors are able to verify the requirements are met and show proof of compliance to the requirements. MIL-STD-471 [9] is a good reference for all vendors to use to verify their maintainability requirements. The maintainability engineer must be involved in the generation of all specifications, SOWs, and contractual documents to the vendors if maintainability requirements flow down to them. The maintainability engineer must also be involved in the verification of and compliance with those requirements by reviewing the analysis and test reports from the vendors.

2.5.8 Task 8: Integration

Maintainability must be integrated with other engineering disciplines, business operations, and logistics support functions in order to be successful. The communications and data exchange between all functions, both engineering and the business, will enable the system or equipment and all its constituent pieces to be designed and integrated correctly for the benefit of the customer over the system/product life-cycle. Maintainability engineers must understand that integration is the connection of different parts of the design (e.g., electronic components and mechanical assemblies) to enable them to work correctly together. Integration is also the connection of different parts of the business operations and logistics support organizations to allow them to work effectively and efficiently together for the benefit of the program and the end-users.

The integration of all components, assemblies, software, equipment (including Government Furnished Equipment [GFE] and Contractor Supplied Equipment [CSE]), and subsystems into the system or System-of-Systems (SOS) is a challenge, which ranges in complexity based on the size of the system and the number of hardware and software item vendors, suppliers, and subcontractors. Just as the system hardware and software interfaces or linkages between elements of the design are only as strong as their weakest link, the same is true for the business side. Any breakdown in a single link results in a breakdown of the system. Thought should be given to how the integration interfaces and linkages should work, and if they could fail, how they can be maintained to prevent failure or to recover from the detrimental effects following a failure of any linkage.

For every linkage within an integration chain, there must be a means to maintain that linkage after the system is developed and deployed. This integration chain involves establishing the interfaces during the development process, which includes, but is not limited to, the electrical hardware design interfaces, the mechanical hardware design interfaces, the software design interfaces, the customer user community interfaces (operational and logistics), and the supply chain interfaces. Obviously, the electrical, mechanical, and software design interfaces are important for system design integration, which require early inputs from maintainability during the development process. The system and equipment developers should request maintainability parameter values from the procuring activity for the subsystems or items within the system's integration chain for those interfaces that the system integrator does not control. These maintainability parameter values from the procuring activity must be included in the data discussed in Task 3, and analyzed as discussed in Task 2, to arrive at the maintainability parameter values for the other items in the system. If the required maintainability parameter values are not available, the contracted system or SOS developer must estimate the maintainability parameter values for use as inputs into the maintainability analysis to develop all maintainability allocations and requirements for every link in the integration chain.

Besides the system design integration during the development process, maintainability must be involved in the integration chain related to the usiness, operational, logistics, and supply chain side of the program. Maintainability should be integrated with procurement personnel to ensure the availability of an uninterrupted supply of parts and assemblies to support maintenance and spares for field applications and services. Effective supply chain integration is necessary to ensure a continuous supply of essential and authentic components and assemblies to replenish field supply inventories as maintenance actions are conducted. Dwindling supplies prompt organizations to allow procurement of components and assemblies from risky sources, which can introduce poor quality or counterfeit components and assemblies into their supply lines and inventories. Counterfeit parts from unknown sources have raised national security issues. "The integration into a system of items over which the contractor has little control is one of the more troublesome contract or system management problems faced today. Early recognition of a problem and good communication between agencies involved can alleviate its total impact" [13].

2.5.9 Task 9: Maintainability Design Reviews

Most programs conduct design reviews to maintain control of the designs during the system or equipment development process. These reviews may be formal reviews involving the customer, or they may be informal internal contractor reviews to determine the status of the development efforts and make sure all tasks are on track for successful completion in accordance with program schedules and cost budgets. Maintainability requirements should receive appropriate consideration during each of these design reviews. It is reliant upon customer and contractor management to expect that maintainability activity status is reviewed and kept on track toward completion. The contractor's organization engineering and logistics elements responsible for implementing and executing the maintainability program must be present at each design review and be made available to present the maintainability program status to management. "In addition to periodic reviews, a final review is required prior to release of drawings and specifications for production. These reviews are important exercises and provide opportunities to insure that maintainability is incorporated into the design. They represent a major assurance aspect of the maintainability program" [13].

Paradigm 8: Understand the maintenance requirements

2.5.10 Task 10: Maintainability Data System

An integrated data collection, analysis, and corrective action system must be established in order to ensure the success of each program. As discussed in Task 3, maintainability data needs to be collected and used as inputs to maintainability analysis, and for other purposes to support the development process as required. The pedigree of the data must be accounted for in the integrated data system. This way, if any questions arise as to the legitimacy or accuracy of the data, the data source can easily be determined and answers provided quickly to resolve any issues. The design of the maintainability database must prevent the entry of bad data, and void "garbage in, garbage out." The maintainability data may be stored in the same data collection system as used for reliability engineering, which is called a Failure Reporting, Analysis, and Corrective Action System (FRACAS). The data collection system should be compatible with other data systems used by the customer to provide maximum benefit for customer entry and retrieval of data.

Paradigm 9: Support maintainability with data

2.5.11 Task 11: Maintainability Demonstration

For any maintainability requirement, there should be adequate analysis to support meeting the requirement within the capabilities of the system or equipment design, but this does not necessarily mean the requirement is verified and validated. A maintainability demonstration is usually the best means to verify and validate that a maintainability requirement was met. MIL-STD-471 [9] is the best reference for conducting a maintainability demonstration. With the successful completion of a maintainability demonstration, the contractor will have proven its achievement of maintainability contractual quantitative requirements.

2.5.12 Task 12: Maintainability Status Reports

Periodic status reports are required by customers for the procuring activity on certain programs, to properly monitor the contractor's maintainability program activities and provide feedback to recommend corrections or adjustments along the development process. These status reports may be written monthly or quarterly, depending on the involvement by each particular customer. Each status report should be reviewed by the contractor's program engineering management before it is transmitted to the customer. These reports may be combined with other system program status reports and submitted to the customer as a Contract Data Requirements List (CDRL) report. The status report usually provides a table to summarize the maintainability parameters, such as allocated, predicted, and observed maintainability values, that have been determined up to that point. "Further, the report should provide a narrative and graphical treatment of trends, problems encountered or anticipated, and actions taken or proposed. The status report will be a key official source of information and communication between the contractor's maintainability group and the procuring activity's maintainability monitor, and should be treated in that light" [13].

2.6 Maintainability Evolution Over the Time Period 1966 to 1978

Following the initial releases of the three military standard documents, MIL-STD-470, MIL-STD-471, and MIL-HDBK-472, there was rapid technological growth in the electronic components and computer industries affecting the retail, consumer, commercial, industrial, aerospace, and military marketplaces. The 12-year period between 1966 and 1978 was a time when the maintainability engineering discipline evolved to align with the technology advancements.

Between 1973 and 1975, MIL-STD-471 experienced several changes to keep pace with the latest technological developments in the electronic systems industry. The first set of changes was released in March 1973 within MIL-STD-471A [18], which was based on the results of the US Air Force Rome Air Development Center (RADC) efforts supporting new maintainability requirements for a wide range of equipment. MIL-STD-473 was released in May 1971 as a special case of MIL-STD-471 for aerospace systems and equipment. After approximately seven years, MIL-STD-471A superseded MIL-STD-471, which was released in Feb 1966. In April 1973, MIL-STD-473 was cancelled, and in two years was replaced by MIL-STD-471A as well. MIL-STD-471A Notice 1 was released in January 1975, which included a title changed from *Maintainability Demonstration* to *Maintainability Verification/Demonstration/Evaluation*. "Notice 1 added four tests for mean/percentile combinations and for preventive maintenance demonstrations" [6].

A new addition was incorporated into MIL-STD-471A, Notice 2, in 1978. This change described testability design characteristics in maintainability demonstrations, along with distribution functions and confidence intervals applied to testability verification test methods. Also, the Poisson distribution was cited as a timing method to evaluate false alarm rates. The new addition was released as an addendum to MIL-STD-471A, which was entitled *Demonstration and Evaluation of Equipment/System Built-In-Test/External Test/Fault Isolation/Testability Attributes and Requirements*. This change reflected a new US Navy initiative. "The Naval Electronics Laboratory Center produced a Built-in Test Design Guide in 1976; the Naval Avionics Facility sponsored a study of standard BIT circuits in 1977; and the Joint Logistics Commanders (JLC) established a JLC Panel on Automated Testing in 1978 to coordinate and guide the Joint Services Automated Testing program. Both the Navy and Rome Air Development Center (RADC) issued BIT design guides in 1979" [6].

During this same period of time, studies were conducted to evaluate maintainability prediction techniques. "The resulting methods used fault-symptom matrices that linked system failure characteristics to its fault signals. Several measures of entropy were used in regression analyses to develop subsequent prediction equations. Although those methods were not formally incorporated into MIL-STD prediction methods, they did foreshadow the types of analysis that would be needed in the next generation of prediction techniques" [6].

2.7 Improvements During the Period 1978 to 1997

"By the end of the 1970s, a substantial collection of maintainability design analysis, prediction, and testing procedures and tools had been developed for application to electronics designs. In contrast to the 1950s, maintainability has become a mature discipline that can be adapt and grow to meet the problems that are a part of the continuing technological evolution of electronic equipment [6]."

MIL-STD-470 was superseded by MIL-STD-470A [19], *Maintainability Program for Systems and Equipment*, in January 1983. This revision was a substantial expansion of the original document. Changes made in MIL-STD-470A include detailed task descriptions for the various maintainability program elements and tasks. MIL-STD-470A stipulates the specific details that customers must provide as requirements to their contractors. MIL-STD-470A also provided additional guidance with a comprehensive appendix to be used by customers or contractors in tailoring maintainability program requirements for program-specific needs.

MIL-STD-470B [20] was released in May 1989. It provided more types of quantitative maintainability requirements. These parameters were Mean-Time-to-Restore-System (MTTRS), Mission-Time-to-Restore-Functions (MTTRF), Direct Man-hours per Maintenance Action (DMH/MA), Mean Equipment Corrective Maintenance Time Required to Support a Unit Hour of Operating Time (MTUT), Maintenance Ratio (MR), Mean-Time-to-Service (MTTS), Mean Time Between Preventive Maintenance (MTBPM), Mean Preventive Maintenance Time (MPMT), Probability of Fault Detection, Proportion of Faults

Isolatable, and Reconfiguration Time. MIL-STD-470B also categorized these parameters into desired and worst-case requirement values.

MIL-STD-471A [18] was not changed during the time period between 1978 and 1983. In June 1995, MIL-STD-471A was reissued and designated as MIL-HDBK-471, and then was immediately cancelled. It was later replaced by MIL-HDBK-470A in August 1997.

MIL-HDBK-472 went through a change in January 1984 with the release of MIL-HDBK-472, Notice 1. Notice 1 included a new maintainability prediction technique with two approaches. The first approach was a quick estimation procedure to be applied to a new design concept or model at the early stage in development when only rough preliminary design data were available. The second approach was a detailed procedure to be applied to a more mature design created in the later stage of the development process when detailed designs and supporting analytical data were available. "The detailed procedure includes the analysis of system failure symptoms to relate them to potential failure candidates" [6].

In 1981, the RADC sponsored a report on the results of a Maintainability Prediction and Analysis Study [21] performed by the Hughes Aircraft Company, Fullerton, CA, that provided the full detailed maintainability prediction procedure. The study effort was initiated to provide a more engineering-oriented maintainability prediction methodology that was less complex and less costly compared with the current military standard procedure. The new maintainability prediction methodology involved a more accurate analysis and tradeoff process that directly quantified the benefits of fault diagnostic and isolation capabilities driven by test equipment requirements to meet maintainability requirements. A paper published by the IEEE Reliability Society in 1981, entitled "A modern maintainability prediction technique" [22], provided a summarized version of the maintainability prediction procedure including the two new approaches.

By the 1980s, integrated circuits (ICs) had reached significant component densities and complexities with reduced size, weight, and costs to warrant added functionality to support improvements in Design for Maintainability from assembly-level designs to system-level designs. IC chips included hardware and software for embedded diagnostics to perform functional testing, fault detection, electronic system health monitoring, and measurements of critical circuit performance parameters for statistical trend analysis. The data collected from the embedded diagnostics could be stored and retrieved from embedded memory and data storage devices for further in-depth data monitoring, analysis, and for historical purposes. Maintainability engineers now had the capability of planning maintenance operations to implement system Built-In-Test (BIT) procedures on a regular basis in a field environment to leverage the IC embedded diagnostics capability to determine system health status, and alert operators and maintainers on the need for corrective maintenance repair actions or preventive maintenance to avoid mission-critical failures in a mission application scenario.

2.8 Introduction of Testability

By 1978, RADC had conducted several studies to develop guidance on specifying and demonstrating testability in electronic systems and equipment. The RADC guidance included how to specify and demonstrate testability, and how to conduct design and cost trade studies involving testability, as described in RADC TR-79-309, entitled *BIT/External Test Figures of Merit and Demonstration Techniques*, and RADC TR-79-327, entitled *An Objective Printed Circuit Board Testability Design Guide and Rating Systems* [6, 23, 24]. RADC published the *Testability Notebook* in 1982 [6, 25], which contained the results of the previous years' testability studies. Shortly thereafter, Analytical Procedures for Testability [6, 26] was published, which discussed the studies of analytical techniques from several government branches of operations research that could become testability applications in the future. A *Testing Technology Working Group Report* was issued in 1983 as part of the Institute of Defense Analyses (IDA)/Office of the Secretary of Defense R&M study to identify high payoff actions in this area [6, 27]. Additional investigations in the testability area in the early 1980s provided payoffs by reducing support problems and enhancing system readiness when deployed. As described in the ARINC Research Corporation's System Testability and Maintenance Program (STAMP) paper at the IEEE AUTOTESTCON conference, October 1982, the integration of testability and maintainability was being realized [6, 28]. STAMP is a testability design and fault diagnostic system modeling tool using first-order test points and component dependencies as inputs to generate higher-order dependencies and their implications. STAMP allows testability assessments through automated identification of component ambiguity groups, redundant and unnecessary test points, and feedback loops.

Paradigm 2: Maintainability is directly proportional to testability and Prognostics and Health Monitoring (PHM)

2.9 Introduction of Artificial Intelligence

Artificial Intelligence (AI) provides quantum improvements in automation of maintenance techniques compared with manual traditional maintenance techniques. AI is pervasive throughout modern military systems that require extensive maintenance procedures using a variety of automated maintenance tools. "These techniques can include natural language processing, inferential retrieval from databases, combinational/scheduling problems, machine perception, expert consulting systems, and robotics. An 'AI Applications to Maintenance Technology Working Group Report' was issued in 1983 as part of the IDA/OSD R&M study to identify potential applications of AI techniques to the DoD maintenance environment" [6, 29]. The Artificial Intelligence Application Committee was tasked to examine the opportunities for applying AI to maintenance, to assess the costs, risks, and development times required, and to provide its recommendations to the DoD for action. The committee's recommendations were to take advantage of the maturity of the current state of the technology for creating maintenance expert systems for high-priority maintenance system applications, to develop versatile maintenance experts for specific domains, to create a tool to automate the creation of system-specific data for developing maintenance experts, to develop smart BIT systems to reduce false alarm rates and improve BIT coverage, to fund applied AI research for maintenance, and finally, to foster an integrated DoD–industry approach with a tri-service working group to coordinate all activity under the existing panel on automatic testing. A top-level focus on AI began in the 1980s and continues to this day, to improve maintenance processes as new automation technological develops and grows, until it is deemed to be mature enough for applications in modern electronic systems and equipment.

2.10 Introduction to MIL-HDBK-470A

MIL-STD-470B [20] was reissued and designated as MIL-HDBK-470 in June 1995. It was almost immediately cancelled in the same month, and then replaced by MIL-HDBK-470A [30] in August 1997. MIL-HDBK-470A was released in two volumes. Both volumes are contained in one electronic file that can be downloaded from the US Government Standards website. The title of MIL-HDBK-470A is *Designing and Developing Maintainable Products and Systems, Volume 1 and Volume 2.* Volume 1 of MIL-HDBK-470A is the main standard. Volume 2 is the design guidelines section of MIL-HDBK-470A, which is also designated as Appendix C. Appendix B of MIL-HDBK-470A Volume 1 describes the maintainability demonstration and test methods that replaced MIL-STD-471A. Appendix A provides guidance for contracting organizations to establish maintainability requirements for acquisition programs.

Unlike previous handbooks, which focused only on maintainability, MIL-HDBK-470A provides information to highlight maintainability in the context of an overall systems engineering effort. The handbook defines maintainability, describes its relationship to other disciplines, addresses the basic elements common to all sound maintainability programs, describes the tasks and activities associated with those elements, and provides guidance in selecting those tasks and activities. Table 2.2 provides a cross-reference of the tasks between MIL-HDBK-470 and MIL-HDBK-470A to show easily the differences and similarities between the two standards.

Due to the many aspects of maintainability and the large number of related disciplines, the depth

Table 2.2 Task cross-reference from MIL-HDBK-470 to MIL-HDBK-470A [30].

MIL-HDBK-470A section	Tasks from old MIL-HDBK-470											
	101	**102**	**103**	**104**	**201**	**202**	**203**	**204**	**205**	**206**	**207**	**301**
4.2 Management approach	×	×	×									
4.3 Design for M(t)			×	×	×	×	×	×	×	×		
4.4.1 Analysis				×	×	×	×	×	×		×	
4.4.2 Test				×								×
4.5 Data collection and analysis				×		×	×				×	
Appendix B												×
Appendix C										×		
Appendix D					×		×					

Tasks:		
	101 - Program plan	203 - Maintainability predictions
	102 - Monitor and control subcontractors	204 - Failure modes and effects analysis
	103 - Program reviews	205 - Maintainability analysis
	104 - Data collection, analysis, and corrective action	206 - Maintainability design criteria
	201 - Maintainability modeling	207 - Maintenance plan and LSA inputs
	202 - Maintainability allocations	301 - Maintainability and testability demonstration

Source: US Department of Defense, *Designing and Developing Maintainable Products and Systems, Volumes 1 and 2, MIL-HDBK-470A*, Department of Defense, Washington, DC, 1997.

Table 2.3 Scope of key topics from MIL-HDBK-470A.

Topic	Scope limited to
Availability and readiness	Basic concepts, effect of maintainability.
Life-cycle costs	Basic definitions, description of effect of maintainability on various cost elements.
Manufacturing	Description of impact of manufacturing on maintainability.
Human engineering	Description of human engineering discipline and relationship to maintainability.
Safety	Description of relationship to maintainability.
Testability	Definition as subset of maintainability, description of concepts, general information on key issues, design techniques and guidelines, definitions of metrics, and demonstration testing (Appendix B). Testability is covered in more detail in other handbooks and standards such as MIL-HDBK-2165.
Logistics support	General discussion with emphasis on how it is affected by maintainability.
Reliability-centered maintenance	Introduction with general procedure outlined.
Predictions	Description of applications with the most-used method from MIL-HDBK-472 included in Appendix D.

of coverage for some topics in MIL-HDBK-470A is limited. Table 2.3 summarizes the scope of key topics extracted from MIL-HDBK-470A [30].

2.11 Summary

This chapter has traced the origins and evolution of maintainability and maintenance engineering, examining several critical standards that set maintainability in motion and drove its growth in the electronics industry. The chapter explored a few research studies that contributed to the development of maintainability and testability practices and techniques. "As maintainability engineering developed in the 1960s, equipment repair times were driven by disassembly, interchange, and reassembly tasks. As electronic technology evolved from tube to transistor to various levels of integrated circuits, electronic systems became modular, modules were easy to remove and replace, and little or no alignment was required after replacement. As a result, maintainability designs now emphasize diagnostic capabilities and equipment testability computers have become an integral part of the maintenance process" [6]. Maintainability continues to evolve as new technological innovations materialize and are employed in modern electronic systems and equipment. The MIL-HDBK-470A standard provides a culmination of the beneficial standards, requirements, and lessons-learned over the last sixty years to provide a streamlined contractual path forward for maintainability engineering on systems and equipment programs.

References

1 Dhillon, B.S. (2006). *Maintainability, Maintenance, and Reliability for Engineers*. CRC Press, Taylor & Francis Group.

2 US Department of Defense (1968). *Policies Governing Maintenance Engineering Within the Department of Defense*, DoD Instruction 4151.12. Washington, DC: Department of Defense.

3 US Department of Defense (1980). *Reliability and Maintainability*, DoD Directive 5000.40. Washington, DC: Department of Defense.

4 US Department of Defense (1986). *Defense Maintenance and Repair Technology (DMART) Program*, Volume II: Appendices, Report AL616R1. Washington, DC: Department of Defense.

5 US Department of Defense (1964). *Maintainability Terms and Definitions*, MIL-STD-778. Washington, DC: Department of Defense.

6 Retter, B.L. and Kowalski, R.A. (1984). Maintainability: a historical perspective. *IEEE Transactions on Reliability* **R-33** (1): 56–61.

7 US Department of Defense (1959). *Maintainability Program Requirements for Aerospace*, MIL-M-26512. Ohio: United States Air Force, Life Cycle Management Center, Wright-Patterson Air Force Base.

8 US Department of Defense (1966). *Maintainability Program Requirements*, MIL-STD-470. Washington, DC: Department of Defense.

9 US Department of Defense (1966). *Maintainability Demonstration*, MIL-STD-471. Washington, DC: Department of Defense.

10 US Department of Defense (1966). *Maintainability Prediction*, MIL-HDBK-472. Washington, DC: Department of Defense.

11 Morrow, L.C. (ed.) (1957). *Maintenance Engineering Handbook*. New York: McGraw-Hill.

12 Ankenbrandt, F.L. (ed.) (1960). *Electronic Maintainability*. Elizabeth, NJ: Engineering Publishers.

13 Stanton, R.R. (1967). *Maintainability program requirements*, Military Standard 470. *IEEE Transactions on Reliability* **R-16** (1): 15–20.

14 US Department of Defense (1964). *Human Engineering Design Criteria for Aerospace Systems and Equipment*, MIL-STD-803. Ohio: United States Air Force, Life Cycle Management Center, Wright-Patterson Air Force Base.

15 US Department of Defense (1961). *Guide to Design of Mechanical Equipment for Maintainability*, ASD TR 61-381. Ohio: US Air Force Systems Command, Wright-Patterson Air Force Base

16 US Department of Defense (1961). *Guide to Integrated System Design for Maintainability*, ASD TR 61-424. Ohio: US Air Force Systems Command, Wright-Patterson Air Force Base.

17 US Department of Navy (1962). *Maintainability Design Criteria Handbook*, NAVSHIPS 94324. Washington, DC: US Navy Bureau of Ships.

18 US Department of Defense (1973). *Maintainability Verification/Demonstration/Evaluation*, MIL-STD-471A, also designated MIL-HDBK-471. Washington, DC: Department of Defense.

19 US Department of Defense (1983). *Maintainability Program for Systems and Equipment*, MIL-STD-470A. Washington, DC: Department of Defense.

20 US Department of Defense (1989). *Maintainability Program for Systems and Equipment* MIL-STD-470B. Washington, DC: Department of Defense.

21 Pliska, T.F., Jew, F.L., and Angus, J.E. (1978). *Maintainability Prediction and Analysis Study*, TR-78-169-Rev-A. Rome, NY: Rome Air Development Center (RADC), Griffiss Air Force Base (AD-A059753).

22 Lipa, J.F. (1981). A modern maintainability prediction technique. *IEEE Transactions on Reliability* **R-30**: 218–221.

23 Pliska, T.F., Jew, F.L., and Angus, J.E. (1979). *BIT/External Test Figures of Merit and Demonstration Techniques*, TR-79-309. Rome, NY: Rome Air Development Center (RADC), Griffiss Air Force Base (AD-A081128).

24 Consollo, W. and Danner, F. (1980). *An Objective Printed Circuit Board Testability Design Guide and Rating Systems*, TR-79-327. Rome, NY: Rome Air Development Center (RADC), Griffiss Air Force Base (AD-A082329).

25 Byron, J., Deight, L., and Stratton, G. (1982). *Testability Notebook*, TR-82-198. Rome, NY: Rome Air Development Center (RADC), Griffiss Air Force Base (AD-A118881L).

26 Aly, A. and Bredeson, J. (1983). *Analytical Procedures for Testability*, TR-83-4. Rome, NY: Rome Air Development Center (RADC), Griffiss Air Force Base (AD-A126167).

27 Neumann, G. (1983). *Testing Technology Working Group Report D-41* (ADA137526). Washington, DC: Institute of Defense Analysis (IDA). https://apps.dtic.mil/sti/citations/ADA137526 (accessed 15 August 2020).

28 Simpson, W.R., and Balaban, H.S. (1982). The ARINC Research System Testability and Maintenance Program (STAMP). Presentation at IEEE AUTOTESTCON conference, Dayton Convention Center, Dayton, Ohio, October 1982.

29 Coppola, A. (1983). *Artificial Intelligence Applications to Maintenance Technology Working Group Report,* D-28 (AD-A137329). Washington, DC: Institute of Defense Analysis (IDA).

30 US Department of Defense (1997). *Designing and Developing Maintainable Products and Systems, MIL-HDBK-470A*, vol. **1 and 2**. Washington, DC: Department of Defense.

3

Maintainability Program Planning and Management

David E. Franck, CPL and Anne Meixner, PhD

3.1 Introduction

This chapter describes the tasks involved in designing for maintainability, the need for maintainability program management, the basic elements of a Maintainability Program Plan (MPP), and the relationship of maintainability engineering to other engineering disciplines. The purpose of this chapter is to provide guidance for the development of an MPP that includes:

1. The descriptions of design, test, and management tasks and activities associated with maintainability engineering that may be performed during the system or product development process.
2. The rationale for specifying maintainability requirements for a system or product.
3. The methods used to verify and validate the maintainability requirements.
4. The structural elements of a maintainability program.

The objective of a maintainability program is to create and implement an MPP that defines how to design and manufacture a system or product that can be repaired from a state of nonoperation or dysfunction during its life cycle, or sustained in an operational or functional state to minimize the risk of operational or functional failures during its life cycle. The objective of a sound Maintainability Program, as stated in MIL-HDBK-470A, is "to design and manufacture a product that is easily and economically retained in, or restored to, a specified condition when maintenance is performed by personnel having specified skill levels, using prescribed procedures and resources, at each prescribed level of maintenance and repair" [1].

3.2 System/Product Life Cycle

This chapter provides information to help the reader understand maintainability in the context of an overall systems engineering effort, which includes consideration of the entire system life-cycle process. All products and systems follow a similar life-cycle process, though they may have different names or be divided up by different milestones, gates, phases, or stages. The basic system life-cycle process begins with an idea, concept, or customer need for a system followed by the early creation of initial system requirements, a list of system performance features or options, and an assessment of technological ability to achieve the desired system goals and requirements. This initial stage of the system life-cycle process leads to system design and development efforts including production manufacturing assessments and system life-cycle support modeling. After the design and development stage, the production stage begins. Once production begins, system life-cycle support planning continues towards an advanced state of readiness to provide

Design for Maintainability, First Edition. Edited by Louis J. Gullo and Jack Dixon.

ample life-cycle support for the system and all the associated products to be delivered to customers and deployed for all the users of the system and products throughout the anticipated useful life of the system and products. After the production stage, the useful life stage begins. The useful life stage is also referred to as the Operations and Support (O&S) stage. Depending on the design life or service life of the system or product, the useful life or O&S stage may be very long, sometimes decades long. After the useful life is exhausted, the End-of-Life (EOL) process or disposal phase begins to remove the systems and products from use and dispose of the constituent parts.

This is a gross simplification of a very complex life-cycle process that has many variations. The generic life cycle has been tailored for use in many industries, countries, products, and technologies. Examples include simple or complex commercial products, disposable consumer products, high-tech manufacturing, the telecommunications industry, the medical industry, aerospace and National Aeronautics and Space Administration (NASA) systems, Department of Defense (DoD) equipment, and various types of military, commercial, and industrial systems that are adapted for use in countries around the world.

A life-cycle variant that is prevalent in the commercial consumer products industry involves market research of potential customers instead of direct receipt of customer requirements to drive decisions during the life cycle. The focus of the life cycle in this variant is more on the expected market life of a product rather than on the design, operation, and support of the product. EOL decisions for consumer electronics products are based on marketing trends and projections of technology transitions instead of design life or service life of the product. The commercial product marketplace life cycle often includes six phases: market analysis, product introduction, product growth, product maturity, product saturation, and market decline.

The focus of this book is more attuned to a product or system life cycle based on the design life requirements, service life requirements, and support requirements of the product, equivalent to the practice within the United States (US) DoD, rather than commercial product marketplace life cycle. A life cycle focused on design life, service life, and support of the product is prevalent in many military defense systems, systems or products regulated for and certified by government agencies, and in capital equipment and systems used in assorted non-military circumstances involving a wide array of environmental stress conditions. DoD life-cycle engineering practices involve situations that are typically characterized by systems and products with a long life, with support costs that are a significant portion of the total Life Cycle Costs (LCCs). Often these systems or products have a mission operational life that exceeds the system/product's specified design life requirements, with insignificant original equipment acquisition and purchase costs compared to the total LCCs, and considering that these systems/products come with an Operational Tempo (OPTEMPO) and Concept of Operations CONOPS defining how the systems or products will be used and supported in a predefined user environment and support infrastructure. As of 2019, the US DoD employed a typical life cycle as shown in Figure 3.1 [2].

NOTE: For purposes of simplicity and consistency, the terms "system" and "product" are used interchangeably in this chapter. When one or the other term is called out, the intent is to apply the thought to both terms.

The process in Figure 3.1 is a rather simplistic version of a more complex diagram depicting all the significant events and tasks in the System Engineering Defense Acquisition Process. A fully detailed representation of this process would require many pages full of diagrams and many more pages of explanation. One item of special note in the DoD Life Cycle: there are numerous points in the process where reviews or milestones are in place, such as the significant milestones A, B, and C designated with triangle symbols in Figure 3.1. These are intended to provide points in the process where management at many levels are able to review the product development, evaluate it against its intended baseline, and assess the risks inherent in any product development effort.

Naturally, as products increase in complexity and quantity produced, the process to achieve acceptable results becomes more deliberate. Whether such products are for the commercial or military markets, careful decision-making is encouraged because the consequences of failing to carefully evaluate and consider the impacts of these decisions can be very

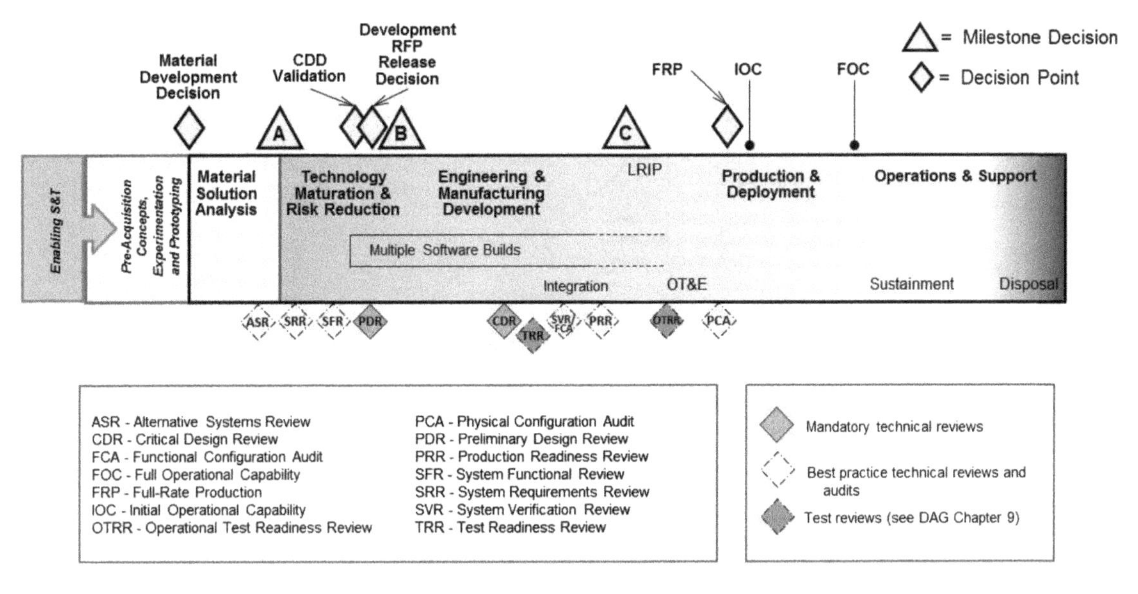

Notes:
- Derived from DoDI 5000.02, Model 5: Hybrid Program A (Hardware Dominant)
- The actual number and type of builds during the program will depend on system type and program strategy.

Additional acronyms: CDD - Capability Development Document; LRIP — Limited Rate Initial Production; RFP — Request for Proposal; DTRR - Development Test Readiness Review; OT&E — Operational Test and Evaluation

Figure 3.1 Weapon system development life cycle [2]. Source: US Department of Defense, *Defense Acquisition Guidebook (DAG)*, Chapter 3 - Systems Engineering, Department of Defense, Washington, DC, 2017.

great. Consider also that the impact of inadequate deliberation can also mean danger or death to human operators and maintainers, and damage to other equipment.

Paradigm 5: Consider the human as the maintainer

In organizations like the military and select commercial situations, where widespread and long-term sustainment of complex products is standard operating procedure, the product development process, such as that depicted in Figure 3.1, must necessarily be detailed and deliberate. Militaries, in particular, apply standardized maintenance planning, procedures, training, documentation, resources and more to products they develop and use. The maintainability and sustainment resources required to operate and support so many different products anywhere and in any situation by organic sources are extremely daunting. A structured forced discipline is required to ensure that products are designed with sustainment and operations in mind. Because these factors must be built into the design from the beginning, heavy emphasis is necessarily on the design and development processes, reviews and decision points to ensure that the required all-aspect considerations are well thought out and fully inclusive of all organizations with interest in the product and its life cycle. To accomplish this goal, the process to field equipment is very deliberate. Other situations may not indicate that such complexity is warranted. Where the product is not expected to be organically supported, product managers may place more emphasis on the manufacturing costs rather than on support costs in the early product tradeoff decisions.

There is a new paradigm emerging in the DoD and other government organizations in an effort to

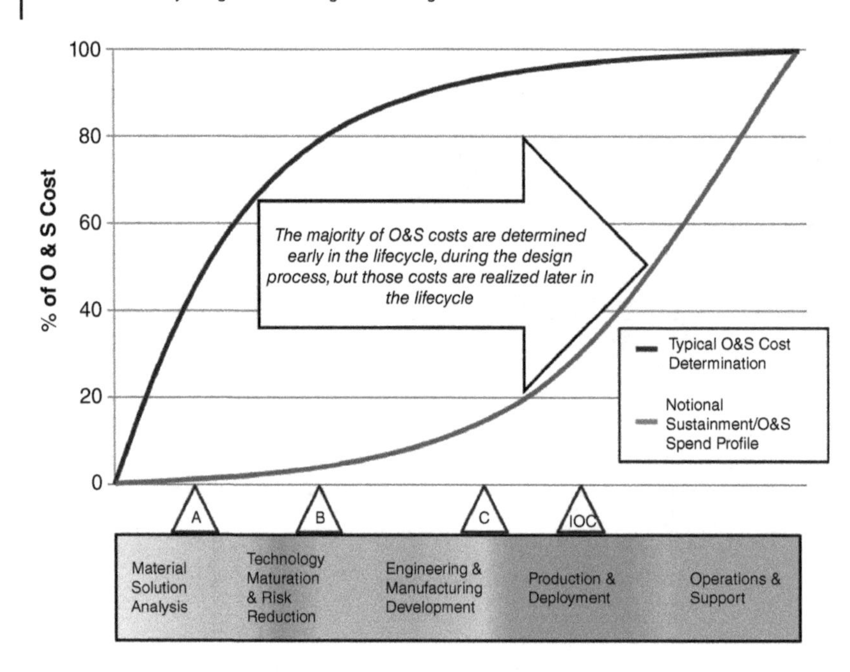

Figure 3.2 Time delay between decisions affecting O&S cost and the realization of those costs [3]. Source: US Department of Defense, *Operating and Support Cost Management Guidebook*, Department of Defense, Washington, DC, 2016.

streamline the system acquisition and development process and provide military users with products quicker. This trend is developing and evolving as ideas and possible alternatives are explored. Changes in this area can be expected to continue as new concepts and development tools mature. The current focus is on streamlining the contracting process and reducing the administrative burdens for government and the contractor. Additional areas of attention appear to be on products and systems that have less non-recurring and Research and Development (R&D) content, and on opportunities to reduce some of the design and development requirements typically imposed by the conventional processes. Continued attention to changes and developments in this matter is warranted.

Since maintainability characteristics, and their inherent cost burdens, are results of the design process, they are best influenced during the design process. Many studies show that 80% of the costs to maintain a product are locked in early in the design process. There are many opportunities during the design process to influence and affect the product maintainability costs, but because their impacts are not realized until the product is in operations, there is a time delay between when the decision is made and when the effects are realized. Since the DoD tracks such cost and relationships, they are able to characterize the typical delays and their contributions to total O&S costs. These are shown in Figure 3.2 [3].

The cost of slowing down the development process and building in pause-points to reflect on alternatives is generally mitigated by the cost savings to be realized by finding and correcting errors or issues early. In addition, getting the product support and user communities involved early provides a chance to identify opportunities for enhancing the supportability, and hence the costs, for the product, and for gaining early insight for planning for that support.

As previously stated, the maintainability characteristics of a product are created during the design process. Lack of effective maintainability characteristics are significant cost drivers over its entire life cycle. Effective mitigation of such costs can only be accomplished in the design of the product. Thus, cost reduction efforts should be initiated early and often throughout the requirements and design phases of the life cycle to address and minimize factors that negatively impact maintainability. Such efforts are especially productive if the design team is educated regarding the impact of maintainability on the life cycle and LCCs, and they are familiar with design actions that improve upon product maintainability.

Integrated Logistics Support (ILS) specialists that are fully engaged members of the design team can help with these efforts and education. The integration of ILS with maintainability specialists is an excellent way to ensure that these maintainability characteristics are considered and accomplished with the overall product design.

Paradigm 8: Understand the maintenance requirements

3.3 Opportunities to Influence Design

Designing maintainability features into the product/system upfront to prevent operator and maintenance issues later is the goal. You can't fix a maintainability issue without changing the design and adding new capabilities. You may have heard of a popular axiom that could be applied to design changes: "If it ain't broke, don't fix it." This axiom can be morphed to apply to maintainability design changes: "If it breaks, then you need to be able to fix it quickly and economically."

Design change is an expensive activity both in terms of engineering time and in terms of dollars.

Consider this scenario: an electronic circuit in a piece of equipment was designed using a cheap unreliable sensor component to identify issues or faults in the equipment. The sensor component failed often, causing excessive false alarms, and needed to be replaced. However, the placement of the sensor made it very expensive to fix due to accessibility to the faulty component, as the equipment's physical frame had to be removed to reach the sensor component. The equipment had to be redesigned by moving the sensor to a location that didn't require disassembly of the whole piece of equipment. If the reliability of the sensor and the Total Cost of Ownership (TCO) had been considered up front, the economic benefits of using a cheap sensor with a lower material cost compared with a more reliability sensor resulting in a lower TCO may not have justified that choice. At the very least, physical placement and accessibility of that sensor could have been recognized as a maintainability design consideration.

An overview of engineering design and common design activities during which the ability to highlight maintainability into the design is discussed next.

3.3.1 Engineering Design

Design of complex systems requires a team of engineers representing the various engineering disciplines involved to design, manufacture, validate, test, and maintain/support the products within a system. In addition, engineers whose role is to take a more holistic approach to the design will participate to provide insights into the impact of decisions made by one engineering team on another engineering team. Engineers responsible for the reliability and maintainability requirements have a more holistic role. Just like testability, maintainability is a characteristic of design; it cannot be fixed or improved without changing the design.

The products that benefit from Design for Maintainability are typically expensive systems ($1 million +), have a long usage (min. five years), and are complex systems with lots of parts (1000+ parts). Design of these large and complex systems requires a large team of engineers, scientists, and technicians to support the design of the product. Product design encompasses the full life cycle of the product, which includes manufacturing, test, acceptance, and maintenance.

There will be lots of meetings, formal discussion of design requirements, drilling down into specifications, design reviews, and working groups to review different components of the design. To understand how maintainability concerns can be best represented and how to identify issues, the time-frame and meeting structure used for the project needs to be understood. In this section we discuss the generally acceptable design process and describe the types of meetings that typically take place and documents that are typically created.

3.3.2 Design Activities

The tradeoffs can be many in a complex design, yet usually all attributes can be mapped to the three "E's": Efficiency, Efficacy, and Economy. The engineer responsible for maintainability should map the maintainability requirements to these Es, to communicate the impact more easily to the other engineering teams. Using the equipment example above, the engineer should have mapped the sensor cost to economy; the placement of the sensor impacts the efficiency of the replacement of the broken sensor, and that maps to the cost of doing a repair.

The ability to influence the other engineering teams relies on the following engineering practices and skills:

- Become involved early in the design process by writing requirements and specifications.
- Regularly attend design meetings and be prepared to point out the impact a design decision has in the maintenance of the product by the customer.
- Apply best-known analysis methods to identify and quantify the maintainability attributes.
- Review previous designs that have similar properties and where maintainability issues may illustrate a concern with the current design.
- Develop relationships with design, manufacturing and test engineers. Influence comes as much from relationships as it does from a set of facts.

Product design is a fluid and progressive confluence of interrelated activities that has a defined end point, which occurs when the design is frozen and becomes the design baseline for production builds. The design process is a collective set of compromises among technical, schedule, cost and supportability considerations. The design teams often are not anxious to stop their work while they pursue the optimum and best solutions within their spheres of influence. In fact, one of the more difficult problems of managing a design process is knowing when to stop engineering. Many times, a product design process is threatened by a design team's pursuit of the best design when good enough is sufficient. This potential runaway pursuit is one reason the project and technical management should pay close attention to the design progress aligned with the contract requirements, and be prepared to stop the design team when those requirements have been met. However, project and technical management must also be cognizant of the fact that the design characteristics and performance capabilities of the product are design features that were intentionally put in place by the design team to meet contract requirements and customer expectations for the product over the life cycle. Once the design is halted and the product moves to the production phase, these characteristics will not change except by design engineering intervention. Management should not halt the design team and freeze the design until it is sure all requirements and expectations are satisfied. This is especially true for maintainability. The support resources needed to maintain the product as well as the actual maintenance tasks themselves are completed and locked in by the product design team when the design activity process flow is completed and the design is frozen. Therefore, the design team should look for and take advantage of all opportunities to influence the product design before the design is frozen and management transitions the program to production.

There are many opportunities to influence the design that can be built into a well-planned product design process prior to the design freeze and production start. A summary of some of the more significant opportunities is provided in Figure 3.3. As seen here, and discussed below, there are many opportunities to influence a design. Additional opportunities to influence the design are discussed throughout this book.

- *Requirements definition and specification documentation*: important design activities to define and formally capture the maintainability design requirements for the product in a robust requirements management tool. These requirements will be turned into the actual product, so appropriate attention and care is necessary during this activity. Specification documentation preparation is

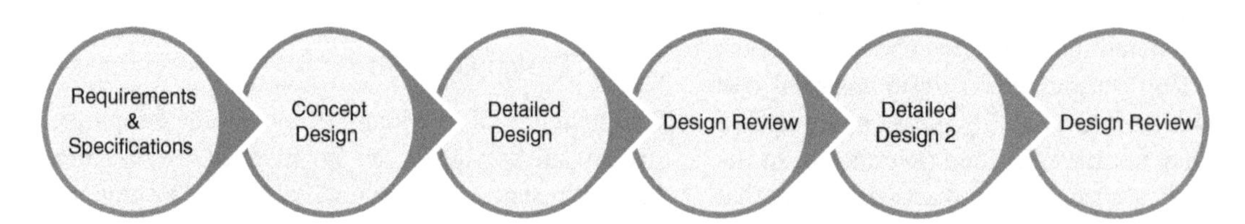

Figure 3.3 Design activity flow diagram.

imperative from the highest level of the design, down to the lowest levels of the design. High-level requirements are flowed down, decomposed, and apportioned to the lower-level requirements. As lower-level elements of the product become refined, product specifications are used to capture all relevant maintainability-related design characteristics and requirements. These specifications will provide continuity as well as the design baseline for producing the products. Enforced use of a single requirements database and management tool, along with an appropriate change and access control tool, is key to meeting the product maintainability requirements. Vendor and subcontractor maintainability requirements and performance should be the same as the prime product requirements for maintainability and for reporting and data sharing. Subcontractor requirements and design activities should be integrated into the overall product design activities as much as is practical. Where commercial products are purchased and used as-is, it is important to verify that the characteristics of the part comply with the allocated requirements. If the part does not comply with all requirements, additional testing and evaluation may be required to allow its use, increasing the effective cost of the part.

- *Concept design*: a key opportunity to think through and capture operational and design-significant elements of how the product is to be used and maintained; what environments it will be subjected to during operations, storage, and maintenance; and what maintainability characteristics are desired or needed in the product. A concept study (e.g., conceptualization or conceptual design) is often a phase of project planning that includes producing ideas and taking into account the pros and cons of implementing those ideas. This stage of a project is done to minimize the likelihood of error, manage costs, assess risks, and evaluate the potential success of the intended project.
- *Detailed design*: many detailed design decisions are made during this stage, which includes initial design activities, prototype development, preliminary design, and design peer reviews. Many of those impact maintainability or dictate the eventual maintainability characteristics of the final product. Maintaining a focus on maintainability requirements and their importance to the life-cycle suitability and costs of the product is suggested. During this stage, many design iterations occur. As shown in Figure 3.3, there is an initial Detailed Design followed by a Design Review and a Detailed Design 2. During the engineering design process, changes occur when an initial implementation doesn't work out. The team then refines that implementation or throws it out and starts with something new. The closer the design is to the end of the design process, the less likely it is to have major changes. The Detailed Design 2 stage is performed to correct the problems identified in the initial design review. The second design review is to ensure all the corrections were effective and the detailed design stage is completed. This is why it is important to participate in design reviews. Typical activities that are performed in the Detailed Design Stage are shown as follows:
  - *Initial design activities*: early design activities provide an important time to capture and explore maintainability requirements and to ensure that the design team is thoroughly indoctrinated into what the requirements are and their importance. Many maintainability-related design decisions and tradeoffs are made at this stage.
  - *Prototype development*: the creation of a prototype provides the design team with actual product-representative hardware and software in which to evaluate the maintainability results of design decisions. It is suggested that the product be handled, operated, and maintained using the tools and resources defined in the support plans in order to discover any issues with tools and maintenance actions.
  - *Preliminary design*: the preliminary design, or high-level design, bridges the gap between design conception and detailed design, particularly in cases where the level of conceptualization achieved during ideation is not sufficient for full evaluation. During this activity, the overall system configuration is defined, and schematics, diagrams, and layouts of the various product designs provide early product and system configurations. During detailed design and optimization, the parameters of the product or system being

created will change, but during the preliminary design phase, the focus is on creating the general engineering framework and design baseline for building the products and the system.

  - *Peer reviews*: regular and informal peer reviews within the design team are ongoing opportunities to visit, revisit, and reemphasize the maintainability requirements and to check that they have been addressed and met.
- *Design reviews*: formal design and customer reviews are important opportunities to verify and demonstrate that the design team knows and understands the maintainability requirements and that the design satisfies them.

3.3.3 Design Reviews

As shown in Figures 3.1 and 3.2, the contribution of maintainability is a large and serious one that deserves attention and focus if LCCs are an issue for the product or system or the user. The DoD system development process is well known for being laborious, while having points in the process where progress is intentionally measured by independent organizations, which helps to maintain focus and keeps the program on track. This development process includes many formal program reviews and design reviews that are part of the process to support major program developments. In all formal design reviews, maintainability should be addressed in the same level of detail and rigor as all other engineering and design aspects of the product. This includes a thorough review of the maintainability requirements and the understanding of them (including all requirements that relate to maintainability, such as maintainer tool sets and skill levels, available test and support equipment, time definitions, reliability, and how the requirements and the product fit into the intended operational environment); review of all requirements allocations and decompositions made; a thorough review of the design and analysis tools being used; discussion of any Logistics Support Analysis (LSA) database analysis tools used; demonstration of how the logistics and design efforts integrate and mutually support each other to achieve the best maintainability solution; a review of the testing, verification and demonstration efforts planned; and identification of any problem areas or issues that exist or are anticipated.

Some of the common formal design reviews that may apply to a product's design process include:

- Conceptual Design Review – generally performed by the originating or development organization to assess the design concept, but sometimes used in large system designs to review the design concept for logic, completeness, and adequacy.
- System Requirements Review (SRR) – a formal review of all of the system level requirements to ensure that the defined requirements accurately reflect the needs and objectives of the customer, user, or market; to identify any requirements that are ill-defined or incomplete, are poorly worded, are not technically possible/feasible, or are not in line with the planned product development limitations; to establish the system requirements as a stable baseline and place them under configuration management version control.
- Preliminary Design Review (PDR) – a formal technical review that establishes the Allocated Baseline (the establishment of Configuration Items [CIs] that comprise the product and the allocation of all system function and performance requirements to the configuration items). The PDR objective is to ensure that the product is operationally effective and that the early design approach, assumptions, and decisions are consistent with the customer's performance and schedule needs. The PDR is a detailed review of the conceptual design to ensure that the planned technical approach has a reasonable prospect of meeting the system requirements within the assigned budget and schedule. Review items for the PDR will include, but not be limited to: requirements allocations and decompositions; analyses and results; system and subsystem specifications, interface specifications; CI specifications and design baselines; drawing plans and status; testing plans; tools to be used (design, drawing, analysis, test, simulation, models, configuration and requirements management, cost and schedule management, etc.); design decisions and alternatives under considerations; and issues. Successful completion of the PDR is generally considered the approval to move into the detailed design phase of development.

- Critical Design Review (CDR) – a formal technical review that establishes the Initial Product Baseline (IPB) for the product (the IPB is placed under formal configuration change management controls following CDR) is complete in all matters and design implementation meets all of its requirements. The CDR is a multidisciplined, detailed product design review whose purpose is to ensure that the product can proceed into fabrication, demonstration, and test and is reasonably expected to meet stated performance requirements within cost, schedule, and risk. The same types of information are presented and reviewed during a CDR as in the PDR, except that they are reviewed in much more detail with all appropriate backup data, analyses, and studies presented or available for review. Following a successful CDR, a product will be cleared for entry into the testing and pre-production planning phases.
- Test Readiness Review (TRR) – a formal review to determine whether the system or its CIs are ready to proceed into formal testing. In the TRR, all test procedures will be reviewed to ensure they are complete and appropriately test the applicable requirements and to verify that they comply with program test plans. A TRR is typically conducted before each major system, subsystem or CI test, and provides management with the confidence that the unit under test has undergone a thorough test process and is ready for turnover to the next test. The Maintainability Demonstration (M-Demo) is one such test.
- Production Readiness Review (PRR) – a formal technical review to ensure that the product design is completely and accurately documented and ready for release to manufacturing. Besides technical matters about the product, the PRR puts considerable emphasis on the manufacturing and production resources, capabilities, and preparedness of the prime contractor and major subcontractors to begin production without experiencing intolerable risks or experiencing any unacceptable schedule, performance, or cost issues. Product maintainability is not normally involved in this review, but maintainability of the manufacturing equipment and correcting manufacturing defects or errors may be addressed.

3.4 Maintainability Program Planning

Product design and development often takes considerable time and is subject to competing priorities and issues that can distract the design and development teams. In addition, staffing is frequently fluid, making it difficult to maintain consistent focus on the original goals. To mitigate the impacts from such organizational and operational issues, an MPP is recommended. A maintainability plan helps keep the design and development team focused on the desired processes, interfaces and requirements. It also stands as a touchstone for both experienced and new team members when issues arise and confusion forms. It is also an important tool for engineering, program, company, and customer management to gain visibility and understanding into the intentions of the product maintainability efforts and how they integrate with the rest of the product plans and teams.

An MPP should include all elements, as found in any reasonable program plan, such as roles and responsibilities of key program members (including the customer and user, if applicable), schedule, management approach, background, key documents, product requirements and interfaces, design and development tools and processes to be used, and validation and testing plans. The content and structure of program plans differ depending on company guidance or customer desires. Regardless of structure, a well-considered maintainability plan should:

- Develop a clear understanding of the customer's maintainability requirements including applicable operational use scenarios, user environments, skills, and tooling.
- Describe how maintainability engineering will be integrated into the product design and development processes, especially with the systems engineering processes.
- Document the product, its interfaces, and its environments sufficiently to provide a thorough understanding of the product and the maintainability characteristics required and desired of it.
- Identify design guidelines and processes needed to create a product in which the defined levels

of maintenance are safe, economical, and easy to perform within the design boundaries established by the product requirements.

- Identify how the product will be validated and tested to measure and demonstrate that the desired levels of maintainability are achieved. Identify simulation methods, analysis tools, and testing efforts to be used.
- Identify if and how product maintainability performance will be tracked during operations and how that information will be used to correct problems, identify improvements, and feed into design guidance for future products.

3.4.1 Typical Maintainability Engineering Tasks

There are many maintainability engineering tasks that should be considered in the development of an MPP. These tasks are summarized in Table 3.1 with indication of which chapter of this book covers each topic.

The maintainability plan, program, and requirements should also be addressed in other related program plans including the Program Management Plan, System Engineering Plan, ILS Management Plan, the Engineering Plan, the Subcontractor/Vendor Management Plan, the Requirements Management Plan, and the Test Management Plan. Oftentimes, the customer has requirements or desires for separate contract deliverable plans or inputs to help them complete their own management plans. It is suggested that attention be given to the needs of the customer for maintainability planning documentation or information.

As the product development efforts proceed, it is advisable to establish a schedule to review the maintainability program efforts, progress, and issues. A review of the MPP should be included to ensure that it is up-to-date and reflects current program management and customer guidance.

3.4.2 Typical Maintainability Program Plan Outline

A suggested generic MPP outline might follow the outline to follow as a start. Of course, each MPP should follow contract or company guidance and should be adjusted for each situation.

- Appropriate title page(s) – title, date, contracting authority, authoring organization, document number, any proprietary declarations.
- Introduction – short paragraph introducing the plan and its content.
- Purpose – a short paragraph or two to describe the reason for the plan and what it is intended to document.
- Objective – a short description of the objective of the plan and what it is intended to achieve.
- Background – a description of the history of the project/product, an explanation of the timeline of activities and tasks up to this time, an account of how this product fits into the customer's operations and how it interoperates with other products/systems, identify the developing organizations over time.
- Scope – discuss the limits of this plan, what is included and what is not, identify any exceptions.
- Roles and responsibilities – identify and describe the significant participants in the program that are related to or involved with the execution of this plan. Identify organizations, organizational hierarchy, describe the roles of each participant, identify the responsibilities and the limits of those responsibilities for each participant, identify communications lines, significant points of contact, email and phone contact details.
- Equipment – describe the product/system in detail, identify the purpose and function of the product and each component within it, include interfaces with other equipment, each major part of the product, what it does, and maintainability features (possibly include technical data such as weight, size, reliability, etc.).
- Schedule – identify the overall program schedule showing where the product and the maintainability program fits into the program schedule, show details of the tasks included in the plan.
- Requirements – identify the source, hierarchy and precedence of the maintainability requirements, explain how the requirements are developed and managed (including the requirements management plan and tool, if any), identify each maintainability requirement and its parent requirements up to the top-level system requirements, identify how

Table 3.1 Suggested maintainability engineering tasks [1].

Type of activity	Tasks and description	Relevant to elements: Understand needs	Understand design	Design for maintenance	Validate maint.	Monitor operations	Chapter where topic is covered
Design	**Testability and diagnostics**. Designing and incorporating features for determining and isolating faults.			×			9, 13
	Design reviews. Formal or informal independent evaluation and critique of a design to identify and correct hardware or software deficiencies.	×		×	×		3, 12
	Environmental characterization. Determination of the operational environment in which maintenance is expected to be performed.		×	×			5, 6, 16
	Supplier control. Monitoring suppliers' activities to assure that purchased hardware and software will have adequate maintainability.	×			×	×	2, 3, 16
	Standardization and interchangeability. Designing for the use and incorporation of common items. Designing so items can be exchanged without alteration or change.			×			5
	Human engineering. Designing equipment so that they may be safely, easily, and efficiently used, operated, and maintained by the human element of the system.			×			10
Analysis	**Testability**. Systematically determining the coverage and adequacy of fault detection and isolation capability. Includes dependency and fault modeling.		×	×			13
	Human factors. Analyzing the design to ensure strength, access, visibility, and other physical and psychological needs/limitations of users, operators, and maintainers are adequately addressed.		×	×	×		10
	Equipment downtime analysis. Determine and evaluate the expected time that the system will not be available due to maintenance or supply.	×		×			6, 7, 15
	Failure Modes, Effects and Criticality Analysis (FMECA). Systematically determining the effects of part or software failures on the product's ability to perform its function. This task includes FMEA.			×	×	×	6, 14

(Continued)

Table 3.1 (Continued)

Type of activity	Tasks and description	Relevant to elements Understand needs	Understand design	Design for maintenance	Validate maint.	Monitor operations	Chapter where topic is covered
	Failure Reporting Analysis and Corrective Action System (FRACAS). A closed-loop system of data collection, analysis and dissemination to identify and improve design and maintenance procedures.			×	×	×	12
	Life cycle planning. Determining maintainability and other requirements by considering the impact over the expected useful life of the product.	×	×	×	×	×	3, 16
	Modeling and simulation. Creation of a representation, usually graphical or mathematical, for the expected maintainability of a product, and validating the selected model through simulation.		×	×			6, 8
	Parts obsolescence. Analysis of the likelihood that changes in technology will make the use of a currently available part undesirable.	×		×	×		3, 16
	Predictions. Estimation of maintainability from available design, analysis or test data, or data from similar products.		×	×	×	×	7
	Repair strategies. Determination of the most appropriate or cost-effective procedures for restoring operation after a product fails.	×		×			4, 7, 9
	Quality function deployment. Determine product design goals (i.e., product maintainability) from the user's operational requirements.	×	×				5
	Allocations. Apportion system-level or product-level maintainability requirements to lower levels of assembly.		×	×	×	×	6
Test	**Functional test**. Verify product is behaving as intended. Of interest to maintainability engineers are issues related to human factors.		×	×			12, 13
	Performance test. Verifying that the product meets its performance requirements, including maintainability.		×	×	×		12, 13
	Verification test. Testing performed to determine the accuracy of, and to update, the analytical data obtained from engineering analysis.		×	×	×		12

Table 3.1 (Continued)

Type of activity	Tasks and description	Relevant to elements: Understand needs	Understand design	Design for maintenance	Validate maint.	Monitor operations	Chapter where topic is covered
	Demonstration. Formal process conducted by product developer and end customer to determine whether specific maintainability requirements have been achieved. Usually performed on production or pre-production items.				×		12
	Evaluation. Process for determining the impact of operational and maintenance and support environments on the maintainability performance of the product.				×	×	7, 9, 12, 16
	Test strategy and integration. Determine most effective and economical mix of tests for a product. Ensure integration of tests to minimize duplication and maximize use of test data.	×		×	×	×	9, 12, 13
Other	**Benchmarking**. Comparison of a supplier's performance attributes to its competitors' and to the best performance achieved by any supplier in a comparable activity.	×	×				3
	Statistical Process Control (SPC). Comparing the variability in a product against statistical expectations, to identify any need for adjustment of the production process.	×					8
	Market survey. Determining the needs and wants of potential customers, their probable reaction to potential products, and their level of satisfaction with existing products.	×					3, 5
	Inspection. Comparing a product to its specifications, as a quality check.				×	×	12

Source: US Department of Defense, *Designing and Developing Maintainable Products and Systems, MIL-HDBK-470A*, Department of Defense, Washington, DC, 2012.

each requirement is to be satisfied (inspection, demonstration, testing, analysis).

- Guidance documents – include contract documents, technical specifications, applicable customer and company directives and standards (i.e., DoD MIL-STDs or corporate design guidance), overarching engineering management documents (i.e. System Engineering Management Plan [SEMP], ILS plan, Program Management Plan). Identify the hierarchy of the documents and which has precedence in case of conflicting guidance.
- Maintainability Program – identify and describe each of the tasks included in the MPP including the purpose and expected results of each task, task participants, the timeline for the task, the actions and steps included in the tasks, any tools or computer models or other resources to be used, and the responsible person or organization for

each task. Identify the engineering procedures that will be applied and where they are applicable. Consider such tasks as: requirements analysis and management, design reviews, design for testability and diagnostics, operational use considerations incorporation, environmental considerations in the design process, supplier control, standardization and interchangeability, human factors application in the design process, Line Replaceable Unit (LRU) definition, the maintenance concept, description of the operators and maintainers, expected tools and test equipment for servicing, maintenance level description, maintenance facilities, troubleshooting plans, modeling and simulation, parts obsolescence (Diminishing Manufacturing Sources and Material Shortages [DMSMS]), maintainability allocations and predictions, repair strategies, quality control processes, market analysis (if applicable), competitor analysis (if applicable), manufacturing interface, transition to maintainability testing.

- Maintainability testing – review the maintainability strategy and maintainability requirements, identify the maintainability testing plans (required, informal, formal, subcontractor development, and maintainability testing), test schedules, test responsibilities, test facilities, development of test-specific test plans, ready for testing plans, test witnessing, test reviews and test reports, resolution of conflicts in data interpretation, re-test rules, and the definition of success and non-success. Clarify and address which test elements are formal and informal or in-process tests, which are functional, performance, verification, demonstration, and analysis tests.
- Action items – provide a list of maintainability program action items, descriptions, equipment involved, action item status, owner, date opened and date closed.
- Conclusions and comments – discuss any appropriate conclusions that develop over the course of the program and any comments about any other topics that relate to or affect the maintainability program.

3.5 Interfaces with Other Functions

Product maintainability characteristics affect, either directly or indirectly, many of the functions associated with the design, manufacture, operation, and support of a product across its entire life cycle. The maintainability engineer is usually assigned to a development program to manage a maintainability program within the development program, to provide the development support for maintainability design tasks, and to interface to the other design functions of the design team. Awareness of what these functions are and an understanding of how maintainability affects them are important in creating a maintainability-effective product. Some of these product-associated functions and interfaces are discussed below.

ILS or sustainment engineering is part of the program design and development team infrastructure that provides cost and operational analysis to ensure the effective use, repair, and supply of a product over its life cycle. This ILS cost and operational analysis is intimately dependent upon the maintainability characteristics designed into a product. The maximum effectiveness of a sustainment program is achieved through early and consistent interactions and liaison between the design team and the support team to ensure the appropriate level of maintainability characteristics are designed into the product. To achieve the best maintainability possible for a product, the experiences and insights of seasoned ILS and maintainability engineers are invaluable in identifying and correcting design issues that will hinder maintenance and support.

The maintainability engineer is also an important design interface between the designers in the program design and development team and the ILS support team that is planning for the support of the product after delivery to the customer. Often, the design team is focused on solving a myriad of high-priority design issues to meet the needs of critical design milestones to complete the development phase, but it is not prone to addressing design factors that are not directly a part of those issues. This creates a gap in the thoroughness of the design. Only those immediate design issues are resolved to ensure short-term success, but long-term design success is not guaranteed. The skill sets of experienced ILS and maintainability engineers bring a focus on maintainability considerations to the design effort to fill in this long-term design area gap and improve the design results over the life of the program.

Achieving a cost-effective support infrastructure as part of the product design and development team's responsibilities requires considerable effort involving maintenance data collection, maintainability predictions, and maintainability analyses. The product's maintainability performance requirements drive these efforts. These maintainability analyses will determine such support needs as the tools required to support the product at all maintenance levels, what repair actions will be conducted at all maintenance levels, where parts will be repaired, the skills and support equipment needed to affect the repairs, the numbers of maintainers needed and their skill sets, and the repair documentation needs at all levels. The determination and development of a support capability takes time and, like the design process, is iterative. The maintainability engineer is the key to collecting the needed data, conducting the right type of predictions and analyses, and providing the data and analysis results to the support team, as well as the overall program design team.

Paradigm 9: Support maintainability with data

Part of product support includes training the maintainers (as well as the operators) in how to troubleshoot and perform maintenance. Courseware development includes identifying the tasks to be performed as well as the skills, tools, and equipment necessary to perform the maintenance. This courseware includes step-by-step troubleshooting, fault diagnosis, and maintenance instructions for every maintenance action for the product, both corrective and preventive maintenance actions. Part of this knowledge to develop this courseware comes from the interface between the maintainability engineer and the ILS maintenance support liaison with the design team.

The maintenance support liaison should be an integral part of the program design and development team, just as the maintainability engineer is. Both roles of a maintenance support liaison and a maintainability engineer may be performed by a single person due to program cost constraints. An appropriate balance in the liaison's roles should be as an active member of the team yet not intrusive into activities in which there is no valuable or legitimate contribution. Active participation in design meetings should be expected, as well as access to all design documentation. Parts lists of candidate electrical and mechanical parts and selected parts are useful for assessing parts commonality for spares provisioning, to maintain a parts supply inventory for replacement parts in the field applications. Access to mockups, prototypes, and test units is also to be expected. The maintenance liaison should be able to create problem or issue reports within the design team, and to make comment on other reports. Participation in change board reviews should be encouraged. A maintainability engineering interface and maintenance support liaison on the design team can, at times, be challenging, but the shared purpose of a common design goal can be advantageous for all involved.

Product maintenance does not start with the user. The process of manufacturing the product may be considered a form of maintenance, which includes correcting any related manufacturing issues during assembly. Similar product design maintainability characteristics that interest the maintainability engineer could be relevant to the manufacturing engineering and operations team for equipment assembly and installation during the program's production phase. The tests that are performed during production manufacturing operations may also be leveraged for maintenance support functions in the customer's use and maintenance applications. Both production and maintenance personnel want to minimize the variety and quantity of tools required to accomplish their tasks, and maximize the use of tools that are already common in the existing manufacturing and maintenance environments. New tools require greater costs to procure, and more training and documentation costs for production. Additionally, new and unfamiliar tools usually mean initially lower production throughput rates, higher maintenance Turn Around Times (TATs), and higher error rates, until production and maintenance personnel familiarization with them grows and matures. The commonality of purpose and experience makes the maintainability liaison and manufacturing team interface important.

Program milestone schedules can be an easily overlooked interface affected by maintainability. One of the requirements for a program schedule is a realistic and complete schedule with all program functions, elements, and influencers included. Another program schedule requirement is to ensure that the data used in the schedule is accurate. The maintainability engineer has a direct impact on early identification of design issues that can be solved before they become actual problems, potentially saving schedule risk. If these design issues are not identified early, they can become major schedule and cost risks later in the program. Also, because maintainability drives so many downstream support functions, their schedule compliance is dependent upon early and accurate access to design data, with easy continuous access to additional design data that may be needed later. Therefore, the maintainability function should be a contributor to the program schedule.

As a design feature that significantly impacts program costs, maintenance interaction with the design team affects the development costs as an additional cost item, but is a potential power factor in avoiding higher costs for manufacturing, operations, and support. The maintenance support liaison activities also tend to reduce other logistics support function development costs by making it easier to get the accurate data needed in a timely manner for all related tasks. It is appropriate to consider maintenance support when laying out the program financial budget.

Because maintainability functions and results impact the cost, technical, and schedule aspects of a program, the management functions should be interested in the performance, functions, results, and impacts of program maintainability efforts. It is reasonable for the maintainability function to report activity and status to program management in the normal course of program element status reporting.

The customer interests for supportability of the product will vary from customer to customer and product to product. The product maintainability function likely will mirror that of the customer to some extent. The level of interaction and the depth of interaction for maintainability concerns should be proportionate and appropriate for the customer and the product. Sharing of maintainability and related design data may span the spectrum from very little data exchange to complete open transparency, depending on the situation. Care should be taken to protect the data, especially where classified or proprietary data are involved.

The product user, depending on the product and the guidance of the customer, may have no interest or involvement in the product design. On the other hand, as in many military product development programs, the user may be heavily involved and have full time representation in the development program. Often, a good relationship with the user community is a valuable help to the maintainability and support teams for their operational perspective and insight into the use and maintenance activities to which the product will be subjected. They also provide valuable input on possible solutions and design ideas.

3.6 Managing Vendor/Subcontractor Maintainability Efforts

Managing a parts vendor or subcontractor brings a second layer of managerial oversight and detailed attention to product management. Often, this function is delegated to functional specialists with engineering and subcontracting expertise. It is advisable that the Product Management Plan (PMP) includes managing vendors and subcontractors. This management plan may be in the form of a separate document called a Program Parts and Process Management Plan (PPPMP), which could be referenced in the PMP. Depending on the extent of their involvement in the program, the PMP might address this as a complete section, as an appendix, or as a stand-alone subcontract management plan.

Perhaps the best way to minimize issues with a product is by integrating a vendor or subcontractor with the prime contractor's design, development, and production processes. The advantage of this option, when executed well, is that technical issues tend to be socialized within the engineering teams, minimizing surprises and miscommunications. This process also has the effect of reducing delays that come from formalized communications between prime contractor and its subcontractor functions.

Vendor integration is often not possible, especially in cases in which the product need for parts is a very small percentage of the vendor's production. For example, it is not realistic to expect a maker of computer chips to go through the costs and efforts to create a team to integrate into a prime's product process if the product need is for 100 chips and the manufacturer is producing a million chips. Additionally, they are not generally amenable to providing extensive part design data, especially proprietary data under such circumstances. Of course, Non-disclosure Agreements (NDAs) could be put in place to acquire and safeguard sensitive data.

When a vendor or subcontractor is agreeable to joining a product design effort, it is important to capture all expectations, restrictions, tasks, assumptions, schedules, and limitations in an appropriate subcontract or purchase agreement. There should be no misunderstandings among the parties, and open communications should be observed. Formal delivery of specific data, products, reports, and so on, should be expected, even if some of that data is provided informally via team interactions. Formal documentation provides a written record of contract compliance and mitigates disagreements or misunderstandings.

It is reasonable for a product prime contractor to request technical data from a vendor or subcontractor regarding their product. Technical performance data is needed by the product design team for full understanding of the part in order to successfully integrate it into the product. Also, it is reasonable to request data regarding any analyses and testing of the part and the conditions of the testing. Of logistical interest, data should be requested about the reliability and maintainability analyses and predictions available for the part, including the assumptions used in the predictions. Background data about the predictions for the part should include whether they are based on actual field performance, in-house testing, analysis, or user-filed data. The conditions surrounding the predictions should be included, such as environment, user activities, length of the testing or operating times, and so on. Knowing the conditions under which the predictions or testing were conducted gives the product support team insight into the pedigree of the data. From that, they can adjust their own analyses to account for differences in test conditions and the expected conditions for the product.

Careful coordination with contracts and engineering specialists of both the vendor/subcontractor and the product owner will go a long way to clearly establishing expectations for each. It also helps if all appropriate points of contact are clearly identified, along with their titles, roles, and contact information.

3.7 Change Management

Among the many tenets of product management is the one tenet that "things will change." Managing that change is among the most difficult parts of product management. Perhaps the best management strategy is to assume that all things will change and plan accordingly. Some changes are expected during the development phase of a program, such as personnel changes, engineering drawing changes, program schedule changes, and changes in program funding. Changes that affect a product's maintainability performance are less obvious and require more in-depth engineering involvement and discipline. A prudent PMP and/or PPPMP will include processes and procedures to guide the design team in creating change requests, preparing for change orders, and adapting to engineering design changes during the course of development.

The maintainability design process combines detailed design knowledge with life-cycle support analysis and planning knowledge. Generally, the design process itself will not change and will be included by reference in the PMP, including the establishment of design reviews and the design baseline freeze point. This plan should also include design guidelines for the product based on user needs and the expected support environment. Such guidelines would, for example, include the tools available to the user. However, the parts chosen by the designers may vary, and parts availability may change.

An effective change management tool is necessary to control all the design changes leading up to and following design requirements lockdown. It is imperative to control changes before the designs are frozen and locked into the design baseline. Establishing a baseline of design requirements is the first step to controlling a design. The design requirements baseline should have the ability for engineers to create decomposed and derived requirements that satisfy the parent requirement. All parent-to-child requirement relationships should be visible in any change management tool. All requirements should be forward and backward traceable through a requirements specification tree structure. The tool should allow for easy identification of defective requirements such as orphan requirements (lower-level requirement that is not traced to a top-level requirement) or parent requirements without children (e.g., top-level requirements that have not been flowed down nor decomposed to a lower-level requirement). An effective requirements management process includes a formal and disciplined method for all vested engineering disciplines to review and approve any changes or additions to the requirement baseline to resolve requirements defects or add new design features.

The product design team designs to a rigid set of requirements established in a design baseline. This rigid set of requirements needs to be managed. The product team should establish a process for controlling and approving the design baseline and changes to it, similar to the process for controlling the requirements baseline. Such mechanisms for controlling requirements and design changes are generally called Change Control Boards (CCBs). When implemented and followed, CCBs are proven management tools for effective control of requirement and design changes.

An important feature of an effective CCB is an impact assessment for all proposed changes. The submitter of a change request must include the impact of the proposed change on the product. These impacts should include the impacts of the changes on the design and manufacturing efforts, on the parts costs, on schedule, on all elements of product support, and any other elements or areas of the product where costs are affected. Because the impacts of a proposed change are so widespread, each leader of a cost area usually provides an impact statement for their particular area.

One of the cost areas that is impacted by most proposed design changes is the logistics area, particularly the maintainability cost component. This maintainability cost component is important because maintainability has a downstream effect on so many other logistics support elements and their associated LCCs, such as tools, training, manuals, and spares. Of particular significance are the definition of LRUs, Shop Replaceable Units (SRUs), a rotatable inventory of common stock hardware for connecting parts and assemblies (e.g., screws, nuts, bolts, wire connectors), and software code implementation that affects maintenance actions.

One of the challenges of long-term and high-volume production and repair operations is the continued availability of the component piece parts and Commercial Off-the-Shelf (COTS) assemblies. Third-tier suppliers may change their design or availability of individual component parts and COTS assemblies at their discretion. The CCB is involved in evaluating the impacts of all such changes and their impact on maintainability and logistics support. A change in a simple lock nut could result in the need for an additional tool size that adds to the maintainer's tool kit, in the cost of that tool kit, in changes to the technical documentation, and in the time required to conduct maintenance.

Part obsolescence is another topic that is a common issue in long-term production situations and should be part of a product CCB process. For complex products and products that closely interact with other products, obsolescence is often a surprise to the design team. To mitigate the impacts and surprise factor of obsolescence, often a special team is created to monitor change notices from parts suppliers and lead times for critical or important components. The term DMSMS has been coined to define the importance of establishing such a team to consider these impacts. Effective execution of a DMSMS team provides early notification where manufacturers are ceasing production of parts so that alternatives can be pursued. DMSMS items also must be subject to the program's CCB processes.

The DMSMS team should maintain a close relationship with parts suppliers to stay current on their product manufacturing plans and new product development plans. Many parts vendors release notices called Product Change Notices (PCNs) to their customers, stating their plans to change their product design or manufacturing line. Some PCNs may provide an early notice of the supplier's intention to discontinue production and delivery of particular parts. These types of PCNs are called EOL notices. This EOL notification allows customers to procure additional quantities to support their future needs until replacement parts are identified and procured. These EOL notices provide a time period for customers to place orders for Last-Time Buys (LTBs). In addition, vendors are generally agreeable to sharing future product development plans (usually with a NDA in place) so their strategic customers can plan ahead for their own future product plans involving new technology insertions and refreshes. Part of a DMSMS team's responsibility may include letting parts vendors know of design characteristics needed in their own future designs. This information helps the parts vendor identify future design characteristics needed by their customers and helps the product developer to select design options that use parts that are early in their production life cycle. This symbiotic information exchange helps to potentially mitigate against shortages in future products and helps in the continued product design evolution process.

Controlling product maintenance costs, along with the subordinate life-cycle support costs, depends upon controlling the changes that are bound to occur in product development. Uncontrolled changes in the design and in production can yield unexpected, uncontrolled costs in the short term, in the early stages of the product life cycle, and unnecessarily higher maintenance and support costs throughout the entire product life cycle. Controlling change does not eliminate the costs from such change, but it does provide a chance to make the most cost-effective change decisions with awareness of the cost implications, and determine whether there is a budget to cover such costs. Some changes may be delayed until a budget is available. For critical changes that don't have a budget, management budget reserves may be used to cover the cost of these changes.

3.8 Cost-effectiveness

Maintainability cost-effectiveness is a series of design tradeoff compromises between competing elements of a product design team to address its support requirements. More time and money should be spent in requirements generation and refinement so that the final design results in a better support position with lower support costs over the life cycle. However, the restrictions of time and budget frequently don't allow a product development team the resources necessary to achieve an optimum design solution and reach its cost-effectiveness goals. Due to the combination of design requirements volatility from constant design changes and a logistics support infrastructure that is unknown or not standardized, the promised support costs and cost-effectiveness goals are generally not achieved. A more realistic way to look at cost-effectiveness is from a point in time snapshot with stated limits, parameters, and understandings. Over time, this snapshot must be updated to reflect the new reality of the moment. This leads to finding bridging factors to help make realistic comparisons between one cost-effectiveness evaluation and the next, thus allowing any trends to be identified and fairly compared.

The subject of cost-effectiveness of a product brings up many considerations and viewpoints. The cost-effectiveness considerations could describe high-level cost benefits, such as a one-time cost avoidance that results in a $1 million cost saving, and low-level cost benefits, such as an incremental cost reduction that decrements the cost of a particular part by 5% per year until the supplier decides that the part should reach EOL and become obsolete. The definition of cost-effectiveness varies depending on who is asked and the context of the question. One question to be answered is "Does the cost-effectiveness result in a cost benefit Return on Investment (ROI) for the customer within one year of implementing a design change?" Maybe the cost-effectiveness is to demonstrate a benefit only to the product OEM. Given the lack of standardized terminology and ill-defined nature of cost-effectiveness measures and processes, special attention must be paid to any efforts to determine or assess the cost-effectiveness of any product. Considerations in assessing cost-effectiveness first begin with defining the scope and boundaries of costs to be considered, such as the high-level cost limits and the lower-level cost constraints. Production costs differ from the cost of ownership by a fairly clear demarcation between the cost to make a product and the sustainment cost to the owner. However, considerations for LCC evaluations differ dramatically depending on the product and the assumptions included. Some LCC evaluations include research and product development costs, whereas others may not. The person conducting the cost-effectiveness analysis should identify, define, and coordinate the terms of cost-effectiveness among all invested parties, and especially gain acceptance from the Finance Department lead on the program team. The financial lead should verify the accuracy of all cost data collected, especially high-level cost data, the validity of assumptions, the potential likelihood that the cost benefits, both lower-level and higher-level, described in the analysis can be achieved. Unless provided with a specific cost metric as a product requirement, cost-effectiveness can be considered as the best result among two or more similar ROI alternatives using a common cost evaluation baseline.

By its nature, evaluating cost-effectiveness automatically dictates a comparison be made against a measure of "effectiveness" in terms of "cost." The "cost" part of the term cost-effectiveness is pretty clear that the assessment be made in terms of some monetary form, usually the currency of the sponsoring country. One must ensure that the currency to be used is defined before the analysis is started. It is also important to determine whether multiple country currencies are to be addressed and included. If this is the case, agreed monetary exchange rates must be determined.

Identifying the "effectiveness" part of the term "cost-effectiveness" is often more elusive or interpretive. Many times, the effectiveness goal that is at the heart of a cost-effectiveness study is not well understood or defined by those wanting the study. Often, extra work and research work are needed to identify, refine, or define effectiveness sufficiently for an analysis team and the consumer of the analysis to have a common understanding of the term. The analysis work to determine cost-effectiveness of a particular course of action or design trade will be of little use to management or the customer if there is disagreement or poor understanding about the most basic underpinning term of the analysis.

Effectiveness may be defined in a requirement by the customer, or it may be an idea from someone's prior experience. Effectiveness may also be an open concept query from a curious management person or it may be a dictate at budget origination. Regardless of how it is defined, cost-effectiveness has an implied level of acceptability and unacceptability. The most important aspects of cost-effectiveness evaluations are that the cost analysis process needs to be thorough, objective, and consistently applied. There must be common analysis measures applied across all options, and the cost analysis results must stand up to scrutiny for fairness and accuracy.

A cost-effectiveness evaluation involves a comparison of multiple options or alternatives to determine the best ROI. The intent of a cost-effectiveness assessment is to objectively compare two or more situations using a common parameter, and generally that common factor is currency. It does not matter whether it is called a Cost–Benefits Analysis (CBA), an Analysis of Alternatives (AOA), a Business Case Analysis (BCA), a trade study or tradeoff analysis, a design alternatives analysis, an options analysis, or any other similar name – the same processes are common to all.

It is important to establish a set of analysis goals and processes early in a cost-effectiveness analysis. Understanding and documenting the purpose of the analysis is critical to establishing and maintaining a baseline throughout the analysis for the team and for reviewers of the results. This is especially important for analyses that are complex or occur over a long time where teams can lose focus. Also, there is nothing to be gained by having an analysis sponsor remark, after reviewing the results of a cost-effectiveness study, "But that is not what I asked you to do." A short cost-effectiveness analysis plan would be appropriate to document all pertinent aspects of a planned analysis to mitigate the potential for such problems.

The cost-effectiveness analysis plan should address all factors that will be evaluated in the analysis. It should address how they will be considered, the length of time the analysis covers, the weights to be given to each factor, the specific options under analysis, the goal of the effort, the team members, and any other relevant information that will help the team's and the consumers' understanding of the results. The cost-effectiveness analysis team should have access to all data they need and to any subject matter experts required, such as financial experts, design staffs, contract specialists, and analysis tool experts.

There are many different methods of comparative analysis. Sometimes the basis of comparison is a fictional or notional product or design option compared to an existing product. Other times the comparison may be against existing competing product designs in the commercial marketplace or an analysis of multiple proposed designs for a particular user problem or mission. Regardless of the subject, often the terms of measure or parameters or other quantified characteristics will be different among the options being considered, and those differences will not always be easy to quantify in common terms. Identifying common factors among the options early will help focus efforts on readily comparable factors.

The remaining, non-common factors will require the cost-effectiveness analysis team to explore ways to quantify features and performance characteristics in common terms and then devise a way to translate this performance into cost.

Some considerations to address in a cost-effectiveness analysis are:

- *Time*: The length of time for the consideration of costs is important to establish and maintain in the plan. The analyst should consider the timing or frequency of the cost analysis, whether the analysis is conducted once or multiple times as more current cost data are collected. The analyst should also consider the time window for the data collected. Are the data based on consistent or irregular time windows (e.g., weeks, months, years)? Are the data based on ancient history with many gaps in the data, or current accurate measurements? An analysis window of two years will likely yield a different view of a product's life costs than evaluating over a 20-year life. When considering this problem, a reasonable approach would be to look at the expected life of the product and the expected life of the mission it is to fulfill. A 20-year cost analysis of a 2-year lifetime product is not very fruitful, whereas a 2-year analysis of a 2-year product would provide more useful results. The utility of a cost analysis increases as the product's life approaches the mission life. The appropriate term should be operationally representative. Too short a term will not capture relevant costs that don't typically occur until later in the lifespan. This could skew the results in an unrealistic and unexpected way. Note that some recent government weapon systems have instituted cost evaluations over a 50-year time span, whereas previous similar systems were evaluated on a 20- or 30-year time-frame. This has, naturally, caused some issues and confusion when comparing life costs among similar systems and when trying to determine which is more cost-effective. The analyst should exercise care when choosing the cost-effectiveness analysis time-frame to evaluate.
- *Countries*: If multiple countries are involved in the product, care needs to be taken to develop a common monetary parameter to use for evaluation purposes. There are two additional significant considerations to address. The first is to identify and define the exchange rates to be applied. The second consideration is to determine the appropriate cost escalation rates for each currency for each of the evaluation years. These factors are crucial in establishing and maintaining consistency and credibility in a changing, multi-currency situation.
- *Scale*: If the intent of the cost evaluation leans more towards a LCC evaluation, the analysis will require considerable rigor, planning, and discipline for all involved. It is advisable to include specialists in financial analysis, cost analysis, and ILS on the evaluation team. The scope of an LCC analysis is large and includes parameters that are often neglected or ignored. Cast a wide net in finding help from experienced professionals who can assist in the analysis or as independent reviewers or sanity-checkers for the effort. The analyst should scale the analysis effort so that the right resources are available when they are needed.
- *Budget*: In the analysis plan, budget restrictions will impact or restrict planned resources needed to conduct the evaluation. Careful planning is necessary to identify what these restrictions are and what resources are affected. The plan will, naturally, need to follow budget restrictions, and also identify how to conduct the evaluation despite the restrictions. This budget restriction issue should be addressed head on in the plan, as well as addressing how people and equipment resource management will be used to maintain the budget. The analyst should consider tools or prior studies that can be leveraged to improve the effectiveness of the resources needed.
- *Weighting*: Not all costs have the same priority or importance in an evaluation or to the customer. Weighting factors need to be established to allow a fair comparison of factors that do not have the same importance. There are several methods for weighting factors. One method is to apply a set of simple 1 to 9 (example) weights to each factor based on relative importance of the factor to each other. Another method is to combine factors where, for example, factor 1 and 2 are twice as important as 3, and 4 and 5 are 50% more important than 7 and 8, and 6 and 9 are 1/3 as important as either 1 or 2. The analyst may perform a pairwise comparison between each of the factors for consideration in a decision matrix to determine the weighting.

Whichever method is used, it is critical to ensure the analysis includes factor weighting.

- *Criteria*: The evaluation criteria and the acceptable parameters should be identified in the evaluation plan. If pass/fail criteria are to be used, the analyst should state this. If a sliding scale of pass/fail scoring is involved, the analyst should state the grades for each evaluation level. It is important to include these items in the plan to prevent errors or misunderstandings, and to ensure the analysis team stays on track.
- *Warranty consideration*: If applicable, the cost of warranty should be addressed as well as the cost of repairs and returns, and the cost of maintaining a repair capability.
- *Product liability*: If appropriate, consider the risks and costs associated with product liability from use and potential costs of litigation. The analyst may want to consult legal counsel authorities for insight into this liability risk topic.
- *Periodic updates*: Long-term cost-effectiveness analysis evaluations should include factors such as the cost of periodic updates to products, planned replacements with new products as part technology insertion cycles, overhaul costs, training of replacement personnel as they enter the support train for the product, parts obsolescence, the introduction of new technologies to the existing product, and the introduction of new technologies to the logistics support infrastructure.

3.9 Maintenance and Life Cycle Cost (LCC)

The classic definition of maintenance, which is typically applied by logisticians, addresses the process of either restoring a product to its original operational state and condition, or keeping the product in an operational state and condition. A product can get into an improper, non-operational condition in many ways, including a regular part failure, a poor design, a maintenance-induced failure, by accident, from acts of war, from handling mishaps, from failures of other equipment, from nature, or even from a "hard" landing when a cargo parachute doesn't open when dropped out of an airplane. Preventive and scheduled maintenance is also included in the classic definition. To more fully represent the scope of what is considered as maintenance, one must also include all maintenance activities involved in a product's lifetime. These activities include assembling a product, teardown and reassembly, removing a product to be able to repair another part, conducting major overhauls, planning incremental product updates, incorporating design changes, demilitarizing the product, or performing EOL disposal activities. Suffice to say there are many, many ways and reasons for maintenance to be required during a product's life cycle.

The life cycle of a product extends far beyond the purchase price, as do the associated costs. As discussed earlier, the life-cycle support costs for a product are typically much higher than the purchase price, and are of considerable interest to those tasked with providing that support. For such situations, maintenance-related costs begin in the very first phases of the life cycle when maintainability and maintenance requirements are being sorted out, validated, and then turned into requirements. They continue to accumulate in the form of costs for early design concepts, contracting costs of buyer and seller, and design, prototyping, and testing efforts. These costs occur before a single product is sold or bought.

Those responsible for life-cycle planning for a product must consider the entirety of its life cycle, identifying and quantifying all activities in which maintenance is involved. Planning for long product time-frames requires one to identify and consider many opportunities to experience costly events and to plan for supporting those events while realizing that products that are used for longer periods of time will wear out and fail more, and therefore need more updating. With existing products, past performance is a good, but often not a great, predictor of future events and costs. With new products, there is no past performance to consider, so the cost estimates are much less precise. This is one reason why so much attention is given to early design efforts and maintainability analyses (along with other related logistics analyses) for new military systems and large commercial products.

LCC analysts try to predict the future for years in advance. Their data products feed into the budgeting processes which eventually result in the approved costs authorized for buying and supporting a product. Budgeting cycles for large companies and governments are measured in years, typically five or more years, and sometimes 10 years. In addition, life-cycle cost estimations for large systems, such as major military weapon systems, have 30–50 year time horizons. As the actual costs in these out years cannot be known with accuracy, life-cycle planning and budgeting by necessity involves predicting support costs for several or many years in the future with the best data available and making judgments about future events. Thus, long-range estimates of future maintenance-related LCC and events is problematic at best. Typically, the further out the estimates go, the far less accurate they are.

One factor in making better maintenance-related LCC estimates is having a thorough understanding of all the cost elements that are associated with all maintenance activities related to support of the product. Proper and accurate maintenance LCC estimates requires a complete understanding of the product, all of its operational uses, and all maintainability related elements for it. It also requires a thorough understanding of its full life cycle. Once the full life cycle is well understood, a good understanding can be achieved of the effects and influences maintenance has on the support costs and where to focus efforts to reduce or at least minimize their impacts. This understanding opens up opportunities to evaluate design options for the product, assess the impacts of those changes, and see whether the results are worth the investment in making the change. Likewise, potential changes in the support concept and plans can be evaluated to see if such changes would yield worthwhile benefits. These analyses and trade studies provide early chances to mitigate maintenance and maintainability-related support issues during the product's design, and to do so when changes are the least costly to make. Later in the life cycle, design changes will be cost-prohibitive. If design changes are not going to be possible or cost-effective, the planning for maintenance support provides an opportunity to adapt to and accommodate the product in the planning stage rather than reacting after the fact.

A relatively recent support concept that might be employed is Performance-based Logistics (PBL). This support option is an alternative that requires an adjustment to the "normal" sustainment LCC assumptions. Under the PBL concept, a vendor accepts maintenance and repair responsibility for a product and warrants an agreed level of operational performance to the user, usually in terms of some operational metric, such as Operational Availability (A_0). Sustainment costs, including maintenance, are borne by the PBL contractor who is also responsible for maintainability performance metrics. Obviously, this paradigm changes the cost equation and the assumptions included in LCC calculations. Any PBL contract requires considerable thought and negotiation, and numerous terms and conditions must be identified in detail. Often, there are specific time-frames that have to be met for the terms to be in effect. Additionally, the veracity of the data that underlie the PBL terms is very critical to an effective PBL effort, and this is of interest to the user as well as the PBL operator. When considering a PBL-type arrangement, one must do considerable due diligence and research and must consult heavily with contracts and logistics specialists throughout the consideration and implementation process.

Trade studies should look at a product and identify and assess the associated costs for all the tools, test equipment, special support equipment, facilities, Personal Protective Equipment (PPE), skills, special certifications and training, special handling or transportation requirements, as well as any other elements of the support infrastructure for cost-saving opportunities. Also, for long lifetime products the costs of periodic upgrades or overhauls need to be identified. Designing in the ability to easily upgrade a product without having to change its basic architecture is a significant cost avoidance opportunity, as is reducing the frequency of overhauls.

Finally, a life-cycle analysis will not be complete if it does not include the maintenance activities and costs associated with EOL, End-of-Service (EOS), demilitarization, declassification, hazardous material handling and disposal, handling of classified and proprietary materials, maintenance involving explosive items, disposal of materials such as batteries, paint and paint removal, and work in difficult, harsh environmental conditions that require nuclear, biological or chemical environment Mission Oriented Protective Posture (MOPP) suits or cold weather gear.

3.10 Warranties

Warranties for product performance are not frequently utilized, except in the case of consumer retail products. Product warranties are inconsistently applied in cost-effectiveness analysis, and variably standardized across many types of industrial, defense, and aerospace products. While not frequently employed, essentially, a maintainability warranty may stipulate that a product will provide a specified level of maintainability performance over a specified period of time. To be in effect, a warranty has to be accepted by two parties (at least), the offeror (usually the product manufacturer) and the buyer (usually the user). In addition, the terms of the warranty need to be specified exactly, including what is offered, how it is to be measured, the conditions under which the warranty is valid and/or invalid, the time-frame for which the warranty is valid, and all exceptions to the warranty. Also, the warranty should specify any penalties that are in force for not meeting the warranty terms and a process by which the buyer/user and the offeror will evaluate and adjudicate warranty claims. Other terms and conditions may apply as well, depending on a number of factors, so it is advisable for the analyst to consult with appropriate legal counsel when considering or working with any warranty contract terms.

Typically, maintainability warranty performance is specified in terms of how long it takes to effect a repair, usually measured in Mean-Time-to-Repair or Mean-Time-to-Replace (MTTR). Since maintainability is a statistical estimation based on many factors, there may be numerous caveats to what MTTR means. This is important to the offeror due to the risk they accept in producing the product and offering the warranty. This is likewise important to the user as they are building a support infrastructure around the performance expectations of the offeror. Both parties will find it helpful to further decompose the MTTR term to provide more definition of such items as the types of maintenance (preventive, corrective, field level, backshop, depot, remove and replace, overhaul, etc.), tools and test equipment to be used, skill levels and training of maintenance personnel, availability of spares, the timeliness of warranty claim reporting, and how the maintenance and time data are to be collected and reported.

Often, especially for military products and systems, the customer recognizes the large LCC involved in supporting the product, and they further recognize the value of early design involvement to mitigate through-life maintenance costs. It is for reasons such as these that, in government and especially military product development programs, one will see heavy involvement of the customer with emphasis on multiple levels of design reviews and progress reporting, including problem reporting. While this may seem, at times, intrusive, it really reflects the inherent risks the customer is accepting to maintain and support a product for, perhaps, 20 years or more. Because the risks and costs are high, the customer involvement is high.

Maintainability warranty discussions should be held early in the product procurement phase, and metrics leading to putting the warranty in force should be reviewed at each formal design review, such as the PDR, the CDR, and the PRR. Typically, a government customer will also specify that specific maintainability predictions be conducted throughout the course of development culminating in a M-Demo, which may be conducted as part of a logistics demonstration.

It should be expected that government users, especially military users, will have an existing data collection and reporting system in place to collect and analyze, among other things, maintenance operations. This data collection capability is an excellent source of maintainability data for warranty purposes, assuming there is agreement on the accuracy and efficacy of the data. Large-scale data collection systems are likely to have inaccuracies inherent in a widespread, many-user system, and this should be discussed and reconciled prior to any agreement to use it as a warranty enforcement source. For example, data collection can be relegated to less than a priority when maintenance is conducted during times of high stress and in theaters of active warfare or during other emergency conditions. This should be understood and agreed in advance by all parties.

One maintainability warranty form that is prevalent, though not often recognized, is the self-warranty situation. Self-warranting is where the user accepts any maintainability performance with the expectation that it will not be too bad. In this situation, the user is taking all the risks. With self-warranting, the upfront costs seen with predictive analyses of maintainability predictions, detailed maintainability design analyses, M-Demos and testing, and field assessments are not in play. This can make a product's development quicker and less expensive up front, but it is an increased risk to the user. This is seen in many commercial situations and is generally acceptable to the user when there is confidence in the manufacturer and prior products. It is applied less often for large products or purpose-built products except when time is of essence and is more important than downstream costs.

If a PBL support contract is part of the warranty provisions for a product, the terms, requirements, limitations, and restrictions should be carefully considered. In some cases, a PBL includes extra costs for items that have been opened or where maintenance has been attempted by the user. The impact of such restrictions on the user could be significant, and it is important that the user's maintainers and the support system are properly trained and notified of such provisions.

3.11 Summary

There are numerous obstacles and problems with which to contend in the planning and execution of a maintainability program. While success is not guaranteed, making smart informed choices can go a long way to implementing a productive, cost-effective path for success. As discussed in this chapter, thorough and thoughtful planning is the most effective tool in the maintainability manager's toolkit. The requisite resources for the program tasks need be identified and arrangements made to have the needed resources ready and available when and where needed. Close, early, and continuing coordination with all organizational elements involved in the product design, development, manufacturing, operations, test, and support is necessary.

It is incumbent upon the maintainability management to ensure that the data, information, understanding, context, interpretation, clarification, and guidance for all team elements are provided and incorporated. It is also necessary to control the information so the entire team is working on the same set and version of information. Configuration control and access control measures will aid this effort considerably.

The development life cycle for a product begins in the concept development phase and progresses through testing and production. Requirements definitions in clear, unambiguous and measurable terms are necessary for the development process to be a success. In-process and formal testing of the design against the requirements should be integral parts of the design and development process. It is necessary to measure and verify that the product being designed actually meets the baseline requirements. The earlier that the design characteristics are measured against requirements, the less time and funding are required to make changes to the design.

The thinking of the design team should include the entirety of the product's life cycle and involve the maintenance and support personnel from the earliest phases of the life cycle. It is advisable, especially early in the life cycle, that the requirements set for the product include sustainability information that can be gained from the actual users and maintainers. Any new insights should be captured and documented and, if appropriate, translated into formal requirements.

Maintainability engineering should be integrated into the overall design team as their requirements are implemented by the design staff. Design alternatives and parts selection should include maintainability, as well as other ILS functions, to review and concur with proposed changes. The design of software should include maintainability engineering to provide insight into user interfaces for ease of use and consistency and for guidance on how to make software maintenance easier and less error-prone. For more mechanical considerations, characteristics such as weights and balances are often neglected until late in the design process. The same occurs for lift and carry ability and transport in passageways, such as the restricted space found aboard ships. Maintenance access and safety considerations are additional important areas for maintainability engineering insight.

One of the key tasks of maintainability and ILS engineers is to work to influence design decisions as a part of the design team. This requires that knowledgeable maintainability and ILS personnel are conversant in design and technology matters, and that they are integral members of the design team. While not always fully appreciated at first, the knowledge and perspective of maintainability engineers is an important element in a successful, complete and requirements-compliant design. It is strongly suggested that the product ILS lead and the various ILS function leads (reliability, maintainability, sparing, etc.) review and sign-off all proposed requirements changes and proposed design changes.

References

1 US Department of Defense (2012). *Designing and Developing Maintainable Products and Systems* , MIL-HDBK-470A. Washington, DC: Department of Defense.

2 US Department of Defense (2017). Chapter 3 - Systems engineering. In: *Defense Acquisition Guidebook (DAG)*. Washington, DC: Department of Defense.

3 US Department of Defense (2016). *Operating and Support Cost Management Guidebook*. Washington, DC: Department of Defense.

Suggestions for Additional Reading

Blanchard, B.S. (2004). *System Engineering Management*. Hoboken, NJ: Wiley.

Blanchard, B.S. (2015). *Logistics Engineering and Management*. India: Pearson.

Maslow, A.H. (1943). A theory of human motivation. *Psychological Review* **50**: 370–396.

Pecht, M. and The Arinc Inc (1995). *Product Reliability, Maintainability, and Supportability Handbook*. Boca Raton, FL: CRC Press.

Raheja, D. and Allocco, M. (2006). *Assurance Technologies Principles and Practices*. Hoboken, NJ: Wiley.

Raheja, D. and Gullo, L.J. (2012). *Design for Reliability*. Hoboken, NJ: Wiley.

US Department of Defense (1988). *Maintainability Design Techniques* , DOD-HDBK-791. Washington, DC: Department of Defense.

US Department of Defense (1997). *Designing and Developing Maintainable Products and Systems*, MIL-HDBK-470. Washington, DC: Department of Defense.

US Department of Defense (2001). *Configuration Management Guidance*, MIL-HDBK-61A. Washington, DC: Department of Defense.

US Department of Defense (2011). *Product Support Manager Guidebook*. Washington, DC: Department of Defense.

US Department of Defense (2013). *Acquisition Logistics* , MIL-HDBK- 502. Washington, DC: U.S. Department of Defense.

US Department of Defense (2017). *Defense Acquisition Guidebook (DAG)*. Washington, DC: Department of Defense.

4

Maintenance Concept

David E. Franck, CPL

4.1 Introduction

The maintenance concept is one of the most basic guiding principles for comprehensive system, subsystem, and component design. The maintenance concept is central to defining and influencing sustainment plans and capabilities, which allow for full life cycle operability. Such life-cycle thinking, by its nature, addresses many system performance, usability, and maintenance factors that are not typically included in design specifications, yet they are critical. It is imperative that the full system capabilities are operable during the critical times within the life cycle when the system needs to operate and lives may depend on it. It is common to find in organizations the philosophy that maintenance, or support in general, has little to do with design. This chapter provides the rationale to reverse this philosophy and justify why maintenance is an important consideration in the system or product design process. Additionally, it is often difficult to be specific about maintenance and other support parameters and expectations through the future, especially when they are unknown, uncontrolled, or not defined. This chapter, and other chapters in this book, provides the specific maintenance and support parameters to be considered and selected for use in defining a maintenance concept for any system or product during its design and development process, prior to production and manufacturing.

A truly inclusive design includes consideration of and accommodation for the maintenance and support elements of the systems or products that are required during their expected useful life experiences. These experiences include factors that go beyond the typical design performance specifications, such as weight, size, data throughput, power utilizations, interfacing, and so on, and beyond how the system or product is to be used. Design inclusiveness includes predicting and anticipating the totality of the system's or product's life experiences. In order to achieve life-cycle effectivity and efficiency, design considerations must include the maintenance environment in which the system or product will spend all or part of its operational life.

NOTE: For purposes of simplicity and consistency, the terms "system" and "product" are used interchangeably in this chapter. When one or the other term is called out, the intent is to apply the thought to both terms.

History demonstrates that the vast majority of a typical system's Life Cycle Costs (LCCs) are post-manufacturing sustainment costs. The maintenance concept establishes the foundation upon which support decisions are based. Sustainment professionals make plans and decisions based on the maintenance concept. They assign resources, create

Design for Maintainability, First Edition. Edited by Louis J. Gullo and Jack Dixon.

capabilities, and identify costs and savings based on the maintenance concept. Therefore, it is prudent to pay attention to these factors. Product designs that do not readily fit into the support infrastructure will experience less than expected operational availability and higher than expected costs later in the life cycle. These costs may be associated with high maintenance times and large product design change costs, usually due to product obsolescence issues.

It is a maxim in the design world that as the design process progresses, the cost of changing the design increases dramatically. Design changes during development trade studies performed early in the design process are simpler to make and far less costly than design changes made later in the design process. The cost of design changes depends on the type of changes and the type of hardware being changed, whether they are complex or simple component, unit or subsystem design changes. It should be obvious, however, that as far as design changes go, the earlier the changes are made, the easier and cheaper they will be.

This "earlier is easier and cheaper" approach also applies to the larger view of the product and its entire support/sustainment environment across the years of its life cycle. The earlier the product maintenance concept is established and baselined, the sooner the resources and capabilities needed to sustain it can be planned and budgeted, and the sooner efforts can be started to put the support sustainment infrastructure in place.

There exists a level of synergy between a product's design efforts and the creation of the sustainment infrastructure needed to support it. The level of this synergy is reflected in the amount of detailed maintenance planning based on the maintenance concept that influences the design requirements. It is like a chicken and egg situation. What came first? Does the product design come before the establishment of the support infrastructure? The ability to maintain a product depends on the design of the product, yet effective support of the product depends on the capabilities and resources of the sustainment infrastructure. In the best of all worlds, the product design parameters should include design features that provide for effective support and maintenance performance. Just as some will say that in the chicken and egg scenario, the egg came first, the same can be said for the product design vs. the support infrastructure conundrum. The support infrastructure must come first. It is impossible to give birth to a new product if the support infrastructure is not already created. A new product design released to the public without an existing support infrastructure is doomed to failure.

Luckily, in many situations, the support infrastructure is fairly well established, such as in the military or the automotive environments. Awareness of this infrastructure helps the design process get started on the right footing, as long as these support infrastructure considerations are made a part of the design requirements. Otherwise, additional upfront work is required to identify and specify the maintenance and support expectations for the design team.

As soon as the maintenance concept is articulated, it is important to document and disseminate the written maintenance concept to all elements involved in the design, development, manufacture, operations, sales, distribution, and product field support. Just as there is a Concept of Operations (CONOPS) for any product design that is produced and deployed to the field, there is a CONOPS for the maintenance and support infrastructure. Socializing the maintenance concept part of the overall CONOPS early in the design process will flush out any issues with the concept and garner feedback from all vested parties, stakeholders, and members of the design team. The earlier in the process this is done the better, as many product design, development, and production decisions rely on this early information to do a thorough job.

The repeated emphasis on "early" is intentional. "Early" matters a lot. It matters in terms of design goodness, time, design costs, production costs, and sustainment costs. For example, the maintenance concept is a major driving force behind design decisions that need to be made before the product's baseline design is confirmed and locked down. After this point, all changes to the design only become harder, take longer, and are more expensive to implement. In addition, early maintenance concept definition provides additional time to conduct cost–benefit design trade studies of alternatives to refine the design.

Consider this point: the product owner is concerned with the product throughout its entire life cycle. In many cases, such as within many government militaries, this support period often lasts for several decades. For long-lived military systems, for example, the associated Operations and Support (O&S) costs typically average approximately 67%, and often are in the range of 70–80% or more of the total LCC. The design and production costs are relatively minor by comparison [1, 2]. A prudent product developer and product owner would do well to search for, identify and consider small design investments early that have positive consequence over decades of product support. In addition, instilling these concepts and focus items on design teams should yield a design which is further optimized for its intended use and support environment.

Another reason to establish the maintenance concept early is that it takes time to put a maintenance support infrastructure in place. Capabilities such as personnel, training (such as operator, maintainer, product support), publications, tools, test equipment, repair locations and capabilities, network troubleshooting, storage capacity, spares inventory, and so on, must all be evaluated and put in place.

Since support capabilities cannot be established without funding, the budgeting process must be considered. The budget cycle of large support organizations, especially governmental organizations, generally begins years before the support capability is needed in place.

Funding requests also require specifics and backup analysis for approval, which similarly drives the need for early and advance analysis of the design and support resource needed. Conducting these analyses takes time and information, including design information. The goal should be to do the analyses in sufficient time for the budgeting cycle such that the defined support capabilities are in place and operational at the same time as the product is made available to the users.

Early support-oriented influence on the design affects the ability to put a support capability in place in a timely manner. A data point to consider: in the US Department of Defense (DoD), typically about 85% of the systems' total LCCs are committed and locked in during the requirements definition phase and before the design baseline is established [3]. This fact should be a guide to design teams in general, as somebody in the product's life-stream is going to pay this cost.

4.2 Developing the Maintenance Concept

The maintenance concept affects all product participants involved in the design, production, operation, maintenance (including repair, calibration, updating, removal, replacement, and reach-back support), transportation, and disposal of the product. By extension, it also affects and provides guidance to: developing training requirements and courseware, creation of technical documentation, tools and test equipment requirements, Line Replaceable Unit (LRU) designation, packaging and storage needs, defining maintenance aids, and the design of Built-In Test capabilities. LRUs are normally defined as electronic circuit card assemblies or printed wiring board assemblies with piece parts and components attached. LRUs are usually the lowest level of a system that can be maintained at the organizational maintenance level. Design reliability dictates how often, and thus how much of these support resources will be required and must be included in the consideration of design alternatives, and their impacts on the support infrastructure and associated costs.

Because the impacts of the maintenance concept are so pervasive, it is critical to an effective design that all members of the design and product teams have a good appreciation of its impact and its relationship to the product design. An effective approach for familiarizing the design and development teams with the maintenance concept and how their design choices affect others is, in a group environment, have the team members list all the potential user and maintainer personnel (or support, if desired) they can think of who would interact with the product, then have them list how each are involved in product maintenance (or support).

Next ask the team how these invested parties are actually going to do their tasks, under what conditions, with what tools, what skills are needed, what facilities are required, and what test equipment is needed for all maintenance. Also, probe the team for other resources that might be needed to accomplish product maintenance throughout its life cycle, including disposal.

Explore the potential conditions the product might be in when these activities occur. The design team should conduct design reviews to consider all the conditions that system operator or maintainer may be expected to work within. Consider the environmental conditions under which the maintenance might take place.

For example, consider the following questions related to different environmental conditions:

- Are the operator or maintainer actions expected to occur in heated or air-conditioned indoor facilities, or on a cold, windy and snowy road or airport ramp or mountainside?
- Can the operator or maintainer do the work in cold weather gear without causing them frustration, with little or no access to the area to be operated or maintained?
- Are the operator or maintainer actions expected to occur in a humid tropical jungle or in a dry arid desert?
- Are the operators or maintainers exposed to mechanical shock and vibrations inside a bouncing, moving vehicle, ship, boat, or plane, or on stable ground?
- Are the physical, emotional, and mental conditions of the personnel being considered?
- Are real-world issues such as tiredness, exhaustion, combat stress, experience levels, and work pressures being considered?
- Are all types of harsh stormy-weather gear, such as rain coats and boots, that the operator or maintainer may be expected to use to get the job done being considered?
- Are dangerous environments, where nuclear, biological or chemical agents are present, being considered for operators or maintainers expected to use anti-contamination suits, such as the military Mission Oriented Protective Posture (MOPP) suits?

Exercises such as these provide all product team members the opportunity to better understand the operational and support environments of the product user and maintainer, and to become more sensitized to the downstream impacts of their design choices.

Developing the maintenance concept requires careful consideration of how the product is to be used and maintained across all life-cycle times and events. Identify the basics – the anticipated lifetime of the product and the major events during its life cycle. Such events might include in-service dates, major overhauls, planned replacement dates, anticipated updates, and end of service or obsolescence dates.

In addition to the environmental conditions described above under which maintenance will be performed, other aspects of the maintenance environment need to be identified and defined. Some considerations that will help to clarify the maintenance environment for a product might include:

- Where will these activities be conducted? Could these activities be conducted out in the open, such as on the flight line, or in a backshop or a shed, like a temporary facility, or in a depot facility or large permanent structure?
- Are the maintenance actions of the nature that requires personnel to be trained and experienced in mechanical-oriented tasks, or electronic hardware tasks, or computer software tasks, or all three types of tasks?
- When will the product be available for maintenance? After a mission? In two years? While in use? Two days next month, then not again for six months?
- What is the work week of maintainers? What holidays are observed?
- Do maintainers work shifts? If so, which ones?
- Who is authorized to do what? For example, are unions or worker organizations involved? If so, what requirements or restrictions apply to their workers?
- What maintenance trades are in place and how do they operate?
- What is the language of the maintainers?
- What facilities and support infrastructure are available for maintenance?
- How will design changes, updates and upgrades be performed during the life cycle? Are regular updates expected? Are periodic upgrades planned? What is their timing? Might they correspond to planned major maintenance events, such as planned depot overhauls?

The intended life cycle must be defined. There is a difference in design approach if the intended lifetime of a product is two years or 20 years or 50 years. Longer lifetimes require consideration of design choices that make upgrades easier, such as open architecture. Are warranty periods planned? Are extended warranty contracts by third-party support providers in the plans for support during the life cycle? The expected life of the product is still relevant; is it 30 seconds or one year? The operational reliability should become a more significant factor for the design team. However, the support team will be interested in design features that impact storage life and conditions.

If the product is disposable and non-repairable, a different set of maintenance concepts and design choices must be addressed. A non-repairable product may follow a "discard on failure" maintenance concept. Another maintenance concept consideration is whether the entire product is disposable or only certain select parts of the product are disposable. Such is the case with a flashlight or a battery. The portable flashlight is a product that should not be disposable if its battery was drained of power. Flashlight design should allow for easy replacement of its battery. The flashlight would not be considered a good design if its battery could not be replaced when the battery failed and the flashlight had to be discarded or thrown away.

Of increasing interest is the matter of environmental disposal impacts. Do any parts of the product require special handling or disposal restrictions? Does the product include Ni-Cd (nickel cadmium) or lithium-ion batteries? Are special military paints, like Chemical Agent Resistant Coating (CARC) paint, being used? Are certain parts or items planned for disposal during maintenance, such as dirty rags, grease, oil-impregnated seals and components, petroleum products, and cleaning materials?

If there are security considerations for the product, they need to be included in the maintenance concept. The conditions and processes for the security levels or proprietary information need to be identified. Additionally, the facilities (including storage), equipment, and personnel requirements must be identified, along with special handling or destruction processes and clearances. Generally, there are established processes, requirements and definitions for compliance with classified equipment design, use, and support. Relevant organizations and personnel are key to compliance in these matters.

Another consideration for the maintenance concept is whether the product is to be used and maintained in different countries. Frequently, the tools included in typical maintainer toolkits will differ from country to country. Also, the use of metric verses American-size tools or connection parts (i.e., threads, bolt heads, lengths) varies. Also, the skill definitions of maintainers and their skill levels vary from country to country. These characteristics need to be part of the maintenance concept.

Also part of the maintenance concept should be a review and identification of any applicable export and import laws and restrictions. Technology transfer is generally the emphasis of such laws, but that also varies from country to country. The transport of certain items through a country may be restricted or prohibited, so a careful review of the laws and the product design is warranted.

Disposal of product parts presents environmental considerations which vary by country, as do laws regarding use, disposal and transport of certain materials. Batteries and battery technologies present an often-overlooked element. If batteries are part of the product, how they are disposed may influence design decisions. Ni-Cd and lithium-ion batteries are the most common sources of current battery disposal concerns, but they are not the only ones. Fluids, gases, carcinogenic materials, and similar materials present potential issues to be addressed in the maintenance concept.

Another often-overlooked maintenance concept component is signage and language requirements for intended users and maintainers. Consider and include the appropriate languages that are needed for safe maintenance and signage standards for the intended user and maintainer cadre.

As examples of some of the above discussion, consider a few of the many maintenance concept-related topics faced in designing and building a large system for the United Kingdom (UK) that was also intended to be used and maintained in another European country. A small sample of considerations might include:

- Who are the maintainers, such as government personnel, military personnel, or contractor personnel?
- What skills do each set of operators and maintainers have?
- Are the maintainers multiskilled or single-skilled?
- What labor laws apply in the different locations?
- Are labor unions involved and what special maintenance definitions apply?
- What are the country's conventions and requirements for electrical wiring?

- Which warning and caution signs and labels are to be used in which location?
- What language would be used for signage, labels, warnings, instructions, manuals, training, etc.?
- What are the environmental considerations of each facility to be designed?
- What are the human factors differences between each facility, such as time/temperature limitations on workers?
- What are the different weather and internal environmental conditions between locations and their impacts on Heating, Ventilation, and Air Conditioning (HVAC) designs?
- What security considerations should be identified and addressed involving the two countries and their militaries, civilian organizations, and commercial data transport resources?

As stated before, the maintenance concept is the basis for defining and establishing the categories of maintenance and support resources needed to support a product. One important item that must be included as a core item of the maintenance concept is how rapidly the defined maintenance needs to be accomplished. This is generally referred to as the systems maintainability, and is typically specified in terms of Mean-Time-to-Repair (MTTR) for active maintenance. The MTTR is a central maintainability metric, but it may come with several caveats and related variations. The intent of the MTTR is to define how quickly repairs are needed to be completed, and it is a statistical average of all potential maintenance actions for the product. There are various aspects to maintainability that may be applicable to a given situation. For example, subsets of MTTR may include preventive maintenance, depot or backshop maintenance, or active versus total maintenance time. There are elements of active or total maintenance time that are not included in MTTR calculations. These elements of time include administrative or logistics downtime.

The MTTR is a significant metric for support planning because it drives what maintenance actions can be performed within the time limits for each of the maintenance levels included in planned support. This in turn drives which maintenance tools, test equipment, training, Built-In-Test (BIT) capabilities, documentation, spares and more are needed at each level of maintenance and location where maintenance is carried out.

The maintenance concept also must include requirements for reliability, typical referred to as Mean-Time-Between-Failure (MTBF) if the system is repairable. If the system is not repairable, the reliability specification should not have an MTBF requirement, but rather may have a Mean-Time-to-Failure (MTTF) requirement. Similar to how the MTTR drives the quantitative maintainability requirements for determining what maintenance is required, the MTBF drives the quantitative reliability requirements of repairable systems and is used to determine how often maintenance resources are required. Naturally, the frequency of maintenance drives the extent to which maintenance-related resources are required to support a given population of products, which drives the cost of providing that support. The relationship between reliability and support costs should be familiar to the design team, since once the design is completed, the reliability of the product does not change. The only opportunity to affect the lifetime support impacts that are driven by reliability lie with the design team.

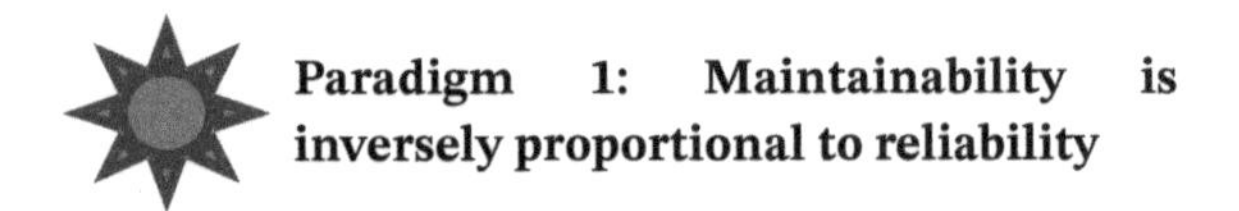
Paradigm 1: Maintainability is inversely proportional to reliability

4.2.1 Maintainability Requirements

The maintenance concept is typically developed from the product's operational requirements. At the high level, these generally include some maintenance-related requirements for the product, its intended use and criticality, planned product support capabilities and infrastructure, and the expected life cycle. These high-level requirements are decomposed into focused, more detailed requirements. Questions that need to be asked to have ample focus and detail to generate requirements to be considered for inclusion in a maintenance concept are:

- What are the operational conditions to which the system will be subjected during its lifetime?
- How will the product be used, handled, stored, and transported?

- What are the plans for handling and disposing of the product at its end of life?
- Is the product's life cycle identified?
- Does the product have a shelf-life? In other words, how long may the product sit on the storage shelf in a non-operating state (and under what conditions) before it is pulled from storage and placed into operation?
- What levels and categories of maintenance will be performed during the product's lifetime?
- Will the product be subject to planned or unplanned upgrades or overhauls?
- What is the duty cycle for the product?
- What are the limitations for conduct of maintenance considering the space needed to perform maintenance at certain locations, and the capabilities and limitations of the maintenance personnel in matching the human to the task?

Paradigm 8: Understand the maintenance requirements

Maintenance concept requirements are translated into system design and support requirements as the system design progresses. As the system design activities are performed, the maintenance concept provides a framework that shapes system design trade study decisions based on maintainability design criteria that lead to detailed maintenance and support requirements. Maintainability design criteria and maintainability requirements will be discussed in greater detail in Chapter 5.

4.2.2 Categories of Maintenance

There are two categories of maintenance depending on the nature of the maintenance. These categories are scheduled maintenance and unscheduled maintenance. Each category can be subdivided further based on the reason for the maintenance:

- Scheduled maintenance
  - Preventive maintenance
  - Upgrades and updates maintenance
  - New capability maintenance
  - Replacement maintenance
  - Removal and Replacement (R&R) maintenance
- Unscheduled maintenance
  - Corrective maintenance
  - Incipient failure
  - Safety of use
  - Unusual conditions

These categories and their subdivisions are discussed in the following sections. There can be some overlap in definition among these groupings, and definitions can sometimes vary from customer to customer. Careful evaluation of the assumptions applied to each situation is appropriate to ensure that all parties are using the same terms and to avoid unwanted "surprises" down the road. It is important to establish a clear understanding of the maintenance design goals and maintenance terms to be used among the design team, support planners, and the customer. Also, forming a close working relationship between the design team and the logistics support (LS) team is an invaluable addition to the overall product development process and should be encouraged. The takeaway from this is that the terms, definitions, and expectations need to be checked and coordinated thoroughly early in the design process, preferably as early as the requirements definition phase. Additional discipline is often needed to make sure the design team is consistent in their use of the terms and definitions.

4.2.2.1 Scheduled Maintenance

As the name indicates, this category of maintenance is conducted in accordance with a defined, predictable metric, generally based on a schedule or operating metric. Some examples upon which scheduled maintenance is based include: hours of operation; calendar days; daily, weekly, monthly, quarterly, or annual schedules; power-on hours; power on/off cycles; take-offs or landings; submergence count (submarines); miles driven; rounds fired; or obsolescence status.

Scheduled maintenance can be further subdivided into additional specialized subcategories described in the following paragraphs.

Preventive Maintenance Preventive maintenance is maintenance that is proactively conducted with the intention of preventing a fault or failure. By its nature, preventive maintenance is intended to find faults, problems, impending problems, or trends that lead to problems before they actually become faults that require corrective actions. Preventive maintenance includes a wide variety of maintenance actions generally based on product usage. Examples include:

- Change oil every 3000 miles
- Change the gun tube after every 2000 rounds fired
- Rotate tires every 5000 miles
- Check system calibration every 100 hours of operation
- Check tires before every flight
- Check hydraulic, oil, and fuel lines for leaks at the beginning of every shift
- Borescope the engine after every 5000 cycles

Preventive maintenance actions run the spectrum from extensive, invasive actions that take days to perform, to cursory visual checks that might take minutes or a few hours to accomplish. The need for and specifics of preventive maintenance actions are design-dependent based on the product issues being addressed. They can be as simple as visual checks for cracks or dirt or part wear. They also include more involved actions, such as running built-in fault detection software or performing calibration procedures, or as extensive as borescoping an engine, checking the inside of an aircraft fuel tank, or manually checking the bilges of a ship.

Starting early in the design process, the design team and support planning team should consider the tradeoffs between design options and their respective maintenance demands. These trade studies should address operational impacts and support consequences throughout the product life cycle. Often, the quantity of final products needed during the life cycle increases as maintenance demands increase, increasing the LCC necessary to achieve operational goals. While design elegance is a desired outcome, it should not come at the expense of an excessive maintenance burden on the user, especially for products with a long expected operational life. Manage the tradeoffs between ease and cost of design production versus design for support costs thoroughly.

Upgrades and Updates Maintenance Products and systems, especially those with extensive software content or with a long life cycle, often require one or more upgrades during their useful life. In its pure form, upgrades and updates typically involve maintenance to replace obsolete parts, to update software, to make the product compatible with other products with which it interfaces, or to add operational capabilities which (usually) do not dramatically change the operational capabilities. Major changes in operations are mostly associated with new, replacement products or major overhaul activities.

In the world of electronic LRU piece parts and components, obsolescence can be a significant issue when trying to manage a moderate or large population of a product over a moderately long time-frame. Many electronic parts become obsolete (out of production or materials no longer available) in two to three years, and 18 months is not unusual for fast-changing commercial parts, with some as quick as 12 months. Designing for ease of upgrade and using parts with established long-term supply history reduces the chances of requiring updates because of obsolescence, thus making the product less of a support burden.

With computer-based systems, it is frequently not possible to achieve this without adding new capabilities, functions or performance. Thus maintenance related to upgrades often overlaps with that in which new capabilities are added.

New Capability Maintenance Maintenance that is intended to add new capability, performance, and function to an existing product is a maintenance subset that involves a unique challenge. Long-life systems present product support management with special challenges to find an optimum solution between competing demands. These include such considerations as: when and where to add in new capabilities; managing new product component demands competing with changing kit needs; competing with existing maintenance resources to conduct the maintenance; working to minimize operational impacts on the customer; and funding issues.

The actual maintenance actions can be as simple as installing software updates, assuming the product and software designs allow this capability in their design architecture. The maintenance might also require complete LRU disassembly and reassembly with design modifications to the host equipment. Replacement of associated electronic part(s) or components on LRUs may be required, which could be straightforward or require extensive disassembly of supporting systems to access the LRUs where parts are to be replaced. On large, complex systems, such as aircraft, ships, tanks, and systems of similar complexity, extensive disassembly is often required, necessitating complex maintenance involving many skill sets and detailed planning.

A design that provides for ease of future adaptability is desired, if possible. Characteristics that aid in achieving this include: ease of access, parts removal, and replacement; use of standard interfaces and parts sizes; use of software and electronic/data interface standards; using the fewest different types of connectors and attaching hardware (screws, nuts, bolts, etc.) as possible; and ease of access *in situ* with minimal removal of other parts or systems to facilitate the maintenance. Additionally, designing-in added margin and capability with excess or spare support capacity is advantageous. Such factors might include: extra capacity in the power supply and heating/cooling sections; extra slots for additional electronic boards; beefed-up support structures to absorb anticipated weight, speed or handling increases; increased structure mounting strength and attachment locations where future capabilities can be accommodated; and using open architecture where possible.

Replacement Maintenance There comes a time in every system's life when it has to be replaced due to failure or accident, to make way for the next-generation replacement, or to make room for new systems. In such situations, the ease of maintenance is not as important a design consideration. Instead, a much more relevant consideration is how to extricate the old equipment from its location and how to safely dispose of the equipment or to make it unusable. During the design phase, design considerations for replacement typically include destruction or retention of proprietary information or designs, proprietary parts, classified software or parts, environmentally sensitive materials (chemicals, radiation sources, poisons), carcinogenic or hazardous materials in the product (batteries, acids, lead, corrosives), carcinogenic or hazardous materials required for dispose of the product (CARC paint, asbestos), or precious metals. Also, consideration is required to address any special handling, storage, shipping, and transportation requirements for the final disposal. For example, air transport of lithium-ion batteries is severely restricted, disposal of lead acid batteries is controlled, and radioactive materials require special handling. Careful reviews of local and country disposal, handling and transport restrictions and processes is mandated.

Removal and Replacement (R&R) Maintenance R&R maintenance is typically performed to remedy a system failure by exchanging a failed LRU with a known good LRU. Sometimes R&R maintenance is performed to allow maintenance on another part of a system, especially in systems with multiple subsystems, and not because of a fault in the LRU being removed. Where this situation can be anticipated, it is advantageous to design-in characteristics that facilitate LRU removal and reattachment easily and without causing any damage to adjacent LRUs. Coordination with the product owners of equipment to which the product connects could identify issues or opportunities to accommodate further R&R actions.

Trade studies might be appropriate where the frequency of R&R actions can be estimated to assess the practicability of making design choices that facilitate R&R activity. Some design choices are low cost, while others are more costly. Examples include: use more robust mounting hardware; use securing and mounting hardware that is rated for more removes and inserts; use mounting hardware that either requires no tools or uses the same tools as those required for the equipment being fixed (lowering the burden on maintainers and the support infrastructure); ensure the weight makes removal easy on tired maintainers; design the operational components to allow more frequent shutdown and restart without harming the product or requiring an extensive reboot processes; ensure more frequent shutdown and removal does not reduce the effective life of the product or any life-limited components; eliminate any special tools required to remove and replace the product; and ensure that any electrical or signal wiring connectors are protected from impacts, dust, contamination, or shorting.

4.2.2.2 Unscheduled Maintenance

Unscheduled maintenance is typically critical to users because it means that the product is not available to perform its intended function when it may be needed.

Unscheduled maintenance impacts the schedule for the allocated maintenance resources for other products waiting for repair. By their nature, unscheduled maintenance activities disrupt ongoing operations. Product users who rely on the product to complete a job or mission or assignment take these interruptions seriously. To be sure, whether it be military operations with the obvious impacts from systems not being available, or the commercial airline's revenue-generating aircraft that cannot depart on time, or the retail consumer flashlight that is needed when the power goes out, or the radio needed for operational communications, unscheduled maintenance is disruptive and gets high priority and attention.

Unscheduled maintenance actions result from a wide variety of causes: part failure, maintenance mistakes, other equipment causing faults, human error, lightning, flood, weather extremes, poor or inadequate preventive maintenance, negligence, maliciousness or criminal action, or misuse. Good design practice would lead a design team to consider such situations during the design process and design mitigations into the product where feasible.

Of course, no product lasts forever: equipment wears out, parts fail, or things break. To mitigate the effects of product failures and to establish realistic expectations, design teams use component and product reliability metrics to estimate how frequently the product is likely to fail. They also use wear data to estimate useful life in parts to predict (estimate) when a product will wear out. The distinction between these two subjects can be found in extensive depth in many technical sources on reliability engineering.

In short, mechanical parts wear out and their performance typically degrades as the wear increases, as in the tires on a car or gears in a transmission. It is a characteristic of mechanical components that the more they are used, the more they wear. The more they wear, the more likely they are to fail. Predicting these failure rates with precision is not an easy task and is rather specialized.

Electronic parts have entirely different failure characteristics with different failure rate prediction functions. Some component failure rates may be characterized by an exponential distribution function, having a constant failure rate. This means that electronic parts have the same likelihood of failing in their first hour of operations (burn-in aside) as they are at any other time in their life.

Where electronic components cannot be predicted with certainty, they can and are predicted based on statistical methods within select confidence bounds. Thus, using the term MTBF identifies the average time between failures. Said another way, if the failure data are normally distributed, half of all specific parts will fail before the stated MTBF. Using this knowledge, design and support teams estimate the frequency of different failures within a design to predict the frequency of expected unscheduled maintenance. These metrics drive support planning and costs.

It is advisable to establish a common understanding among all stakeholders of the terms and assumptions used to estimate MTBFs and what constitutes a failure. Once the design process has completed, it is too late to productively change a product's reliability, and hence its contribution to maintenance demands. Establishing common terms and understandings should start in the operational and functional requirements phase and then be assigned to the appropriate design authorities. In this way, there is a clear linkage established between performance expectation and the design team.

Typically, the user community will want to know, and reasonably expects, information on the predicted failure rates of a product, how they were determined, and what the data pedigree is (i.e., based on published prediction data, actual field performance data, and similar equipment performance test data). Environmental and stress factors will also be important factors for the design team to apply, as well as electronic and mechanical derating, safety factors, and design margin.

Reliability is important to corrective maintenance because reliability directly impacts the expected frequency for conducting corrective maintenance actions. Since each corrective maintenance action is a perturbation to the operational planning of the product user, as well as a source of support costs, product reliability directly affects user mission completion and LCC parameters (e.g., spares, people, tools, test equipment, facilities, and training).

Paradigm 1: Maintainability is inversely proportional to reliability

Thorough design analysis and well thought-out design decisions produce better products. As a design goal, maintenance actions that require an abundance of maintenance time and resources should be designed so they occur infrequently. Likewise, failures that occur more frequently should be designed such that they can be corrected easily and quickly.

Unscheduled maintenance can be further subdivided into additional specialized subcategories described in the following paragraphs.

Corrective Maintenance Corrective maintenance is required whenever a part breaks or fails to perform its functions properly. Maintenance actions must be taken to correct the fault and return the product to its proper functionality. In select situations, full functionality is not possible and degraded functionality is acceptable. In still other situations, switching to an alternate functional path may be required or allowed by the customer. Each situation is governed by operational requirements and design tradeoffs.

Situations requiring corrective maintenance result in unplanned interruptions to planned service and operations. The product not being available for planned use impacts the user, sometimes seriously, and should be taken seriously by the design team. Proper design efforts consider and address all contingencies, operational scenarios, and emergency situations anticipated to be encountered when in use. Design reviews need to address the spectrum of operational use and user interaction. Testing should, likewise, be thorough, especially for all software-based elements. If appropriate, sharing test plans and results with the user community should be considered. This not only builds confidence with them, but provides an opportunity to gain additional insights to improve the testing or the product.

The design team should identify all possible maintenance actions that their design will require, define the resources that will be required for each action, how often they are anticipated to be required, and how long each one takes. A close working relationship and communications with the support team and use of Logistics Support Analysis (LSA) tools is highly recommended. LSA tools provide a mechanism to collect and quickly analyze failure and maintenance metrics and assess their support impacts.

Also, test the corrective maintenance procedures identified – all of them. Consider such tools as Failure Modes, Effects, and Criticality Analysis (FMECA), tabletop walk-throughs of repair actions, and Maintainability Demonstrations. The FMECA is a standardized logistics tool in which every failure mechanism is identified along with its effects on product operations or other equipment. Then each element in the analysis is assessed for the criticality of those effects to mission operations, personnel safety, and equipment safety. Tabletop walkthroughs are useful to identify failure modes and the resources and procedures that are required to address them. Bring design, support and user participants together and talk through each part failure to identify how it is noticed, its impact, what steps are needed to correct the fault (including tools, test equipment, skills, etc.), and then restore full operations.

A Maintainability Demonstration is often called for in military contracts. The Maintainability Demonstration is a formal demonstration with customer observers in attendance wherein select product faults are corrected using defined repair processes and tools. Repair times are captured for each repair action. The purpose of the Maintainability Demonstration is to demonstrate that the predicted corrective maintenance processes and times are realistic. These results are important to build confidence that the planned support posture will, in fact, meet operational needs.

Thorough testing of software elements, especially those that identify faults or are employed in troubleshooting faults, should be integral to the entire process discussed above.

Incipient Failure An incipient failure is the condition where a part is on the verge of failing but has not yet actually failed. It is generally preferred to replace a part before it has failed, rather than wait until it actually does fail. The impact on operations is reduced and there is an opportunity to plan the corrective maintenance action around available support resources. The catch to replacing parts based on an anticipated incipient failure is determining when to initiate the maintenance. Waiting too long results in excessive actual faults, and initiating replacements too soon wastes residual operational life in the part.

There is no golden rule to guide this decision process. Each product is different because of its criticality, the importance of its mission, safety concerns, part costs, and design and historical evidence. These are different for every product, use, customer, and need. For example, knowing that the brakes on a particular airplane are expected to fail sometime in the coming three months is of limited use. Knowing they are anticipated to present a serious safety risk in the next three days is helpful. But what is the right answer? It depends. If the plane flies regularly to defined locations, which all have the needed spare parts and maintenance capabilities, and extra time is allocated to replace the brakes on the flight line, then less advance notice of impending maintenance might be acceptable. If this same plane does not have a comprehensive support infrastructure in place, more advance notice might be required before maintenance can be performed.

These decisions are not binary; rather, there are usually many more than two simple options available. Understanding the user and customer mission, operational considerations, and support plans helps in making and evaluating maintenance plans in concert with design tradeoff decisions that best fit the user and customer needs and expectations. The design team should consider and document the customer position and the reasons for design tradeoff decisions. Tradeoff options should include the support costs required to implement the options. For example, vague long-lead notice of a pending failure means additional costs for excess just-in-case spares, and for the added storage costs (probably in multiple locations). Vague identification of incipient failures also means that the support and operational planners don't really know when or where a fault will actually occur. To ensure minimum interruption to operations, extra unnecessary spare parts are needed as well as additional costs for shipping, handling, and personnel costs to expeditiously get replacement parts to where they need to go. These costs do not include the costs of damage, safety, or inconvenience.

It should be evident that accuracy in incipient fault prediction is critical. Support managers do not like uncertainty. Because of the high costs of life-cycle support, support planning is heavily optimized to identify what is needed when and where. Customers do not have funding to pay for parts and people to sit around waiting for failures that may or may not happen.

What role does the design team have in helping the customer minimize the unknowns and maximize product operations? The proliferation of electronics, computer technology, sensor technology, and communications technology unlocks design options that offer capabilities never before possible. Encourage and motivate the design team to consider new technologies and to think out of the box. The design team has the ability to affect the user and support communities for years to come.

Does the product allow for sensors for such parameters as power-on time, operating time, calendar time, speed, current, voltage, distance, temperature, cameras, voice, magnetics, movement, GPS, and so on? Some examples include: chip sensor detectors that measure vibration in engines, tire pressure, current draw in motors, firearm rounds shot (shot counter), thermal heating and cooling cycles, cool-down time, over-temperature conditions, mechanical shock, motion detection, angle measurement, direction, location, wear indicators, and presence of personnel. Parameters such as these provide the ability to measure, track, and report important information, not only regarding the proper operation of the product, but also if incipient problems require special attention. More of these proactive design features are discussed in detail in Chapter 9.

Does the product have a design feature that provides users with the capability to manually or automatically transmit maintenance data, either wirelessly or wired? Consider available communications technology for alerting the operator or maintainer of pending problems. Two methods are used in this approach.

The first method is to measure, track, and retain performance data. These data would be manually downloaded during operational periods or non-operational maintenance checks. Data downloads can be performed by maintenance diagnostic device plug-ins, using near-field wireless connections, reading performance parameters off a data display on the product, or reading from the attached maintenance diagnostic device. In this situation, trend or status analysis would occur either locally, or at remote locations at a long distance from the product, thus initiating appropriate actions from a centralized maintenance location.

An alternative in this scenario would be to build the trend or status analysis into the product so the user/maintainer would see only when a parameter trend has reached the point where near-term maintenance is indicated. This approach generally uses BIT or monitoring capabilities common in complex systems where the user needs to know about incipient or actual failures, but does not have the time to monitor the data directly. Aircraft, tanks, cars, refrigerators, unmanned vehicles, and ships are only a few examples of where this technology is common. Further discussion of BIT capabilities is discussed in Chapter 13.

The second basic method to consider for providing warnings of incipient failures is by transmitting the data automatically to the user and responsible support personnel. Such a capability is becoming more and more common in complex, computer-based systems, such as military and commercial aircraft. Where communications exist, it is becoming more common to find designs in which the system monitors itself and self-reports its status and problems detected to a support entity, to provide notice of impending maintenance need. An example of this type of method is in common use on commercial automobiles in which vehicle use is monitored and reported to centralized stations based on defined criteria. Examples include automatically recognizing when the vehicle has had an accident, initiating communications with the vehicle (voice and data), and reporting the accident and its location to authorities for appropriate response. Such systems also monitor maintenance parameters and provide reminders of oil changes that are coming due, tire rotations due, other scheduled maintenance due, recalls that may be in effect, and more. The vehicle's maintenance status is also often shared with local maintenance sources, such as the dealer from whom the vehicle was purchased, to let them know of potential maintenance needed and to help them to prepare for maintenance they may be doing.

The data can then be used to inform the operators of the problem (if appropriate) and suggest actions to take to mitigate any consequences. Consider an airliner over the middle of the ocean getting a message that a piece of navigation, power, hydraulic, or engine equipment is not operating properly. At the same time, the airline support infrastructure is notified of the issue and the proper technical support, spares, and personnel are waiting at the flight's destination to effect the repair. While not well publicized, this is actually a common occurrence. Not surprisingly, engines used on commercial airliners have extensive performance and fault analysis capabilities that automatically report performance data to the operator, the airline, and the engine manufacturer.

Safety of Use One critical baseline design consideration for any product should be designing it for safe operation and maintenance. The first goal for the design team is to design a product that does not cause unsafe situations for the user, the maintainer, other people, or other equipment.

There may be times when absolutely safe maintenance is not possible. In such cases, the design needs to identify unsafe conditions. The operator and maintainer need to be aware so they can take appropriate actions to protect themselves and other equipment. Oftentimes, systems are required or commanded to shut down partially or completely to prevent or mitigate a hazardous, dangerous, or otherwise unsafe situation. Other considerations include notifying the user of an unsafe situation (flat tires, excessive leaking fluids, backup systems outage), automatically initiating emergency actions in case of an incapacitated user, or securing equipment in case of a crash, which may include fuel line shutoff and power disable.

Unusual Conditions Infrequently, and generally quite unpredictably, situations arise in which maintenance is required under conditions that are not "baseline." These include situations that are unavoidable and present maintenance challenges that are out of the ordinary. The genesis for such conditions are often combat actions, extreme weather, accidents, explosions, fire, floods, vehicle motion or upset conditions, or damage from other equipment malfunction. Unusual maintenance occurs when maintenance must be conducted under these extreme conditions and absolute safety is degraded or compromised. Regardless of cause, the primary considerations of the design team include:

- How to identify and classify the situation.
- How to report the situation and equipment condition to the user and upper levels of oversight and control, such as a controlling computer system, operations management, support management, or the user.
- Identifying the appropriate actions to render the equipment and personnel safe.
- How to mitigate unsafe conditions to allow repair activities.
- Identifying information important to users and maintainers to allow safe operations and maintenance.

The design team should also explore and address additional design options as they relate to the product performance in usual operations or conditions where an accident may occur. The design team might add a design option to the product, which proactively identifies a potentially unsafe situation before it occurs, and alerts the users, thus preventing an accident. This product design option might include software that initiates an automatic emergency shutdown or takes control away from the user. An example of this type of product design option is software programmed in some jet fighter aircraft that monitors the flight, senses close proximity to the ground, and identifies when flight into the ground is imminent. The software automatically initiates a series of actions including warnings, shaking the pilot controls, taking control of the aircraft away from the pilot, initiating a direction pitch-up to avoid crashing into the terrain, and, in some situations, initiating crew ejections if a crash is unavoidable. The crew ejection decision may be initiated by the pilot manually or automatically by the flight computer. In this situation, if the crew does not or cannot eject on their own, the system is able to take control and self-initiate crew ejection.

For many modern high-performance military aircraft, they cannot fly if the flight computer is not working. Even though the pilot can still fly the aircraft without the flight computer, the failure of the flight computer will cause the aircraft to be grounded at the flight line until maintenance can be performed. The flight computer performs various airworthiness and safety-critical functions, so it must be fully operational before the aircraft is allowed to take off and fly again.

A reasonable question for the design team to address, if a safety of operation situation cannot be avoided, is: how could the consequences of an unsafe situation be mitigated? Unsafe and unusual situations often result in investigations to determine what happened, why, and what can be done to prevent a recurrence. It is prudent for design teams to consider options for preserving incident data for analysis after the fact. For example, can operations and performance data be recorded and preserved to allow for reconstruction of the event after it occurs? This capability is seen in flight data recorders (the Black Box); data buffering in recording devices, such as security cameras, television recorder boxes, and automobiles; data burst transmission of location and status data during the last XX seconds (e.g., 30 seconds or whatever is specified in the design requirements) of operational data; and data containing the last XX days (XX referring to a certain amount of time specified in design requirements) of maintenance performance and access information.

It might also be advantageous for the design team to explore options by closely examining single-function design threads and exploring the threads in the full scope of non-standard operational situations. Some suggestions include: what happens in an unsafe situation because of external reasons, such as: vehicle crash, fire, explosion, earthquake, acts of war, lightning, and flood? How does the product act and react? Can it become dangerous? For example, does it have flammable fluids or gases under pressure, explosives, or hazardous materials that make a bad situation worse?

The design team should review and identify the appropriate responses, actions, tools, protections, Personal Protection Equipment (PPE), and mitigations necessary to allow maintainers to safely work under unusual situations. Are emergency shutdown procedures possible? Is automatic shutdown an option for unsafe situations? Can the product sense conditions to initiate shutdown or does it come from an external source? What source? Can power be remotely, automatically or easily removed from the equipment? Can warnings be displayed to protect innocent bystanders, users, and maintainers in case of an unsafe situation? Is the product designed to allow maintenance under unexpected or unsafe conditions, such as under attack, extreme cold/heat, no support equipment, inadequate personnel, tired personnel, or personnel suffering from exposure to the elements? Can the system display warning information to users and maintainers, either on the equipment or transmitted to external devices, that an unsafe condition exists and perhaps provide alternate maintenance procedures? Does the design minimize any impact to other equipment, systems, people, structures, or the environment in unsafe conditions?

Paradigm 5: Consider the human as the maintainer

4.3 Levels of Maintenance

There are many definitions and descriptors used to attempt to define maintenance and the different levels of maintenance. Many times, the definitions vary from country to country or try to adapt to new technologies and work standards. Typically, maintenance of systems, especially complex, high-tech equipment containing multiple technologies, requires several levels of maintenance complexity and maintainer specializations and skills. These levels may have different names depending on the organization or country hosting the customer or support operations, or whether the user is military or civilian. The design team needs to understand the maintenance capabilities and limits the customer maintenance organization plans to employ, where they are performed (indoors or outdoors), and their locations. The product should be designed to facilitate ease of maintenance and reduce the demand for that maintenance at all levels. This may require some design tradeoffs to optimize the total maintenance burden across all maintenance levels. Robustness and fault-tolerance are desired characteristics that contribute significantly so maintenance actions are minimized in terms of frequency, repair time, tools and support equipment demands, new skills, and special equipment or facilities.

The maintenance levels typically used by the US military are commonly used elsewhere and have been adopted for use in this book. The levels of maintenance are:

- Organizational
- Intermediate
  - o Direct
  - o Indirect
- Depot

These levels of maintenance are described as follows:

Organizational maintenance – This includes maintenance that can be performed by the operator or maintainer while the equipment is in its operating location and environment. Generally, removal of equipment is not required to perform this level of maintenance, and the maintenance tasks performed are relatively simple, such as high-level trouble-shooting, adjusting, calibration, rebooting, perhaps some software loading, running fault detection and isolation software (i.e., BIT), power and connectivity checks, or battery replacements. Organizational maintenance is sometimes subdivided into operator and maintainer maintenance. Organizational maintenance may also include corrective or preventive maintenance.

Intermediate maintenance – This maintenance is typically performed with the equipment removed from its operating location. Typically, this maintenance means simply removing the part from an equipment rack or detaching it from higher-level equipment, but also includes removing it from an aircraft or vehicle to conduct maintenance. This level of maintenance typically includes maintenance that involves some disassembly of the equipment, R&R of parts and cables/wiring, select software modification or loading, more intensive troubleshooting, calibration, mechanical checks or adjustments, and decisions for disposal. Intermediate maintenance may be performed at the operating location (i.e., direct) or in more robust maintenance facilities providing more extensive maintenance capabilities (i.e., indirect).

Depot maintenance – This level of maintenance capability is considered extensive and complex, requiring specialized skills, tools, and support equipment. It is usually reserved for only the most intense, detailed and difficult maintenance. Typically, depot maintenance is capable of performing any maintenance that the manufacturer can, including completely tearing down and rebuilding the product. The depot, in some cases, may be the actual factory. Depot maintenance is infrequently performed because it is so extensive, time-consuming, and expensive. If possible, equipment should not be designed such that repairs require this level of maintenance on a routine basis.

4.4 Logistic Support

A complete maintenance concept addresses all topics related to sustainment of a product. Typically referred to as Logistics Support (LS), Integrated Logistics Support (ILS), or Integrated Product Support (IPS), product sustainment encompasses all aspects of supporting an equipment or system during development, production, testing, use, and retirement from use. ILS involves all related activities to achieve the totality of a support infrastructure, from training to transportation to technical documentation, from spares and warehousing to tools and test equipment and PPE. To those who are not familiar with the scope and breadth of comprehensive ILS, there are numerous resources available for familiarization. It should be noted that the totality of ILS encompasses many topics and subtopics related to product sustainment, each of which could require an entire career to master. As defined by the Defense Acquisition University (DAU) in S1000D, there are 12 elements of IPS [4]. These IPS elements are:

1. Product support management
2. Design interface
3. Sustaining engineering
4. Supply support
5. Maintenance planning and management
6. Packaging, Handling, Storage, and Transportation (PHS&T)
7. Technical data
8. Support equipment
9. Training and training support
10. Manpower and personnel
11. Facilities and infrastructure
12. Computer resources

Often, design teams are confused about what logistics support has to do with the design of a system or product. On the surface, this seems like a reasonable question, especially with the emphasis product companies and engineering educators have to place on the fundamentals of engineering and design. Pulling back from the single-focus design considerations of pure engineering, the scope of engineering considerations expands to include a wider view of related equipment, interfaces, and functionalities. This expanded view naturally begins to include the user and operating environment as design parameters and, with further expansion, includes the entire life cycle of the product. Establishing an adequate ILS posture takes considerable planning, coordination, and work on the part of many people and skill sets. Implementing a robust support posture requires even more effort.

If an equipment or system is intended for commercial use, the totality of the extensive ILS considerations of most military organizations will not be applicable. While all ILS elements are applicable to a commercial product, the provision of those support elements are spread out among many elements, not all of which are directly involved in its support. Component manufacturers are not usually focused on the issues of the prime product developer, users and supporting organizations support many products as a business, and transportation services transport many products for many customers. The focus of these ancillary service elements is their particular expertise, available to anyone. This contrasts with military systems where the support infrastructure is designed for, and focused on, supporting one client base. Regardless, it is still incumbent on the design authority to ensure that the product is designed for the support concept, plans, and capabilities of the customer, whether it is a civilian using the product in their home or a trade worker using it in the field.

The design team should also consider and document the expected life of the product. A short product lifetime with a customer who has little or no support capability requires a different design approach than the case where there is essentially no customer support (such as with retail products) and the product has an expected lifetime that is fairly long, but the customer does not have a mature support capability established. It is not infrequent, in the latter case, that an educational process is required for both the design team and the customer to be able to answer the questions that arise in trying to support a product.

4.4.1 Design Interface

The vast majority (approximately 90%) of the total LCC for a typical system used by the military are locked in by design decisions before the Full Rate Production (FRP) decision is made [2]. It should be obvious from this data point that the product's design is a tremendous determining factor and impactor upon how much the product is going to cost its customer and user. Design decisions from the early and middle design phases lock in the logistics support costs of the system or equipment for as long as it remains in use. This is the reason why, in any mature ILS planning effort, so much emphasis is placed on involvement during the earliest design phases. These efforts, often referred to as design interfaces, are intended to gain early knowledge of design activities and to influence design decisions to improve product supportability and reduce LCC.

The design interface efforts are most effective when begun during the requirements definition phase (or even the concept development phase) of the product's life cycle. It is at this beginning point that the product's future performance and supportability are first shaped. What the product does, how it does it, what it weighs, how easy it is to maintain, and much more are directly traceable to the system-level requirements, to supporting subsystem level requirements, and further decomposed to component requirements.

If a supportable system is the desired goal, this is the starting point where the design begins. It is common for a supportability specialist (or more than one) to be embedded within the product design team. This specialist should bring the right experience and skill set to intentionally influence the design team regarding decisions and options that impact on the product's supportability, reduce support costs, and improve the system's fielded availability. It is a common byproduct of this design/support integration that all participants learn new aspects of the system and what product support entails.

4.4.2 Design Considerations for Improved Logistics Support

During the design process, it is sometimes difficult to identify opportunities or decisions that can actually affect logistics and LCC costs. Generally, the design team is hard at work focused on engineering-centric issues and don't have time to solve "down the road" problems. The problems in front of them are hard enough. This is why a supportability specialist is an important part of the team. They can look out for such matters and make suggestions or conduct analyses that add to the totality of the product design. It is their ability to understand the bigger picture and the multiple areas of support interest, with design awareness, that contributes to a better product.

Most companies have detailed design guidance for their products, and that is an excellent first step. Also, general industry design guidance is readily available. People who work in the DoD environment are familiar with the myriad, extensive, and detailed design guides used in the design and development of DoD systems and equipment. People who are not familiar with the DoD community can still get access to the same military, DoD, Federal, NASA, Department of Energy, and government specifications, standards, handbooks, and publications and design guides from several public information sources. Some suggested sources are listed at the end of this chapter.

4.4.2.1 Tools

Effective maintenance relies upon the right tools for the right job with adequate access. Optimized maintenance requires the fewest different tools possible, maximizes the use of existing tools with which maintainers are familiar and have on hand (always check on the tool inventory of the maintainer's standard toolkits), provides room for tool and maintainer access, provides room for ingress and egress of the failed part or replacement parts, requires no new training, and is able to be quickly performed. Where possible, use common attaching hardware. The fewer types and sizes of screws, nuts, bolts, washers, cotter pins, Dzus fasteners, and so on, the easier the maintenance will be.

The use of quick-disconnect connectors simplifies the tools required and enhances maintenance actions. Be mindful of the customers' needs regarding metric or standard or torx tools (torx was developed by Camcar Textron). Minimize the need for special tools or test equipment. Where test or support equipment is necessary, try to use equipment with multiple functions and equipment that are technology and industry standards.

4.4.2.2 Skills

The skills needed for product maintenance is one of the largest cost drivers of long-term support. Skills definition early in the design process is important for two significant reasons. First, it drives the design team to address this subject when making design decisions. Second, developing new skills and the training that goes with it takes time. Early identification of new training or modified training material helps the support system be ready when the product needs it.

Skills are expensive to develop and maintain, and they must be constantly refreshed and renewed as support personnel change over time. Managing the development of support skills is a constant focus item to be sure they are available and ready when needed. When a design team pays attention to minimizing, or at least identifying, any special skills, especially unique skills, the ability to support the product is enhanced.

The team should be extra mindful if the product is to be used and maintained internationally. The skills available and their definitions vary in different countries, perhaps extensively. Research the details of these skills and expect differences in basic definitions.

The design team should also pay attention to regulatory restrictions, limitations, or definitions incorporated in law on skills, and their implementations. This also includes situations where organized labor is in effect or where governmental controls of labor are prevalent.

4.4.2.3 Test/Support Equipment – Common and Special

Maintenance often requires equipment to test, troubleshoot, diagnose, or confirm a repair in addition to the normal toolkit. These items are generally referred to as test and support equipment and are further categorized as common and special. Test equipment is considered "common" when it is widely used and available (e.g., digital multimeter). Test or support equipment is considered "special" when it has limited application or is used for specialized technology or applications.

For effective maintenance, the design should require a minimum of test and support equipment. This can be achieved by letting the product itself do the test and diagnostic work. This is not always possible or feasible. In the case where it is not feasible, the product should minimize the need for test equipment, especially for specialized test equipment.

If possible, product design might contain all the troubleshooting and operational health capability possible or feasible. Capabilities such as BIT software and screen-based computer-led diagnostics and maintenance aids result in faster maintenance (more operational time for customers) and lower support costs. Design reviews should include verifying that the need for additional tools, especially special tools and support equipment and unique support equipment, are fully documented and justified.

If future support costs and LCC are concerns for the product, the design team should also be concerned about tool requirements and selection. Added tools and support equipment, especially special and unique items, are expensive cost drivers of support costs. They must be purchased and maintained, training (initial and recurring) is required for each item, recurring calibration is often required, spares are required if they are prone to failure, upgrades and replacements are needed over time, spares are needed, and operational and storage space is required.

A digital multimeter is an example of common test equipment, whereas an engine borescope would qualify as special test equipment. A jet engine stand would be considered special support equipment. The jack in a car might be considered special support equipment because it is not in the regular toolkit and is generally applicable to that particular car for special situations. A customized laptop that is plugged into a system, or an aircraft, that allows the maintainer access to system health and operational parameters not otherwise accessible is special test equipment.

4.4.2.4 Training

Training is a critical element of an effective support capability, and the decisions of the design team will drive the need for training. Creating new training requires extensive planning and preparation, including needs and skills analyses, identifying training objectives and methods, as well as facilities, training aids, and training equipment. Decisions to include new technologies or capabilities in a product should be weighed against the cost to develop new training for the technologies, knowledge, or skills that will be required.

Modifications to existing courseware are sometimes an alternative. While less resource-intensive than developing new courseware, it still demands considerable analysis and effort to prepare revised courseware. Additionally, training materials and mockups may need revision or development, adding to the concerns involved in preparing for product support. Often the training capabilities are being developed in parallel with the product development. In such cases, the synergy between the product design and the training system design is so intertwined that any design changes are likely to have significant impacts on the cost and availability of training.

Training facilities are always a concern to support planners, especially if new facilities are needed. Budget planning processes, in conjunction with contracting and construction processes, mean that new training facilities often take years to put in place. To be ready when the product is, early identification of training needs and facilities is mandatory.

In some cases, training is conducted using simulation technologies. Simulations are effective training tools, but they require special development knowledge and skills. Some simulation capabilities can be hosted on laptops, but large simulation systems require specialized facilities and sometimes their own buildings.

4.4.2.5 Facilities

The maintenance concept should address the facilities needed to house, operate, store, maintain, dispose of, and transport the product and all elements ancillary to it. The existing facilities available in the time-frame of expected product introduction should be assessed and confirmed with the authorities that control those facilities. Once definitive facility requirements are identified, gaining commitments from the facility stakeholders for those facilities should begin. Facility impacts and needs should be addressed for all areas of product support, both direct and indirect. Consider where all levels of maintenance will occur and what resources and capabilities will be needed by each, including *in situ*, field, backshop, and depot maintenance.

Also consider all aspects of facilities design, such as: HVAC, electricity (types and amounts), floor bearing and loading, dust control, humidity control, air pressure, egress and ingress needs (examples include door sizes to move equipment, or limited passageways as in a ship), fire detection and response, security systems, personnel support facilities, and hydraulic or pneumatic distribution. Some facility concerns to address include:

- Storage facilities for equipment and spares, short-term and long-term, local and central.
- Facilities for personnel to include working and non-working time and personnel care, such as food, sleep, washing, clothes maintenance, etc.
- If security is a consideration, the need to develop and maintain secure facilities, especially where additional personnel are required to operate or support such a facility.
- Transportation facilities are needed to move equipment, spares, tools, test equipment, documentation, and personnel.
- Facilities for operating the product including power, roads, noise, personnel access, security, safety.
- Maintenance requirements such as: benches, anti-static, clean air, special power, sound attenuation, pneumatics, and security.

4.4.2.6 Reliability

One of the most significant logistics support elements that is driven by product design decisions is product reliability. The entire support infrastructure and its costs hinge on the reliability of the product. The elements of support are integrally linked, not unlike a spider's web where a tug on one piece affects all the others. The need for maintainers is linked to the levels of maintenance, which is linked to the tools and equipment needed, which is linked to the training for those maintainers, which is linked to the documentation needed to perform the maintenance. All these things are linked to the facilities to provide the training and house the personnel and store the spare parts, which are linked to transportation and shipping needs, which are linked to the documentation systems to track and manage parts, activities and usage. The quantities of all these elements and more are driven by the frequency at which they are needed. A design that is highly reliable has a lower demand on support resources and costs the product owner less.

Reliability is typically measured as the average operations between bad things happening. This very basic description is provided to summarize the general understanding of reliability across almost any application. However, there is not really any one definition that can apply across all situations except the classic definition: the ability of an equipment or system to perform its purpose or function without degradation or failure.

Reliability is most often referred to in terms of MTBF or the probability of equipment completing a defined mission without degradation or failure. There are also many other variations of reliability terms that apply to different scenarios. Sometime, the miles between failures is a more appropriate metric, or landings between failures, or mean number of rounds fired between failures. The proper application of appropriate terms and statistical methods is critical to obtaining meaningful metrics for each particular situation. Choosing inappropriate terms will be detrimental to determining the appropriate support needs for a product. The design team should include reliability expertise to ensure that reliability estimates, predictions, and terms are properly applied.

Calculating reliability is often a matter of details. It is prudent, indeed necessary, for the design team and the user community to be in agreement on what reliability means for a product. Reliability can be measured and analyzed and used in many different ways. All product parties need to have a common understanding to prevent misunderstandings or disagreements later in the design, development, testing, and operational phases. A real-world example of how a simple word can cause issues comes from a weapon reliability testing program where meetings of the oversight management of the test team agreed to biweekly meetings. The confusion came from exactly "biweekly" means. Is it twice a week or every other week? This question got resolved after much discussion, but it is a small example of how details matter in the reliability world. (By the way – the dictionary said it is either.)

Terms that should be addressed and carefully defined by all involved parties include:

- **Mission**: Define a typical mission or profile to use as a design goal and against which to measure product performance. Consider defining alternate missions or mission extremes that may apply for scoring purposes.
- **Failure**: Not all failures are alike. Carefully define levels of failures that may occur. Suggestions include catastrophic, maintenance-induced, sympathetic, relevant and non-relevant, non-test, scorable, mishandling, accident, and more.
- **Acceptable operations**: Identify what functionality or performance functions the product is expected to perform. This defines the acceptable operations, and failure to meet these parameters is consideration for a failure. Often, the inability to meet these parameters results in the equipment being shut down.
- **Minimum operations**: In many cases, there is a minimum set of performance that is acceptable without the product being considered failed. This is a common situation in complex systems where not all functionality is fully available, but partial functionality is sufficient to keep the item operating. This characteristic requires additional work to define partial success and how to evaluate occurrences within the definition of failure.
- **Time**: There are many forms of time to consider: power-on time, operating time, passive non-operating time (i.e., missiles on aircraft wings), and calendar time. Alternate equivalents include power on/off cycles, landings, rounds fired, miles driven, and so on.
- **Test conditions**: Identify and define the conditions under which the product will be tested. Generally, these are derived from the requirements documents and should address normal as well as extreme conditions. Conditions often include environmental factors, but may include vibration, temperature rise or fall, atmospheric pressure, salt environment, water submersion, design or customer operators or maintainers, lab or field locations, and so on.
- **Who**: Define who will operate the equipment during testing and what training, skills and tools they will require. If testing is in-house manufacturer testing, do the same, but address whether user representatives will be present or participating.
- **Documentation**: Agree on the documentation needed to identify the facility used as a test facility, the Unit Under Test (UUT), actual operating environment, operational logs, maintenance logs, failure or anomaly documentation forms, recording of dates and times or readings from Elapsed Time

Indicators (ETIs), and all other related data. It is too late after the fact to realize that certain data are needed but were not collected.

As a basic design and support characteristic that drives all other LCC considerations, reliability should be a prime design factor in any product. A product that is allowed to complete the design process with poor reliability will fail frequently, will likely disappoint the user, and will have an unnecessarily high demand for maintenance and related support resources. The additional LCCs caused by the poor reliability will far exceed the cost of designing in better reliability. Alternatively, a product which is highly reliable and requires infrequent maintenance has lower support costs and generally better user satisfaction.

Paradigm 1: Maintainability is inversely proportional to reliability

Each product and situation is different, and the end result is always a compromise among a myriad of competing criteria. Design choices are not simple. Some compromises are technical, such as weight or size or power demands. Others are less direct, such as the design budget, manufacturing target cost, time allowed to finish the design, or design guidance limitations.

One maxim of product design and support is: reliability is designed in. It cannot be improved once the design is finished.

4.4.2.7 Spares Provisioning

One direct result of product reliability is the cost associated with spare parts and spare complete units. A complete maintenance concept should include the entire spares provisioning topic in sufficient detail for the design team to clearly comprehend the types and importance of spares provision planning for the product and to allow sufficient detail for support planners to develop detailed plans for the acquisition and provision of an adequate spares posture. The maintenance concept should address such spares-oriented topics as:

- What are the replaceable parts of the product at each level of maintenance?
- Where can the entire product be replaced?
- What is the expected failure rate of each LRU or part?
- What is the product's reliability requirement?
- Who will provide the spare parts?
- When will the spares be provided at each needed location?
- How will the spares be transported to storage sites?
- How much space/volume is needed to store the spares in each location?
- What special handling or handling equipment is required for the spares?
- Do the spares require periodic inspections and maintenance during storage?
- What environmental requirements or shelf-lives do the spares have?
- Do the spares require calibration, maintenance, or software loading prior to installation?
- What is to be done with the packaging material after the spare is installed?
- What is to be done with the removed LRU or part?
- What documentation, tools, support equipment, or safety items are needed for the spares?

The above is not a complete list of considerations. The list gets longer and more detailed as planning morphs into specific plans and then into actual actions. The design team needs to have an appreciation of these factors and be comfortable discussing them with support personnel on the product team.

4.4.2.8 Backshop Support

An element of product support that is frequently given inadequate and timely attention is the support needed away from the in-place product. The maintenance concept needs to address all levels and forms of maintenance to be complete, and the design team needs to include these considerations in their design process.

A backshop is generally considered any maintenance or technical support capability that is not involved in the basic field-level direct maintenance of the product. Some military organizations refer to backshops as Intermediate, or I level, maintenance or shore maintenance, or real echelon maintenance. More and more, depot- and manufacturer-level maintenance organizations are providing backshop support, especially for very complex or intricate systems. Regardless of what it is called, backshop personnel generally have more training, skills, and maintenance authorities than field maintainers. The backshop personnel are able to perform repair capabilities that are more in-depth and complex than organizational-level or intermediate-level maintainers.

As product experts, backshop maintainers are sometimes called upon to assist in problems at the operational site. The tools, test and support equipment, and facilities for backshop support are fairly specialized and take time to properly define, plan, and establish. Early identification of the resources needed at the backshop level will help the support teams responsible for setting up their facilities and capabilities achieve their goal on time.

As new technologies come into play, the design and support teams should consider technological options for achieving the desired support. With the advent of ubiquitous data and communications capabilities, it is now feasible to provide backshop technical support via web chat, interactive video chat, and strategic or tactical reach-back assistance involving technical research or training support. The product design might be such that backshop personnel can access the product and initiate remote diagnostics and troubleshooting, directing on-site personnel to perform actions via virtual support.

The operational environment of a product plays a big part in determining support opportunities. For example, for a remote system with little access by maintainers or telecommunications, extensive redundancies, self-diagnostics, and self-repair capabilities might be appropriate design goals to mitigate support costs. Artificial intelligence developments offer design teams the possibility of new capabilities for remote systems self-diagnostics and repair.

Some operations are remote, but still accessible physically or via communications methods. An example is airlines where pilots in flight need technical assistance, but are in a remote area with nowhere to land and get help. For such situations, they use radio telecommunication reach-back to request help from technical experts to diagnose and correct an issue in-flight. This process also allows for maintainers and spares to be made available at the destination location when the plane lands.

4.5 Summary

A key factor for successful product design is planning ahead. This should come as no surprise to those trained in or experienced in the product design process. It is the forcing function that says, paraphrasing, you cannot design what you cannot define, that helps avoid vague and generic requirements early in a product's life cycle. In fact, it is the same saying that makes us pay attention to requirements at all. This is also why, in today's enlightened design environment, considerable time, energy, and resources are invested early in a product's life cycle to identify, define, and refine the myriad technical factors that will eventually become a finished product. Many hard lessons have been learned, and relearned, on this path to enlightened product design and development to reach the point where early design planning and documentation is considered not just "best in class," but mandatory for a product's success.

An increasingly significant element in the early product definition is defining how the product is going to be maintained or even if the product is going to be maintained. For products that have a significant expected lifetime, it has become accepted that the users' maintenance cost burden as part of the total product cost is lessened by implementation of design attributes that were adopted as design considerations early in the design process as the maintenance concept was defined and matured. It is also accepted that maintenance and support design characteristics present opportunities and competitive advantages. As demonstrated in this chapter, if and how a product is to be maintained influences the design, and the design influences how the product is to be maintained. Applicable maintenance and design personnel need to work hand-in-hand, starting early in the design process, to explore, identify, define, refine, model, and socialize all aspects of a product's intended operational life with a full, common understanding of every aspect of the maintenance concept across all elements of the design team and process.

References

1 US Department of Defense (2014). *Operating and Support Cost-Estimating Guide*. Washington, DC: Department of Defense.
2 US Department of Defense (2016). *Operating and Support Cost Management Guidebook*. Washington, DC: Department of Defense.
3 United States General Accounting Office (2003). *Best Practices: Setting Requirements Differently Could Reduce Weapon Systems' Total Ownership Costs*, GAO-03-57. Report to the Subcommittee on Readiness and Management Support, Committee on Armed Services, US Senate. Washington, DC: United States General Accounting Office.
4 Defense Acquisition University. (2019). Integrated Product Support (IPS) Element Guidebook. https://www.dau.edu/tools/t/Integrated-Product-Support-(IPS)-Element-Guidebook- (accessed 18 August 2020).

Suggestions for Additional Reading

AeroSpace and Defence Industries Association of Europe. (2014). *International Procedure Specification for Logistic Support Analysis (LSA)*, ASD/AIA S3000L. Brussels, Belgium: ASD/AIA. www.s3000l.org/.

Blanchard, B.S. (2003). *Logistics Engineering & Management*. Pearson.

Raheja, D. and Allocco, M. (2006). *Assurance Technologies Principles and Practices*. Hoboken, NJ: Wiley.

Raheja, D. and Gullo, L.J. (2012). *Design for Reliability*. Hoboken, NJ: Wiley.

UK Ministry of Defence (2016). *Integrated Logistic Support. Requirements for MOD Projects*, DEF STAN 00-600. London: UK Ministry of Defence.

US Department of Defense (1993). *Logistics Support Analysis (LSA)*, MIL-STD-1388. Washington, DC: US Department of Defense.

US Department of Defense (1995). *Definitions of Terms for Reliability and Maintainability*, MIL-STD-721. Washington, DC: US Department of Defense.

US Department of Defense (1995). *Reliability Prediction of Electronic Equipment, MIL-HDBK-217F*. Washington, DC: US Department of Defense.

US Department of Defense (1996). *Maintainability Prediction*, MIL-HDBK-472. Washington, DC: US Department of Defense.

US Department of Defense (1997). *Maintainability Verification/Demonstration/Evaluation*, MIL-STD-471. Washington, DC: US Department of Defense.

US Department of Defense (2011). *Logistics Assessment Guidebook*. Washington, DC: US Department of Defense.

US Department of Defense (2012). *Designing and Developing Maintainable Products and Systems, MIL-HDBK-470*. Washington, DC: US Department of Defense.

US Department of Defense (2012). *Design Criteria Standard, Human Engineering*, MIL-STD-1472. Washington, DC: US Department of Defense.

US Department of Defense (2013). *Acquisition Logistics*, MIL-HDBK-502. Washington, DC: US Department of Defense.

US Department of Defense (2016). *Product Support Manager Guidebook*. Washington, DC: US Department of Defense.

5

Maintainability Requirements and Design Criteria

Louis J. Gullo and Jack Dixon

5.1 Introduction

As discussed in Chapter 1, designers should spend a significant amount of time in the early stages of system design and development on system requirements generation and analysis. Developing and understanding requirements for new systems or products is a significant challenge. There are several reasons for this challenge. "First the customer typically has not thought through all of the product features and needs. There are functional requirements which jump out first. Other requirements seem to take time to discover. Then there are requirements of what the product should not do and how the product should behave when presented with off-nominal inputs (product robustness). Additionally, there are requirements to meet business and regulatory needs. Then there are questions of infrastructure, which often are not specified. They are just typically assumed to be present and only noticed in their absence" [1].

Davy and Cope [2] indicated that "it is clear that requirements elicitation has not been done well and that failure causes considerable problems." In a 2006 paper, Davis, Fuller, Tremblay, and Berndt [3] stated "Requirements elicitation is a central and critical activity in the systems analysis and design process." They went on to declare, "It is well known that requirements elicitation is one of the most important steps in systems analysis and design. The difficulties encountered in accurately capturing system requirements have been suggested to be a major factor in the failure of 90% of large software projects." An earlier work by Lindquist [4] similarly observed that: "Analysts report that as many as 71 percent of software projects that fail do so because of poor requirements management, making it the single biggest reason for project failure – bigger than bad technology, missed deadlines or change management fiascoes." The costs associated with requirements failures are substantial. Browne and Rogich [5] found that failed or abandoned systems cost $100 billion in the USA alone in 2000. Therefore, it is important to develop good requirements early to avoid the potential for system failures later. Most system failures originate from bad or missing functional requirements in specifications. The cause of most of these requirement defects is incomplete, ambiguous, and poorly defined specifications. These requirement defects can result in expensive engineering changes being required later in a program.

5.2 Maintainability Requirements

In addition to performance, there are other significant system design considerations including reliability, safety, and maintainability among many others. This chapter is focused on maintainability requirements and design criteria.

Good design requirements ensure a good design. The proactive incorporation of maintainability considerations in the design of products, equipment,

Design for Maintainability, First Edition. Edited by Louis J. Gullo and Jack Dixon.

and systems is important to ensure that they will be maintainable by the user. Total system performance effectiveness recognizes that system performance is a function of both the equipment and the user. For a system or product to be its most effective, it must be both usable and maintainable. Therefore, it is imperative that how the system/product will be maintained must be considered early and throughout the entire design and development process.

System/product requirements must address the expectations of the customer in terms of efficient and safe system/product operation and maintenance. Inadequate specifications and requirements plague the system engineering discipline in all industries. If generic specifications with vague or inadequate requirements are provided to the designers, the result will be faulty designs and, ultimately, dissatisfied customers. For any system or product, the most effective means of ensuring customer satisfaction in the area of maintainability is to implement an organized maintainability engineering effort beginning in the conceptual design phase and continuing throughout its development, fabrication, testing, production, use, and ultimate disposal. A systems-based approach to maintainability requires the application of scientific, technical, and managerial skills to ensure that maintainability considerations are addressed throughout the life cycle of a system.

Maintainability is a key element of supportability. "Supportability is the degree to which system design characteristics and planned logistics resources meet system … requirements. Supportability is the capability of a total system design to support operations and readiness needs throughout the system's service life at an affordable cost. It provides a means of assessing the suitability of a total system design for a set of operational needs within the intended operations and support environment (including cost constraints). Supportability characteristics include many performance measures of the individual elements of a total system. For example: Repair Cycle Time is a support system performance characteristic independent of the hardware system. Mean Time Between Failure (MTBF) and Mean Time to Repair (MTTR) are reliability and maintainability characteristics, respectively, of the system hardware, but their ability to impact operational support of the total system makes them also supportability characteristics" [6]. These types of requirements are crucial to the supportability of the product or system.

Maintainability engineering provides requirements and methodologies for the optimization of maintainable products or systems during use. In addition to system/product usage during normal operations, maintainability engineering must provide requirements and methodologies on how to efficiently and safely operate and maintain a system or product during periods of extreme conditions. Requirements must be provided to the users so they know the boundaries of system operation. These user-oriented requirements may be provided to the user or maintainer through training materials, technical courses to certify operators and maintainers, and through operator or maintainer manuals that describe system operation and maintenance. When operators and maintainers don't follow the established procedures and work instructions provided to them, equipment damage or accidents may occur. The means to prevent equipment damage or accidents is through well-defined and effective maintainability requirements.

Maintainability analysis should be conducted and support requirements should be defined early in the design phase. A Level of Repair Analysis (LORA), which is typically conducted early in the design of a system, provides a method to analyze the system and develop maintenance requirements. A LORA is paramount to understanding the maintenance requirements upfront and planning for them. "Level of Repair Analysis (LORA) is a prescribed procedure for defense logistics planning. Level of Repair Analysis (LORA) is an analytical methodology used to determine where an item will be replaced, repaired, or discarded based on cost considerations and operational readiness requirements. For a complex engineering system containing thousands of assemblies, sub-assemblies, and components organized into several levels of indenture, and with a number of possible repair decisions, LORA seeks to determine an optimal provision of repair and maintenance facilities to minimize overall life cycle costs. Logistics personnel examine not only the cost of the part to be replaced or repaired but all of the elements required to make sure the job is done correctly. This includes the skill level of personnel, tools required to perform the task, test equipment required to test the repaired product, and the facilities required to house the entire operation" [7].

5.2.1 Different Maintainability Requirements for Different Markets

Top-level maintainability requirements may be imposed by the customer. This is particularly true when government organizations contract with private industry to produce system or product designs. The Department of Defense (DoD), the National Aeronautics and Space Administration (NASA), the Federal Aviation Administration (FAA), and other government entities typically will dictate the high-level maintenance requirements, along with performance and other requirements, in their Request for Proposal (RFP). Customer imposition of requirements, including maintenance requirements, is also prevalent when large industrial customers, such as airlines, energy sector companies, and the automotive industry procure products or systems.

In situations where consumer products or smaller-scale industrial products are supplied to the general market, the producer of those items will establish the maintainability requirements. These maintainability requirements are determined by considering the market, the customer desires and capabilities, the product's use and life, and the competition. Establishing the proper level of maintainability requirements can be a tricky business – maintainability requirements that are too demanding may price the product out of the market, whereas if they are set too low, or are non-existent, the product may be looked at unfavorably, resulting in lost sales.

5.3 The Systems Engineering Approach

This section is meant only to serve as a brief overview of the systems engineering process. It is not intended to be an exhaustive, detailed treatment of systems engineering. The reader is invited to review the books recommended in the suggested reading section at the end of the chapter for more in-depth treatises on systems engineering.

Whether you are a government entity contracting for a complex weapon system, the prime contractor who is going to develop it, the subcontractor who is developing an important complex subsystem that will be part of it, or a small business manufacturing a complicated widget to be sold directly on the open market, the systems engineering approach has been proven to be the best, most efficient, and cost-effective process for successful development of complex systems and products.

The INCOSE 2004 *System Engineering Handbook* [8] defines systems engineering as: "an interdisciplinary approach and means to enable the realization of successful systems." Systems engineering integrates many areas of expertise in order to design and manage a system over its entire life cycle. Systems engineering provides a cradle-to-grave methodology that begins at the concept stage, developing and analyzing requirements, and continues through systems design and development into production, culminating with delivery to the customer. Systems engineering also integrates all of the various support functions as well as end-of-life disposal considerations into the development process.

The life cycle of a system comprises the following stages: concept, development, production, utilization, support, and retirement, as defined in ISO/IEC/IEEE 15288 [9]. In the concept stage the system is defined from a user perspective reflecting the system performance needs for the customer in a set of initial operational requirements. Also created early in the concept phase, usually when system acquisition begins, is the Concept of Operations (CONOPS), which was discussed in Chapter 4. The CONOPS defines how the system will be used and supported in a defined user environment and support infrastructure. Initial requirements documents and architecture diagrams are created during the concept stage and refined during the development stage. During the development stage, detailed design, implementation, integration, test, verification, and validation processes are performed. During this process, design evaluations and design reviews assess the system's ability to meet the needs of the customer/user. The production stage, in which the system is manufactured, occurs after successful validation of the system. The utilization, support, and retirement stages occur after the system is delivered to the customer. Operations and maintenance occur in these later stages.

Functional analysis is conducted early in the systems engineering process. This effort is used to define the system in terms of the functions it must perform. Functions are initially determined while defining the needs and top-level requirements of the system. The system operational requirements and the maintenance concept are then defined and functional analysis is conducted. The functional analysis will lead to the identification of resources such as hardware, software, people, facilities, and the various elements of support, which will also include maintainability criteria, and all of which will be needed to ensure a successful system design. Evolving from these activities will be the establishment of design criteria that must be incorporated into the design so that it will ultimately meet the overall requirements for the system.

The systems engineering process may need to be tailored to fit the development project at hand, and may be scaled up or down to match the size and scope of project. "The systems engineering process is continuous, iterative, and incorporates the necessary feedback provisions to ensure convergence" [10] to an optimal design.

5.3.1 Requirements Analysis

During the early concept phase, after operational needs are assessed and the CONOPS and high-level concept definition are completed, the next step is the initial creation of requirements, usually by the customer. These general, top-level requirements will cover various aspects of the system, including such things as user needs and capabilities, operating environments, quality, support requirements, and so on. Further analysis of the requirements is typically performed by the systems engineers, usually by the prime contractor in the case of government or large, complex procurements. This analysis will address any gaps in the requirements that may have been overlooked in the top-level system specifications and will transform the users' needs into system requirements. The analysis will also decompose the requirements and allocate them to lower-level subsystem specifications.

Requirements analyses is an iterative process that progressively translates the stated and implied customer operational needs into an implementable set of requirements which:

- define major system functions
- allocate these functions to lower-level system elements (i.e. subsystems, components, etc.)
- derive lower-level functional requirements
- allocate derived requirements to increasingly lower-level elements.

Top-level system requirements typically use formal "shall" requirement statements in the top-level system specification documents. The use of the term "shall" forces the product/system developer into the mandatory implementation of a described capability in order to meet the "shall." A more discerning use of "shalls" can be beneficial in situations in which the needs are likely to evolve over time. Disproportionate use of "shall" statements can overly constrain the designers and can contribute to excessively expensive solutions. Requirements that employ less formal requirement statements that use the term "should" rather than "shall" can be used to illustrate the customer's preference, but permit more freedom in the development of design solutions.

As this process of requirements analysis progresses, it is critical to ensure consideration is given to the maintainability of the end product. It is important to define and formally capture the maintainability design requirements for the product. Specification documentation is prepared for the highest level to the lowest levels of the design. High-level requirements are flowed down, decomposed, and apportioned to the lower-level requirements. As lower-level elements of the design are refined, their specifications are used to capture all design characteristics and requirements, including those related to maintainability. These specifications will provide traceability of requirements as well as the design baseline for producing the products.

5.3.1.1 Types of Requirements

There are many different types of requirements used in the systems engineering discipline. The systems engineering process decomposes the top-level requirements into lower-level system functional flows, interface requirements, performance requirements, and constraints. A brief description of some of the many types of requirements follows:

System requirements – These are the top-level requirements typically provided by the customer.

Operational requirements – These are requirements imposed by the customer that describe where, when, and under what conditions the system will be used. They will also describe what the system is expected to accomplish and what functions it will be required to perform. They will also define the life cycle of the system. The maintenance concept is typically developed from the product's operational requirements.

Functional requirements – These are design requirements, largely derived from the operational requirements, which define the functional capabilities, such as what functions the system is required to perform and when, where, and how it will perform these functions.

Performance requirements – These are usually quantitative values that are associated with a particular functional requirement. Performance is generally associated with quantity, quality, or timeliness.

Interface requirements – These requirements define the interfaces (i.e., electrical, mechanical, communications, etc.) between systems, subsystems, or components.

Verification requirements – These requirements define the verification methods and success criteria for all of the design requirements.

Maintenance and support requirements – These requirements are derived from the operational requirements and define how the system will be maintained and supported throughout its life. These requirements will address such items as spares, test and support equipment, transportation and handling equipment, personnel skills and training, facilities, manuals, and other data.

Operations and maintenance training requirements – These requirements define the training programs required for the operators and maintainers of the system.

Derived requirements – These requirements are those derived from higher-level requirements that make it possible for the system to operate effectively but which are not directly imposed by the user. Many of the requirements listed here are considered to be derived requirements.

5.3.1.2 Good Requirements

The generation of good requirements is a critical function of systems engineering. If the requirements are wrong, incomplete, or ambiguous, they will have a negative ripple effect throughout the development process. If the requirements are not correct from the beginning of the system design process, costly rework will be required and/or the system will fail to perform as desired.

According to MILSTD-961E [11], a good specification should do four things: (i) identify minimum requirements; (ii) list reproducible test methods to be used in testing for compliance with specifications; (iii) allow for a competitive bid; and (iv) provide for an equitable award at the lowest possible cost. The last two items refer more to project management and the acquisition process than to engineering. Focusing on requirements that are more relevant to the systems engineering process brings us to the attributes of "good" requirements:

- Clear – easily understood, unambiguous
- Complete – contains everything pertinent
- Consistent – free of conflicts with other requirements and compatible with other requirements
- Correct – specifies what is actually required
- Unique – only stated once to avoid confusion and duplication
- Feasible – technologically possible
- Objective – no room for subjective interpretation
- Need-oriented – states the problem only, no solutions
- Singular – specific and focuses on only one subject
- Succinct – free of superfluous material, avoids over-specification
- Verifiable – can be measured (tested) to show need is satisfied
- Traceable – traceable to its source such as a system-level requirement, which in turn is traceable back to an operational need.

Each of these characteristics contributes to the quality of the requirements and to the system being developed. Imposing these characteristics throughout the requirements development or flow-down activity will help to keep the entire development effort organized and will help to avoid issues later in the life cycle. The goal of a requirements process is to define a system that meets the needs of the users and to ensure that when it is built it will satisfy those needs and will be delivered on schedule, within cost, and possesses the desired performance characteristics, including characteristics such as quality, reliability, and maintainability.

5.3.2 System Design Evaluation

An essential part of the systems engineering process is that of system design evaluation. System design evaluation typically occurs iteratively at numerous times during the development process as the design evolves. It should be done early as the design is conceptualized and throughout the development process to evaluate design alternatives. It may involve either formal or informal design reviews.

The first step is to establish a baseline against which a given design alternative is to be evaluated. This baseline is created using requirements analysis consisting of identifying the needs, determining feasibility, defining operational requirements along with the maintenance concept. The functional requirements of the system are described first at the system level and later allocated to lower levels; these will lead to determining the characteristics that will be incorporated into the design. Both operational functions and maintenance and support functions must be determined at the top level. These functional requirements should be prioritized so that they can then influence the design process by placing varying emphasis on specific design criteria. Once specific design criteria are identified, analysis and tradeoffs are conducted on possible design alternatives. Design synthesis can then be accomplished and the design evaluation process is repeated. This process is conducted at the system level, the subsystems level, and lower levels as necessary to meet the overall customer requirements.

Assuming that maintainability is an important, top-level requirement, then maintainability design criteria such as accessibility, interchangeability, standardization, modularization, packaging, and so on, along with all the other important operational criteria, should be incorporated into the design through the system design evaluation process. Details concerning maintainability design criteria will be discussed in greater detail later in the chapter.

5.3.3 Maintainability in the Systems Engineering Process

Supportability and life cycle costs are generally driven by technical design decisions occurring during the concept and the development phases of the systems engineering process. It is therefore important to understand the impact that technical decisions have on the usability, maintainability, and supportability of the system. The development team must take into account the logistics implications of their decisions as part of the systems engineering process.

Maintainability ultimately determines how well and how rapidly a deployed system can be restored after a failure. When developing maintainability requirements, consideration should be given to preventive maintenance, condition-based maintenance, Built-In-Test (BIT), level of repair, sparing strategies, maintainer skill levels, maintainer training, maintenance manuals, and identification of required support equipment. It is easy to see why requirements derived from these considerations and the dependencies they impose on logistics support will have a direct impact on sustainment and on life-cycle costs. While some maintainability requirements are considered the upper level, overarching requirements, it is often necessary to decompose these upper-level requirements into lower-level design-related quantitative requirements such as MTBF and MTTR. These lower-level requirements are specified at the system level and then must be allocated to subsystems and/or components.

5.4 Developing Maintainability Requirements

With a functional analysis completed, the designers along with operators can begin to define the maintainability requirements for not just the system, but for all of the various subsystems. The maintainability requirements must be defined to ensure that the system will meet not only the functional needs, but also any supply chain and cost requirements. These requirements will ensure that the system can be supported wherever it may be located, without undue burden to the system owner/operator.

Table 5.1 Quantitative vs. qualitative requirements.

Quantitative requirements	*Qualitative requirements*
Mean-Time-to-Repair (MTTR)	No safety wire or lockwire shall be used
Mean Corrective Time (MCT)	All maintenance will be performed at existing facilities
Mean Man-hours per Operating Hour (MMH/OH)	No more than 15% of all access panels will be designed to require more than 4 fasteners per side (or a total of 12 per perimeter)
Maximum time to repair ([at a specified confidence level, *F*] ($M_{Max}(F)$))	90% of all corrective task times must be less than 60 min at the 95th percentile

Paradigm 8: Understand the maintenance requirements

Maintainability requirements can take the form of quantitative and qualitative requirements as demonstrated in Table 5.1. Quantitative requirements are specific measurable requirements that can be determined during analysis or based on system performance. Quantitative requirements may specify the required MTTR or Mean Corrective Time (MCT) of the entire system along with the individual subsystems or components.

Qualitative requirements are not as easily measured, or may not be measurable at all, but nevertheless are defined as specific requirements. Qualitative requirements may identify the need for fault isolation, workspace requirements, or modularization. Both types of requirements should be defined, as it may be easy for a designer to achieve certain quantitative requirements, while neglecting qualitative requirements that may have a drastic impact on the total cost of ownership.

A good maintainability requirements document will contain both quantitative and qualitative measures, along with a breakdown of specific requirements for individual systems, if required. The maintainability requirements documents may also contain requirements on the following:

- Standardization
- Modularization
- Interchangeability
- Accessibility
- Identification of components
- Fault isolation
- Fault notification
- Use COTS components to improve supply chain performance

The maintainability requirements document should provide the designers with enough information to design a system or subsystem in a way that will support the functional requirements of the system while considering the need for maintenance.

5.4.1 Defining Quantitative Maintainability Requirements

Quantitative maintainability requirements should be defined using a consistent and repeatable process that is used to develop other system design requirements such as reliability or quality. The process often used is the Quality Function Development (QFD) process. The QFD is a tool that is used for deriving customer requirements into appropriate design requirement at each stage of design and development. This method uses a matrix known as the House of Quality [12] (see Figure 5.1).

Breaking down the House of Quality, we can walk through the customer requirements using a methodical approach to arrive at the quantitative requirements:

1) All of the customer requirements are listed here. For example, these may be the required operating or flight hours in a given period where the asset will be operating. The requirements may be broken down into primary, secondary, and tertiary requirements.
2) Prioritization of what is most important to the customer. This will enable the designers to evaluate potential tradeoffs. This should be done on a numerical scale to identify the most important to least important requirements to the customer.

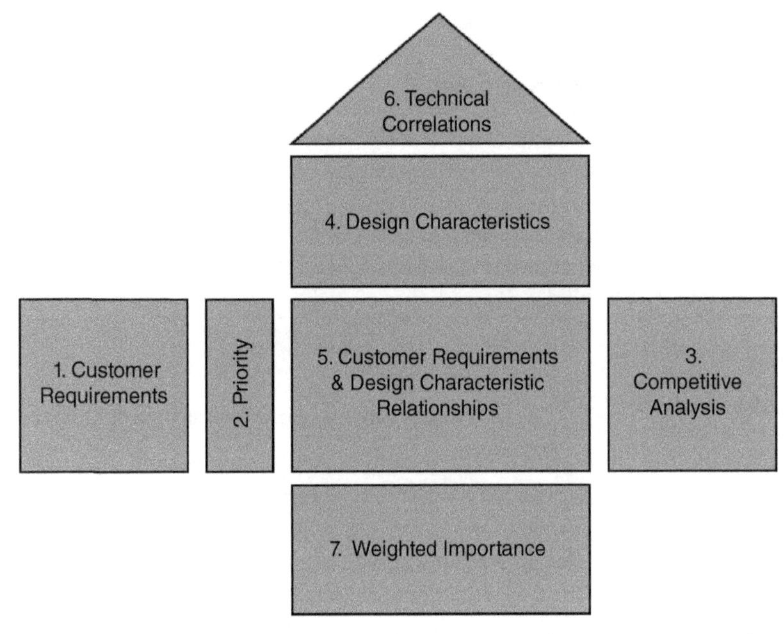

Figure 5.1 House of Quality [12].

3) What are the competitors offering? Can the competitors meet or exceed the requirements? These should be ranked as: Does Not Meet, Meets, Exceeds. You may do this for multiple competitors.
4) What design characteristics must be in place to meet the customer requirements? All design characteristics are listed across the top. For example, this may be the availability requirement to meet the number of flight hours over a given period.
5) Identify the strength of the relationship between a customer requirement and a design characteristic. These should be ranked using Very Strong, Strong, or Weak, or a numerical value.
6) Identify the strength of the technical interrelationships between the design characteristics. These should be evaluated as Very Strong Relationship, Strong Relationship, or Weak Relationship, or a numerical value may be used.
7) Evaluate the importance of each design characteristic on how it will impact the customer priorities and requirements. To do this, multiply the customer priority ranking by how well the design characteristic meets the need. Continue for each customer requirement in the vertical column. Then sum up all of the values for each design characteristic. Also consider the weight of each design characteristic against the competitive analysis.

By using the House of Quality at the system level in Figure 5.2, specific design characteristics can be prioritized. As each additional subsystem level of the system is evaluated, an additional House of Quality can be used to ensure the right design characteristics are incorporated [12]. Once the House of Quality process is completed, the specific quantitative maintainability requirements can be finalized. There are numerous quantitative maintenance requirements. Many of the most common are listed below:

- Maximum downtime per operating cycle
- Mean Man-hours per Operating Hours (MMH/OH)
- Mean Active Maintenance Time (MAMT) per maintenance action
- Corrective Maintenance Time (CMT) [13]
- Mean Corrective Maintenance Time (MCMT) [13]
- Median Active Corrective Maintenance Time (MACMT) [13]
- Preventive Maintenance Time (PMT) [13]
- Mean Preventive Maintenance Time (MPMT)
- Median Active Preventive Maintenance Time (MAPMT)
- Mean-Time-to-Repair (MTTR)
- Maximum Time to Repair (at a specific confidence level, F) ($M_{Max}(F)$) [12]
- Mean Time Between Preventive Maintenance (MTBPM)
- Probability of Fault Detection (PFD)

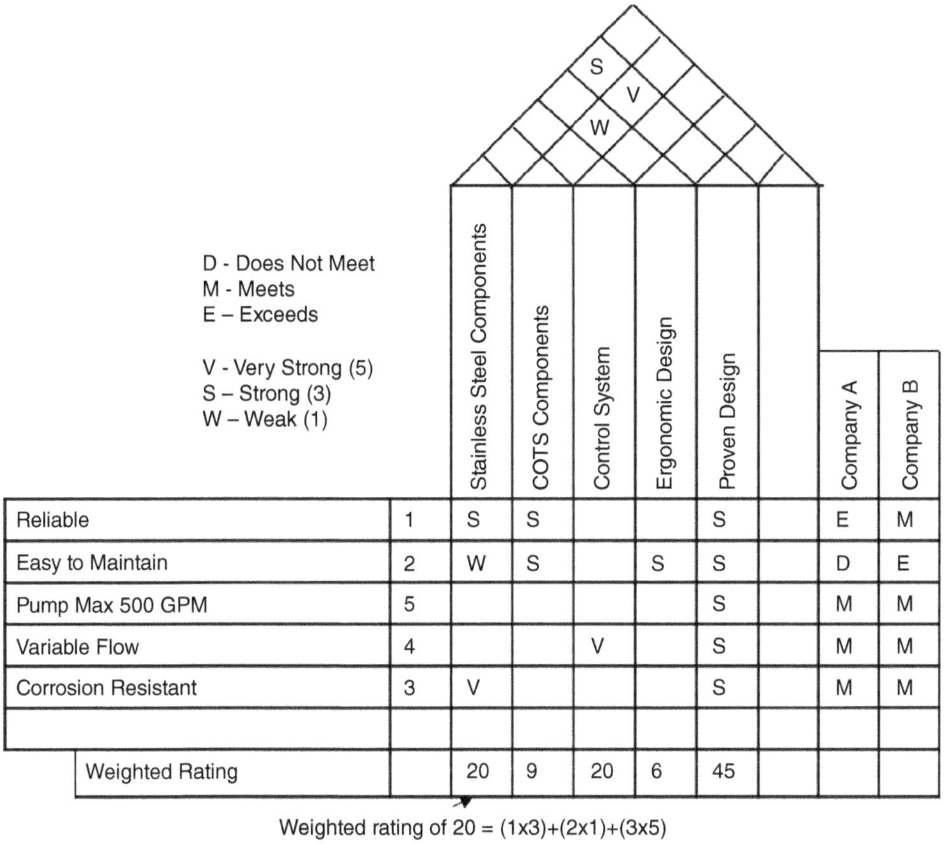

D - Does Not Meet
M - Meets
E – Exceeds

V - Very Strong (5)
S – Strong (3)
W – Weak (1)

Roof correlations: S, V, W

		Stainless Steel Components	COTS Components	Control System	Ergonomic Design	Proven Design		Company A	Company B
Reliable	1	S	S			S		E	M
Easy to Maintain	2	W	S		S	S		D	E
Pump Max 500 GPM	5					S		M	M
Variable Flow	4			V		S		M	M
Corrosion Resistant	3	V				S		M	M
Weighted Rating		20	9	20	6	45			

Weighted rating of 20 = (1x3)+(2x1)+(3x5)

Figure 5.2 Completed House of Quality.

Quantitative maintainability requirements often fall into one of two categories – preventive maintainability requirements and corrective maintainability requirements.

In addition, many quantitative requirements are focused on the mean, median, and maximums. This is done to better understand how the asset will truly perform in the field. If the distribution of repairs times is normally distributed, then the mean and median would be the same. Unfortunately, in the real world, the repair times are not normally distributed and the mean vs. the median must be understood in order to address outliers in the repair times (Figure 5.3). In addition, the maximum repair time with a given confidence interval (usually 95%) is required to ensure that the maximum downtime requirements can be met.

5.4.2 Quantitative Preventive Maintainability Requirements

Preventive maintainability requirements are used primarily to ensure that the system can be maintained in a given amount of time to meet the availability requirements of the asset or system. Often the primary measure used to define the preventive maintainability requirements is MPMT.

MPMT is usually expressed by the formula:

$$\text{MPMT} = \frac{\sum(fM_P)}{\sum f}$$

where f is the frequency of occurrence of preventive maintenance actions per 10^6 hours, and M_P is the time in hours to perform a preventive maintenance task.

Given the times and frequencies to perform the preventive maintenance task shown in Table 5.2, we can determine the MPMT.

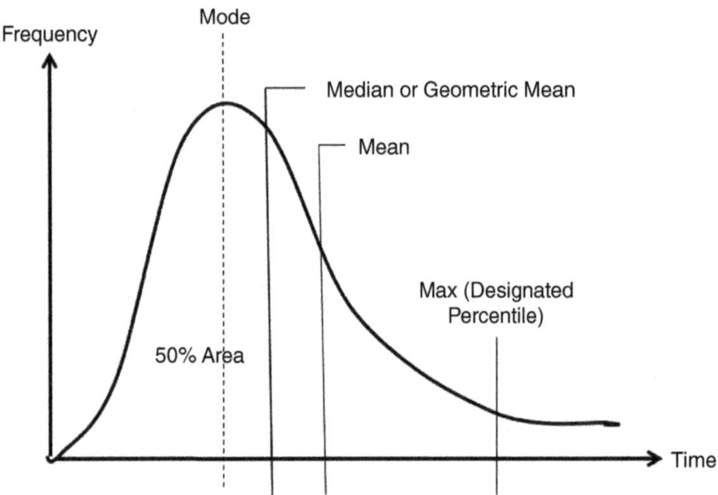

Figure 5.3 Mean, median, and maximum.

Table 5.2 Preventive maintenance activities.

M_P	f	fM_P
0.5	1388.89	694.44
1	1388.89	1388.89
4	694.44	2777.78
7	231.48	1620.37
2	462.96	925.93
9	114.16	1027.40
4	462.96	1851.85
9	1388.89	12500.00
3	694.44	2083.33
6	114.16	684.93
8	231.48	1851.85
1	694.44	694.44
0.25	2976.19	744.05
1	2976.19	2976.19
5	5952.38	29761.90
7	231.48	1620.37
2	462.96	925.93
6	925.93	5555.56
10	114.16	1141.55
2	231.48	462.96
$\Sigma = 87.75$	$\Sigma = 21\,737.97$	$\Sigma = 71\,289.73$

$$\text{MPMT} = \frac{71289.73}{21737.97} = 3.28$$

Given the information, it can be concluded that the MPMT for the system will be 3.28.

5.4.3 Quantitative Corrective Maintainability Requirements

Corrective maintenance activities are often more of a concern for organizations as they are often more unpredictable and less consistent compared with preventive maintenance, since there are a wide range of unknowns and variables. The unknowns and variables could be related to troubleshooting, finding additional repairs during corrective activities, and many more. In order to account for these variables, the corrective maintenance requirements usually define the need for MCMT, MACMT, and Maximum Time to Repair (at a specific confidence level, F) ($M_{Max}(F)$), with the confidence level set at 95% [13]. These requirements ensure that a true understanding of the corrective maintenance requirements is achieved.

MCMT, which is also known as Estimated Repair Time (ERT), is used to determine the average repair

Table 5.3 Corrective maintenance actions.

R_p	λ	λR_p
0.67	83.49	55.66
1.18	43.64	51.64
1.25	114.08	142.60
1.12	105.29	117.57
1.43	103.00	147.63
0.97	41.93	40.54
0.87	40.82	35.38
1.07	150.38	160.40
0.68	61.07	41.73
1.23	60.24	74.29
0.72	47.25	33.86
0.92	56.05	51.38
0.80	75.91	60.73
0.90	51.82	46.64
0.85	84.72	72.01
0.75	139.28	104.46
1.07	186.15	198.56
0.65	52.63	34.21
0.50	45.95	22.98
1.20	176.99	212.39
$\Sigma = 18.82$	$\Sigma = 1720.69$	$\Sigma = 1704.68$

time over 1,000,000 hours.

$$\text{ERT} = \frac{\sum(\lambda R_P)}{\sum \lambda}$$

where λ is the frequency of occurrence of preventive maintenance actions per 10^6 hours, and R_P is the repair time required to perform a corrective maintenance action in hours.

In our example, the system has 20 different repairable components, each with their own average replacement times and failure rates (see Table 5.3). Given that we want to understand the MCMT over an operating time of 8760 hours, we can use the formula given above to calculate the MCMT as follows:

$$\text{ERT} = \frac{1704.68}{1720.69} = \text{MTTR}$$
$$= 0.99$$

Now we can conclude that the system will experience an MCMT of 0.99 hours assuming that the data are normally distributed. If the data follow an exponential distribution, then the ERT equation is described as follows:

$$\text{ERT} = 0.69\ \text{MTTR}$$

Building off the example above, if the data follows an exponential distribution, the ERT would be:

$$\text{ERT} = 0.69 * 0.99$$
$$= 0.6831$$

When repair times follow a log-normal distribution, the formula is:

$$\text{ERT} = \frac{\text{MTTR}}{\text{antilog}(1.15\,\sigma^2)}$$

where σ is the standard deviation of the logarithms, to the base 10, of repair times. The average value of σ is approximately 0.55, in which case:

$$\text{ERT} = 0.45\ \text{MTTR}$$

Continuing with the example above, we can conclude that if the data follow a log-normal distribution, the ERT would be:

$$\text{ERT} = 0.45 * 0.99$$
$$= 0.4455$$

As you can see, it is vital to understand what distribution your data follow to get an accurate ERT or MCMT.

MACMT [13] is used to divide all corrective maintenance values so that 50% are equal to or less than the median of the dataset, and 50% are equal to or greater than the dataset. If the data were normally distributed, the value would be the same as MCMT. However, if it is not, this metric will provide insights to what is likely to occur in the field with the system or asset. To calculate the MACMT, the formula is [14]:

$$\text{MACMT} = \frac{\sum(\lambda_i)(\log\ \text{MCMT})}{\sum(\lambda_i)}$$

There are also times when there needs to be a defined Maximum Time to Repair (at a specific confidence level, F) ($\text{M}_{\text{Max}}(F)$) [14]. Usually the confidence level used is 90% or 95%. This metric specifies what percentage of corrective maintenance actions is expected to be completed below a specific value of downtime. The formula used to calculate ($M_{Max}(F)$) is [12]:

$$M_{M_{st}} = \text{anti ln } [\overline{t'} + z(t'_{1-\alpha})\,\sigma_{t'}]$$

where:

- $z(t'_{1-\alpha})$ is the value from the normal distribution function corresponding to the percentage point $(1-\alpha)$
- t is the repair time for each failure
- $t' = \ln t$
- st = standard deviation of repair times

Specific industries and customers may have additional quantitative requirements over and above those listed. It is imperative that, regardless of the quantitative measures used, the design team understands each requirement and how it should be calculated and tested for.

5.4.4 Defining Qualitative Maintainability Requirements

Any maintainability requirements that cannot be given a specific measurable value such as MTTR or CMT will generally fall into the qualitative requirements category. These qualitative requirements can cover a vast range of design outcomes, and they are critical to ensuring the asset can not only meet the needs of the operator, but also make it cost-effective. These requirements are often related to tools, training, support equipment, or facilities. Qualitative maintainability requirements are usually derived from maintainability design criteria. Common qualitative maintainability requirements include:

- No less than 75% of maintenance actions will perform with tools available in the maintainer's standard tool kit.
- No safety wire or lockwire shall be used.
- Existing skill levels must be used at the operational level.
- Existing depot level facilities must be used.
- No more than 25% of all access panels will require more than three fasteners per side.
- All lubrication activities will take place while the asset is operating with guarding in place.
- All guarding for belts and chains should be accessible and easily removed.
- All condition monitoring actions can be completed safely from outside of the guarding while the asset is operating.
- All workspaces must include limiting clearances required for various body positions according to NAVSHIPS 94324, *Maintainability Design Criteria Handbook for Designers of Shipboard Electronic Equipment*.

Often the design may not be able to translate the qualitative requirements into specific design criteria, such as the use of existing skill levels at all maintenance levels. Therefore, the design team must consider this requirement and come up with design guidelines or rules to support the achievement of the objective.

The design team should also consult the operator of a previous generation of similar assets to uncover other maintainability issues with the asset, which can then be incorporated into the design. It is only when the design and potential operator of the asset come together that meaningful maintainability requirements can be defined and the asset designed for maintainability.

5.5 Maintainability Design Goals

The purpose of maintainability engineering is to develop a product or system that can be easily maintained to ensure its continued cost-effective use, and maximum availability for the customer. In order to accomplish this, maintainability design goals must be established early in the product's development. Some typical maintainability goals during system design phase may include:

- Minimizing the need for support resources
- Minimizing the skill requirements for maintenance personnel
- Minimizing maintenance training requirements
- Maximizing safety for operating and maintenance personnel
- Maximizing modularity
- Ensuring easy accessibility
- Optimizing commonality and interchangeability of components

- Ensuring that the cause of defects can easily be determined and corrected
- Ensuring that all items requiring maintenance are clearly labeled
- Minimizing support equipment requirements
- Minimize requirements for special tools
- Minimizing calibration and alignment requirements
- Ensuring ease of testing

Once the goals are established, they need to be converted to clear, concise maintainability criteria, and requirements that can be incorporated into a specification for product development.

5.6 Maintainability Guidelines

"Although quantitative measures are used extensively to evaluate the maintainability of a design as it evolves, much of the 'art' of designing for maintainability involves the application of tried and true ... guidelines" [12]. Merriam-Webster defines a guideline as "an indication or outline of policy or conduct" [15]. Wikipedia defines guideline as, "a statement by which to determine a course of action" [16]. Guidelines are typically general rules and are not absolutes that must always be followed. Design guidelines are often used to perform tradeoffs among the various design alternatives (i.e., maintainability, performance, reliability, safety, etc.).

Guidelines can come from numerous sources. For example, MIL-HDBK-470A [12] contains a large number of guidelines (over 7000) covering many categories, topics, and different types of equipment. Another useful source is MIL-STD-1472 [17], which is replete with potential guidelines related to human factors. Many companies develop their own sets of guidelines for the various types of equipment they develop.

Paradigm 5: Consider the human as the maintainer

Appendix A of this book contains a maintainability design verification checklist, with a collection of guidelines drawn from numerous sources and references. Use these, and other, guideline compilations to create maintainability guidelines tailored to your particular products.

In order to make guidelines useful for a particular product or system development effort, some screening is typically needed:

- Tailor the guidelines to the specific type of product being developed.
- Screen guidelines for applicability to the project at hand prior to initiating conceptual design efforts.
- Review guidelines on an iterative basis as the design evolves.
- Pay particular attention to guidelines that might help prevent accident or loss, personnel death or injury, or collateral damage.
- Focus on guidelines that might prevent the product or system from performing its function or mission.
- Revise, expand, and update guidelines to keep them current as technology changes.

Once the appropriate guidelines have been identified for the type of product being designed, each guideline should be translated into specific quantitative or qualitative design criteria.

An example of translating a guideline into a design criterion is as follows:

Avoid swivel type connectors and fittings for air, fuel, and hydraulic line interfaces due to their history of low reliability. This guideline does not prohibit the use of swivel type connectors and fittings. However, if used, some action must be taken to avoid the problem of low reliability encountered in the past. Also, if a trade is to be made, whatever advantages might be obtained through the use of swivel connectors would have to be weighed against its historically low reliability (and the correspondingly high maintenance rate) [12].

5.7 Maintainability Design Criteria

The following definitions for design criteria provide insight to what they are and how they are used:

- "Design criteria constitute a set of 'design to' requirements, which can be expressed in both qualitative and quantitative terms. These requirements represent the bounds within which the designer must 'operate' when engaged in the iterative process of synthesis, analysis, and evaluation. Design criteria may be established for each level in the system hierarchical structure" [18].

- A second, simpler definition is that design criteria are "a group of characteristics provided as input to the design process (e.g. factors which dictate equipment design)" [19].
- Another definition for maintainability design criteria is that they "describe or reference specific design features (i.e. criteria) that are directly applicable to the system being developed. This may relate to quantitative and qualitative factors associated with equipment packaging, accessibility, diagnostic provisions, mounting, interchangeability, section of components, and so on. These criteria constitute an input to the design process" [20].

Maintainability design criteria are developed to ensure that maintainability considerations prevail through the design and manufacturing of the system. As was discussed earlier, the systems engineering process is used to decompose the customer or user performance requirements and allocate them to lower-level subsystems, components, and so on. This process also includes allocation of maintainability-related requirements that have been provided as part of the top-level requirements, the maintenance concept, standards, and guidelines. These requirements must be transformed into specific maintainability design criteria. These design criteria will be implemented to ensure that the existing design meets the specified maintenance concept and system performance requirements. Examples of a few of the more common top-level quantitative performance requirements that might be specified are:

- MTTR is one of the most common methods of specifying maintainability requirements.
- MMH/OH is a requirement that is a combination of the time to perform a task and the number of persons required to perform each task.
- BIT or testability requirements ensure the proper level of BIT to reduce the maintenance time required. It also helps to reduce the amount of test equipment required.
- Fault Detection/Fault Isolation (FD/FI) requirements define the percentage of faults that must be detected.

To achieve established maintainability goals, maintainability design criteria are developed from the previously established goals, requirements (such as those above), and guidelines. These design criteria are then used as a basis for evaluating the proposed designs in terms of their maintainability, testability/diagnostics, and so on, as the design evolves. As the design matures, design criteria are used to design in the necessary maintainability features and to perform tradeoff studies to optimize design solutions. Some of the areas to consider in developing maintainability design criteria include:

Standardization: Maximizing the use of standard parts simplifies maintenance and reduces the spares inventory.

Modularization: Designing equipment using modules facilitates repair and replacement of failed components.

Simplicity: Using the simplest, least complicated design will streamline maintenance.

Interchangeability: Physical and functional interchangeability will facilitate component removal and replacement, reduce downtime, and minimize inventory.

Accessibility: Providing reasonable access to components will allow for easier diagnosis, repair, and replacement while reducing maintenance time.

Anthropometric considerations: Designing equipment while taking into account human factors such as the capabilities and limitations of maintenance personnel to accomplish the assigned tasks, displays and control requirements, lighting requirements, and so on, will facilitate maintenance activities while protecting the health and safety of the worker.

Fault recognition, detection, and isolation: Implementing fault recognition, detection, and isolation techniques will simplify maintenance and lower corrective maintenance time.

Testability and test points: Facilitating testing will expedite diagnosis and repair time.

Environmental conditions: Defining equipment exposures to temperature/humidity, vibration, weather conditions, and so on, early will ensure the design meets the requirements imposed.

Supportability requirements: Considering spares, supply chain, levels of maintenance, and so on, during the design will ensure that the system can be properly supported when deployed.

Tools and support equipment: Defining necessary tools and support equipment, using what will be available at the assigned maintenance locations, and minimizing specialized tools and equipment will provide the best maintenance capability.

Maintainer requirements and skill levels: Design the equipment to use the personnel resources that are available and to be useable and maintainable by personnel with the specified skill levels.

Training requirements: Defining the amount and scope of training that will be provided to maintenance personnel will help in personnel selection and ensure that sufficient training will be available.

Packaging, handling, and storage: Defining how equipment is packaged, what material handling equipment will be needed, what storage space will be required, and what environments the equipment will be exposed to will ensure that the equipment will be designed to meet these needs.

Facilities: Defining the size, location, and requirements for needed facilities early will ensure their availability when required.

Paradigm 6: Modularity speeds repairs

There are many maintainability design criteria that are typically developed during the design of a system. A few examples are provided here to illustrate the concept of how the considerations in the above list can be translated into detailed maintainability design criteria:

- Use quick-release cables and locate cables to make removal and replacement easy and to avoid having to remove one cable to gain access to another. Provide adequate space for cables, including sleeving and tie-downs, and adequate service loops for ease of assembly/disassembly.
- Use positive locking, quick-disconnect electrical connectors to save man-hours.
- Avoid using identical electrical connectors in adjacent areas.
- Provide hoist fittings or hard points for hoist fitting attachments that are readily accessible.
- Use anthropometric measures in the man–machine interface to satisfy the wide range of personnel body measurement spectrum from the 5th percentile female to the 95th percentile male.
- Avoid special handling or shipping requirements of repair materials.
- Ensure electrical, electronic, and coaxial interfaces between fixed and moveable surfaces contain quick-release and quick-disconnect fasteners and connectors to simplify replacement of the moveable surface or the electronic module.
- Ensure the removal or replacement of electronic equipment does not require the removal of any other piece of equipment.

5.8 Maintainability Design Checklists

Checklists are used in our everyday life to remind us of either something to consider or something that requires action. We often use checklists in our daily lives without even thinking about them. Almost everyone has a "to do" list to keep track of the things that must be done, although we may not always get everything done that we would like. Your grocery shopping list, your holiday shopping list, and your daily calendar are all examples of familiar checklists.

Checklists are another tool the maintainability engineer can use. Checklists play an important part in the world of systems engineering in general and maintainability engineering in particular. Checklists can help to ensure maintainability is incorporated into design of new products and systems, in their production, and in their maintenance and support operations.

A maintainability design checklist is typically used as an aid to jog the analyst's memory and can be used to support any number of maintainability functions. Maintainability design checklists can provide a source of material for identifying maintainability requirements and design considerations in the early stages of design.

The maintainability engineer should not rely exclusively on checklists to help them influence design efforts. Other methods should be used to supplement the use of checklists. These methods may include brainstorming with a group of experts knowledgeable of the system being developed, analysis of similar systems, or lessons-learned databases. However, a high-quality maintainability design checklist will provide a good starting point.

Another important use of checklists is in the support of the system evaluation process described in Section 5.3.2. As the system design and development process proceeds, there may be occasions when an evaluation of the design is appropriate. As the system design matures, these evaluations, or design reviews, may be either formal or informal, and they may occur as ad hoc reviews or at discrete times in the development process, as required by the customer or by company policy. In support of the evaluation process, design review checklists are often used to highlight areas that should be addressed during the review.

There are numerous sources the maintainability engineer can use to create checklists that may be of help when developing a new product or system. For example, some common sources include:

- Requirements from contracts, specifications, and/or standards
- Regulations
- Company policies
- Checklists from similar systems

Checklists should be tailored to support the effort at hand. They should be tailored to reflect the type of equipment being developed, the subsystems involved, and/or the specifications and standards being used. When using a design checklist to assist in designing a new system or product, the maintainability engineer must remember that no one checklist can be all-encompassing. Design checklists are not the end-all for solving all problems; they are only meant to be a job aid to help stimulate the engineer's thinking.

Appendix A contains a checklist that can be used as a starting point for the reader to create their own checklists. It is intended to provide the reader with a list of items that may help in generating requirements for the maintainability design aspects of the product, to facilitate trade studies, or to verify the implementation of design requirements in the product design. The checklist topics in Appendix A are:

Section 1: Requirements Management
Section 2: Accessibility
Section 3: Tools
Section 4: Maintenance
Section 5: Software
Section 6: Troubleshooting
Section 7: Safety
Section 8: Interchangeability
Section 9: Miscellaneous Topics

There are many other sources of checklists, and a few are provided in the suggested reading at the end of this chapter.

In summary, the checklist is a valuable tool that can be used for many purposes, but checklists must be used cautiously and must not be relied upon as the only tool.

5.9 Design Criteria that Provide or Improve Maintainability

"In DoD, the 21st century challenge will be improving existing products and designing new ones that can be easily improved. With the average service life of a weapons system in the area of 40 or more years, it is necessary that systems be developed with an appreciation for future requirements, foreseen and unforeseen. These future requirements will present themselves as needed upgrades to safety, performance, supportability, interface compatibility, or interoperability; changes to reduce cost of ownership; or major rebuild. Providing these needed improvements or corrections form the majority of the systems engineer's postproduction activities" [21].

While this statement refers to military systems, the same concept applies to many commercial systems, in that they too have long lives, and often the products are in service well beyond their intended useful life. For example, according to the Bureau of Transportation Statistics, the average age of aircraft in use today is 25.49 years, with some aircraft from the 1950s and 1960s [22]. Therefore, often product improvements will need to be made. These service life extension programs may refurbish and/or upgrade systems to increase their service life.

Potential reliability and maintainability upgrades that make it less expensive to use, maintain, or support a system or product may include:

- The development of new supply support sources
- Improved fault detection systems or software
- Replacement parts having longer life or simpler replacement processes
- Improvements based on data captured in Failure Reporting and Corrective Action System (FRACAS) on failures and improvements to correct failures
- Improvements based on warranty data which typically track costs associated with post-fielding problems
- Software maintenance activity such as patches, upgrades, and new software revisions
- Periodic inspections of critical components or subsystems to identify the subsystems or components of the system that need improvement

The maintainability engineer, like the systems engineer, must strive to ensure the life of future systems can be extended and improved with minimum cost and effort.

5.10 Conclusions

This chapter has focused on developing, implementing, and using maintainability requirements, guidelines, criteria, and checklists. Maintainability engineers use design criteria in the evaluation process and as input to the design process and equipment selection. Application of the design criteria by the design team will ensure compliance with maintainability requirements and will support the optimization of the system's sustainability. The importance of good requirements cannot be overemphasized. They are the key to developing a successful system that meets the needs and desires of the customer or user, and they will allow for the efficient and cost-effective maintenance of the system or product that has been developed. The reader is encouraged to use the checklist in Appendix A to develop and tailor their own unique maintainability design requirements, criteria, and design guidelines, and then verify the presence of those criteria and requirements in the design after development.

References

1 Raheja, D. and Gullo, L.J. (2012). *Design for Reliability*. Hoboken, NJ: Wiley.

2 Davy, B. and Cope, C. (2008). Requirements elicitation – what's missing? *Informing Science and Information Technology* **5**: 543–551.

3 Davis, C.J., Fuller, R.M., Tremblay, M.C., and Berndt, D.J. (2006). Communication challenges in requirements elicitation and the use of the repertory grid technique. *Journal of Computer Information Systems* **46**: 78–86.

4 Lindquist, C. (2005). *Fixing the Software Requirements Mess*. CIO. https://www.cio.com/article/2448110/fixing-the-software-requirements-mess.html (accessed 20 August 2020).

5 Browne, G.J. and Rogich, M.B. (2001). An empirical investigation of user requirements elicitation: comparing the effectiveness of prompting techniques. *Journal of Management Information Systems* **17** (4): 223.

6 US Department of Defense (1997). *Handbook Acquisition Logistics*, MIL-HDBK-502. Washington, DC: Department of Defense.

7 Defense Acquisition University (2011). *Integrated Product Support (IPS) Element Guidebook*. Fort Belvoir, VA: DAU.

8 INCOSE (2004). *Systems Engineering Handbook*. Seattle, WA: INCOSE.

9 ISO/IEC/IEEE (2015). *ISO/IEC/IEEE 15288, Systems and Software Engineering: System Life Cycle Processes*. Geneva: IOS/IEC/IEEE.

10 Blanchard, B.S. (1998). *Systems Engineering Management*. Hoboken, NJ: Wiley-Interscience.

11 US Department of Defense (2014). *Military Standard: Defense and Program-Unique Specifications Format and Content*, MIL-STD-961E. Washington, DC: Department of Defense.

12 US Department of Defense (1997). *Designing and Developing Maintainable Products and Systems, MIL-HDBK-470A*. Washington, DC: Department of Defense.

13 Dhillon, B.S. (2006). Corrective and preventive maintenance, Chapter 12, . In: *Maintainability, Maintenance, and Reliability for Engineers*. Boca Raton, FL: CRC Press.

14 Blanchard, B.S. (2004). *Logistics Engineering and Management*, 6e. Upper Saddle, NJ: Pearson Education Inc.

15 Merriam-Webster Dictionary. "Guideline." https://www.merriam-webster.com/dictionary/guideline (accessed 20 August 2020).

16 Wikipedia."Guideline." https://en.wikipedia.org/wiki/Guideline (accessed 20 August 2020).

17 US Department of Defense (2012). *Department of Defense Design Criteria Standard: Human Engineering*, MIL-STD-1472G. Washington, DC: Department of Defense.

18 Blanchard, B.S. and Fabrycky, W.J. (1981). *Systems Engineering and Analysis*. Englewood Cliffs, NJ: Prentice-Hall.

19 Blanchard, B.S. (1974). *Logistics Engineering and Management*. Englewood Cliffs, NJ: Prentice-Hall.

20 Blanchard, B.S., Verma, D., and Peterson, E. (1995). *Maintainability: A Key to Effective Serviceability and Maintenance Management*. Hoboken, NJ: Wiley-Interscience Publication, Wiley.

21 US Department of Defense (2001). *Systems Engineering Fundamentals*. Fort Belvoir, VA: Department of Defense, Systems Management College.

22 Bureau of Transportation Statistics (2019). *Average Age of Aircraft 2019*. Washington, DC: US Department of Transportation https://www.bts.gov/content/average-age-aircraft (accessed 20 August 2020).

Suggestions for Additional Reading

Blanchard, B.S. (1976). *Engineering Organization and Management*. Englewood Cliffs, NJ: Prentice-Hall.

Knezevic, J. (1997). *Systems Maintainability: Analysis, Engineering, and Management*. London, UK: Chapman & Hall.

Raheja, D. and Allocco, M. (2006). *Assurance Technologies Principles and Practices*. Hoboken, NJ: Wiley.

Additional Sources of Checklists

US Department of Defense (1997). *Designing and Developing Maintainable Products and Systems, MIL-HDBK-470A*. Washington, DC: Department of Defense.

US Department of Defense (2012). *Department of Defense Design Criteria Standard: Human Engineering*, MIL-STD-1472G. Washington, DC: Department of Defense.

US Department of Energy (2001). *Human Factors/Ergonomics Handbook for the Design for Ease of Maintenance*, DOE-HDBK-1140-2001. Washington, DC: Department of Energy.

US Department of the Navy (1972). *Maintainability Design Criteria Handbook for Designers of Shipboard Electronic Equipment, NAVSHIPS 0967-312-8010 (formerly NAVSHIPS 94324).* Washington, DC: US Department of the Navy.

6

Maintainability Analysis and Modeling

James Kovacevic

6.1 Introduction

Maintainability must be designed into the asset or system with intention. Maintainability is not something that just happens; therefore, it must be thought out and included in the design. Maintainability is a function of design, and the designer must plan for and understand the impact of various designs on maintainability. Therefore, the design must be modeled to understand whether the asset will meet the needs of the customer. The level of maintainability or lack thereof cannot be overcome in the field with better training, a good supply chain, or better tools. The rigor with which maintainability is considered will set the absolute minimum amount of downtime the asset requires. This is why maintainability must be analyzed and modeled in the design phase.

In order to understand the levels of inherent maintainability required for a design, the designer of the asset needs to understand what the functional requirements are, along with the mission profile of the asset. With a firm understanding of the requirements, the designers can then use a variety of analysis methods and models to validate and ensure that the asset will meet the customer requirements. The ability to meet the customer needs is based on the unique design of the asset, and there may be multiple design iterations which work to balance maintainability, reliability, and cost in given operating environments.

Initially, a designer will use maintainability allocations to determine what levels of maintainability are required at the subsystem and component levels during the development phase of a program. During this phase, the designer may have to sacrifice maintainability in one subsystem to improve maintainability in another. It is this delicate balance that will result in the design meeting the needs of the customer. This system requirements balance will set the design requirements for each component or subsystem of the asset, which will serve as a guide for the lower-tiered design teams.

With the design complete, maintainability modeling techniques are used to validate the design in various configurations or operating environments. This modeling will enable the design to be evaluated along with reliability to understand how the system will operate in the real world. In addition, the maintainability models will set the framework for how the asset will be supported in the long term by determining such things as the number of staff, their skills, the tools needed, and the downtime required. Armed with this information, various techniques such as Level of Repair Analysis (LORA) can be used to help establish the expected costs for maintaining the asset over its life, which will feed into a Life Cycle Costing Plan. This enables organizations to justify the added upfront costs vs the inflated Life Cycle Cost (LCC) of not designing for maintainability.

Design for Maintainability, First Edition. Edited by Louis J. Gullo and Jack Dixon.

6.2 Functional Analysis

Since maintainability must start from the customer requirements, it is important that the designer of the asset conducts a thorough analysis of the functional requirements of the asset and any potential maintainability impacts to the functional requirements. The designer cannot solely rely upon any maintainability requirements supplied by the customer, as there is no way to confirm that the maintainability requirements are realistic or attainable.

The first step to design for maintainability is to start with a functional analysis. This functional analysis will review all of the functional requirements for the asset and its subsystems to identify where maintenance (both planned and corrective) may be needed. This allows the design team to arrive at realistic and specific maintainability requirements for the asset. The functional analysis is based upon the primary and secondary functions of the asset. A function is the specific requirements that the asset must deliver. For example, a pump has a primary function of moving liquid at a given flow rate over a given distance. There may be additional functions such as to operate without leaking, only require maintenance once per year, and so on.

A functional analysis is a method to drill down from the functional requirements of a system or asset. This analysis enables designers to understand what the system is required to do, along with the mission profile. As the mission profile is further drilled down, the specific needs of the system components or subsystems can be defined. The analysis does not just look at the operational requirements, but also at what maintenance actions may take place for each functional failure of the system. In this way, the functional requirements for each component or system can be defined, along with maintainability and reliability requirements. The functional analysis usually takes a form of a functional block diagram, as shown in Figure 6.1.

The functional block diagram can have many different levels, starting off at the system level, which describes at a high level the functional requirements of the system. The next level down, known as the subfunctional level, describes the specific ways in which each function is performed. This can be broken down into further levels as determined by the complexity of the system, as seen in Figure 6.2. Often the functional block diagram includes the use of Go and No-Go lines, which would state which maintenance action would take place, should a function fail. Each function block should be numbered sequentially and linked to their parent block.

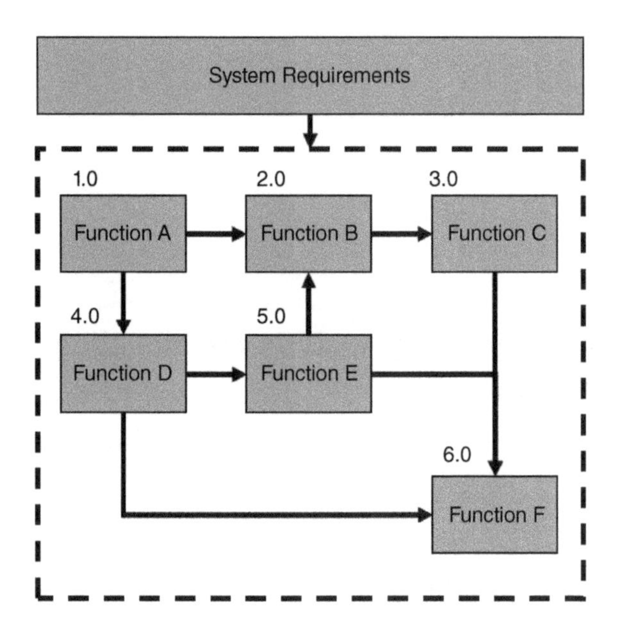

Figure 6.1 Example of a functional block diagram.

The use of the No-Go lines enables the initial development of maintenance requirements. As a No-Go event [1] is determined, the designer can consider the required corrective maintenance to restore the function or a preventive maintenance action to prevent the functional failure. The maintenance requirements may contain additional steps such as "transport to depot," "replace *xyz* component," "rebuild component," or "return to inventory" [1]. The level of detail is dependent upon the team's understanding and vision of what the asset may be like or on past experience. While this is not designed to develop an entire maintenance program, the functional analysis does allow the design team to determine what maintenance may be required, which can then be used in the maintainability allocation analysis. The identification of maintenance actions during this analysis also enables the design team to begin evaluating the need for supportability elements and to begin to evaluate the costs associated with maintenance and develop alternatives.

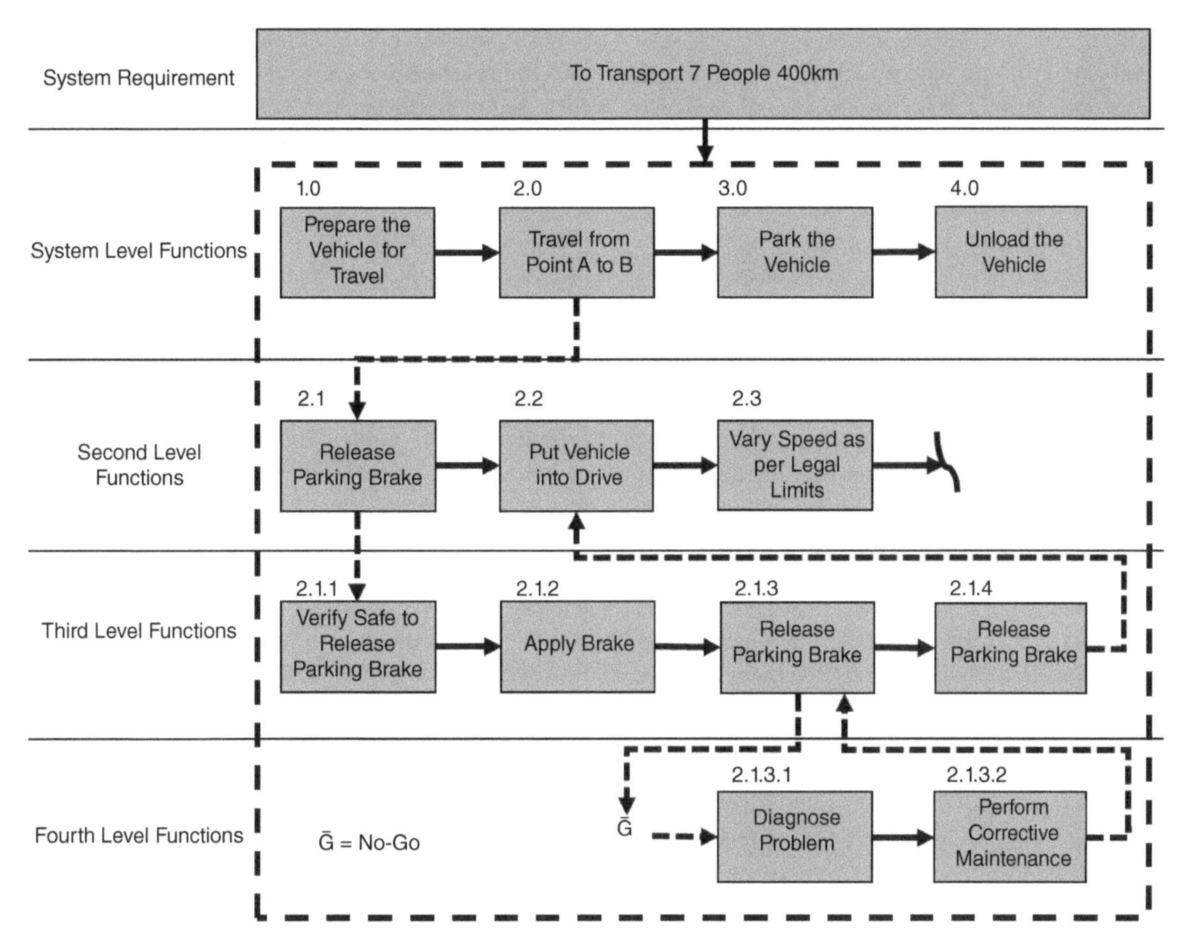

Figure 6.2 Multi-tiered functional block diagram.

6.2.1 Constructing a Functional Block Diagram

To create a function block diagram, a defined methodology should be used to ensure that it is completed thoroughly and completely. A properly completed functional block diagram will enable the designers to properly evaluate maintainability, along with reliability of the system, and allocate the requirements to all subsystems. The diagram will also enable a supportability analysis to be performed by the designers and the operators/maintainers. It is for this reason that it is imperative that the diagram be constructed completely and accurately. A suggested methodology to build the functional block diagram is:

1. **Create the function block diagram**: Begin by constructing the blocks at the system level, identifying each of the unique functions that the asset is expected to perform. Each separate function to be performed by the system should be represented by a single block in the diagram. Each functional block is a specific action that the system (hardware and/or software), the personnel, or a combination thereof, is required to perform at some point or various points in time during the system's life cycle. As each function is identified, it may be necessary to build the various sublevel functional diagrams.
2. **Number the function blocks**: After the functional block diagram has been completed, the boxes must be numbered in a sequential manner (1.0,

2.0, 3.0, 4.0, etc.) for the system level. Each sublevel should take a parent identifier along with a sequential decimal; for example a second level may take a sequence such as 2.1, 2.2, and so on. This is continued for all blocks in the diagram.
3. **Draw connections**: With all of the blocks numbered, the connection lines can be constructed for the diagram. These lines may indicate both series and parallel pathways in the functional flow of the system. The connections should be drawn with an arrow to reflect the sequence of the system.
4. **Add gates**: In some instances, the system may have "AND" or "OR" gates in the functional flow. "AND" gates are used to identify where two functions must be performed for the system to continue on in the process. "OR" gates are used to identify where a single function out of many functions can initiate the next function in the sequence. These gates help the designers to understand the true relationship of the system functions, which will enable them to better understand and identify No-Go paths.
5. **Identify Go and No-Go paths**: The symbol G is used to identify a No-Go path that indicates what should occur when a functional failure has occurred. This is usually where maintenance actions are identified. To keep the diagram clean, it is recommended that only No-Go paths are labeled and the remaining connections are assumed to be "Go" paths.

In the author's experience, it can be very helpful to use "post-it" notes to create the functional block diagrams, as often the functions will be moved around many times during the initial development. Post-it notes enable the diagram to be reconstructed with ease, and also increases the engagement and collaboration with the team.

6.2.2 Using a Functional Block Diagram

The functional block diagram is used by the design team to understand how the system will function and to begin to develop a conceptual design. The designers may also use the diagram to identify what subsystems are required and whether or not Commercial Off-the-Shelf (COTS) components may be used or if new components have to be developed.

With an understanding of how the system functions and the functional requirements, the designers can then begin to understand what the potential reliability and maintainability constraints are for the entire system. If the system requirements specify a specific Mean-Time-Between-Failure (MTBF) or Mean-Time-to-Repair (MTTR), then the designers can determine what specific subsystem requirements may be necessary. This is known as Maintainability Allocation, which is covered in Section 6.4.5.

In addition, after the functional block diagram is completed, the designers can begin to pull together operational and supportability requirements. The operational requirements may be the number of staff required to operate the unit, the specific operator training, and ongoing competency development. As for supportability, it may define the type of maintenance facilities required, who will perform the maintenance, specific tools, software, or diagnostic capabilities required, along with the number of maintenance staff, and training requirements. The development of these operational and supportability requirements will enable the designers and operators of the asset to evaluate where potential change could be made to reduce the total LCC of the system.

6.3 Maintainability Analysis

Maintainability analysis is an essential activity of the system engineering process and is conducted in parallel with the development process. It is used to transform top-level operational requirements into design criteria, to evaluate design alternatives, and to provide input into the supportability analysis processes, including the identification of spares, training and support equipment requirements, and to verify that the design complies with the system requirements.

The purpose of the maintainability analysis is to translate the maintenance concept, requirements, and constraints, into detailed quantitative and qualitative maintainability requirements. The maintainability engineer uses maintainability analysis to help achieve the specified level of maintainability in accordance with top-level requirements. The products of the maintainability analyses are used to generate the design criteria that will later be used to address the ease, accuracy, and cost-effectiveness of maintenance activities that have been integrated into the system/product design. Another benefit of the maintainability analysis is the identification of potential maintenance and support problems early enough in the design cycle to allow tradeoffs to be made. To ensure that the design will ultimately include those maintainability criteria, numerous top-level documents and factors are considered and provide input to the maintainability analysis process:

- Operational and support concepts
- Maintenance concepts
- Operating requirements
- Maintenance levels
- Testability and diagnostics concepts
- Environmental conditions where maintenance is to be performed
- Skill levels of maintenance personnel
- Maintainability requirements
- Logistics support requirements
- Facility requirements
- Support equipment and tools
- Software maintenance considerations

Like most of systems engineering, maintainability analysis is an iterative process that is performed at all levels within the design and throughout the design process. The process of maintainability analysis is typically performed following the guidance of MIL-HDBK-470A [2], and much of this section is adapted from that handbook.

6.3.1 Objectives of Maintainability Analyses

Maintainability analyses have five main objectives:

- To establish design criteria that will provide the necessary maintainability features
- To support the evaluation of design alternatives and tradeoff studies
- To provide inputs to the process of identifying and quantifying support requirements (e.g., spares, training, support equipment, etc.)
- To evaluate the effectiveness of the support concept and maintenance policies and to identify needed changes to the concept and policies
- To verify that the design complies with the maintainability design requirements.

6.3.2 Typical Products of Maintainability Analyses

The products of performing maintainability analyses include, but are not limited to:

- Mean and maximum times to repair (at various levels of maintenance)
- Inputs to LORA
- Maintenance time or labor hours per task or operating hour
- Inputs to maintenance personnel requirements (e.g., number required, existing or special skills, etc.)
- Inputs to spares requirements
- Support equipment requirements
- False alarm rates, methods of fault detection, and effectiveness of Built-In-Test (BIT)
- Mean time between scheduled and preventive maintenance
- Maintainability models and block diagrams

6.4 Commonly Used Maintainability Analyses

The following sections present and highlight common types of maintainability analyses that are used to meet the overall supportability objectives in the most cost-effective way. The various types of maintainability analyses include, but are not limited to:

- Equipment downtime analysis
- Maintainability design evaluation
- Testability analysis
- Human factors analysis
- Maintainability allocations
- Maintainability design trade study
- Maintainability models and modeling
- Failure Modes, Effects, and Criticality Analysis – Maintenance Actions (FMECA-MA)
- Maintenance Activities Block Diagrams (MABDs)
- Maintainability prediction
- Maintenance Task Analysis (MTA)
- Level of Repair Analysis (LORA)

The depth and scope of any of these analyses will vary with the design detail available and the complexity of the equipment. A description of each of these analyses is provided in the following sections.

6.4.1 Equipment Downtime Analysis

Equipment downtime analysis is used to evaluate the expected time that a piece of equipment is not available (i.e., it is down) due to maintenance or a supply backlog. This value is the sum of elapsed maintenance time, awaiting parts time, and awaiting maintenance time. It is a primary measure of merit that considers reliability, maintainability, support system attributes, and operational environment. The results of this analysis may be used to calculate other equipment measures of merit, such as mission capable rate and equipment availability. The results of the analysis indicate those areas driving non-availability of the equipment and can be used to evaluate alternative designs and support concepts.

Equipment downtime analysis may be used at any time during the program or product life cycle. Early use of downtime analysis will provide criteria to influence design for supportability, while later use will point out corrective actions that can be taken through changes in the design or support system. The depth of this analysis will increase as the system becomes better defined in the later phases of development.

Paradigm 7: Maintainability predicts downtime during repairs

Equipment downtime analysis results in a figure of merit called "equipment downtime," measured in hours, days, or other time cycle appropriate for the equipment evaluated. It can be used to identify areas driving system non-availability, to compare alternate design or support system concepts, and as input to other equipment capability measures. More details on availability and downtime can be found in Chapter 15.

6.4.2 Maintainability Design Evaluation

Maintainability design evaluation is the process of analyzing the maintenance implications of an evolving design and providing feedback to the design team in a timely manner. A major goal of this evaluation is to ensure that maintainability is designed into the product from the start.

The process starts with a set of system documents that are available to the designer and maintainability engineer. These normally consist of a preliminary description of how the system will be used, maintenance concepts, top-level qualitative and quantitative maintainability requirements, and lessons learned. Design evaluations are used to refine the maintenance concepts that will later form the basis for the maintenance elements of logistics support analysis. The depth of this analysis will depend on the phase that the design is in and the complexity of the equipment being designed. Design criteria will provide a basis for evaluating a design for maintainability.

6.4.3 Testability Analysis

Testability analysis is important at all levels of design and can be accomplished in a variety of ways. For instance, when designing complex integrated circuits (ICs), such as Application Specific ICs, or ASICs, it is important to develop test vectors that will detect a high percentage of "stuck at" faults (i.e., signal stuck at logic "1" or "0").

For non-digital electronics, fault detection efficiency is typically determined with the aid of a Failure Modes and Effects Analysis (FMEA) as described in the previous section. The FMEA will identify those faults that result in an observable failure and can therefore be detected. The test engineer then must develop a test that will verify operation and detect any malfunctions as identified in the FMEA. This process can occur at all levels of design.

Testing and testability is thoroughly discussed in Chapter 13.

6.4.4 Human Factors Analysis

One of the most basic, and important, maintainability requirements is that the system is easy to maintain by human personnel. Maintainability analysis of a system typically involves identifying maintenance tasks required for repair or removal and replacement of a part or subassembly. Maintenance tasks usually involve the disassembly of the equipment in order to gain access to the component in need of repair or replacement.

Paradigm 5: Consider the human as the maintainer

Human factors analysis is performed to identify problems related to the interaction between maintenance personnel and the design in performing each maintenance task. This analysis is used to verify that each required maintenance task can be performed by humans. This analysis often deals more with qualitative requirements than with quantitative requirements. As with many analyses, it is important that this analysis be done in the early stages of design.

Human factors analysis involves three major considerations:

- Strength – ability to carry, lift, hold, twist, push, and pull objects in various body positions
- Accessibility – ability of maintenance personnel to access the work area
- Visibility analysis – ability to clearly see the work area, labels, displays, and controls

There are a variety of modern, animated, computer-aided design (CAD) tools, and virtual reality techniques available to assist the maintainability engineer in effectively and efficiently performing these analyses. When problems are discovered during the human factors analysis, the proposed design modifications can be quickly verified for their effectiveness using these same tools and techniques.

A more detailed coverage of human factors analysis can be found in Chapter 10, and a general discussion on the use of virtual reality in maintainability design can be found in Chapter 17.

6.4.5 Maintainability Allocations

This section is adapted from *Designing and Developing Maintainable Products and Systems*, MIL-HDBK-470A Section 4.4.1.6.2 [2].

With the system or asset maintainability requirements defined, the designers must decide on how maintainability should be allocated or budgeted for each subsystem. The combination of all subsystems should not result in a system maintainability that does not meet the system requirements. This allocation is a balancing act that must be performed by the designer. If there is a significant focus on one subsystem or component, it may sacrifice maintainability in another or it may result in a higher than expected cost to build the system. Maintainability allocations are a type of a maintainability model.

Maintainability allocation is the process of apportioning the system-level maintainability requirements to lower levels of assembly. In other words, the system requirements are apportioned to each subsystem; each subsystem's requirements are apportioned to components and equipment within the subsystem; and, finally, the component and equipment requirements may be apportioned to modules.

Maintainability allocation requires a detailed analysis of the system architecture and knowledge of the characteristics of various types of systems, subsystems, and so forth. Allocations are made primarily for corrective maintenance requirements. Historically, system-level requirements have been difficult to fully assess without a prototype or first-production version of the system. So allocations have been used to assess the progress being made toward achieving the system-level maintainability requirement.

Maintainability allocations are a natural management tool. They are used by the customer, prime contractor, and subcontractors and suppliers to:

- Derive "not-to-exceed" maintainability values (i.e., maximum MTTR) for the system's lower level indentures of assembly.
- Provide designers and maintainability engineers with a standard for monitoring and assessing compliance with stated maintainability objectives.
- Identify areas needing additional emphasis (regarding maintainability) and areas where improvements in maintainability will have the greatest effect on the system.

Maintainability allocations provide a "budget" of maintainability values which, if met, will ensure with a high degree of confidence that the system-level requirements will be achieved. This budget is the standard against which subsequent maintainability predictions and demonstrated (i.e., measured) values are compared. The allocation of maintainability requirements must be completed and the results made available to the designers and any subcontractors early in the program.

Allocation is an iterative process. The feasibility of achieving the initial set of allocated values must

Table 6.1 Typical types of "in-place" repair and maintenance.

Type of maintenance action	*Performed on*
Repair	Hydraulic, pneumatic, lubrication, and fuel lines
	Electrical cables and wiring
	Structural components
	Control cables
Calibration and adjustments	Subsystems, components, or items
Fueling and servicing (includes lubrication)	Product, components, items

be evaluated and, if the allocated values are not reasonable, the allocation must be revised.

One final note regarding allocations: as discussed thus far, and will be shown in the specific methods that follow, the maintainability values allocated to subsystems, components, and so on, are expressed in the same terms as used for the product (MTTR, for example). However, an item may simply be removed and replaced to repair the product. Repair of the item itself would then be done off the product. For example, if an aircraft (the product) had an engine fail internally, the engine would be removed and replaced. It would then be sent to the engine shop or the engine manufacturer for repair. For complex products that are mobile (wheeled and tracked vehicles, aircraft, railroad engines and cars, and, to a lesser extent, ships), many "repairs" consist of removing and replacing the failed item or component. Table 6.1 shows the types of repairs and maintenance that are made on the product (i.e., in-place).

6.4.5.1 Failure Rate Complexity Method

In this method, the most stringent maintainability requirements (that is, the lowest MTTR values) are allocated to the subsystems and components having the lowest reliability; and conversely, the least stringent maintainability requirements are allocated to the subsystems and components having the highest reliability. The assumption is that the most complex items will have the highest failure rates. For that reason, the method is referred to as the Failure Rate Complexity Method (FRCM). The procedure for the method is as follows:

Step 1. Determine N_i, the number of each item in the product for which the allocation is being made.
Step 2. Identify λ_i, the failure rate for each item (constant failure rate is assumed).
Step 3. Multiply λ_i by N_i to find C_{fi}, item's contribution to total failure rate.
Step 4. Express each item's MTTR, M_i, as the product of (λ_H/λ_i) and M_H, where H is the item having the highest failure rate.
Step 5. Multiply each result from Step 4 by the corresponding λ_i. The result is C_{M_i}.
Step 6. Using the following equation, solve for the MTTR of the item having the highest failure rate.

$$\text{MTTR}_{\text{Product}} = \sum C_{M_i} / \sum C_{fi}$$

where $C_{M_i} = M_i C_{fi}$

Step 7. Solve for the MTTR of the other items by multiplying the MTTR found in Step 6 by λ_H/λ_i.

Table 6.2 illustrates an example of maintainability allocation using the FRCM, for the subsystems shown in Figure 6.3. The same method was used to allocate the MTTRs found for subsystem B to its components.

6.4.5.2 Variation of the Failure Rate Complexity Method

A method used by Blanchard and Fabrycky in their text [3] is a variation of the FRCM. In this approach, an initial MTTR is assumed for each item and the product-level MTTR, M_{Product}, is calculated. If the result is equal to or less than the required M_{Product}, the allocation is complete. If it is not, then new values of each item's MTTR are selected and the process repeated until the calculated M_{Product} is equal to or less than the required M_{Product}. The initial values for the items' MTTRs can be selected based on similar items already in use or engineering estimates.

6.4.5.3 Statistically-based Allocation Method

A well-documented, statistically sound methodology for performing a maintainability allocation may be found in IEC-706-6 [4]. The key underlying assumption is that, within a product, item maintainability is inversely proportional to item complexity.

Table 6.2 Allocation using failure rate complexity method.

Item	Step 1.	Step 2.	Step 3.	Step 4.	Step 5.
	Determine no. of items per product (N_i)	**Identify failure rate $\lambda_i \times 10^{-3}$ f h^{-1}**	**Calculate contribution to total failure rate $C_{fi} = N_i\, \lambda_i \times 10^{-3}$ f h^{-1}**	**Express each MTTR (M_i) as $(\lambda_H/\lambda_i) \times M_H$**	**Calculate contribution to system MTTR $C_{Mi} = M_i\, C_{fi}$**
A	1	5	5	M_a	$5\,M_a$
B	1	1.111	1.111	$4.5\,M_a$	$5\,M_a$
C	1	0.833	0.833	$6\,M_a$	$5\,M_a$
			$\Sigma C_{fi} = 6.944$		$\Sigma C_{Mi} = 15M_a$

Step 6. Solve for M_a

$$\text{MTTR}_{\text{Product}} = \Sigma C_{Mi}/\Sigma C_{fi} \Rightarrow 1.44 = 15\,M_a/6.944 \Rightarrow M_a = 0.67\text{ h}$$

Step 7. Solve for M_b and M_c

$$M_b = 4.5\,M_a = 3\text{ h};\ M_c = 6\,M_a = 4\text{ h}$$

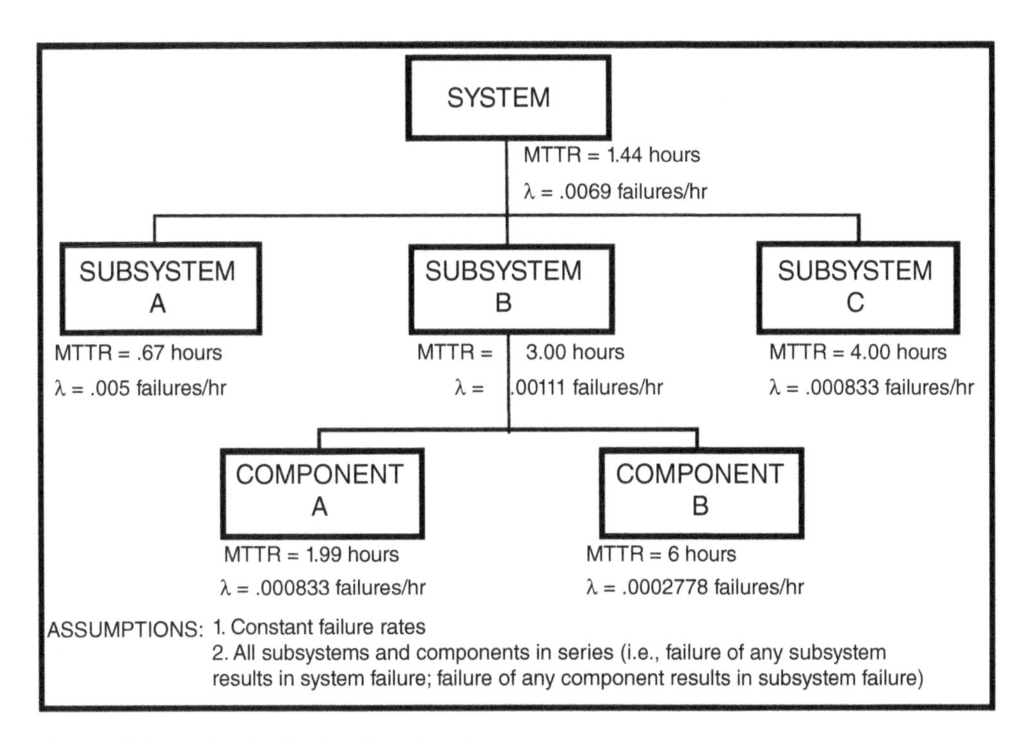

Figure 6.3 Example of maintainability allocation.

This method is based upon the frequently used assumption that the maintenance times, and especially the active corrective maintenance part of them, which is generally under the control of the supplier, can be adequately described by a log-normal distribution with mean active corrective maintenance time (MACMT) and 95th percentile maximum active corrective maintenance time (ACMT95; also called $M_{Max}(95)$). ACMTs longer than ACMT95 are also determined so as to provide the complement to the accumulated MACMT specified for the item.

6.4.5.4 Equal Distribution Method

This method is applicable when the items have equal, constant failure rates. The Equal Distribution method simply allocates the product-level value of maintainability to each lower indenture item. As shown in Table 6.3 for the product depicted in Figure 6.3, using the product-level MTTR for each item does indeed result in an allocation that supports the product-level requirement. The assumption underlying this method is that repair times are unrelated to the failure rate (i.e., MTTR is not affected by complexity). The method is identical in principle with the Equal Distribution method used for reliability allocations.

6.4.6 Maintainability Design Trade Study

As designers strive to improve maintainability, they must make sacrifices, mainly in terms of cost. While it is possible to improve both reliability and maintainability of the components, subsystems, and systems, it often comes at a substantial cost. Therefore, many designers balance the design performance requirements with maintainability, reliability, and costs, as a minimum set of design trade study criteria. As one example, a design trade study could be conducted to determine the best course of action to obtain an availability requirement. A decision could be made after analyzing a design that spending equipment design and development funding to improve equipment design reliability may have little impact on overall equipment availability, but the cost of improvement in maintainability may have a significant improvement in availability with a desirable Return on Investment (ROI) when comparing the dollars spent on the design versus the dollars saved in the equipment life cycle with an availability improvement. Design trade studies seek to achieve a delicate balance between cost, reliability, and maintainability to achieve the optimal availability.

In order for the designer to make the proper tradeoff during any particular design trade study, the designer needs to review the requirements of the asset, system, or equipment, and determine what the priority is for the asset purchasers and operators. In many cases, the designer must perform a tradeoff to balance the overall availability with the total cost to operate the asset over its life. This cost to operate over the asset's life is known as the Life Cycle Cost (LCC). LCC takes into account the cost not only to purchase the asset, but also to operate, maintain, refurbish, and dispose of it.

Determining the LCC of an asset will allow designers the opportunity to see how the changes in maintainability impact the costs throughout the asset's life and not just the upfront cost. The designer may have many LCC models, with each model reflecting a potential change in design for maintainability or reliability. Life Cycle Cost Analysis (LCCA) is a proven, practical approach to evaluating the various design tradeoffs, but it is both part science and part art. The late Paul Barringer was a big advocate of LCCA, and provided extensive resources on the topic. In summary, Mr Barringer provided an outline for LCCA, which has been

Table 6.3 Example of equal distribution method.

Item	No. of items per product (N_i)	Item failure rate $\lambda_i \times 10^{-3}$	Contribution to total failure rate $C_{fi} = N_i\,\lambda_i \times 10^{-3}$	MTTR (M_i) (each set equal to product-level M)	Contribution to system MTTR $C_{Mi} = MC_{fi}$
A	1	5	5	1.44	7.2
B	1	1.11	1.11	1.44	1.6
C	1	0.833	0.833	1.44	1.2
			$\Sigma C_{fi} = 6.943$		$\Sigma C_{Mi} = 10$

CHECK

$$\text{MTTR}_{\text{Product}} = \Sigma C_{Mi}/\Sigma C_{fi} = 10/6.943 = 1.44\,\text{h}$$

summarized and adapted below to reflect the Design for Maintainability process [5]:

1. **Define the problem requiring LCCA**: In terms of maintainability, this would be defining the various courses of action in which maintainability can be improved against losses in reliability, changes in availability, or increases in cost.
2. **Define alternatives and acquisition/sustaining costs**: For each option, the acquisition costs such as design, testing, and purchasing must be developed and estimated for each alternative. The sustaining costs may include the staff, training, facilities, parts, and disposal costs to operate the asset over its life. In addition, any downtime or lost opportunity costs should be captured for each alternative, as this will usually play a large role in truly understanding the benefits of design for maintainability. All elements should be defined for each alternative. Often, one alternative is the "Do Nothing" alternative, where the current design is analyzed to provide a baseline for the other alternatives.
3. **Chose analytical cost model**: The cost model may range from a spreadsheet to an advance LCCA specific tool. The cost model should merge the financial data, engineering data, and good maintenance practices. For example, if it is good practice to change the filter along with the oil, then costs for both parts and labor would be estimated for that activity along with estimated occurrence over the life of the asset. The financial calculation usually considers the Net Present Value (NPV) of the costs.
4. **Gather cost estimates and cost models**: With the cost model built, the designer must capture costs for each alternative. This is where the art comes in. The designer may not have all of the sustaining cost data, so they must make assumptions based on similar assets and systems. All of the cost elements should be determined, with assumptions documented for each alternative.
5. **Develop cost profiles for each year of study**: Each alternative should be evaluated for the total costs to operate over its life, as well as any lump sum costs at the beginning, middle, or end of its life.
6. **Develop break-even charts for the alternatives**: In order to better evaluate the true costs of each alternative, it is advisable to plot the costs over the life of each alternative against each other. This will show the designer where the ROI occurs in the asset's life for investment in maintainability improvements. This ROI may be the determining factor for some customers as they may have a specific hurdle rate in which investments must be returned.
7. **Determine vital few contributors**: Armed with the LCC data, the designer can then perform a Pareto analysis to determine whether there are additional options to reduce the LCC. In addition, the high contributors should be selected for additional analysis to ensure the model will represent real-life application.
8. **Perform sensitivity analysis of high costs**: The high cost contributors should be subjected to a sensitivity analysis to understand the impact of uncertainty of those contributors. This analysis may be used to provide a worst-case scenario, best-case scenario, and a middle of the road scenario. Various mathematical models can be used for the sensitivity analysis, with more advanced analysis using Monte Carlo simulation.
9. **Perform risk analysis of high-cost items and occurrences**: With the sensitivity analysis performed for the high-cost items, each cost factor can be analyzed for risks, and preventive or corrective measures can be put in place to reduce the risks to an acceptable level.
10. **Select preferred course of action**: With cost models, sensitivity analysis, and risks assessed for each alternative, the designer, in collaboration with the asset owner/operator, can evaluate the alternatives and determine what course of action would be best for the design.

The end result of the LCCA should not only be a cost analysis, but it should also provide insight into how the various alternatives contribute to, or detract from, maintainability, reliability, availability, and cost. This enables the designer and the asset owner/operator to select the right level of tradeoffs.

LORA (see Section 6.4.12) is another method that lends itself to evaluating the potential changes in maintainability with the LCC. Regardless of the approach chosen, the designer must evaluate the cost of improved maintainability against the initial cost of the asset as well as its impact throughout its life.

As a practical example, there may be a requirement by the customer to utilize existing maintenance facilities and to utilize organization level for 50% of all maintenance activities. This means reducing 50% of maintenance activities that are performed at the operational or product site. Maintenance at this level is normally limited to periodic performance checks, visual inspections, cleaning, limited servicing, adjustments, and removal and replacement of some components (i.e., constituent module, part, item, or component of the product). As a result of this requirement, maintainability should be improved to enable the existing staff to perform all routine maintenance, which means reducing complexity, or adding intelligent systems to the asset. This will drive up the cost of the asset sustainably, and it is ultimately up to the designer to evaluate the costs of doing so versus what the customer will pay for the product.

6.4.7 Maintainability Models and Modeling

Maintainability modeling activities are considered *de rigueur* for modern design and development programs. A design process that does not have accurate models to predict and assess product performance and provide for tracking test results should be considered suspect. A maintainability model should be expected to be an integral part of the product development efforts. A maintainability model should include a similar amount of rigor as would a reliability model or any number of system performance and predictive models. The maintainability model is updated frequently throughout the design and development process to reflect the system or product design, analysis results, and testing results. The primary purposes of maintainability tests are to verify the accuracy of the design model and provide opportunities to update the models to accurately reflect the maintainability design performance collected from test results, to find instances where the design is inadequate to allow for cost-effective design changes, and to provide confidence that the product meets the maintainability requirements.

Maintainability Models and Modeling is a key maintainability analysis technique, which further allows the designers to refine and enhance the asset design. Maintainability Models and Modeling are used mainly to:

- Establish design criteria
- Support design evaluation and enable evaluation of alternatives, such as LCCA
- Establish and quantify support requirements for the design, such as spare parts, training, support, and equipment [2]
- Evaluate the effectiveness of the support structure and maintenance plans [2]
- Verify that the design complies with the design requirements [2]

The maintainability models will consider the functionality of the system, or asset, and all of the supporting elements to establish how likely it is to perform in real life. As such, it is a vital contributor to performing LCCA or LORA. The maintainability model will provide the design and owner/operator with vital system information such as MTTR, maximum time to repair, and how much maintenance time or labor hours will be required per hour of mission time. In addition, the model will provide a forecast as to what support elements will be needed, such as spare parts requirements and a forecast, maintenance staffing numbers along with skill requirements, and what maintenance will be performed at the organizational, intermediate, or depot level.

Some of the most common type of maintainability models are the FMECA-MA and Maintainability Block Diagrams. These models will verify that the allocation has been performed correctly and validate that the design will in fact meet the requirements, both quantitative and qualitative.

6.4.7.1 Poisson Distribution in Maintainability Models

Given that maintainability models are generally based on the failure frequency, or number of occurrences over a period of time, the Poisson distribution is well suited for use in maintainability models. The Poisson distribution models rates such as the number of repair events or preventive events. The Poisson distribution expresses the probability of a given number of events

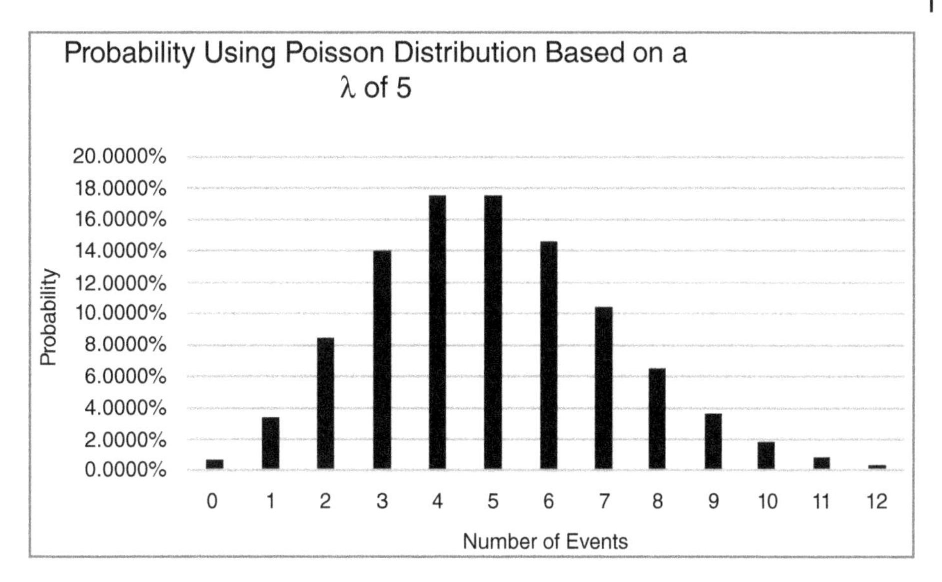

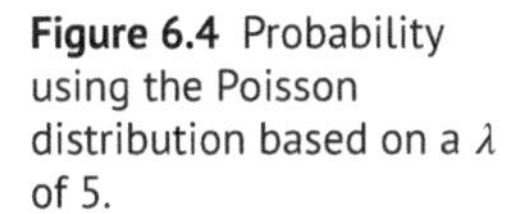
Figure 6.4 Probability using the Poisson distribution based on a λ of 5.

occurring in a fixed interval of time, if these events occur with a known constant mean rate and independently of the time since the last event. For example, if we want to consider the probability of an event occurring with a constant mean rate of 5, we can see the probability values for each variable from 1 to 12 in the plot in Figure 6.4. We can also see how the probability is highest with four or five events occurring over the time period, and how the probabilities diminish as we move away from the median.

The Poisson distribution is related to the exponential distribution such that if x is an exponential random variable, then $1/x$ is a Poisson-distributed random variable. Like an exponential distribution, the chance per interval of time or space provided is equal. In a maintainability perspective, if there is a 1/100 chance of a delayed maintenance action due to logistical delays (e.g., not having a part) over 10 corrective maintenance events, then there would be a 1/10 chance of having a logistical delay over 100 corrective maintenance events.

For example, let's consider that on average we experience six stock-out delays per year. Using the Poisson distribution, we want to determine what the probability is that we will experience zero stock-out delays over the next year.

$$\mathrm{P}(x) = (\mathrm{e}^{-\lambda} * \mathrm{a}^{x})/x!$$

$$\mathrm{P}(0) = (\mathrm{e}^{-6} * 6^{0})/0!$$

$$\mathrm{P}(0) = 1.487\%$$

This indicates that the probability of having zero stock-outs over the one-year period is 1.487%, whereas the probability of have four stock-outs would be 13.385%. Knowing this, we can adjust our maintainability models to reflect the stock-outs that may be experienced in the field.

The Poisson distribution may also be used to consider the mission success rate with backup systems. In a simple example, consider a partially redundant system with 6 components. An average of λ failures per hour can be expected if each failure is instantly repaired or replaced. If we assume that the λ is 0.002 per hour, $t = 100$ hours, and the number of elements is 6, then we can determine the probability of the system surviving with 0 failures, or having 1, 2, or 3 failures in the system. This assumes that the system can operate with a minimum of 3 components.

Taking the general Poisson expression of [6]:

$$f(x) = \frac{(\lambda t)^{x}\mathrm{e}^{-\lambda t}}{x!}$$

We must expand the expression to reflect the number of components within the system, and the average number of failures in the number of mission hour. Taking this into account, we would have a new Poisson expression of [6];

$$f(x) = \frac{(n\lambda t)^x e^{-n\lambda t}}{x!}$$

With $\lambda = 0.002$ per hour, $t = 100$ hours, for $n = 6$, then, $m = n\lambda t = 6(0.002)100 = 1.2$

$$f(x) = \frac{(1.2)^x e^{-1.2}}{x!}$$

$$f(x = 0) = 0.301 = \mathrm{P}(0)$$
$$f(x = 1) = 0.361 = \mathrm{P}(1)$$
$$f(x = 2) = 0.217 = \mathrm{P}(2)$$
$$f(x = 3) = 0.087 = \mathrm{P}(3)$$

Using the Poisson distribution expression, we can see that the system has a probability of success of 0.301 over a 100-hour mission with zero component failures. The probability of success with 1 or fewer failures is simply, $P(0)+P(1)$ which is 0.663. If the system can successfully complete the mission with up to three failures at one time, the probability of success is $P(0)+P(1)+P(2)+P(3)$, which is 0.966 23. This model can be used to highlight the advantage to online repairs, which allows for failure to occur without sacrificing mission success. It can also be used to determine what level of redundancy needs to be planned for to enable mission success.

6.4.8 Failure Modes, Effects, and Criticality Analysis – Maintenance Actions (FMECA-MA)

The FMECA-MA is used to determine what the actual maintenance actions will be for both preventive and corrective maintenance. The FMECA-MA is based upon on the FMECA methodology, but it is focused towards the actions required by maintenance and ignores actions such as redesign or process changes. The FMECA-MA is a very robust model for establishing very specific requirements in terms of the big six elements of supportability, such as spare parts requirements, facility requirements, personnel requirements, training and skills development, support and test equipment, and technical manuals [2].

The Failure Modes, Effects, and Criticality Analysis (FMECA) is a common tool in reliability used to identify the failure modes, the effects on the operation of the asset, as well as a criticality ranking. The criticality ranking is typically based on the potential consequence (C) and probability (P) of each failure mode, as well as the ability to detect (D) the failure. The product of these three risk factors is known as the Risk Priority Number (RPN) and enables the designers to sort and focus on the most critical failure modes in the design. In many instances, designers and owners/operators will only focus on a portion of the failure modes which are based on the RPN. The FMECA-MA defines what would cause each functional failure, as well as the effect that the individual failure mode will have on the system, along with any maintenance requirements needed to prevent or correct the failure mode.

In Table 6.4, an example of a pump, the FMECA-MA evaluates each of the failure modes, ranks them, and identifies each of the corresponding maintenance actions.

As can be seen in the example, the preventive and corrective maintenance actions all drive supportability requirements and can enable the designer to evaluate the design and provide input into the Maintainability Model and other analysis. The level of detail required in the FMECA-MA will be driven by the end user and their requirements for maintainability; however, the more detailed the FMECA-MA, the more accurate the models and the other analyses will be.

6.4.9 Maintenance Activities Block Diagrams

Maintainability modeling may also take the form of an MABD (see Figure 6.5). The MABD is a very robust process that provides a graphic representation of maintenance tasks, which enables the designer to assess the length of a maintenance action, as well as the big six areas of supportability. As with many other modeling techniques, designers can develop alternative models to reflect proposed design change to evaluate the impact on maintainability of the system. Ideally, the MABD is used thoroughly and throughout the design process as changes are made to the design and alternatives are evaluated [2].

In an MABD, there may be series and parallel paths that represent what activities occur sequentially or which activities occur at the same time. This is important, as the ability to add parallel activities with additional manpower can dramatically reduce

Table 6.4 FMECA-MA output.

Function	*Functional failure*	*Failure mode*	*Failure effect*	*C*	*P*	*D*	*RPN*	*Maintenance actions*
Deliver 100 gpm of water at 25′ of head	Deliver less than 100 gpm	Impeller worn due to cavitation	Reduced output of pump, slow-down of process. 2 h to repair.	7	4	2	56	PM: monitor flow rates and trigger alarms based on lower limit and rate of decrease.
								CM: requires the following; – impeller (P/N 12345), seal (P/N 17548), and a gasket (38573).
								– One level-two mechanic
								– Repair performed in-place
	Not deliver any water	Bearing seized due to lack of lubrication	Pump failed, process stopped, 3 h to repair.	8	5	3	120	PM: monitor vibration via installed sensors. Lubrication requires;
								– One level-one lube tech for three minutes per lube point
								– Ultrasonic lubrication tool
								– NLGI two grease
								CM: requires the following;
								– 2 × bearing (P/N 49502), seal (P/N 39112)
								– One level-two mechanic
								– Requires bearing heater, bearing puller, and bearing press tools
		Impeller jammed due to foreign material	Pump failed, process stopped, 2 h to repair	7	2	9	126	CM: requires the following; – impeller (P/N 12345), seal (P/N 17548), and a gasket (38573).
								– One level-two mechanic
								– Repair performed in-place

the amount of downtime required, which may be an option to improve the MTTR.

As an example, we will use an MABD to evaluate the maintenance activity of replacing a set of belts that drive a pump:

Task Description: Replace cooler pump belts
Number of maintenance personnel: 2
Equipment/parts: Belt tensioner, 8′ × 2″ nylon rigging strap, 3/4″ drive torque wrench, 3/4″ socket, 1″ socket, 8× Size D belts (P/N MC88071)

Activities:

1. Gather tools and parts – 1 person 5 minutes
2. Lock Out/Tag Out (LOTO) the pump and driver – 1 person 5 minutes
3. Verify LOTO – 2 people 10 minutes
4. Remove belt guard – 2 people 5 minutes
5. Loosen the driver bolts and remove tension from belts (8) – 1 person 8 minutes
6. Remove old belts – 1 person 5 minutes
7. Install new belts – 1 person 3 minutes
8. Align pulleys – 2 people 8 minutes
9. Adjust tension to 18–22 lb across all belts – 2 people 5 minutes
10. Torque driver bolts to 150 ft lb – 1 person 7 minutes

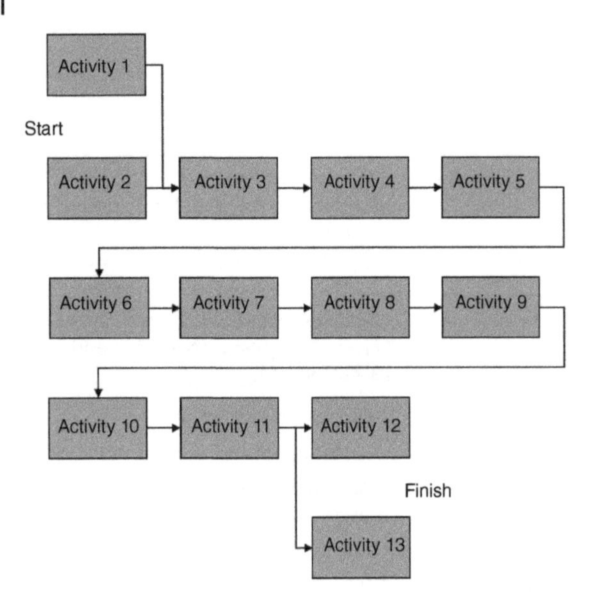

Figure 6.5 Maintenance Activities Block Diagram example.

11. Reinstall belt guard – 2 people 10 minutes
12. Start-up and verify proper operation of pump – 1 person 10 minutes
13. Put away tools and equipment – 1 person 5 minutes

The MABD should be constructed for each maintenance activity, both preventive and corrective actions, to develop a deep understanding of each activity and all supporting requirements for each. With the MABD complete, the design can evaluate each step and determine how the time to complete could be reduced. In the above example, reinstalling the belt guard is one of the most time-consuming activities. The designer could ask why and determine that it is large and bulky and requires the staff to align the bolt holes. Through a proposed design change, the guard could be redesigned to be self-aligning by resting on a ledge or with the use of locating pins. Both options will add cost to the design but could reduce the amount of time required to perform the activity.

Regardless of the approach taken, the MABD will spell out the supporting requirements as well as provide an estimated time for the completion of each task, which is vital to building various models and predictions, such as RAM Analysis, LCCA, or LORA.

6.4.10 Maintainability Prediction

Maintainability predictions are estimates of design performance from a maintainability perspective [7]. The predictions are used as another method to compare and evaluate design alternatives, as well as determining whether the asset can achieve the maintainability requirements defined by the asset owner/operator. Maintainability predictions often are performed early in the system design to help establish the high-level alternatives that the designers can take.

Maintainability predictions are not precise and often contain a degree of uncertainty because they are full of assumptions. These assumptions must be documented as they will have a significant influence on the outcome. As with maintainability allocation, maintainability prediction is an iterative process. As such, maintainability predictions should be used with other analyses to establish the best course of action for a design. Where the maintainability prediction excels is in its ability to provide the designer with the following information:

- Identifies design weaknesses
- Supports maintainability tradeoff analysis
- Determines whether the design is ready to move to the next level, based on the selected approach(es) [7].

The maintainability predictions are used to validate that the maintainability allocation is feasible or to develop supportability planning. The outputs of the maintainability predictions are usually the common maintainability metrics [1] such as:

- Mean-Time-to-Repair (MTTR)
- Mean Time Between Maintenance (MTBM)
- Mean Corrective Maintenance Time (MCMT)
- Mean Preventive Maintenance Time (MPMT)
- Maintenance Labor Hours per Operating Hour (MLH/OH)

Maintainability predictions are covered in detail in Chapter 7. It is vital to understand maintainability prediction as it directly feeds into the LORA.

6.4.11 Maintenance Task Analysis (MTA)

The MTA is another method available to the design team to determine the system's ability to meet the maintainability requirements. A maintenance task is defined as the maintenance effort necessary for retaining an item in, changing to, or restoring it to a specified condition [8]. MTA describes and assesses all maintenance tasks and associated logistical tasks to ensure maintenance requirements are satisfied. MTAs collect the data to verify all the maintenance and logistical requirements that are needed to perform all required maintenance task activities and actions (corrective and preventive). The MTA is a way to collect maintenance task data to be used for maintainability predictions. It serves as a basis for a maintainability prediction model when enough historical maintenance data have accumulated.

The MTA, along with the associated logistical information, enables organizations to not only prepare for the asset and the ability to maintain the asset, but also enables the team to investigate potential design changes to reduce any logistical requirements. As such, the MTA should be conducted early in the design phase of the asset and be revised as changes are made to the design. This will ultimately allow the designers to make specific changes to the system design that will improve and allow the system to meet the maintainability requirements.

The MTA is a tool that allows the designers to validate the existing design and its ability to meet the maintainability requirements, as well as provide an information package on all of the logistical requirements that the asset operator will require. The MTA arrives at the estimated time to perform a maintenance task, based on a detailed step-by-step analysis of the task. By analyzing each task, along with the associated frequency of each task, the design team can accurate model the systems maintainability. Further information on MTA is available in Chapter 7.

6.4.12 Level of Repair Analysis (LORA)

One of the most common methods used to evaluate and provide decision support to maintainability tradeoffs as well as to supportability analysis is the LORA. LORA is an analytical methodology used to establish what items will be replaced or repaired based on cost considerations. The LORA is also used to establish at what level the maintenance will be performed based on staffing, capabilities, costs, and operational readiness requirements. This enables the owner/operator to strike the right balance of frontline staff, intermediate staff, and depot-level staff, as well as what level of designers are required to ensure the maintenance can be performed.

Before diving too deep into LORA, one must understand the different levels at which maintenance is performed. Typically, organizations will have three different levels of maintenance – organizational, intermediate, and depot level [2]. There are instances, such as with the United States Coast Guard (USCG) where the intermediate and depot levels are combined [9].

The organizational level of maintenance is generally performed at the operational site. Maintenance at this level normally consists of performance checks, basic care such as cleaning, inspection, lubrication, and may include limited servicing such as adjustments, and removal and replacement of certain components. Most routine maintenance is conducted at this level. Most components, even if repairable, are not repaired at this level, and may go on to the intermediate level for diagnostics and repairs. The goal of the organization level is to keep the asset in a ready condition while being able to quickly restore the asset to operational status with low to moderately skilled staff [2].

The intermediate level can easily be defined as a shop location, such as an electronics shop. The intermediate level is usually performing the repair on a repairable component, by diagnosing the fault, and replacing specific parts or modules. The intermediate level may also perform larger maintenance activities such as rebuilds or overhauls. The skill level at the intermediate level is usually higher than the organizational level, and as such, the tasks associated with the intermediate level are more complex than the organizational level [2].

The depot level is the most specialized level of maintenance and repair. The depot level maintenance may be in control of the asset owner/operator or be part of the asset manufacturer. The depot level usually performs extensive rebuilds, overhauls, and very complex diagnosis. The depot level has the highest level of skill and usually includes the most sophisticated test and repair equipment. As such, it is usually the costliest level to perform maintenance. Each level of maintenance may include a mix of in-house staff and contractors.

LORA is ultimately an analysis that seeks to understand the economic and non-economic factors of the maintenance that must be performed over the life of the asset, while seeking to establish criteria for the system design to reduce and optimize the costs. Economic factors typically include the cost elements of performing the repair such as parts, support equipment, labor, contracted labor, training costs, and so on. Non-economic factors normally consist of criteria established to determine if there is an overriding reason why the maintenance should/should not be performed, which may include maintenance constraints, operational requirements, or environmental restrictions. The non-economic drivers may include workforce proficiency, maintenance philosophy, tools, environment, organizational policies, support contracts, training, supply chain, and facilities [9]. Both economic and non-economic factors must be considered during a LORA.

In its simplest terms, LORA is a decision-making tool, a form of a design trade study, which defines when, how, and where the various maintenance activities will be performed.

6.4.12.1 Performing a Level of Repair Analysis

While the thought process behind a LORA is simple, the analysis can quickly grow complex and require extensive amounts of data and information. In order to reap the benefits of a LORA, the designer must perform a thorough enough analysis to influence the design of the asset while establishing the most cost-effective manner to maintain the asset. Figure 6.6 has been adopted from the works of Lucas Marino and his dissertation [9] and is designed to walk the designer and analysis team through the LORA process.

The first step to performing a LORA is to understand the problem that is trying to be solved. For example, the LORA may not need to be performed at the asset level but at a subsystem level, as it is the primary driver of maintenance costs. Understanding how the asset is expected to perform based on the

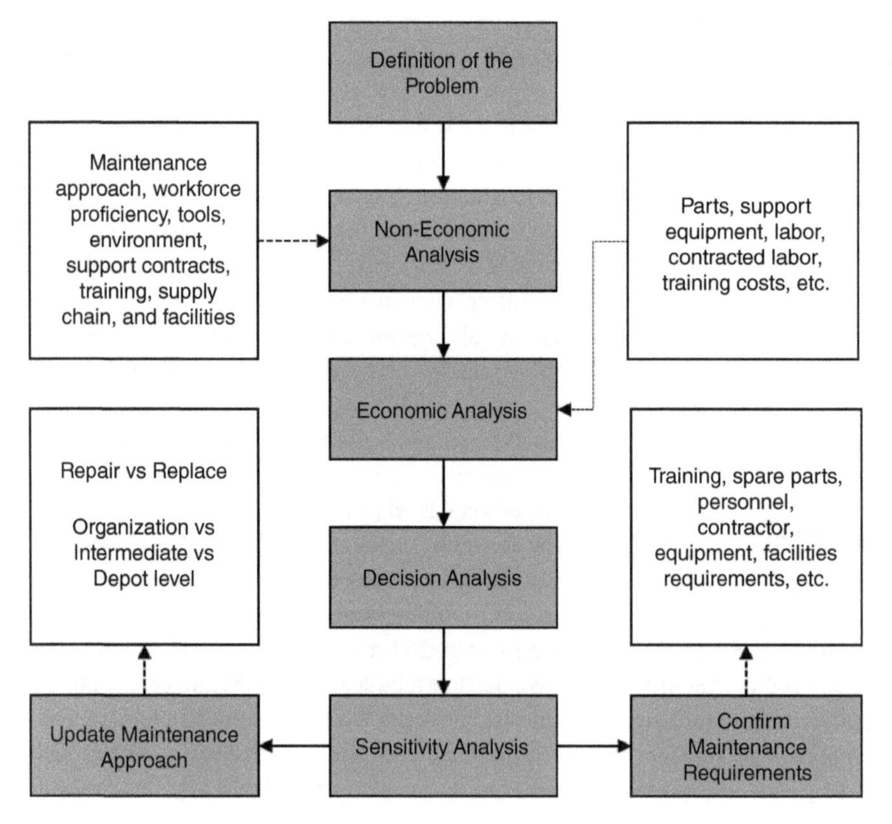

Figure 6.6 Level of Repair Analysis process.

maintainability prediction will allow the analysis to limit the scope of the LORA, allowing the analysis to be more effective and efficient. For example, a ship's turbine is the largest cost contributor for a ship, and therefore, the approach to maintenance of the turbine must be carefully analyzed. Once the scope has been properly defined, the team will need to gather relevant information such as drawings, maintenance schedules, maintenance activity block diagrams, and maintainability predictions. This data will support the team in defining the non-economic and economic elements of the analysis.

With the scope of the LORA defined, the analysis team can begin to factor in the non-economic elements and drivers. The non-economic elements will be both qualitative and quantitative in nature [9]. The identification of current personnel, skills, tools, and so forth, will allow the analysis team to identify what elements will need to be supplemented which can then be passed on to the economic analysis as quantitative data. Other qualitative elements such as operational requirements and the environment can be used in the decision analysis and sensitivity analysis phases as risks are evaluated.

The economic analysis is focused on identifying the cost to both repairs and replacement of the various components within the system at the various levels (organizational, intermediate, and depot), and the use of in-house personnel and contractors. This economic analysis is designed to evaluate all potential options along with their associated costs. The associated costs should include all specialized tools, facilities, recruitment, and training costs. The economic analysis should result in a detailed cost breakdown for all expected maintenance activities within the scope of the LORA.

Armed with the economic and non-economic analysis, the LORA team can then begin to determine what the best course of action is for determining the maintenance approach. The use of a Decision Analysis form (see Table 6.5) [8] is recommended for conducting this type of design trade study.

The use of the Decision Analysis Form enables decision-makers to have a complete understanding of the organizational impacts and risks for the decision made in terms of the maintenance approach [9]. Based on the answers to the questions in the form, the analysis team can then decide on the best course of action for each maintenance activity. Each decision should have the risks documented and evaluated as they will be used as an input to the sensitivity analysis.

Table 6.5 Decision analysis form [9].

	Decision analysis form	Y (good)/ N (bad)
1	Is the cost of maintenance acceptable?	
2	Are support elements in place to ensure success?	
3	Is the proposed maintenance appropriately funded?	
4	What life cycle phase is the asset in?	
5	Do non-organizational resources have the capacity to meet the demand?	
6	Are external maintenance resources satisfactorily responsive?	
7	Does the organization have the capability to complete the maintenance tasks?	
8	Does the maintenance demand meet operational requirements?	
9	What is the lowest estimated maintenance cost?	
10	What is the highest estimated maintenance cost?	
11	How do the analyses impact the existing ILSP?	
12	What is the most undefinable cost at this point?	

Source: From Marino, L., *Level of repair analysis for the enhancement of maintenance resources in vessel life cycle sustainment*, Dissertation, George Washington University, 2018.

At the completion of the decision analysis, the analyst will have determined one of two options.

The first is where each assembly is to be repaired (at the operational level, intermediate, or depot level), and which assemblies are to be discarded at failure. With this approach, the analyst should review the interactions that occur and result in the lowest possible cost. In essence, each assembly is evaluated individually based on certain economic and non-economic drivers. The results are then reviewed in the context of the whole and any possible feedback is assessed to ensure that there is no significant impact resulting from the decision [1].

The second is the determination of the least cost-burdened approach for the entire system if it was treated an entity. If this is the case, all assemblies would default to the least-cost approach, for example, repaired at the intermediate level [1].

With the decisions made as to the maintenance approach to be used, the decision should be analyzed using a sensitivity analysis which is used to understand the impact of uncertainty on those decisions. This analysis may provide a worst-case scenario, best-case scenario, and a middle of the road scenario. Various mathematical models can be used for the sensitivity analysis, with more advanced analysis using Monte Carlo simulation. The sensitivity analysis may determine that there is significant risk or costs associated with a particular approach which may require the analysis team to review and revise the LORA. In addition, the sensitivity analysis may include "what-if" scenarios, which may be accepted if a mitigation plan were put in place. For example, there could be a risk that access to depot-level facilities will be reduced dramatically if the manufacturer of the asset wins a new contract which would then increase the non-available time of the asset. To overcome this, the team may decide that they will pay a premium to reduce the wait time, or they will invest to bring more of those activities to the intermediate level.

Once the decisions are made, the maintenance approach for the asset or subsystem must be updated so that the LCCA can reflect the new changes. In addition, any design changes that came out of the LORA should be referred to the designers right away to be incorporated. Lastly, once the design has been finalized and the LORA updated to reflect the as-build, the maintenance requirements should be communicated to the supportability team, so that they can begin their preparations to support the asset.

Since the LORA is based on maintainability predictions, it is usually conducted early in the design of the asset because the outcomes may suggest design changes. As the design matures, the LORA model should be updated to reflect the current design and to verify that the improvements have been captured. Lastly, the LORA model should be compared to the actual maintenance performed on the asset to identify any opportunities for improvement in the supportability of the asset.

6.4.12.2 Managing LORA Data

Data is a vital part of the LORA, and as such, the quality of the data can greatly affect the outcomes. While in an ideal world the LORA team would have complete and accurate data, it is not always the case, especially with completely new designs. Therefore, the lack of data and assumptions must be carefully managed.

Paradigm 9: Support maintainability with data

Much of the data that the LORA team will use will be task data. Task data typically contain the task duration, skills, frequency, tools/equipment, and parts information. Ideally, the task data will include a bill of materials along with pricing information. If this data are not readily available, the LORA team can pull data from similar pieces of equipment but must be sure to document any assumptions and changes to the data-set based upon the new use case. Without these assumptions being documented, the LORA may not be repeatable or believable to those that will use the information.

In addition, the LORA team must also acquire the maintenance organization's data, such as personnel, historical performance, skill levels, support facilities information, and so on. Often the data coming from the maintenance organization, especially in terms of historical performance, are suspect and must be taken with a grain of salt. Their actual duration times will likely include logistical delays and not true maintenance time.

Perfect data are not required for a LORA to be successful, but the data should be subjected to a review, and documentation of all risks and assumptions should be captured. The quality of the data, along with the risks and assumptions, will drive much of the swing in the sensitivity analysis.

6.4.12.3 Level of Repair Analysis Outcomes

Once the LORA has been completed, the maintenance planning and supportability process can begin. Often, organizations do not begin the supportability planning until the first asset arrives on-site. This puts the asset and organization at risk because the organization may not have the right parts, tools, or personnel and skills in place to support the asset. The LORA analysis allows organizations to get well ahead of the process to ensure that assets are maintained properly from the beginning.

In addition, the LORA analysis can then be used by the procurement team to establish the right support contractors for the intermediate and depot level maintenance. Having a true understanding of the requirements will ensure that the support contract reflects the actual needs of the asset owner/operator and will allow the contractor to plan for the upcoming work properly. This will ensure a high level of service and improved turnaround times for the repairs.

Lastly, the LORA may be used to drive future design changes as more is learned about how the asset behaves out in the field. Often the first system in a fleet is put into operational use while much of the fleet is still a plan or just starting the construction process. By using the LORA after the first asset becomes operational, additional opportunities can be identified and can further reduce the cost of maintenance for the next generation of assets.

6.5 Summary

The ability to model an asset during the design phase is critical to ensuring that the design will meet the defined performance requirements. The modeling tools are iterative and should be used to drive changes to the design, resulting in an asset that meets the maintainability requirements. While there are many modeling tools (both quantitative and qualitative), they are unique in their approach, and should often be used in concert with each other.

Oftentimes there is a balance that must occur, and where improvements in maintainability occur, there may be a sacrifice in reliability or cost. In addition, performing the repairs, along with supplying all of the required support equipment, may not be the most cost-effective approach, and contractors may be used. It is this tradeoff that takes place during the design phase that ensures the asset and its support system is designed correctly. When modeling is combined with the various prediction tools, the design can be validated and the asset owner and operator can rest assured that the asset will meet the operational availability requirements.

References

1 Blanchard, B.S. (2004). *Logistics Engineering and Management*, 6e. Upper Saddle, NJ: Pearson Education Inc.

2 US Department of Defense (1997). *Designing and Developing Maintainable Products and Systems*, MIL-HDBK-470A. Washington, DC: Department of Defense.

3 Blanchard, B.S. and Fabrycky, W.J. (1981). *Systems Engineering and Analysis*. Englewood Cliffs, NJ: Prentice-Hall.

4 IEC 706-6 (1997). *Guide on Maintainability of Equipment - Part 6: Section 9: Statistical Methods in Maintainability Evaluation*. Geneva, Switzerland: International Electrotechnical Commission.

5 Barringer, P. and Weber, D. (1996). Life Cycle Cost Tutorial. *Fifth International Conference on Process Plant Reliability*. Houston, TX: Gulf Publishing Company.

6 Morris, S. (2011). *Poisson Distribution*. Reliability Analytics Blog. http://www.reliabilityanalytics.com/blog/2011/08/31/poisson-distribution/#more-312 (accessed 21 August 2020).

7 US Department of Defense (1966). *Maintainability Prediction*, MIL-STD-472. Washington, DC: Department of Defense.

8 Dhillon, B.S. (2006). Corrective and preventive maintenance. In: *Maintainability,*

Maintenance, and Reliability for Engineers, 143–160. Boca Raton, FL: CRC Press.

9 Marino, L. (2018). Level of repair analysis for the enhancement of maintenance resources in vessel life cycle sustainment. Unpublished dissertation. George Washington University.

Suggestion for Additional Reading

US Department of Defense (2015). *Level of Repair Analysis, MIL-STD-1390*. Washington, DC: Department of Defense.

7

Maintainability Predictions and Task Analysis

Louis J. Gullo and James Kovacevic

7.1 Introduction

This chapter describes the methodology for performing maintainability predictions based on the key standard that launched maintainability predictions as an important engineering process. That standard is MIL-HDBK-472 [1]. Some of the prediction techniques discussed in MIL-HDBK-472 are described in this chapter. The purpose of the maintainability prediction is to facilitate an assessment of the maturity of the system or product design with regards to the maintainability design criteria, features, and parameters to satisfy the maintainability design requirements. The maintainability predictions enable design trade study decisions concerning the compatibility of proposed design options with specified maintainability requirements. The maintainability predictions provide assessment results of design criteria to determine the best design option to choose, by prioritizing the option choices after differentiating the different design alternatives compared to trade study scoring criteria.

This chapter also describes the method for performing a Maintenance Task Analysis (MTA). An MTA is an engineering analysis of maintenance efforts that are necessary for sustaining an item or asset in the customer use environments, and restoring it to an operational condition following an occurrence of a failure. An MTA defines all the logistical requirements that may be required to sustain and maintain an item, system, product, equipment, or asset after it has been deployed. The MTA is a complete data package that includes all scheduled and unscheduled maintenance tasks and sequences of tasks that are likely to occur with the asset, along with all associated task times and logistical information. The MTA enables organizations to prepare for the operation and maintenance of an asset after installation in a customer field application, and to investigate potential design changes to reduce any logistical requirements. Also, the MTA is a way to collect maintenance task data to be used for maintainability predictions of new designs, and it serves as a basis for a maintainability model when enough historical maintenance data have accumulated.

7.2 Maintainability Prediction Standard

MIL-HDBK-472 is the military handbook entitled *Maintainability Prediction*, which offered guidelines for maintainability analysis techniques to ensure maintainability requirements are designed correctly prior to the verification and testing of the particular system. The initial release of MIL-HDBK-472 presented four procedures (I–IV) for conducting maintainability predictions. Maintainability prediction procedures I and III are applicable to electronic systems and equipment. Maintainability prediction procedures II and IV are applicable to all types of systems and equipment. MIL-HDBK-472

Design for Maintainability, First Edition. Edited by Louis J. Gullo and Jack Dixon.

Change Notice 1 (CN1), released in 1984, adds a new Procedure V, which is explicitly used for military avionics, ground-based systems, and shipboard electronics. Procedure V is used to predict maintainability parameters at the organizational, intermediate, and depot levels of maintenance. A summary and highlight for each of these procedures are described for consideration by the reader in the following section on maintainability prediction techniques. The specific details of these methods and the differences between these methods are not fully described in this book. The reader is encouraged to research these methods further if they are interested to learn more beyond the bounds of this chapter.

7.3 Maintainability Prediction Techniques

Maintainability predictions are assessments of the quantitative maintainability allocations similar to maintainability analyses to assure that maintainability requirements will be met. Maintainability predictions give credence to the design option selection following a maintainability trade study.

Each maintainability prediction procedure depends upon the use of recorded reliability and maintainability data, which have been collected and compiled from similar predecessor systems and products that operated under similar use scenarios and environmental conditions. Also, each maintainability prediction procedure depends upon the knowledge of skilled engineers who have experienced operating and maintaining similar predecessor systems and products. In order for the predecessor system and product data and engineering knowledge of skilled engineers to be deemed useful for a new design, the principle of transferability must apply. This principle allows empirical data accumulated from one system or product that was deployed in a field application to be applied to another similar system or product to predict the maintainability of this comparable system or product. The principle of transferability is justifiable when the degree of commonality between the systems or products can be established and the differences between their designs are minimal. With a high degree of commonality, a high level of positive correlation can be established between the design functions, features, operating procedures, maintenance task times, and levels of maintenance.

All maintainability prediction procedures are dependent on at least two basic parameters:

- Failure Rate (FR) of an item or asset derived from a reliability prediction
- Repair time of the item or asset at a particular maintenance level derived from actual time studies or historical models as found in MIL-STD-472

There are many sources for FR data. Some of these sources are organizational maintenance history composed of actual failure occurrences, failures experienced under controlled test environments, supplier data sources, and traditional military or industry handbook methodologies. FR data are functions of usage and environmental conditions. FR data are measured in terms of Failures Per Million Hours (FPMH) or Failures in Time (FIT) with units of failures per billion hours.

Paradigm 9: Support maintainability with data

Maintenance task times are collected from organizational maintenance history composed of actual system or equipment maintenance data logs resulting from either preventive or corrective maintenance actions. Preventive maintenance actions do not include repair times since maintenance is not being performed to resolve a failure occurrence, but rather is an action to prevent a failure occurrence. Preventive Maintenance (PM) actions can be categorized into six types of PM task subcategories:

- Maintenance Preparation Time (MPT)
- Inspection Time (IT)
- Calibration and Adjustment Time (C&AT)
- Remove and Replace Time (R&RT), or Component Swap-Out Time
- Service Time (ST)
- Functional Check Out (FCO), or Final Test Time

The following equation is used to calculate the total times for each PM task:

$$\text{Total PM time} = \text{MPT} + \text{IT} + \text{C\&AT} + \text{R\&RT} + \text{ST} + \text{FCO}$$

Corrective maintenance actions include repair times since maintenance is being performed to resolve a failure occurrence. Repair times will be accumulated and analyzed to develop resultant metrics. These resultant metrics will usually be Mean-Time-to-Repair (MTTR) and Maximum Maintenance Time (M_{max}). M_{max} can be separated into the Maximum Time for Preventive Maintenance (MTPM) or Maximum Time for Corrective Maintenance (MTCM).

Repair times associated with corrective maintenance tasks are usually divided into the subcategories of repair times listed below. The sum of each of these repair times is used to calculate the total Time-to-Repair (TTR) for each corrective maintenance task. The average of the TTRs for all corrective maintenance tasks is used to calculate the MTTR. The individual data points from the TTR values determine the distribution function to calculate the probabilities of repair within a certain time parameter. Maximum Time for Corrective Maintenance may be calculated from a distribution function for the time parameters in the 95th percentile of the distribution.

- Maintenance Preparation Time (MPT)
- Failure Verification Time, or Fault Detection Time (FDT)
- Fault Localization Time (FLT), or Fault Isolation Time
- Remove and Replace Time (R&RT), or Component Swap Out Time
- Functional Check Out (FCO), or Final Test Time

The following equation is used to calculate the TTR for each corrective maintenance task.

$$TTR = MPT + FDT + FLT + R\&RT + FCO$$

7.3.1 Maintainability Prediction Procedure I

This procedure from MIL-HDBK-472 is used to predict downtime of airborne electronic systems involving modular replacement of items at the flight line using the organizational maintenance level concept and aerospace maintenance operations. This procedure is used to collect the data and track all the fundamental elements of system downtime as discrete maintenance activities for each repair action. This procedure considers all the time elements for each maintenance activity that are accumulated to reach a total system downtime metric for each corrective maintenance task. The total system downtime metric considers several categories of active repair times. These categories are:

- Preparation
- Failure verification
- Failure location
- Part procurement
- Repair
- Final test

The discrete maintenance activities that may be associated with each of these categories are listed and explained.

7.3.1.1 Preparation Activities

The following corrective maintenance activities are associated with the category of preparation. Each of these separate activities under the category of preparation should have a time measured. These individual time measurements will be summed to a total preparation activity time metric for the category of preparation.

- Turn on power to the failed system, adjust controls, set dials and counters as necessary, and allow system to warm up.
- Record any visual indications as the system powers-up and begins to function; and record any unusual sounds, vibrations, or smells.
- Review past maintenance records.
- Obtain test equipment, technical orders, and maintenance manuals.
- Gain access to test connectors by removing doors or panels.
- Install test equipment to the system, as required by maintenance manuals or technical orders.
- Turn on power to test equipment, adjust controls, set dials and counters as necessary, and allow test equipment to warm up.

7.3.1.2 Failure Verification Activities

The following corrective maintenance activities are associated with the category of failure verification. Each of these separate activities under the category of failure verification should have a time measured. These individual time measurements will be summed to a total failure verification activity time metric for this category.

- Run all relevant tests for troubleshooting the system.
- Observe and document all test messages and indications.
- Observe and document all failure symptoms.
- Use additional test equipment to verify test indications.
- Test for pressure leaks, if applicable.
- Perform a visual and physical integrity check of the test setup.

7.3.1.3 Failure Location Activities

The following corrective maintenance activities are associated with the category of failure location. Each of these separate activities under the category of failure location should have a time measured. These individual time measurements will be summed to a total failure location activity time metric for this category.

- Determine if fault is self-evident from the symptom observations.
- Interpret symptoms to analyze potential causes, using additional test meters or devices.
- Gather data from different combinations of control settings and test messages or indications.
- Determine which unit or assembly is the most likely cause of the system failure.
- Determine the location of the unit or assembly in the system.
- Remove the unit or assembly that is suspected of causing the system failure.
- Observe the unit or assembly following procedures to visually inspect, check condition, and test separately from the system. The unit or assembly may be tested, either immediately on the flight line or later at a maintenance shop, as allowed by maintenance manuals and technical orders.
- Determine location and availability of a replacement unit or assembly with a substitute unit or assembly.

7.3.1.4 Part Procurement Activities

The following corrective maintenance activities are associated with the category of part procurement, whether a spare unit or assembly is or is not readily available at the flight line. Each of these separate activities under the category of part procurement should have a time measured. These individual time measurements will be summed to a total part procurement activity time metric for this category.

- Determine means to transport replaced unit or assembly to a local flight line spares storage facility or maintenance facility with a substitute unit or assembly, if a spare unit or assembly is readily available to swap from the storage facility. (Note: the time measurement for this activity may take several hours depending on the proximity of the system on the flight line to the storage facility.)
- If the unit or assembly is not readily available, determine whether a replacement unit or assembly could be removed from another system that is down for repairs and that the downed system will be waiting to be repaired for a long time (e.g., weeks or months). If a unit or assembly is to be removed from another system that is down for repairs, steps should be taken to verify that the unit or assembly in the downed system is operating properly, and is not faulty (contributing to the reason for the system being down for repairs, or perhaps damaged by the faulty system).
- If a replacement unit or assembly is not available from a downed system awaiting parts for repair, attempt to order a replacement unit or assembly through normal maintenance supply logistics channels and determine the fastest means to expedite the shipment. (Note: the time measurement for this activity may take several days depending on the proximity of the system on the flight line to the nearest storage facility with an available unit or assembly, and the shipment method, such as air transport or ground transport.)

NOTE This category of part procurement within the possible corrective maintenance activities is not considered a time parameter for calculating MTTR. It is rather a time parameter of administrative logistics downtime. Administrative logistics downtime is described in detail later in the book.

7.3.1.5 Repair Activities

The following corrective maintenance activities are associated with the category of repair. Each of these separate activities under the category of repair should have a time measured. These individual time measurements will be summed to a total repair activity time metric for this category.

- Replace unit or assembly with a substitute unit or assembly using documented installation procedures from the maintenance manual or technical order.

- Use all parts provided for installation that may be included in an installation kit.
- Ensure all electrical wire and cable connections are properly routed, mated, and tight.
- Ensure the replacement unit or assembly fits properly as the original unit or assembly that was removed.
- Clean up the area and discard any excess material left over from the installation procedure.

7.3.1.6 Final Test Activities

The following corrective maintenance activities are associated with the category of final test. Each of these separate activities under the category of final test should have a time measured. These individual time measurements will be summed to a total final test activity time metric for this category.

- Perform functional check-out test to determine whether the system failure was eliminated or is still present.
- If system failure is still present, repeat steps in sequence starting with preparation activities in Section 7.3.1.1, and record all times for each category on the second sequence combined with the times for the initial sequence.
- Repeat the sequence as necessary until the final test is completed with passing results to signify removal of the system fault.

7.3.1.7 Probability Distributions

The prediction procedure includes time distributions for maintenance activities that fit into three types of probability distribution functions:

- The fitted normal distribution (Gaussian distribution)
- The fitted log-normal distribution
- The corrected time log-normal distribution

Based on past experience, maintenance time data are most often found to be normally distributed. It is not unusual to see maintainability predictions make this assumption. However, when maintenance time parameter data are plotted on a curve and an equation of the distribution created, based on a standard deviation and mean, it may result in the data fitting poorly to a normal curve. In this case, the other two options involving a log-normal distribution are available for use to provide a better fit of the data and a more accurate prediction.

7.3.2 Maintainability Prediction Procedure II

This procedure from MIL-HDBK-472 is used to predict corrective, preventive, and active maintenance parameters. Active maintenance combines both corrective and preventive maintenance parameters. This procedure is used to collect the data and track all the fundamental elements of system downtime as discrete maintenance activities for each repair action, as for Procedure I, except it excludes all times associated with part procurement and any tasks related to administrative downtime or logistics downtime. This procedure considers all the time elements for each maintenance activity that includes only the actual repair time, which is the period when repair work is in progress. All actual repair times are accumulated to reach a total system downtime metric for each corrective maintenance task. The total system downtime metric considers seven categories of active repair times, which are similar to those used in Procedure I. These categories are:

- Localization
- Isolation
- Disassembly
- Interchange
- Reassembly
- Alignment and calibration
- Check-out

This procedure also includes two different measures for corrective maintenance: actual elapsed time and man-hours. The metric "man-hours" is a measure of manpower required to complete a maintenance activity in a given time.

Paradigm 7: Maintainability predicts downtime during repairs

7.3.2.1 Use of Maintainability Predictions for Corrective Maintenance

Corrective maintenance predictions using Procedure II follow the Equipment Repair Time (ERT) method in terms of hours to perform a particular repair task.

The ERT is defined as the median of individual repair times and their corresponding failure rates. ERT is usually calculated with different equations based on the applicable distribution function established from the measured repair time data. For a normal distribution, the median is equal to the mean.

The equation used to calculate the Mean Corrective Maintenance Time (MCMT) or ERT or MTTR when the distribution function is a normal distribution (Gaussian distribution) is:

$$\text{MCMT} = \text{ERT} = \text{MTTR} = \sum(\lambda \times \text{TTR}) / \sum(\lambda)$$

where λ is the Failure Rate (FR) and TTR is the time in hours to perform a corrective maintenance task.

When the repair times follow an exponential distribution, then the ERT equation is described as follows:

$$\text{ERT} = 0.69 \;\; \text{MTTR}$$

When the repair times follow a log-normal distribution, then the ERT equation is described as follows:

$$\text{ERT} = \text{MTTR}/\text{antilog}(1.15\,\sigma^2)$$

where σ is the standard deviation of the logarithms of repair times, to base 10.

The average value of σ is approximately 0.55. This approximation results in the simplification of the ERT equation as shown:

$$\text{ERT} = 0.45 \;\; \text{MTTR}$$

When the repair times follow a log-normal distribution, then the geometric MTTR occurs at the median. In this case, the MTTR equation is described as follows:

$$\text{MTTR} = \text{antilog}\left[\sum(\lambda \times \log \text{TTR}) / \sum(\lambda)\right]$$

where λ is the Failure Rate (FR) and TTR is the time in hours to perform a corrective maintenance task.

7.3.2.2 Use of Maintainability Predictions for Preventive Maintenance

The equation used to calculate the Mean Preventive Maintenance Time (MPMT) expressed in man-hours is:

$$\text{MPMT} = \sum(\text{f} \;\; \text{Mp}) / \sum(\text{f})$$

where f is the frequency of occurrence of preventive maintenance actions per one million hours, and Mp is the man-hours required to perform a preventive maintenance action.

An alternative method with a different equation used to calculate the Median Preventive Maintenance Time (Median PMT), when the distribution function is a log-normal distribution for the set of PMTs in units of hours, is [2]:

$$\text{Median PMT} = \text{antilog}\left[\sum(\lambda \times \log \text{PMT}) / \sum(\lambda)\right]$$

where λ is the Failure Rate (FR) and TTR is the time in hours to perform a preventive maintenance task.

This alternative method is not part of MIL-HDBK-472, Procedure II. It is provided for comparison purposes. When the preventive maintenance task time data are plotted and determined to be normally distributed, then the median value calculated from the set of PMT data is equal to the mean value of this same MPT data-set. The decision to select one method over the other method is dependent on the distribution function of the maintenance time data that are collected following a MTA.

7.3.2.3 Use of Maintainability Predictions for Active Maintenance

The Mean Active Maintenance (MAM), in terms of man-hours, which includes corrective and preventive maintenance parameters, is determined from the equation as shown:

$$\text{MAM} = \left[\left(\sum(\lambda)\,\text{Mct}\right) + \left(\sum(\text{f})\,\text{Mpt}\right) / \sum(\lambda) + \sum(\text{f})\right]$$

where λ is the Failure Rate (FR) based on operating time, f is the frequency of occurrence of preventive maintenance actions per one million hours based on calendar time, Mct is the Mean Corrective Maintenance Time and Mpt is the Mean Preventive Maintenance Time.

7.3.3 Maintainability Prediction Procedure III

This procedure describes a method of performing a maintainability prediction of ground electronic systems and equipment by utilizing the basic principles of random sampling. The method involves selecting random samples of replaceable items (e.g., units, assemblies, or components) from the total complement of items comprising the system, subdividing this sample into smaller subsamples by discrete classes of items, and conducting a maintainability analysis for every replaceable item in the subsample.

Corrective and preventive maintenance parameters are used in this procedure. The basic parameters of measure for this procedure are:

- Mean Corrective Maintenance Time (MCMT), also defined as Mct in this procedure
- Mean Preventive Maintenance Time (MPMT), also defined as Mpt in this procedure
- Mean Downtime (MDT), also defined as Mt in this procedure
- Maximum Time for Corrective Maintenance Time (MCMT), also defined as Mmax in this procedure

The fundamental approach to obtaining a maintainability prediction for corrective and preventive maintenance parameters is to randomly select samples for maintenance analysis and evaluation. The samples are representatives from a population of replaceable items, which comprise the system. This total sample is termed sample size N. The N sample is then subdivided into a number of subsamples of size, *n*, also termed "task samples." Each of these task samples represents a specific class of parts, such as resistors, capacitors, motors, and so on. The size of each *n* sample (task sample) is determined by considering the relative frequency of failure for a particular class of replaceable items. The classes of items with a relatively higher failure rate, as compared to the mean failure rate, would be represented by a larger subsample than the classes of items with a relatively lower failure rate than the mean value. The technique for determining the sample sizes, accuracy, and confidence levels are explained in detail in MIL-HBDK-472, Procedure III [1]. The sample size equation in MIL-HDBK-472 is based on the confidence level, the population standard deviation, the population mean, and the desired accuracy of the prediction as a percentage of the mean. This is not the only method to calculate sample size. Another equation that is widely used is based on only two parameters: confidence level that the sample size is an accurate representation of the population, and reliability of the item to be sampled.

The use of sampling in predicting downtime is justified based of uniformity of design with respect to replacement and repair times for like categories of replaceable items. This implies that, on average, it should take the same time to replace any resistor or capacitor as is required for any other resistor or capacitor. This implication assumes that the methods of mounting, fault location, adjustment, calibration, and final test are similar for replaceable parts of the same class. Therefore, these maintenance actions are referred to as samples of maintenance tasks. These samples are selected randomly in order to provide a status of universal applicability. The item to which these maintenance actions are applied is called the maintenance "task sample." On this basis, if sufficient maintenance task samples are randomly selected from each class of replaceable items, these should suffice to provide a prediction of downtime for that specific category.

The downtime is calculated by performing a maintainability analysis of the maintenance tasks, which entails a step-by-step accounting of a logical diagnostic procedure. This results in numerical scores, which are assigned by following certain scoring criteria of applicable checklists. These numerical scores are then translated into a quantitative measure of downtime in hours. This data translation is accomplished by entering the scores into a linear regression equation, which was developed from past studies and experience with comparable systems. For details of this regression equation and how to use it, refer to MIL-HDBK-472, Procedure III [1].

7.3.4 Maintainability Prediction Procedure IV

This procedure is based on the use of historical experience, subjective evaluation, expert judgment, and selective measurement for predicting the downtime of a system or equipment. The procedure uses as much existing data as possible to the maximum extent available to benefit the accuracy of the prediction results. This procedure integrates preventive and corrective maintenance task data similar to Procedure II. Task times to perform various maintenance actions are estimated and then combined to predict overall system/equipment maintainability. The intrinsic maintainability of the system/equipment is predicted under the assumption that optimal utilization of specified maintenance test support equipment and personnel will occur once the system/equipment is deployed.

The basic parameters of measurement for this procedure are:

- Mean Preventive Downtime (MPDT)
- Mean Corrective Downtime (MCDT)
- Total Mean Downtime (TMDT)

Preventive MTA uses all available preventive maintenance task time data from predecessor systems/equipment for a particular preventive maintenance task to predict a similar type task time for a new designed system. The task time data from predecessor systems/equipment for a particular preventive maintenance task are entered into an equation to calculate Preventive Downtime (PDT) for each maintenance task. The equation to calculate PDT is the sum of all preventive maintenance activity times to accomplish a maintenance task for one item within a system/equipment asset. A distribution of the individual task times within each action for a single item maintenance task performed for multiple system/equipment assets can be developed into a model to compare to other data-sets. The model will include the Mean Preventive Downtime (MPDT), which is calculated by averaging all PDT data for a particular maintenance task as performed in Procedure II. There would potentially be separate models with their own unique MPDT for each maintenance task for a type of similar system/equipment asset items.

The comparison of an observed data-set from a new similar system/equipment to the prediction model may result in a finding where the preventive maintenance task time measured for a similar newly designed system/equipment is much larger than the model. In this situation, the disproportionate correlation of the observed data to the model could justify the need to identify critical design improvements before deploying any assets of this new system/equipment design to customer field sites. If the observed data closely matches the model, a high degree of prediction accuracy can be assessed, and the new data could be added to the prediction model's data-set for an updated accuracy of the prediction model when it is used again in the future.

Just as the case for preventive maintenance tasks, corrective MTA uses all available corrective maintenance task time data from predecessor systems/equipment for a particular corrective maintenance task. This procedure leverages the system/equipment detectable preventive maintenance action data and maintenance data during the process to restore an operational function, to predict a corrective maintenance task time for a newly designed system. The fault isolation capability is analyzed during preventive maintenance as well as for corrective maintenance. Troubleshooting, repair, and verification times for repairable and non-repairable assets are determined and compared. The task time data from predecessor systems/equipment for a particular corrective maintenance task are entered into an equation to calculate Corrective Downtime (CDT) for each corrective maintenance task. The equation to calculate CDT is the sum of all corrective maintenance activity times to accomplish a maintenance task for one item within a system/equipment asset. A distribution of the individual task times within each action for a single item maintenance task performed for multiple system/equipment assets can be developed into a model to compare to other data-sets. This model is different from the other models since downtime is a prime resultant and it considers failures over the entire system, distinguishing between items that are detectable and items that can be isolated, but non-repairable. This prediction model includes the Mean Corrective Downtime (MCDT), which is calculated by using all CDT data for a particular maintenance task and entering the data into the equation as shown below.

$$\mathrm{MCDT} = \left[\sum(\lambda_{\mathrm{im}} \times \mathrm{T}_{\mathrm{im}}) + \sum(\lambda_{\mathrm{ijm}} \times \mathrm{T}_{\mathrm{ijm}})\right] \Big/ \left[\sum \lambda_{\mathrm{im}} + \sum \lambda_{\mathrm{ijm}}\right]$$

where λ_{im} is the Failure Rate in Failures per Million Hours (FPMH) of detectable failures from the "ith" asset, T_{im} is the time in hours to perform a fault detection maintenance task and correct a failure of the ith asset, λ_{ijm} is the Failure rate in Failures per Million Hours (FPMH) from the "ith" asset in the "jth" non-repairable group which can be isolated during the maintenance actions, and $\mathrm{T}_{\mathrm{ijm}}$ is the time in hours to perform a fault isolation maintenance task and correct a failure of the "ith" asset in the "jth" non-repairable group.

There would potentially be separate models with their own unique MCDT for each maintenance task for a type of similar systems/equipment asset items. The comparison of a data-set to the model may result in a finding where the corrective maintenance task time measured for a similar newly designed system/equipment is much larger than the model, which can justify the need to identify critical design points where improvements are needed before deploying any assets of this new system/equipment to customer applications.

7.3.5 Maintainability Prediction Procedure V

This procedure permits analyzing the maintainability of electronic systems/equipment including fault detection, fault isolation, and test capabilities. Only active maintenance time is analyzed. Administrative and logistical delays and time to perform clean-up are excluded. There are two prediction methods in this procedure. The first method is an early prediction method that uses estimated design data, but no actual time measurements. The second method is applied later in development, which uses actual detailed design data to predict maintainability parameters. The procedure requires monitoring the overall system/equipment maintainability performance parameters during the system/equipment development phase. The analyst identifies whether or not the specified maintainability design requirements will be met before the system is completed and assets are deployed to customer field sites. If it appears that the system maintainability design requirements will not be met before the system design is completed, the designers are informed of the issues and the design changes made immediately before the costs make design changes prohibitive later in the program.

The basic maintainability design parameter is MTTR, which is predicted using this procedure. Other maintainability design parameters that can be predicted using this procedure are M_{max} at a particular percentile (e.g., 95th percentile), percentage of faults isolated to a single replaceable item, percentage of faults isolated to multiple replaceable items, mean maintenance man-hours per repair, mean maintenance man-hours per operating hour, and mean maintenance man-hours per flight hour. Some of these parameters have been explained in previous procedures in MIL-HDBK-472, and some are new, such as the percentage of faults isolated to a single replaceable item and the percentage of faults isolated to multiple replaceable items. These two parameters are common parameters with testability metrics. The reader may find more information on this testability parameter in Chapter 13.

Paradigm 2: Maintainability is directly proportional to testability and Prognostics and Health Monitoring (PHM)

7.4 Maintainability Prediction Results

The maintainability prediction results should provide assurance that the quantitative maintainability requirements will be satisfied. If the maintainability prediction results conclude that the maintainability requirements will not be met, the prediction should provide recommendations for design changes that could be performed during the remaining time in the development program to correct the maintainability design issues that have surfaced. The results of maintainability predictions are compared to Maintainability Demonstration (M-Demo) results to further ensure the maintainability requirements are met prior to system or product deployment. Results of maintainability predictions are compared to maintainability allocations to determine whether the allocations were correct, or should be changed.

Initially, a designer will use maintainability allocations to determine what levels of maintainability are required during the development phase of a program for the different levels of hardware that can be repaired. For example, the system is the top level, and the component or piece part is the lowest level. The allocations are apportioned or flowed down from the top level to the lowest replaceable item. For this reason, this process is called a top-down approach. Flow down from the lower level perspective means the lower levels of replaceable hardware use the same allocation number as the allocation at the next higher level of assembly. From the top-level system perspective,

Table 7.1 Modified excerpt of allocation example from Table 6.3.

Item	No. of items per product	Item failure rate $\lambda_i \times 10^{-3}$	Contribution to total failure rate, $\lambda_i \times 10^{-3}$	MTTR	Contribution to system MTTR
A	1	5	5	1.44	7.2
B	1	1.11	1.11	1.44	1.6
C	1	0.833	0.833	1.44	1.2
Total			6.943		10

the top level allocates its number to each of the items at the next lower level of hardware. Apportion means allocation numbers at the top level are decomposed to the lower levels. Each hardware level would decompose the allocation differently to the next lower levels of hardware, depending on the design features and complexity of each item. The lower level numbers are derived from the top level number. The lower level numbers could be rolled up to the top level using an additive model. The allocations, no matter if they are flowed down or apportioned at the lowest hardware levels, may be used as maintainability requirements for the specifications of each of these replaceable items.

Once allocations are determined, the allocation results are used later in the development process to compare with the prediction results, when they are available. An example of the results from a maintainability allocation is shown in Table 7.1, which shows a modified excerpt of the data originally contained in Table 6.3 following one of the allocation examples from Chapter 6. Table 6.3 is an example of the equal distribution method for maintainability allocations. This table is modified and reprinted here for convenience.

In this example, Table 7.1 shows the product-level MTTR allocation results, which became maintainability requirements for Items A, B, and C. Each of these MTTRs is 1.44 hours using the equal distribution method. These MTTR allocations, along with the Item Designation and the Number of Items columns, are copied into a new table (Table 7.2). Table 7.2 compares allocation results from Table 7.1 to the results of a prediction. This prediction example is designed to show how to interpret the differences between allocations and predictions. These predicted MTTR values would have been generated from a separate maintainability prediction with MTTR values. For the sake of this example, no maintainability prediction report is included in this example.

In keeping with this example, the allocation data from Table 7.1 was copied into the three columns starting from the left side of Table 7.2. Three new columns were added to the right side of the table. These three columns are MTTR prediction, percentage difference, and comments/actions recommended.

The comments column in Table 7.2 reflects the thoughts of the analyst when a design change is recommended or some other action is needed based on the percentage difference between an item's allocation and prediction. When allocations are very different from the prediction results, a design team decision is usually needed. The design team should decide on recommended actions, whether the maintainability allocations should be altered or a design improvement be made, as shown in Table 7.2. Comments related to slight allocation adjustments mean the allocations are fairly close to the predictions (relatively small percentage difference) and no design change is necessary. However, allocations that are far from the

Table 7.2 Example of maintainability allocations compared to predictions-modified excerpt of allocation example from Table 6.3.

Item	No. of items per product	MTTR allocation (hours)	MTTR prediction (hours)	Percentage difference	Comments/actions recommended
A	1	1.44	3.23	+124%	Recommend a design change and redo prediction
B	1	1.44	1.71	+19%	Adjust allocation up slightly
C	1	1.44	0.55	−89%	Adjust allocation down

predictions (relatively large percentage difference) may not be enough to warrant a design change, but may require some action to readjust the allocations to align with the predictions. In the case of Item A, the design team recommended a design change. A new prediction should occur after the design update to demonstrate the effectiveness of the design change, bringing the prediction closer to the allocation and decreasing the percentage difference. In the case of Item B, an adjustment of the allocation could go up to bring the two numbers closer, but the difference is small, and the design team might have been ok if no action was recommended. For Item C, the design team requested the allocations be adjusted down to bring the percentage difference closer to 0%.

Sometimes, one set of data to determine predictions is simply not enough to convince designers to make a change to improve maintainability features in the design. A combination of data-sets may be collected and equated to each other using a variety of statistical and probabilistic methods. One such method to collect multiple data-sets and build a probabilistic maintainability model that is constantly improving its accuracy as more data is collected is called a Bayesian methodology.

7.5 Bayesian Methodologies

Bayesian methods use historical and recent maintenance data representing the current state of knowledge of the statistical performance and status of a system or product to determine probabilities as reasonable expectations of the occurrence of a future state, or as the quantification of a personal belief. Bayesian methods are a means to quantify uncertainty rooted in behavioristic considerations. Behavioristic considerations are confined to the study of observable and quantifiable aspects of behavior, excluding subjective phenomena, such as feelings and emotions. Bayesian methods are important for maintainability and reliability predictions because they provide treatment of data such that there is always an inference of a probability of failure for a system or product at any given time. Just as there is a probability that something will fail, there is a probability that something will not fail. If there is a probability that something will fail during a specific period of time, then there is a probability that maintenance will need to be performed during that same time period. And this probability is likely to change as more data is collected. Since Bayesian methods assess the past and current conditions using continuous data measurements to determine the way things are, the data are constantly updated to revise the assessment of the way things will be.

Bayesian methods are probabilistic methods, as opposed to deterministic methods. In deterministic methods, there is a 100% certainty that an output or event will occur. Deterministic methods state that if a specific set of inputs are received, under established conditions and controls, then a specific output will occur. For probabilistic methods, if a specific set of inputs are received, under established conditions and controls, there is a probability that a specific output will occur and an inverse probability that the output will not occur. As an example of a deterministic method: $2+2 = 4$. As an example of a probabilistic method: $1.75+1.75 = 4$ with a 60% probability and 3 with a 40% probability depending on decimal truncation or rounding up. In this example, the output must be a whole number. If this example allowed the output to be a decimal with 2 significant digits, then $1.75+1.75 = 3.50$, which would be a deterministic method.

Bayesian methodologies are evidential probabilities. Bayesian methods are used to derive probability distribution functions that can be referred to as Bayesian probabilities, which are used to evaluate the probability of a hypothesis. The Bayesian probability always begins with the specification of a prior distribution based on the assessment of all available historical and current data. The prior distribution is the initial posterior distribution to estimate future outcomes. As new data are collected, these form the initial likelihood distribution. This likelihood distribution is combined with the prior distribution to form a new posterior distribution. A Bayesian posterior distribution is updated as new, relevant data are unveiled. The Bayesian methodologies provide an interpretation of data-sets using a standard set of procedures and formulae to perform distribution function calculations. The Bayesian method or rule is sometimes referred to as the perfect inference engine. The Bayes rule is not intended to provide the correct answer.

Rather, the Bayes rule provides the probabilities that any numbers of alternative answers are true. These probabilities determine the best answer that is most probably true.

The term "Bayesian" refers to Thomas Bayes, who proved a special case of a theorem (now called the Bayes Theorem) in a paper entitled: "An Essay towards Solving a Problem in the Doctrine of Chances." In that paper, Bayes described the prior and posterior distributions as beta distributions using data that came from Bernoulli trials. Pierre-Simon Laplace introduced a general version of the theorem. He used this version of the theorem to solve problems in several fields, including celestial mechanics, medical statistics, and reliability. Early Bayesian inference methods used uniform prior distributions following Laplace's principle of insufficient reason. This inference method was later called "inverse probability." Inverse probability infers backwards from observations to probabilistic modeling parameters. The inverse probability method is a means to collect failure effects data to determine failure causes [3].

7.5.1 Definition of Bayesian Terms

When performing Bayesian methodologies related to maintainability predictions, the following terms are commonly used:

Prior, P (A) = Probability of an event, A, based on past history (aka prior probability).

Likelihood, P (B|A) = Conditional probability, which is the degree of belief in event B, given that event A occurs and is true based on the latest collection of failure or maintenance action data.

Margin likelihood, P (B) = Probability of an event, B, using the latest failure or maintenance action data.

Posterior, P (A|B) = Probability of an event A given that B occurs and is true based on failure or maintenance action accumulated from past history and likelihood data (aka posterior probability).

Equation: $P(A \mid B) = [P(B \mid A) \times P(A)]/P(B)$.

Posterior, $P(A \mid B) = (\text{prior} \times \text{likelihood})/\text{marginal likelihood}$.

The posterior distribution equals the prior distribution when there is no likelihood data.

7.5.2 Bayesian Example

As an example of the use of these terms, assume that the past failure history of an equipment (type A) used within a system tested over the last year estimates a prior failure distribution with a probability of equipment type A failure equal to 0.001. This equates to 5 equipment type A failures for 5000 systems tested in this time period. There is one piece of this equipment type A in each system tested. There is 100% probability that an equipment type A failure will cause a system failure. The latest collection of recent failure data reveal 9 equipment type A failures out of the last 10 systems tested, or a probability of 0.9. No other causes of system failure occurred. Based on this recent system test data, for any sample of 10 systems tested, 9 will fail due to equipment type A. The overall system test failure rate for the month is 0.01, which is based on 9 failures for 900 systems tested. This is called the marginal likelihood. The current equipment failure probability estimate, or posterior, is determined as:

$$\text{Probability that a system will fail due to the equipment (A)} = (0.9 \times 0.001)/(0.01) = 0.09$$

The initial prior was 0.001. The new posterior distribution is 0.09. This means the new prior is 0.09. It is this prior that will be combined with the next set of likelihood data to determine the next posterior.

7.6 Maintenance Task Analysis

As discussed earlier, a maintenance task is defined as the maintenance efforts, activities, or actions necessary for retaining or sustaining an item, changing or improving an item, or restoring an item to a previous state of functionality given a specified operational performance condition [4]. A MTA describes all maintenance tasks and associated logistical tasks to ensure maintenance requirements are satisfied. MTAs collect the data to verify all the maintenance and logistical requirements that are needed to perform all required maintenance task activities and actions. Some typical maintenance requirements have already been discussed in the earlier chapters of this book, so they won't be repeated here. Sample maintenance task efforts, activities, or actions have been listed and associated with different maintainability prediction procedures earlier in this chapter.

The MTA is a way to collect maintenance task data to be used for a maintainability prediction. It serves as a basis for a maintainability prediction model when enough historical maintenance data has accumulated. As such, the MTA should be conducted early in the design phase of the asset and be revised as changes are made to the design.

Oftentimes, the full data content of a MTA are not transferred to a maintainability prediction. Many of the logistical requirements may be considered administrative logistic support requirements that are not within the scope of the maintainability predictions, depending on the type of prediction procedure selected.

When the maintainability predictions require consideration of the administrative logistic support requirements, to analyze all sources and causes of system/equipment downtime, then the MTA becomes an integrated part of the supportability and logistics analysis. The MTA will be a complete information package that includes all scheduled and unscheduled maintenance tasks that are likely to occur with the asset, along with all associated logistical information. This combined maintenance and logistical information includes:

- All preventive maintenance task actions along with their associated frequency
- All corrective maintenance task actions along with the failure rates of each repairable item
- Personnel requirements such as the skill level, and estimated duration for each task
- Spare parts and inventory requirements broken down by task
- Support or test equipment required by task
- Packaging, handling, storage, and transport requirements
- Facility requirements to perform the maintenance tasks
- Manual and technical information requirements
- A summary of the maintenance plan for the asset

When an MTA includes the associated logistical information, it enables organizations to not only prepare for the fielding of the asset and the ability to maintain the asset, but also to investigate potential design changes to reduce life-cycle costs tied to any unnecessary logistical requirements.

The MTA should be structured in a way that enables the asset-owning/operating organization to develop their support systems and requirements to ensure that the design chosen can meet the operational requirements. The MTA also allows the asset-owning organization to acquire the required support, either through developing skills internally, or organizing contracts to support the asset. Lastly, the MTA enables the organization to prepare for the maintenance of the asset, by identifying critical spare parts, tools, test equipment, and training that can be procured as part of the larger asset purchases, which normally reduces the costs. The MTA ensures that the organization is ready to support the asset the moment it arrives for operation, and there are no unexpected surprises.

The MTA should address all repair items identified through the Level of Repair Analysis (LORA), as described in Chapter 6. The MTA should analyze maintenance levels for all repairable items that are complex and time-consuming to repair, costly to procure, and warrant an additional level of analysis to define the logistical requirements to complete required maintenance tasks. Each of the tasks should be documented with an analysis performed for each of the repairable item levels corresponding to the required maintenance levels. If a repairable item has a maintenance task involving troubleshooting, removal and replacement, inspection, and calibration activities, then an individual analysis should be completed for each activity within this maintenance task. Once all maintenance tasks and associated activities for the repairable item have been analyzed, the maintenance and logistical requirements are summarized at the repairable item level. Once all repairable items have been analyzed and summarized, a final MTA report will be created for design, maintainability, and logistical benefits. This MTA report should be used by the logistics organization owning the assets to acquire the resources needed to support the system assets in the field (see Figure 7.1).

After all these tasks are identified, they are consolidated into a maintenance plan, which will be used to determine and verify that the asset will meet the operational requirements. The maintenance plan will include a summary that, based on the analysis,

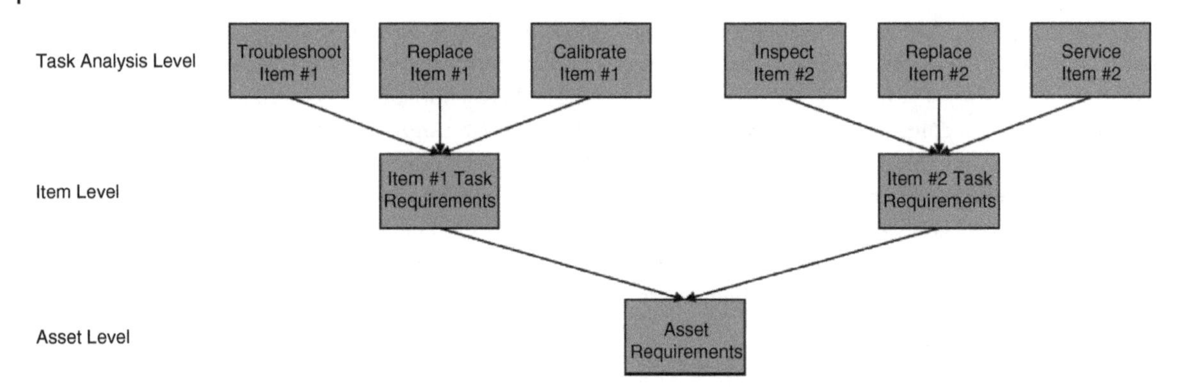

Figure 7.1 Sample Maintenance Task Analysis process.

will highlight the asset's ability to meet the defined maintainability requirements such as Maintenance Labor Hours per Operating Hour (MLH/OH), which takes into account the time to perform each task, plus the associated probability that it will occur. The maintenance plan may also be used to support availability calculations.

Often the final MTA will include a number of summary forms or reports that consolidate all of the individual task requirements into a summary that the organization can use to procure spares, identify personnel and skill requirements, and generate facility requirements.

7.6.1 Maintenance Task Analysis Process and Worksheets

Performing a MTA should be a structured process; however, there is no fixed format for the analysis, and it should be based on the level of detail required by the support team and designers. As such, there is flexibility for the analysis team in how the MTA can be completed. However, the MTA must be able to identify the required resources and supporting elements for each task. In order to work through each task and identify those resources and supporting elements, it is recommended that a structured approach be taken. In addition, the MTA should not be conducted in isolation and should be performed, or at minimum reviewed, by a team that has various levels of experience in performing those maintenance tasks. The MTA requires that the analyst is able to put themselves in the place of the technician and visualize what steps and resources are required to perform each maintenance task.

The MTA process should be defined by the analyst up front, along with any associated forms or documentation. This will ensure that analysis runs smoothly. Figure 7.2 shows a generic MTA process that may serve as a guide for the analyst.

Each of the repairable/replaceable items in the asset needs to be analyzed using the MTA process for each task type that is applicable to the repairable item. Often a FMEA can be used as a source document to identify what preventive and corrective actions may take place for each repairable/replaceable item. If a FMEA is not available, the analysis team can use past designs and experience as a starting point. Once all of the repairable/replaceable items have been identified, the analysis may begin.

The MTA may consist of three worksheets. These worksheets include the Task Analysis Worksheet (Figure 7.3), the Supporting Element Worksheet (Figure 7.4), and the Item Summary Worksheet (Figure 7.5). Examples of each type of worksheet are provided for guidance to the reader and to serve as a template.

The header information for the sheets can be populated before the analysis takes place. The header information in the Task Analysis Worksheet (Figure 7.3) consists of the following:

- System: The name of the system, product, or asset that the component belongs to. Be sure to use the proper description from any drawings or functional concepts.
- Component Name/Part No.: Enter the name of the component and any associated subassembly numbers as per the drawings or other engineering specifications.
- Parent System: Enter the parent system that that the component belongs to (the immediate higher-level in the hierarchy above the system, product, or asset

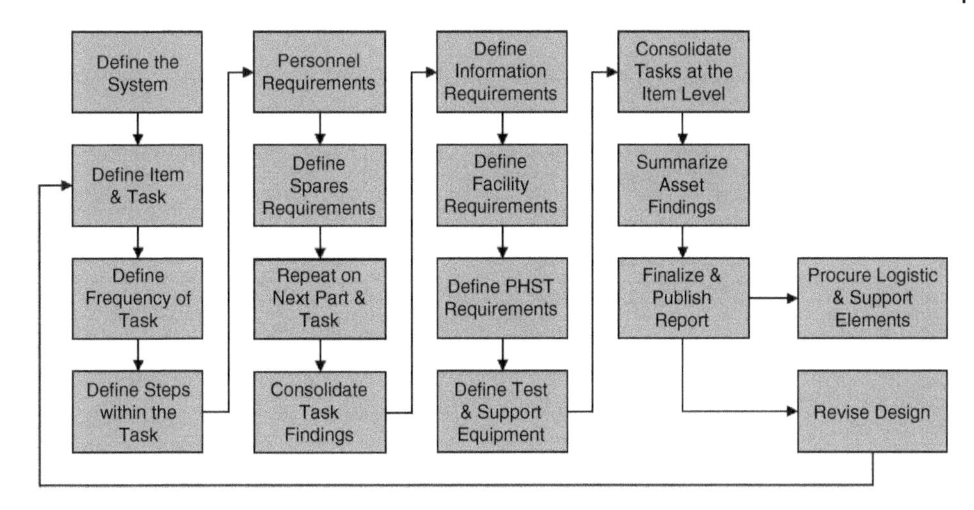

Figure 7.2 Levels of analysis in an MTA.

System	Component Name/ Part No.	Parent System	Description of Requirement:				
Component Task No.		Task Type:	Maintenance Level:		Frequency of Task		
Step Number	Task Description	Task Time (min)	Frequency of Task	Personnel & Skills (minutes)			
				Basic	Intermediate	Expert	Total
		Totals					

Figure 7.3 Task Analysis Worksheet.

Component Name / Part No.		Component Task No.	Task		Task Frequency		Maintenance Level	
		Replacement Parts		Test and Support/ Handling Equipment			Description of Facility Requirements	Instructions
Step Number	Qty Per Assy.	Part Nomenclature Part Number	Quantity Required	Quantity Required	Test & Support Equipment Description & Number	Use Time (min)		

Figure 7.4 Supporting Element Worksheet.

System	Component Name/ Part No.	Functional Description:						
Maintenance Description:								
Component Task No.	Task Type	Maintenance Level	Frequency of Task	Task Time	Skills	Personnel Time (mins)	Parts Required	Test & Support Equipment Required
Totals								
Notes:								

Figure 7.5 Item Summary Worksheet.

being analyzed); there may not be a parent system so this data field is optional.

- Description of Requirement: A description of the maintenance task that is being analyzed. If it is corrective in nature, include the fault to the system that will be observed. If preventive, include the failure mode that it is designed to prevent or detect.
- Component Task No.: If the component has multiple tasks associated with it, enter a unique identifier to this task.
- Task Type: Define the task type from a list such as troubleshooting, remove and replace, repair, servicing, alignment and adjustment, functional test and check-out, inspectional, calibration, overhaul, and so forth.
- Maintenance Level: Describe at which level the maintenance will take place, such as organization, intermediate or depot.
- Frequency of Task: What is the estimated frequency at which this task will be performed?

7.6.2 Completing a Maintenance Task Analysis Sheet

After completing the header information, the next step in the MTA is to define the task that is being performed on the specific asset. The task type should be recorded in the header of the analysis sheet. With the task type defined, the analyst must define each element or activity of the task. This entails looking at each and every step that the maintenance team must perform on the asset to complete the identified maintenance task. In most maintenance tasks, there are a number of elements that make up the task. The elements of an MTA might follow the activities selected for a maintainability prediction, if a prediction analysis exists. Options for the selection of activity elements and the procedure used in a maintainability prediction were discussed earlier. The data entered in the MTA should match data from past maintainability predictions so that the new MTA data can fit into the same infrastructure to further refine the maintainability prediction models.

The analyst populates the MTA sheet with each step in the proper sequence of activities that are required to accomplish the particular maintenance task. Each step within the task should be identified with a unique number and listed on the left-hand side of the worksheet, along with a description of the task as shown in Figure 7.6.

7.6.3 Personnel and Skill Data Entry

Once the task and header information has been defined, the analyst needs to determine the personnel required to perform the task. This is more than simply stating 1 or 2 people, for example. It requires the analyst to determine what level in the organization will be performing the maintenance. Remember there are typically three levels [5] at which maintenance will be performed:

- Organization maintenance is performed at the operational level. Maintenance at this level normally

System	Component Name/ Part No.	Parent System	Description of Requirement:				
#2 Seal Water Pump	Coupling #OICU812	Main Transfer System	During routine maintenance, the coupling has been found to be damaged, or displays excess wear. The damage has determined that the coupling must be removed and replaced.				
Component Task No.		Task Type:	Maintenance Level:		Frequency of Task		
OICU812-01		Remove and Replace	Organization / Intermediate		0.0005		
Step Number	Task Description	Task Time (min)	Frequency of Task	Personnel & Skills (minutes)			
				Basic	Intermediate	Expert	Total
0010	Lock-Out / Tag Out	15	0.0005				
0020	Remove guard	5	0.0005				
0030	Disassemble coupling	10	0.0005				
0040	Remove coupling	15	0.0005				
0050	Inspect pump/motor shaft	5	0.0005				
0060	Install new coupling	15	0.0005				
0070	Align coupling	30	0.0005				
0080	Install guarding	5	0.0005				
0090	Remove Lock-Out / Tag Out	10	0.0005				
0100	Start-up pump	5	0.0005				
	Totals	115	0.0005				

Figure 7.6 Maintenance Task Analysis Worksheet with header and step information.

consists of performance checks, basic care such as cleaning, inspection, lubrication, and may include limited servicing such as adjustments, and removal and replacement of certain components.

- Intermediate maintenance is performed within specialty shops. The intermediate level is usually performing the repair on a repairable component or assembly, by diagnosing the fault, and replacing specific parts or modules. The intermediate level may also perform larger maintenance activities such as rebuilds or overhauls.
- Depot maintenance is the most specialized level of maintenance and repair. Depot-level maintenance may be in control of the asset-owner/operator or be part of the asset manufacturer. The depot level usually performs extensive rebuilds, overhauls, and very complex diagnosis.

In addition, the analyst must determine what skill levels will be required at the specified level of maintenance. This may include using a mix of expert and basic skills to complete a task. Typically, skills are broken up into three levels as well. The use of different skills enables organizations to plan resources and determine how many junior vs senior people may be required. The skill levels [6] are as follows:

- A basic skill level is used to indicate someone that is typically a general laborer. They usually have a high school diploma, and maybe some minimal training.
- An intermediate skill level may be thought of as an apprentice or technologist. They have additional training from a college or university along with 2–5 years of experience in the specific field.
- Expert skill is the highest, with 2–4 years of formal training along with 5–10 years of experience in the specific field.

As the analysts begin to assign the resources to the tasks, they must define the minimum acceptable skill which would be required to perform the task correctly. If a specific step requires two or more staff, then they should define the minimum acceptable skills for that step. For example, if they require an expert to perform the step, but the expert requires a second set of hands, then a basic skill level may be used as the second set of hands. Building upon the example in Figure 7.7, it can be seen how the personnel are added to the worksheet. In the example, while the total task time is 115 minutes, the task actually requires 110 minutes of basic skill, and 90 minutes of intermediate skill levels. In addition, some analysis sheets will enable the analyst to define which tasks are completed in series or in parallel with one another.

7.6.4 Spare Parts, Supply Chain, and Inventory Management Data Entry

With the task defined, the analysts must then define what is needed to perform the task in terms of spare

System	Component Name/ Part No.	Parent System	Description of Requirement:		
#2 Seal Water Pump	Coupling #OICU812	Main Transfer System	During routine maintenance, the coupling has been found to be damaged, or displays excess wear. The damage has determined that the coupling must be removed and replaced.		
Component Task No.		**Task Type:**	**Maintenance Level:**	**Frequency of Task**	
OICU812-01		Remove and Replace	Organization / Intermediate	0.0005	

Step Number	Task Description	Task Time (min)	Frequency of Task	Personnel & Skills (minutes)			
				Basic	Intermediate	Expert	Total
0010	Lock-Out / Tag Out	15	0.0005	15	15		30
0020	Remove guard	5	0.0005	5			5
0030	Disassemble coupling	10	0.0005	10			10
0040	Remove coupling	15	0.0005	15	15		30
0050	Inspect pump/motor shaft	5	0.0005		5		5
0060	Install new coupling	15	0.0005	15	15		30
0070	Align coupling	30	0.0005	30	30		60
0080	Install guarding	5	0.0005	5			5
0090	Remove Lock-Out / Tag Out	10	0.0005	10	10		20
0100	Start-up pump	5	0.0005	5			5
							0
							0
							0
	Totals	115	0.0005	110	90	0	200

Figure 7.7 Maintenance Task Analysis Worksheet with personnel information.

parts, consumables, and so on. This is crucial as it will feed into the spare part stocking requirements of the organization, and enable the organization to support the assets in the field.

Typically, the analyst will define the specific part(s) required for each step, along with how many would be needed for the particular step. Some analysts prefer to define how many of a specific part are in the assembly. This would enable the organization to decide whether they would like to replace items in pairs or sets.

To populate the worksheet, the analysts will take the header information from the Task Analysis Worksheet (Figure 7.6) and populate the header of the Supporting Elements Worksheet (Figure 7.8). As the analyst works through each step, they need to include the specific part description, along with the part number. This will ensure that the correct parts are incorporated into the stocking plan. Building on the example of the coupling replacement in Figure 7.7, we can see which parts are required in Figure 7.8.

In some instances, it may be wise to call out parts that may not be expected to be replaced, but may need to be if additional damage is found. In the example, a guard could be found to be damaged, but it is not guaranteed, so in the worksheet it is called in the quantity in the assembly, but not in the quantity required field.

Component Name / Part No.		Component Task No.	Task		Task Frequency		Maintenance Level	
Coupling #OICU812		OICU812-01	Remove and Replace		0.0005		Organization / Intermediate	
		Replacement Parts		Test and Support/ Handling Equipment				
		Part Nomenclature						
Step Number	Qty Per Assy.	Part Number	Quantity Required	Quantity Required	Test & Support Equipment Description & Number	Use Time (min)	Description of Facility Requirements	Instructions
0010								
0020								
0030								
0040								
0050	1	Shaft #OICS8645						
0060	1 1	Coupling #OICU812 3" x 3" Shim Pack # CNS3391	1 1					
0070								
0080	1	Guard #OICG324						
0090								
0100								

Figure 7.8 Supporting Elements Worksheet with parts.

7.6.5 Test and Support Equipment Data Entry

With the steps, personnel, and part requirements defined, the analyst must begin to determine what tools or test equipment will be required to perform each step. This upfront review enables organizations to purchase any specialized tools with the asset, which usually results in a significantly lower cost.

The analyst must determine what test equipment, or specialty support equipment, will be required to perform the maintenance tasks. This may include electronic test equipment, overhead cranes, or even test stands. Each tool or test equipment should be defined with the number required, the description along with any specifications (i.e., capacity of a crane or test stand), along with the estimated use time per task. This allows the analyst to determine the total quantity of test equipment or tools that may be required to support the entire operation or fleet. The analysts must also consider whether the test equipment will require multiple units in parallel to perform the operational checks properly. Lastly, the analyst may need to consider any calibrations required, and if that will impact the availability of the test equipment. The calibration routine may warrant additional test equipment to be procured and on-site to ensure the required number of units can be made available to the staff.

In the example shown in Figure 7.9, the test and supporting equipment requirements are defined for each individual step. Using the estimated frequency of the tasks, the estimated use time per task, along with the fleet size, the analyst can determine the final amount of the equipment required in the facility.

7.6.6 Facility Requirements Data Entry

The facility requirements outline any special facility requirements needed for the maintenance to take place. For a ship or aircraft, there may be requirements for a dry dock capable of supporting the ship, or a hangar with a specific size door. While these seem like obvious examples, many organizations do not have the proper facilities to support many of the maintenance activities. For example, in an industrial environment, many organizations rebuild pumps or valves in a general maintenance shop, which is not clean and can lead to contamination and damage to the assembly being rebuilt, which has an adverse effect on the availability of the asset.

Facility requirements will identify specialty shops such as a machine shop, an instrumentation shop, or even clean rooms for rebuilds. The specifications for these rooms should also include any lighting requirements, space requirements, the size of the doors, or the weight capacity of the work benches. The specifications for the shops should also include any support services such as the voltage and capacity of the electrical system, along with any other support services such as water, compressed air, or other specialty gases.

In the example of a pump in Figure 7.10, the description of a machine shop has been identified along with the need for a specific lathe. The analyst should be careful to ensure that facility or specialty shop requirements are consolidated into a single facility or specialty shop for efficiency and cost-effectiveness.

Component Name / Part No. Coupling #OICU812		Component Task No. OICU812-01	Task Remove and Replace		Task Frequency 0.0005		Maintenance Level Organization / Intermediate	
		Replacement Parts		Test and Support/ Handling Equipment				
		Part Nomenclature						
Step Number	Qty Per Assy.	Part Number	Quantity Required	Quantity Required	Test & Support Equipment Description & Number	Use Time (min)	Description of Facility Requirements	Instructions
0010								
0020								
0030								
0040				1	Torch Kit #EQ8462	15		
0050	1	Shaft #OICS8645						
0060	1 1	Coupling #OICU812 3" x 3" Shim Pack # CNS3391	1 1	1	Torch Kit #EQ8462	15		
0070				1	Alignment Kit #EZL440XT	60		
0080	1	Guard #OICG324						
0090								
0100								

Figure 7.9 Supporting Elements Worksheet with test and support equipment.

Component Name / Part No. Coupling #OICU812		Component Task No. OICU812-01	Task Remove and Replace		Task Frequency 0.0005		Maintenance Level Organization / Intermediate	
		Replacement Parts		Test and Support/ Handling Equipment				
Step Number	Qty Per Assy.	Part Nomenclature / Part Number	Quantity Required	Quantity Required	Test & Support Equipment Description & Number	Use Time (min)	Description of Facility Requirements	Instructions
0010								
0020								
0030								
0040				1	Torch Kit #EQ8462	15		
0050	1	Shaft #OICS8645					Machine Shop with a 14" x 40" Lathe	Manufacture a new shaft
0060	1 1	Coupling #OICU812 3" x 3" Shim Pack # CNS3391	1 1	1	Torch Kit #EQ8462	15		
0070				1	Alignment Kit #EZL440XT	60		
0080	1	Guard #OICG324						
0090								
0100								

Figure 7.10 Supporting Elements Worksheet with facility requirements.

7.6.7 Maintenance Manuals

The last item of consideration during a MTA is the documentation requirements. Often the documentation does not arrive until after the equipment is in place and operating. Even when the documentation is received, there is no standard way to catalogue and store the information. While the role of the MTA is not to establish how the documentation will be stored, the MTA should identify what information may be required to perform the task. There are various forms of information or documents that may be required, which may include but not be limited to:

- Standards may include any national, state, or corporate standards for how a task may be completed or provide performance levels that must be achieved. An example of a standard could be an alignment standard that indicates to what tolerance the alignment must be performed given the various operating speeds.
- Drawings such as assembly, mechanical, electrical, or logic diagrams may be provided. These are often used by the maintenance personnel to perform the maintenance task.
- Maintenance manuals are manuals that include maintenance procedures, troubleshooting guides, and potentially a Bill of Materials.
- Operating manuals are manuals that include operating procedures, specifications, and parameters. The operating manual typically includes the Theory of Operation, which explains how the equipment operates.

Ideally, the MTA will also identify what format the documents should be in, such as electronic, hard copy, or both. The MTA should also identify how many may be needed to perform the task being analyzed. By identifying the documentation requirements upfront, a proactive approach to managing the documents can be implemented.

7.6.8 Maintenance Plan

The Maintenance Plan is an outcome of the MTA which includes all corrective and preventive actions that will be performed while the equipment is operating. The Maintenance Plan will provide an estimate of the number of personnel, parts, tools, and so on, that will be required. This estimate is used to appropriately staff and train the right number of personnel, along with stocking the right parts and tools. The Maintenance Plan is essentially a summary of the MTA that has been performed for each task at a component level (see Figure 7.11), and is displayed using the Item Summary Worksheet.

Using these Item Summary worksheets, the analysts and the supporting team can begin to develop stocking plans, training plans, and so on. The summaries may also be used to bulk-buy spare parts as part of the equipment or for asset acquisition. The MTA and the Item Summary Worksheets are then

System	Component Name/ Part No.	Functional Description:
#2 Seal Water Pump	Coupling #OICU812	To provide power transmission between the #2 seal water pump and the driver (a 25hp motor).
Maintenance Description: All preventive and corrective maintenance actions that will be performed at the organization and potentially the intermediate level.		

Component Task No.	Task Type	Maintenance Level	Frequency of Task	Task Time (mins)	Skills	Personnel Time (mins)	Parts Required	Test & Support Equipment Required
OICU812-01	Remove and Replace	Organization	0.0005	115	Basic/ Intermediate	110 / 90	Shaft #OICS8645 Coupling #OICU812 3" x 3" Shim Pack # CNS3391 Guard #OICG324	Torch Kit #EQ8462 Alignment Kit #EZL440XT
OICU812-02	Inspection	Organization	0.0015	30	Basic	30		
OICU812-03	Alignment & Adjustment	Organization	0.0008	45	Intermediate	45	3" x 3" Shim Pack # CNS3391	Alignment Kit #EZL440XT
Totals				190		275		
Notes:								

Figure 7.11 Item Summary Worksheet.

used to assist in the development of an Integrated Logistic Support (ILS) plan, and a Logistics Support Analysis (LSA) that will be covered in detail in Chapter 16.

The MTA is a great tool to further understand the requirements needed to support an asset or piece of equipment once it has been deployed. The MTA can be used to further identify issues with the overall asset design and assist the designers in improving the maintainability of certain products, items, components, or assemblies.

7.7 Summary

Maintainability predictions and a detailed MTA are crucial steps to ensuring that the asset will be able to achieve the desired performance objectives over its life cycle. Through a means of statistical analysis and detailed task-building, the design of the asset can be validated. While not foolproof, the techniques used in this chapter will go a long way to demonstrating that a design is ready. In the event that the maintainability prediction or MTA provides evidence that the design will not meet the requirements, the design team has the ability to correct any issues before the asset is built. However, if the design is changed, the predictions and MTA will need to be reviewed and revised to reflect the design changes. This ensures that the design changes are validated. Using this iterative process, the designer and asset owners can ensure that the best possible design is put forth within the financial constraints. In conclusion, the use of MIL-HDBK-472 in concert with this book will facilitate the reader's ability to design, develop, and produce equipment and systems that require a high degree of maintainability and maintenance preparation for the equipment and system life cycle.

References

1 US Department of Defense (1966). *Maintainability Prediction*, MIL-HDBK-472 (updated to Notice 1 in 1984). Washington, DC: Department of Defense.

2 Dhillon, D.S. (2006). *Maintainability, Maintenance, and Reliability for Engineers*. Boca Raton, FL: CRC Press.

3 Wikipedia. Bayesian probability. https://en.wikipedia.org/wiki/Bayesian_probability (accessed 23 August 2020).

4 US Department of Defense (1997). *Designing and Developing Maintainable Products and Systems, MIL-HDBK-470A*. Washington, DC: Department of Defense.

5 Marino, L., (2018). Level of repair analysis for the enhancement of maintenance resources in vessel life cycle sustainment. Unpublished dissertation. George Washington University

6 Blanchard, B.S. (2004). *Logistics Engineering and Management*, 6e. Upper Saddle, NJ: Pearson Education Inc.

8

Design for Machine Learning

Louis J. Gullo

8.1 Introduction

Traditional Design for Maintainability (DfMn) approaches, as discussed throughout this book, have been impactful to date, but modern Artificial Intelligence (AI) techniques with Machine Learning (ML) offer a dramatic leap forward in the design and benefit of maintainability, as shown in this chapter. Machine Learning is necessary for the future of designing systems and products for maintainability. DfMn with Machine Learning capabilities is a paradigm shift moving away from today's strict dependence on scheduled maintenance events. Machine Learning is moving DfMn towards designing and planning for maintenance events based on the system's or product's need to conduct real-time corrective and preventative actions, as determined by their exposure to actual environmental and mechanical stress conditions and the past historical data of accumulated physical fatigue from similar exposures to stress conditions resulting in failures.

Machine Learning (ML) is the science of how computers learn to make decisions using data. The premise of Design for Machine Learning (DfML) in the context of maintainability is to augment the capabilities of a system operator or maintainer to enhance what can be done to maintain a system and keep a system in operation without downtime. ML technologies are designed into a system to allow skilled users and maintainers of the system to scale up their capabilities as demand for their skills increases. ML is a skill multiplier for the human in the loop. Now, imagine a world where ML is deployed and executed in all systems with which we interface, and ML continuously progresses into Deep Learning (DL) and beyond. One can only imagine a world where system operators and maintainers will be able to do more with the systems of the future with far less stress and impact on their wellbeing compared with similar systems of today.

This chapter discusses the meaning of Machine Learning and Deep Learning, and the differences between Machine Learning (ML), Artificial Intelligence (AI), and Deep Learning (DL). This chapter explains what Machine Learning is and how it supports DfMn activities that facilitate Preventative Maintenance Checks and Services (PMCS), Digital Prescriptive Maintenance (DPM), Prognostics and Health Management (PHM), Condition-based Maintenance (CBM), Reliability-centered Maintenance (RCM), Remote Maintenance Monitoring (RMM), Long Distance Support (LDS), and Spares Provisioning (SP). ML provides value-added benefits when used with PHM, CBM, and RCM to reduce or eliminate the need for corrective maintenance (CM) actions. This means that ML potentially has the capability to drive corrective maintenance costs and activities to zero. When a machine learns from the system's performance data, and understands the precursors of a failure mode, the machine makes effective and efficient preventive CBM decisions. These machine

Design for Maintainability, First Edition. Edited by Louis J. Gullo and Jack Dixon.

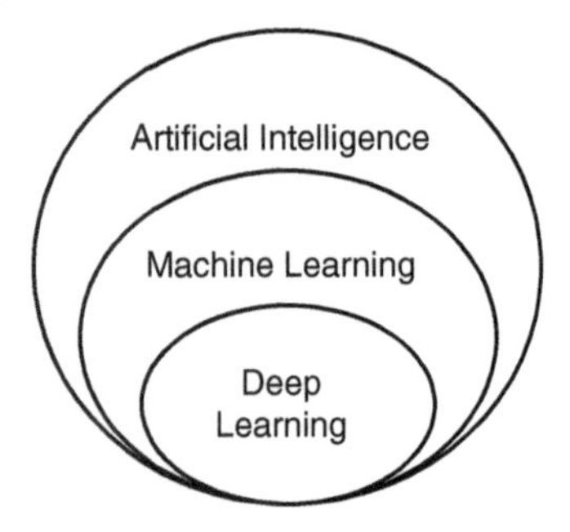

Figure 8.1 Relationships between DL, ML, and AI.

decisions are usually made faster and more reliably than those made by humans, which results in fewer errors and less expense to operate and maintain the system.

Simply stated, Machine Learning is a method used in Artificial Intelligence, and Deep Learning is a specialized case of Machine Learning. It is not necessary to conduct DL in order to perform ML. Likewise, it is not necessary to perform ML to perform AI. Figure 8.1 illustrates the relationship between DL, ML, and AI with a Venn diagram.

8.2 Artificial Intelligence in Maintenance

Artificial Intelligence is used to replace human intelligence with machine intelligence in a wide variety of applications, especially for AI Support Systems. AI Support Systems apply to large electronic systems with self-initiated maintenance modeling to select optimal repair methods. An AI Support System is different from a traditional support system in that it shifts the maintenance focus from a reactive and schedule-based preventative maintenance model to one that is condition-based and transparent to the end customer. RMM and LDS are examples of AI support structures that unlock value in providing a higher level of autonomous logistic support services to customers. Autonomous maintenance, which was developed in the automotive and medical industries, relies on the system capabilities to employ adequate AI and ML techniques. The success of AI Support Systems relies heavily on the capabilities for Knowledge-based Systems (KBSs), Expert Systems, and Machine Learning employed on a system.

Paradigm 4: Migrate from scheduled maintenance to Condition-based Maintenance (CBM)

KBSs are closely linked to Expert Systems that utilize rule-based reasoning. Initial interest in applying AI techniques in maintenance started with KBS in the 1980s. "The early paper by Dhaliwal (1986) argued for the use of AI techniques in maintenance management of large scale systems. Kobbacy (1992) proposed the use of KBS in this area and that was followed by a paper, Kobbacy et al. (1995) detailing 'IMOS', an Intelligent Maintenance Optimization System that uses KBS." "Batanov et al. (1993) presents a prototype KBS system for maintenance management (EXPERT-MM) that suggests maintenance policy, provides machine diagnosis and offers maintenance scheduling. Su et al. (2000) presents a KBS system for analyzing cognitive type recovery for preventive maintenance (PM). Gabbar et al. (2003) suggests a computer aided RCM based plant maintenance management system. The adopted approach utilizes a commercial Computerized Maintenance Management System (CMMS)" [1]. Since then, interest in KBS Expert Systems used for maintenance automation has varied over the years since 1980, but it has remained relatively constant in terms of moderate interest as a rolling average over this period.

Fuzzy Logic (FL) is a popular AI approach that has been frequently employed for maintenance systems in the past. Many system designers were interested in FL for modeling the uncertainty related to the solutions for various causes of maintenance problems. "Mozami et al. (2011) uses FL to prioritize maintenance activities based on pavement condition index, traffic volume, road width, and rehabilitation and maintenance cost. Al-Najjar and Alsyouf (2003) assessed the most popular maintenance approaches using a fuzzy Multiple Criteria Decision Making (MCDM) evaluation methodology. Derigent et al. (2009) presented a fuzzy methodology to assess component proximity based on which opportunistic maintenance strategy can be implemented. They used component proximity fuzzy modeling, which targeted the system components 'close to' a given reference component, on which a maintenance action was planned. Sasmal and Ramanjaneyulu (2008) developed a systematic procedure and formulations for condition evaluation of existing bridges using Analytic Hierarchy Process (AHP). Khan and Haddara (2004) presented a structured risk-based inspection and maintenance methodology that used fuzzy logic to estimate risk of failures by combining (fuzzy) likelihood of occurrences with (fuzzy) consequences, which was applied to oil and gas operations. Singh and Markeset (2009) presented a methodology for a risk-based inspection program for oil pipelines using a fuzzy logic framework" [1].

ML refers to an inference method used in Expert Systems for AI based on an existing a priori derived model of the physical world by a human who is a Subject Matter Expert (SME) in a specialized field. Inference is the rational process of deriving logical conclusions from information or knowledge that is either known to be true or assumed to be true. A primary focus of ML application development is developing inference techniques that significantly reduce or potentially eliminate reliance on the SME for decision-making. This primary focus that drives a design to adapt ML techniques for inference generates opportunities to reduce manpower and sustainment costs of the system or product over its life cycle, as well as to reduce the possibility of errors introduced by the SME. The same primary focus is applicable for KBS and Big Data Analytics (BDA). The goal is to develop the processing power and intelligence of a system to be smart enough to automatically process data and be able to autonomously generate the need for certain types of actions related to the operation or maintenance of the system involving people or drive operations and maintenance actions without a person in the loop. AI development involves automated inference systems to emulate human logic and reasoning. Before delving into the design of a KBS or Expert System, it is important to understand what is meant by the term "reasoning."

"Reasoning is the process of using existing knowledge to draw conclusions, make predictions, or construct explanations" [2]. Reasoning can take several forms. These forms include inductive reasoning, deductive reasoning, and abductive reasoning. Inductive reasoning is the logical process to reach a conclusion by assessing premises, and deciding which premises supply the strongest evidence for the truth of the conclusion. Premises are propositions presupposing the truth or the bases of arguments or inferences. The truth of the conclusion of an inductive argument is probable, based upon the evidence given. Deductive reasoning is a top-down approach, which starts from a general hypothesis and draws a conclusion about something more specific. Deductive reasoning is the logical process of concluding that something must be true because it is a special case of a general principle that is known to be true. Deductive reasoning may involve using the process of elimination by ruling out all the premises that are known not to be true and leaving only the premises that could be true. Deductive reasoning is the easiest reasoning approach to implement, but is the most error-prone and unreliable reasoning technique. Deductive reasoning is well known to lead to very erroneous conclusions. Abductive reasoning is a logical approach, which starts with an observation, and then progresses to a theory, which accounts for the observation. Abductive reasoning seeks to find the simplest premise using existing knowledge as the most likely explanation for the observation to draw a conclusion and make a decision. In abductive reasoning, the premises do not guarantee the conclusion. Simply stated, abductive reasoning is inference to the best explanation [3].

KBS and Expert Systems have been used for AI for many years with and without ML methods. Designers of KBS and Expert Systems begin the development process by identifying and asking questions of SMEs to collect data, information, and knowledge that answer the questions. The knowledge gathered from the SMEs is used in the KBS or Expert System to develop an AI capability where the machine will make decisions and draw conclusions just as a SME would do, but in place of a SME. Depending on the questions and answers from the SME, the logical flow of the data processing ultimately leads to a logical conclusion, which may or may not be able to solve a problem at hand. The logical conclusion may lead to a deterministic answer or a probabilistic answer. The deterministic answer results in a definitive "YES" or "NO" answer, while a probabilistic answer results in a probability that an answer is "YES." The probability is a number between

"0%" and "100%" or between 0 and 1, with infinite decimal values in between. For each probability (P) that an answer is "YES," there is a $(1 - P)$ value that the answer is "NO." For instance, a probabilistic answer for the question "Will it rain tomorrow?" may result in a 75% probability that it will rain, and a 25% probability that it will not rain.

ML may be applied for fault detection, fault diagnosis, fault isolation, and prognostic capabilities in all forms of electronic systems. ML may apply deterministic or probabilistic approaches to these capabilities. The development of deterministic or probabilistic approaches to ML will result in certain types of models that describe how the system will perform with ML. The models will be different depending on the types of systems that use them and how the system will apply ML capabilities. Two types of systems are widely distributed systems and embedded systems. Widely distributed systems apply ML capabilities on multiple subsystems and platforms across a network (such as a Wide Area Network [WAN] or on the internet's World Wide Web [WWW]) where processing power is pervasive and redundant. Embedded systems will apply ML capabilities within a Real-time Operating System (RTOS). Embedded systems will apply ML capabilities within the designs of its embedded test functions. These embedded test functions, such as the software that processes fault diagnostics and isolation logic, will be described as part of the embedded test system architecture. Embedded test system architecture applies methods and approaches to fault detection and diagnostic designs that will either include or not include explicit models. When models of the observed system are used for anomaly detection, fault detection, and failure diagnosis, this is often referred to as model-based testing. Since model-based testing is very important in DfMn, as well as design for testability, some explanation of model-based testing is warranted before continuing with ML designs for maintainability.

8.3 Model-based Reasoning

Model-based testing is an application of model-based design for designing test artifacts, such as executable test software, and executing these test artifacts to perform software testing or system level testing. Models can be used to represent the desired behavior of a System Under Test (SUT), or they can be used to represent testing strategies and a test environment [4]. Model-based testing may apply ML techniques in considering the potential for automation that it offers, but it depends on how effectively the model-based testing is applied in the system.

A model may be translated to or interpreted as a finite state diagram or a state transition diagram. These diagrams may be used to develop test cases. The test cases may also be developed from use cases. This type of model is usually deterministic, or it could be transformed into a deterministic model. These models represent performance of the observed system, which would be useful in design of AI to automate the system beyond its current capabilities. "When models of the observed system are used as a basis for fault detection and diagnosis, this is often referred to as 'model-based reasoning'" [5]. There are several types of model-based reasoning:

- Normal operations models or abnormal operations models
- Quantitative models (e.g., based on numbers and equations) or qualitative models (e.g., based on cause and effect models)
- Causal or non-causal models
- Compiled models or models based on first principles
- Probabilistic or deterministic models

An Ishikawa Fishbone diagram [6], which is widely used in Statistical Quality Control (SQC) and Statistical Process Control (SPC), is an example of a qualitative cause and effect model that leverages the data from model-based testing. Fishbone diagrams are used to develop ML capabilities in a newly designed automated system based on the results of defect detection and corrective actions experience gained from predecessor systems. These ML capabilities embedded into the system's control functions that are constructed from Fishbone diagrams allow the system to predict imminent failures and alert the user or maintainer that action is needed at some point in time to take the system down gracefully when it is convenient for the user to enact preventive repairs by the maintainers and avoid mission-critical system downtime.

SPC is an SQC methodology to further leverage results of defect detection and corrective actions experience gained from predecessor systems for use in monitoring and controlling process variables in a manual or automated system, to determine when maintenance should be performed prior to a process failure occurrence. SPC can be used to detect anomalous behaviors, subtle drifts, and significant changes in system performance. Some of this behavior may be a pattern in sensor data outputs that can be ignored if it is varying within nominal ranges, or it may be cause for concern if the output veers outside control limits. If one-time random occurrences are found outside control limits, they may be classified as outliers and be monitored for evidence of repeat occurrences in the future. Outliers are behavior that occurs outside a nominal case or outside control limits. Outliers can be considered anomalies when they first occurred.

In order to define fully any model used for testing, it is important to define terms that are used in test development and in the model development for testing purposes. The terms to be defined in this case are diagnosis, health monitoring, and prognostics.

Paradigm 2: Maintainability is directly proportional to testability and Prognostics and Health Monitoring (PHM)

8.3.1 Diagnosis

Diagnosis is the process of identifying the symptoms of the cause of unusual system or product performance behavior, to classify the behavior as a type of fault or failure. The primary function in diagnosis is fault detection. Diagnostics is a fault detection approach that usually involves fault identification, and may also include fault isolation. Once a fault is detected, the diagnosis process to identify and isolate a fault may be performed with automated test diagnostic routines or performed manually.

8.3.2 Health Monitoring

Health monitoring is the function of estimating a system's state of health. Health monitoring is accomplished by measuring state variables and identifying if the states of these variables indicate an off-nominal condition. Health monitoring may be included with diagnostics, or it may be performed as a complementary or supplementary capability to diagnostics.

8.3.3 Prognostics

Prognostics is a type of predictive analytical process for predicting when the occurrences of degradation events or system failures are expected to occur based on predictable time domain assessment, such as number of stress cycles, miles traveled, operational hours, and so on. Prognostics is the process of predicting a system's Remaining Useful Life (RUL) by predicting the progression of a fault given the current degree of degradation, the load history, and the anticipated future operational and environmental conditions to estimate the time at which the system will no longer perform its intended function within the desired performance specifications.

Further descriptions of these terms are included elsewhere in this book (see Chapters 9 and 13).

8.4 Machine Learning Process

Machine Learning is the design and application of algorithms that can learn from and make predictions based on data. These algorithms interpret inputs in order to make predictions, rather than following static program instructions. Machine Learning involves computational statistics, heuristics, and rule-based expert systems. Machine Learning includes data analytics, data grouping, classifiers, feature selection, and feature extraction. Algorithms may be developed by ML methods such as Support Vector Machines (SVMs), Bayesian Belief Networks (BBNs), Bayesian Entropy Networks (BENs), Neural Networks, Boosted Trees, and so on. A ML process was developed for documenting a step-by-step approach for developing these ML algorithms.

The ML Process Flowchart shown in Figure 8.2 describes a flowchart for requirements generation, data collection from predecessor repairable systems that were designed for maintainability, selection and performance of several ML methods, and data analysis to lead the ML practitioner towards development of a predictive analytics model for use in Preventive Maintenance (PM) vs. Corrective Maintenance (CM) decisions. This process is iterative with a closed-loop feedback path between assessment of the algorithm results and collecting more repair data from similar systems to improve the ML algorithms. As the fidelity of the ML algorithm improves, more maintenance actions will be preventative in nature and fewer actions will be corrective in nature. The goal is to drive the ML design towards zero CM effort and minimal PM that is only performed when the conditions warrant such actions to prevent system failures.

The first step in the process shown in Figure 8.2 is to generate requirements at a system level to quantify and qualify the amount of maintainability design effort that is needed and to specify these in a system-level

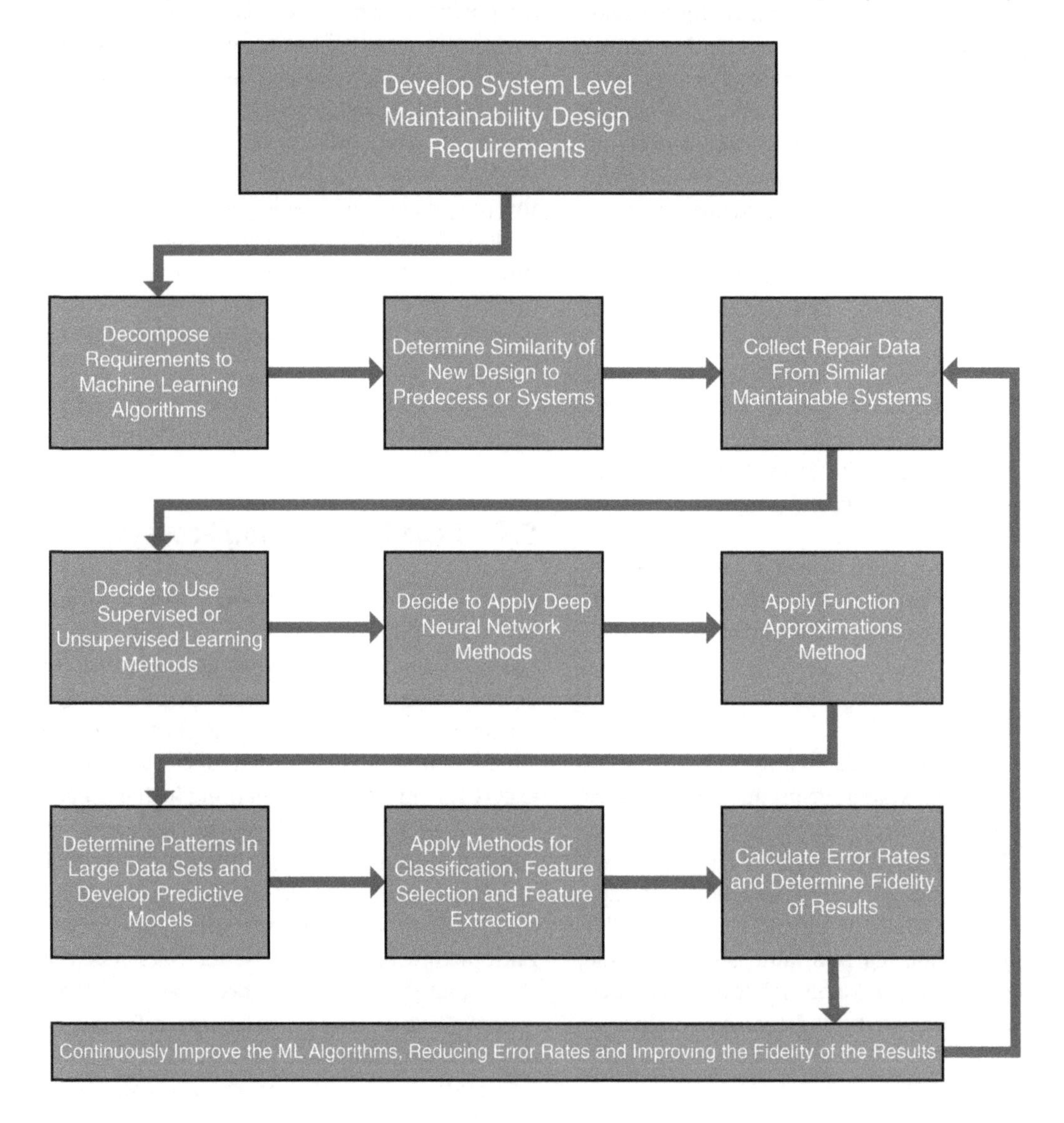

Figure 8.2 ML process flowchart.

requirements document. The next step is to decompose or flow down the system-level maintainability design requirements to specification documents at the lower levels of the hardware and software design hierarchy to describe the requirements for the ML algorithms. This is a critical step in the process, to properly document all lower-level requirements in such a way as to make the requirements intuitive for designers to understand and to do their jobs effectively and efficiently. The requirements should be documented to avoid any ambiguities and defects in the requirements that may be detected later in the development of the ML algorithms such that would require development rework involving cost and schedule impacts.

The next step in the process is to determine similarity of the newly designed system to predecessor systems, and to collect all available repair data from these systems' performance in field applications. These repair data are then formed into data groupings based on the different types of repair actions to resolve a system failure that caused system downtime. These data groupings are then brought to SMEs for data analysis and data tagging. Once SMEs examine the data groupings using statistical data sampling methods and complete their analysis, they will develop correlations in the data both within and between data groupings. Then they will create data labels that contain words, or are associated with a string of descriptive words, about the data contained in the groupings. These descriptive words may include nominal system performance conditions or extreme environmental stress conditions of the system at the time of a detected failure, such as cold start-up, rapid warm-up, high or low temperature peaks, mechanical shock, power overload, electrical current fluctuations, ripple current/voltage, and low voltage spikes, to name just a few conditions.

By using manual or automated statistical data sampling methods in the data analysis of data grouping subsets, with their corresponding data labels, the SMEs can determine the relative consistency or purity of the groups, from which data classifiers can be derived. These data classifiers may be derived using supervised or unsupervised learning methods as described in the following section. The data analysis should result in a quantification and statement about the amount of data consistency or data uncertainty that may exist in each of the data groupings and between the groupings.

8.4.1 Supervised and Unsupervised Learning

Machine Learning can be supervised or unsupervised. Supervised learning means a person is telling the machine what to look for. Unsupervised learning means a person is not telling the machine what to look for, but rather, the machine learns what to look for. "In supervised learning algorithms such as ensemble classifiers like bagging and boosting, Multilayer perceptron, Naive Bayes classifier, SVM, Random Forest, and Decision Trees are compared. In case of unsupervised learning methods like Radial base network function, clustering techniques, such as K-means algorithm, and K nearest neighbor are compared against each other" [7]. The reader is encouraged to research these methods further in the cited reference [7] and other references related to learning classifiers and function approximations.

An unsupervised learning method partitions data into groups and enables a supervised ML algorithm to discover rules that explains the distinctions between the data groups. A supervised learning method with a person-in-the-loop can use the results from unsupervised learning to decide which existing, already classified data group the new data should be classified with. Unsupervised learning partitions data into groups of similar entities that resemble each other in some form or another. Unsupervised learning will assess measurable differences between data groupings to draw characteristic distinctions to separate collections of entities that resemble each other from entities that are widely different. Without knowing the nature of a selection of data groupings, it is possible for a person to select any two or more data groupings and determine whether they are similar or different. A person can train a supervised ML algorithm using the unsupervised learning machine-generated data grouping descriptors.

SMEs may discover that certain kinds of meaningful labels were assigned to several different groups. A post-processing of the data groups may be warranted to expedite the extraction of data grouping features later on in the ML algorithm process, to help explain why members of different data groups could be joined together into a new single larger grouping. Once these new labels and additional features are derived by the SME, a new ML algorithm classifier is trained as post-processor to try to explain the new regrouping of entities from the first self-supervised auto-partitioning. This tiered approach to processing the information flow through a series of different partitioning algorithms followed by classifiers where new features or additional information is brought to bear is called Deep Learning.

"For supervised learning tasks, Deep Learning methods obviate feature engineering, by translating the data into compact intermediate representations akin to principal components, and deriving layered structures that remove redundancy in representation [8]."

DL algorithms can be applied to unsupervised learning tasks. "This is an important benefit because unlabeled data are more abundant than labeled data. Examples of deep structures that can be trained in an unsupervised manner are neural history compressors and Deep Belief Networks (DBNs)" [8].

If the machine applies "trial and error" to reach an objective, then it is applying Reinforcement Learning. Reinforcement Learning uses Deep Neural Network (DNN) methods. A DNN is an Artificial Neural Network (ANN) with multiple layers between the input and output layers [9].

8.4.2 Deep Learning

Deep Learning (DL) is an ML method that is based on learning to select the best data features and make the right decisions based on data representations, as opposed to generating task-specific algorithms. DL is also known as deep structured learning or hierarchical learning. DL inspires deep thinking, both for the human-in-the-loop and the machine.

DL is a class of ML algorithms that use multiple layers composed of an input layer, several layers of nonlinear data processing elements, and an output layer within a network for feature selection and feature extraction. Each layer of nonlinear data processing elements or nodes within the network accepts the output from the previous layer of nonlinear data processing elements. DL is not a specific algorithm, but rather a process flow distributed across multiple methods that offers different techniques and different sets of features to contribute to problem-solving, decision-making, and refinement of solutions in stages. If each of these stages is automated, then the Deep Learning architecture or decision architecture embodied by an instantiated algorithm is called a Deep Learning (DL) algorithm.

Deep Learning architectures are based on models such as DNNs, Convolutional Neural Networks (CNNs), Deep Belief Networks (DBNs), and Recurrent Neural Networks (RNNs). These DL architectures have been applied to system or product designs including speech recognition, natural language processing, image processing, audio and visual recognition, bioinformatics, genetic engineering, physical material inspection, modeling and simulation, and computer games. DL architecture applications have produced automated results comparable to the results produced manually by SMEs, and in some cases have produced results that are far superior to human experts.

Many DL models are based on an ANN methodology, such as DNNs and CNNs. ANNs may also include propositional formulae or latent variables organized layer-wise in deep generative or statistical models such as the nodes in DBNs. Propositional formulae are well formed and have a truth value. Deep Learning is a special type of Machine Learning since it involves the creation of a relatively large number of layers through which the input data are transformed.

In DL applications, such as closed-circuit camera image processing or visual face recognition tools, each level of a neural network learns to transform its visual frame input data into a slightly more abstract and composite representation of the input data. The camera's digital image input data may be a subset matrix of pixels grouped on an image frame that is provided to the neural network in multiple layers, where the first representational layer in the neural network may abstract the frame's image pixels and encode edges. The second layer of the neural network may formulate and encode arrangements of edges. The third layer may encode a portion of a human facial feature, such as a person's nose, mouth, and eyes. The fourth layer may compare the facial features to known images of a face to be able to recognize that the image contains a human face. The fifth layer may be a process of comparing the facial features in the input data to known faces of specific individuals contained in a large extensive database of human facial images

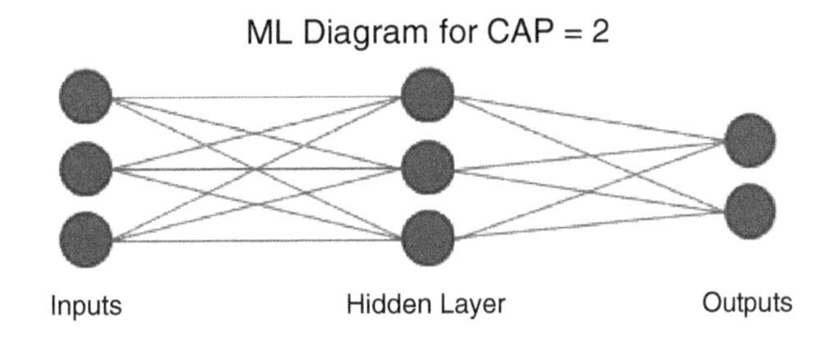

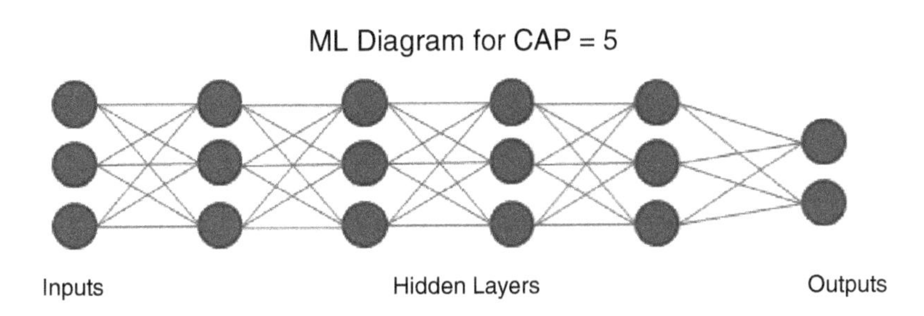

Figure 8.3 ML diagrams for CAP = 2 and CAP = 5.

to identify the human face by name. A Deep Learning process can automatically learn which image features to optimally place in the appropriate level of the neural network without manual intervention.

Deep Learning systems have a substantial Credit Assignment Path (CAP) depth compared with less complex forms of machine learning. "The CAP is the chain of transformations from input to output. CAPs describe potentially causal connections between inputs and outputs. For a feedforward neural network, the depth of the CAPs is that of the network. The CAP is calculated as the number of hidden layers plus one, which includes the output layer. For recurrent neural networks, in which a signal may propagate through a layer more than once, the CAP depth is potentially unlimited. No universally agreed upon threshold of depth divides shallow learning from deep learning, but most researchers agree that deep learning involves CAP depth > 2" [9]. Layers with CAP depth greater than 2 may not add to the effectiveness of the function approximator in the network. Deep models with a CAP greater than 2 are able to extract better features than models with a CAP of 2 or less. This means that models with a CAP greater than 2 provide extra layers that improve the ability of the neural network to learn features.

Examples are provided in Figure 8.3, showing diagrams for CAP equal to 2 and CAP equal to 5.

8.4.3 Function Approximations

Engineers and scientists learn by experimentation or from data collected from normal system usage and system test events. These data are categorized in different ways for function approximation and modeling, and then archived as historical data for future usage. Machines learn function approximations from these historical data. Linear regression is a function approximation method used in manual analysis as well as for ML. Linear regression is good for predicting a continuous-valued output. Classification is another function approximation method used in the development of ML algorithms that includes logistic regression, SVM, random forest, decision trees, and Naive Bayes. SVMs are considered by some as the best supervised learning algorithms. Other forms of function approximation for ML algorithms are collaborative filtering such as Alternating Least Squares, dimensionality reduction such as Principal Component Analysis, and clustering such as K-Means and Linear Discriminant Analysis (LDA). Logistic regression is a classification algorithm usually limited to two-class classification problems. If you have more than two classes, then LDA is the preferred linear classification technique.

"In general, a function approximation problem asks us to select a function among a well-defined class that closely matches ("approximates") a target function in a task-specific way [10]."

The following are two major classes of function approximation problems:

- Case 1, the target function is known: To approximate the actual target function as closely as possible, polynomials or rational (ratio of polynomials) functions are normally used. A generalized Fourier series is one example of a function approximation where the target function is known. A Fourier series is a periodic function composed of harmonically-related sinusoids, combined by a weighted summation [10].
- Case 2, the target function is unknown: Instead of an explicit formula, only a set of points of the form $(x, g(x))$ is provided, where g is the target function and x is the input to the target function. There are several ways to approximate an unknown target function. The selection of the best technique to approximate the target function depends on the structure of the domain and codomain of g. The domain is a function defined by the set of all "input" or x argument values. The codomain is the range or target set of a function, also called the set Y. The range or target set are all the values of the output of the function g that are expected. It is this range of values for Y that constrains the output of the target function [10].

"To some extent, the different problems (regression, classification, fitness approximation) have received a unified treatment in statistical learning theory, where they are viewed as supervised learning problems [10]."

8.4.4 Pattern Determination

ML algorithms are trained on large data-sets with various types of data patterns. These data patterns model performance variation over time. ML techniques use these data patterns to conduct statistical analysis of the data that are well beyond the abilities of any human brain. The result of applying ML techniques to data patterns using statistical analysis methods is an ML model or algorithm. Unsupervised learning methods rely heavily on pattern determination.

8.4.5 Machine Learning Classifiers

A mathematical function that is implemented by a model or algorithm that categorizes or classifies data is a classifier. A classifier or classification algorithm is a model of known states and labels that maps input data to a category, and then applies this category to similar current and future data groupings. A classifier detects and classifies data groupings that it was trained for. Classifiers assign data groupings to known classes within a training set. A training set contains data groupings whose categories are known and readily identifiable. Once a decision is made to train a classifier on an input data grouping or set of data groups, the result is a class of data or several classes of data. A classifier is able to rapidly detect the simplest boundary between two identified classes of data groupings. A classification algorithm receives a new data grouping and compares this data grouping with its database of known categories of past history of data groupings to determine which category the new data grouping fits into. Supervised learning methods rely heavily on classification algorithms.

Learning classifiers, as used in ML, are rule-based learning methods that combine a discovery component with a learning component. This learning component may be the result of performing supervised learning, reinforcement learning, or unsupervised learning. "Learning classifier systems seek to identify a set of context-dependent rules that collectively store and apply knowledge in a piecewise manner in order to make predictions (e.g. behavior modeling, classification, data mining, regression, function approximation, or game strategy). This approach allows complex solution spaces to be broken up into smaller, simpler parts. The founding concepts behind learning classifier systems came from attempts to model complex adaptive systems, using rule-based agents to form an artificial cognitive system (i.e. artificial intelligence)" [11].

As stated earlier, supervised learning methods rely heavily on classification algorithms. For this reason, classification is considered an instance of supervised learning when performing ML techniques, or any type of learning for that matter. This means that a training set of correctly identified observations is available.

The corresponding unsupervised learning methods in machine learning are known as clustering. This method involves grouping data into categories. The data grouping categories will be based either on some measure of inherent similarity or some measure of inherent difference or separation between groupings. "Often, the individual observations are analyzed into a set of quantifiable properties, known variously as explanatory variables or features. These properties may variously be categorical (e.g. 'A,' 'B,' 'AB,' or 'O,' for blood type), ordinal (e.g. 'large,' 'medium,' or 'small'), integer-valued (e.g. the number of occurrences of a particular word in an email), or real-valued (e.g. a measurement of blood pressure)" [12].

It is appropriate at this point to mention a few words about terminology that is used here, since the terms are quite variable across different fields. For instance, in the field of statistics and probabilities, classifications are often accomplished using logistic regression methods. Logistic regression is a class of regression analysis where an independent variable is used to predict a dependent variable. An analyst may start with linear regression and then evolve to logistic regression to achieve a logistic model to solve a classification problem. Regression usually refers to continuity and is used for predicting continuous variables depending upon features. Logistic regression is a special method of regression that is used for predicting binary variables when the target variable is categorical, not continuous. The properties of observations are termed explanatory variables, independent variables, or regressors. The categories to be predicted are known as outcomes. These outcomes are possible values of a dependent variable. When the dependent variable has two categories, then it is a binary logistic regression. When the dependent variable has more than two categories, then it is a multinomial logistic regression. "In machine learning, the observations are often known as instances, the explanatory variables are termed features (grouped into a feature vector), and the possible categories to be predicted are classes. Other fields may use different terminology: e.g. in community ecology, the term 'classification' normally refers to cluster analysis, i.e. a type of unsupervised learning" [12].

8.4.6 Feature Selection and Extraction

Data reduction is a necessary task to simplify the data analysis, whether it is performed manually or automatically. Data reduction is performed to eliminate data redundancies and shorten data processing time. When a data-set or data grouping is too large to be efficiently input by an ML algorithm and it takes too long to be processed by the ML algorithm, it is possible that the data grouping includes many data redundancies. These data redundancies may involve repeated measurements of the same numerical value, or multiple copies of the same images represented as pixelated or binary coded decimal data. In these cases of data redundancies, the large data-sets should be transformed into a reduced data-set containing unique and specific data features. Large amounts of data can be removed from the data grouping without loss of valuable information. The data features are called feature vectors. "Determining a subset of the initial features is called feature selection. The selected features are expected to contain the relevant information from the input data, so that the desired task can be performed by using this reduced representation instead of the complete initial data" [13].

Feature selection, also known as variable selection or attribute selection, is the process of selecting a subset of relevant features or variables for use in ML algorithm or model construction. Feature selection techniques are used for:

- Model simplification for better SME understanding
- Variance reduction
- Data generalization to reduce data overfitting
- Shorter ML training times

Feature selection is different from feature extraction. Feature selection is often used in domains where there are many features and comparatively few data samples or data points. "Feature extraction creates new features from functions of the original features, whereas feature selection returns a subset of the features. Archetypal cases for the application of feature selection include the analysis of written texts and DNA microarray data, where there are many thousands of features, and a few tens to hundreds of samples" [14].

When machine learning techniques are applied for pattern recognition and image processing, feature extraction will be used. Feature extraction starts from an initial set of measured data and builds derived values (features) intended to be informative and non-redundant, facilitating the subsequent learning and generalization steps. In some cases, feature extraction leads to better human interpretations. Feature extraction is a dimensionality reduction process, where an initial set of raw variables is reduced to more manageable groups (features) for processing, while still accurately and completely describing the original data-set [15].

Feature extraction reduces the amount of resources required to describe a large data-set. This can be extremely valuable when working with Big Data and a large number of variables. When performing analysis of Big Data, one of the major problems to deal with is the number of variables involved. Analysis with a large number of variables generally requires a large amount of memory, data storage, and data processing. It may also cause a classification algorithm to over-fit model parameters to training data samples and generalize poorly to new input data groupings. Feature extraction describes methods of constructing combinations of variables to circumvent model over-fitting problems while still describing the data with sufficient accuracy. Properly optimized feature extraction is an important key to effective model construction [15].

8.5 Anomaly Detection

An anomaly is an undesirable system event that has not been classified as a system fault or failure based on past system performance data. System behavior that has not been seen before is called an anomaly the first time it is encountered. When an anomaly or anomalous system state occurs and it cannot be classified as a system fault or a failure mode due to its new or anomalous behavior, then it requires a SME to investigate the symptoms of the behavior and determine a root cause for this behavior. After a SME analyzes the anomalous system state and determines whether a root cause exists, the occurrence is classified as a fault. If the SME cannot identify root cause, the anomaly is classified as a fault without a known root cause and flagged in the data infrastructure as a fault that requires further data collection and investigation. For anomalies that are classified as faults with known root causes, they are labeled in the data infrastructure as such, so that any future occurrences are no longer considered anomalies and the data are available for rapid classification the next time it occurs. The anomaly becomes a "known-unknown" fault, while true anomalies will be the "unknown-unknowns" since they are unexpected and occur for the first time.

Anomaly detection is the discovery of unusual or unexpected system performance behavior that does not conform to previously observed system performance data patterns. The process of analyzing performance data and distinguishing "acceptable" behavior from "suspect faulty" behavior enables a classifier to be created to detect and recognize them as being "faulty" behavior should they be encountered again in the future. The classifier will be created so that it will model specific instances of "faulty" behavior. The classifier is modeled after the anomaly has been observed and recognized in the system as acceptable or faulty, based on the functional requirements of the system.

A classifier is different from an anomaly detector. Anomalous data has labels initially assigned by the ML algorithm based on the system's self-supervised numerical group labels. Once the data labels are analyzed by a SME, the data labels may be changed to more meaningful labels for future SME reference and ease of understanding. The newly classified observations or data groupings that represent the initial anomalies that were detected will now identify a precursor of an impending fault condition and be included as a data grouping associated with a classification algorithm. This classification algorithm may be used to establish a prior distribution as used in a Bayesian methodology, such as a BBN, which could prove to be the initial algorithm providing valuable predictions of future failure occurrences as part of a PHM system.

8.5.1 Known and Unknown Anomalies

ML and predictive analytics, generally speaking, provide data classification into two categories: Data Category 1 are "known-knowns"; and Data Category 2 are "known-unknowns." As stated earlier, "known-known" data are anomalies that have been classified as faults with known root causes. The "known-unknown" data are anomalies that have been classified as faults with unknown root causes. Predictive analytics with anomaly detection features allows data classification into three categories: "known-knowns," "known-unknowns," and "unknown-unknowns." It is the "unknown-unknowns" that cause lots of added costs and missed opportunities to correct future problems. The "unknown-unknowns" data classification means that data are available, but no one has taken the time to find the data related to the problem symptoms, and analyze the data. True anomalies are identified from "unknown-unknown" data, which are anomalies that have not been classified as faults, and therefore are without known root causes. These are anomalies that are waiting to be found. These data contain potential latent defects that have yet to be discovered. As more and more data come into the data repositories, there is less chance of people spending the time to data-mine and discover the gold nuggets that tell them something about system behavior that they did not know before. Big Data and Internet of Things (IoT) technology allows people to discover the unknown-unknowns in the data to predict potential future outcomes. Once these unknowns are learned, then they are classified and modeled using a ML algorithm. When the data are modeled and embedded in an algorithm, and available for reuse applications, it usually proves to be invaluable for multiple future uses.

Unknown anomalies are the opposite of defined or known anomalies. Defined anomalies are modeled using the comparison of expected versus unexpected system behavior. SMEs observe the nominal and worst-case conditions of system performance and behavior. The SMEs conduct these observations by monitoring the frequency of occurrence of any types of faults or anomalies causing unexpected system behaviors. SMEs are aware when the statistics associated with expected faults or known anomalies change. SMEs are mostly concerned with a high frequency of fault and anomaly occurrences. Even though SMEs worry less about faults or anomaly occurrences that occur in expected ranges of times and locations, they still keep a watchful eye on the data to determine the increasing rates of occurrences, whether they are gradually increasing or rapidly increasing. SMEs may observe symptoms of anomalies, but do not understand their root cause. The goal is identifying and alerting operators to any anomalous behavior that falls outside that envelope (nominal behavior with a tolerated envelope of known anomalies). Perturbations to the system can trigger alarms of anomalies, but could be considered non-relevant in terms of effects on the system performance. The challenge in recognizing anomalies is to generate an understanding of the anomaly and determine whether action is needed. The meaning of the anomaly should be understood by discerning subtle differences between anomalies, and deciding how to use such distinctions to derive classifiers and validate the meaning of the indicators that lead to the anomaly. Anomalies may be a possible early warning indicator of degraded functionality leading to a system failure, or an undesired system state change, which could damage the equipment or cause a safety-critical hazard with a catastrophic effect. Classifiers are used to detect known warning states and provide the signals to early warning indicators. Recognizing an anomaly by an early warning indicator means the anomaly is no longer an anomaly, but rather a fault detection of a known class.

8.6 Value-added Benefits of ML

As stated earlier, Machine Learning supports DfMn activities that facilitate PMCS, DPM, PHM, CBM, RCM, RMM, LDS, and Spares Provisioning (SP). One value-added benefit of ML and DPM when used with PHM, CBM, and RCM is reduced corrective maintenance actions, and potentially, zero corrective maintenance costs and activities. When a machine makes maintenance decisions in place of humans, the decisions are made rapidly and more reliably, which results in fewer maintenance errors and less expense to operate. ML with Prescriptive Maintenance (ML/PM) leverages software for pattern recognition to identify specific stress points mapped in a three-dimensional (3D) space to explicitly diagnose root cause issues, followed by message alerts to operators or maintainers to initiate precise and timely actions to change the inevitable outcome of a system or product failure. Industrial firms are using ML/PM programs and predictive analytics software to reduce or eliminate unplanned downtime and maximize profitability and equipment reliability. AI-enabled ML/PM is unique in that it strives to produce outcome-focused recommendations for operations and maintenance using predictive analytics instead of just predicting impending failure without recommending timely preventive actions.

8.7 Digital Prescriptive Maintenance (DPM)

DPM allows manpower tasks related to preventive and predictive maintenance to become digital automated tasks. Maintenance digitization is becoming the new norm. The digitization of maintenance that is prescriptive is achievable through Dynamic Case Management or Adaptive Case Management. The DPM reference [16] provides highlights of a number of use cases from key industries that are leveraging the IoT or the Industrial Internet, Internet of Everything, and Machine-to-Machine. DPM's Dynamic Case Management is analogous to digitizing the logic from a detailed Failure Modes, Effects, and Criticality Analysis (FMECA), Fault Tree Analysis (FTA), and Probability Risk Assessment (PRA), and embedding the functionality into processors for autonomous real-time decision-making.

The prescriptive digitization approach to maintenance starts with descriptive maintenance, moves to predictive maintenance, then to prescriptive maintenance with digitized decisions, decision cases, and IoT technology applied to each decision case that is intelligent, timely, and responsive. This IoT technology is being applied effectively in multiple industries, including aerospace, defense, automotive, energy and utilities, agriculture, mining, and consumer products, such as home appliances, lighting, thermostats, televisions, healthcare devices, and consumer wearables. This approach provides continuous online, real-time data collection for data-driven decisions with innovative insights.

"A disruptive model harnesses the power of connected devices and the Internet of Things in a way that changes the dynamics of conventional Total Productive Maintenance (TPM), which defines maintenance as simply minimizing machine downtime. Incorporating intelligent software into these connected devices (Things) is proving to be a key enabler for diagnostics and proactive maintenance. Each layer of device and software creates a greater level of control and efficiency [16]."

8.8 Future Opportunities

Many businesses usually miss opportunities because they don't know what they don't know. They collect large amounts of data to perform maintenance on a single system, but don't analyze the larger data-sets collected for a single system or compare the data between multiple systems of similar types. These businesses usually suffer from data overload. They collect more data than their capability to quickly analyze it. The data analysis tasks seem too daunting for any one person to perform. They collect plenty of diagnostic data that are analyzed by individual electrical system trouble-shooters and maintainers to perform system repairs. Large companies don't have systems to collect data for the purpose of learning what they need to know. They typically don't collect data on a single platform in a form that could be easily read into a ML algorithm as part of BDA. Also, they don't know how to find the data from multiple customers that use the data to make their own repairs. These maintenance data are generated by customers for repairs that they perform on their own, which only further increases the size of the data population, and adding to the data analytics complexity. These external customer-generated maintenance data may be difficult to obtain, but is well worth the effort in terms of increasing the potential value of the results from the overall data population, and improving the accuracy of predicted failures. All these issues can be resolved if these businesses employ ML techniques such as those described in this chapter.

Large companies typically use predictive analytics for predictive maintenance to classify data into two top-level classifications or categories: known data (such as known failure causes) and unknown data (such as failure symptoms where the cause is not yet understood and to be determined). The failure causes usually fit into ambiguity groups, such as with "shotgun trouble-shooting," where we remove and replace three assemblies to correct a system failure code, instead of replacing one assembly at a time until the failure code clears and the system operation is restored. Applying the shotgun method, a trouble-shooter does not know the single assembly that caused each failure mode, but rather knows only that the group of assemblies that were replaced includes the culprit. There is no cause and effect conclusion from these repairs. This also means that there is wasted cost in replacing assemblies that were not defective. There is added cost in the repair activity from the waste of removing two out of three "potentially good" assemblies in trying to find the one bad assembly. If a sequential trouble-shooting method were applied, instead of the shotgun method, there is the potential that less good assemblies would be mistakenly replaced, thus decreasing the cost for each repair. However, it would probably take longer to conduct each repair.

Paradigm 3: Strive for ambiguity groups no greater than 3

Besides reducing the cost, a sequential trouble-shooting method would result in actual failure rate data to calculate the assemblies with the highest failure rates, and use of these data in predictive maintenance models that could be fed into ML tools for creating ML algorithms. Just as the maintainer would know RCM of each assembly using calculated assembly failure rates, so the ML algorithm would know the same and use these data in the decision on automating the maintenance remove-and-replace process for assemblies in the system. ML algorithms would be able to handle the data based on what is known and what is unknown about the defect causes within assemblies that cause system failure modes and failure effects. These ML algorithms can predict degraded system performance and future problem occurrences of systems. Maintenance actions can be planned to prevent system failures from the data modeled from ML classifiers and through the use of anomaly detection. With a library of ML classifiers and models of big data-sets, a company is better prepared to handle the maintenance challenges for new business opportunities and ensure customer satisfaction in the future.

8.9 Summary

Machine Learning moves DfMn capabilities from antiquated manual tasks to automation of maintenance events. ML techniques take the decisions away from a person where errors may be made, and puts the decision responsibility on the machine itself. The machine makes the decision solely based on the machine's need to conduct corrective action, whether it is immediately needed or can be scheduled as preventive maintenance in the future. The decision to know when to schedule preventive maintenance is determined by the machine's exposure to environmental and mechanical stress conditions accumulated as physical fatigue from similar exposures to stress conditions that resulted in failures in the past. Offloading the burden to determine exactly when is the best time to perform preventive maintenance can save hours of unnecessary maintenance. Machine Learning can save customers much time and cost in maintaining their systems over the system life cycle.

References

1 Kobbacy, K.A.H. (2012). Application of artificial intelligence in maintenance modelling and management. *IFAC Proceedings Volumes* **45** (31): 54–59. https://doi.org/10.3182/20121122-2-ES-4026.00046.

2 Butte College (n.d.). *Deductive, Inductive and Abductive Reasoning*. TIP Sheet, Butte College, Oroville, CA. http://www.butte.edu/departments/cas/tipsheets/thinking/reasoning.html (accessed 23 August 2020).

3 Sober, E. (2013). *Core Questions in Philosophy: A Text with Readings*, 6e, 28. Boston, MA: Pearson Education.

4 Wikipedia. (n.d.). Model-based testing. https://en.wikipedia.org/wiki/Model-based_testing (accessed 23 August 2020).
5 Stanley, G. (1991). Experiences using knowledge-based reasoning in online control systems. *IFAC Proceedings Volumes* **24** (4): 11–19. https://doi.org/10.1016/S1474-6670(17)54241-X.
6 Wikipedia (n.d.). Ishikawa diagram. https://en.wikipedia.org/wiki/Ishikawa_diagram (accessed 23 August 2020).
7 Aleem, S., Capretz, L., and Ahmed, F. (2015). Benchmarking machine learning techniques for software defect detection. *International Journal of Software Engineering and Applications* **6** (3): 11–23.
8 Deng, L. and Yu, D. (2014). Deep learning: Methods and applications. *Foundations and Trends in Signal Processing* **7** (3–4): 197–387. https://doi.org/10.1561/2000000039.
9 Schmidhuber, J. (2015). Deep learning in neural networks: An overview. *Neural Networks* **61**: 85–117. https://doi.org/10.1016/j.neunet.2014.09.003.
10 Butz, M.V., Lanzi, P.L., and Wilson, S.W. (2008). Function approximation with XCS: Hyperellipsoidal conditions, recursive least squares, and compaction. *IEEE Transactions on Evolutionary Computation* **12** (3): 355–376. https://doi.org/10.1109/TEVC.2007.903551.
11 Urbanowicz, R.J. and Moore, J.H. (2009). Learning classifier systems: a complete introduction, review, and roadmap. *Journal of Artificial Evolution and Applications* **2009**: 1–25. https://doi.org/10.1155/2009/736398.
12 Alpaydin, E. (2010). *Introduction to Machine Learning*, 9. London: MIT Press.
13 Alpaydin, E. (2010). *Introduction to Machine Learning*, 110. London: MIT Press.
14 Wikipedia. (n.d.). Feature selection. https://en.wikipedia.org/wiki/Feature_selection (accessed 23 August 2020).
15 Wikipedia. (n.d.). Feature extraction. https://en.wikipedia.org/wiki/Feature_extraction (accessed 23 August 2020).
16 Khoshafian, S. and Rostetter, C. (2015). *Digital Prescriptive Maintenance*. Pegasystems, Inc., a Pega Manufacturing White Paper. https://www.pega.com/system/files/resources/2019-01/Digital-Prescriptive-Maintenance.pdf (accessed 23 August 2020).

9

Condition-based Maintenance and Design for Reduced Staffing

Louis J. Gullo and James Kovacevic

9.1 Introduction

Condition-based Maintenance (CBM) is a predictive maintenance technique that enables smart decision-making by monitoring the condition of an asset (e.g., component, product, or system), processing and interpreting data to detect off-nominal behaviors or anomalies, and providing warnings to asset operators and maintainers to perform maintenance when the asset demonstrates signs of an impending failure. These warnings predict future failures. This technique is not just simply reacting to failure, but is a cost-effective approach that has demonstrated improved availability of assets in many different sectors. However, being effective does not mean doing maintenance. Rather it means that the right maintenance tasks are performed at the right time prior to operations being affected to ensure that an asset will operate when needed by the organization that owns the asset. CBM is not Time-based Maintenance (TBM). CBM provides the opportunity for an organization to move away from the traditional approaches to TBM.

Paradigm 4: Migrate from scheduled maintenance to Condition-based Maintenance (CBM)

CBM has the ability to dramatically reduce the man-hours needed to maintain an asset. Instead of lengthy periods of overhauls, maintenance is only performed when needed, and on the specific components of the asset that do need maintenance. This enables organizations to operate with lower staffing requirements. The need to reduce staffing requirements is not just driven by cost pressures, but cost pressures do play a role. The need for reduced staffing comes from a variety of areas:

- Skills shortages is a very big concern in many industries right now. Technical skills are in demand, yet fewer people are entering these fields. According to a recent study from Deloitte, in 2018 there were 508 000 open positions in manufacturing, which will grow to an estimated 2.4 million open positions by 2028. In addition, it is becoming harder to recruit those skills, with the average time of 70 days in 2015 to 93 days in 2018 to recruit to positions posted [1].
- Increased complexity of assets requires more specialized skills to maintain and repair the assets. Considering electrical systems that were designed 50 years ago; electricians needed an understanding of relay logic to be successful. Currently, many industrial electricians,not only need to understand relay logic, but now they need to understand various communication protocols, programming languages, specialized servo-controls, and so on. The cost of acquiring the knowledge of these different systems is extensive and prohibitive to some organizations. The cost and time associated with developing the

Design for Maintainability, First Edition. Edited by Louis J. Gullo and Jack Dixon.

skills necessary to maintain these systems means that there is less of a chance of developing expertise in these systems. Most technicians can specialize in only a few of the systems but not all. So, in order to support the skill development required to maintain the assets, and combat the forgetting curve, organizations need to ensure that the staff are deployed to the locations where they are needed most, since adding headcount dedicated to support each location is often not an option. In order to deploy staff where they are needed most, organizations need to leverage CBM to reduce the total workload of the staff.

For CBM to be practical and effective there needs to be a clear understanding of how equipment is used, how it functions, and most importantly, how it fails. Without this understanding, CBM may not be deployed where it is needed, or may be deployed where there is no value to the organization. In order to accomplish this, there have been many methods developed to help identify what maintenance should be performed on the asset.

The first is a Failure Modes and Effects Analysis (FMEA), which identifies the failure modes of an asset and is used to develop risk mitigation actions, but does not include criteria for which actions should be taken. The second methodology available is called Reliability-centered Maintenance (RCM), which utilizes a FMEA and a logic tree to decide what maintenance or design actions to take. RCM is covered later in this chapter. Another similar approach to RCM used in aviation is Maintenance Service Guidelines 3 (MSG3). Various other approaches have been developed but are not standardized to ensure a consistent approach in developing maintenance activities.

9.2 What is Condition-based Maintenance?

CBM is the application of maintenance actions based on indications of asset health, as determined from non-invasive measurement of operation and condition indicators. CBM allows preventive and corrective actions to be optimized by avoiding traditional calendar or run-based maintenance.

CBM should be used when the failure mode does not have a defined wear-out period. By using CBM, the components or asset can be monitored for indications of a potential failure, and preventive or corrective action put in place only when necessary. This approach enables organizations to plan their downtime, and further, focus on the right work when the asset is down. CBM can take many forms, but the same principle is behind each – to monitor for changes in the asset, or its operating characteristics, and act only when needed.

9.2.1 Types of Condition-based Maintenance

CBM involves monitoring the asset or its operating characteristics for changes, then acting to mitigate the consequence of the failure. These characteristics could include physical changes, process changes, or even changes in the operating parameters of the asset. Generally, organizations monitor a combination of the parameters using various forms of monitoring techniques:

- **Process data**: Variation in the output of a process can be monitored, as changes in quality generally indicate a change to the process. In addition, parameters within the process can be monitored to understand changes. For example, if flow rate is decreasing, and the current draw is decreasing, it may indicate that an impeller is wearing.
- **Condition monitoring**: Involves the use of specialized equipment to monitor and trend the condition of the equipment. The specialized equipment generally monitors changes in vibration, acoustics, or the release of particles into the environment. Other techniques monitor changes in the lubrication oil, changes to the chemical make-up of lubricating oil, or changes and release of chemicals into the environment. Condition monitoring also detects changes in the physical structure of the equipment, in terms of wear, damage, or other dimensional changes. Temperature is also monitored, as are changes in electrical characteristics, such as resistance, conductivity, dielectric strength, and so on.
- **Quantitative measurements**: Are used to capture similar information as condition monitoring, but without the specialized equipment. These activities

may include using Vernier calipers or micrometers to measure and record thickness as an indicator of wear. These data can be trended over time to determine degradation rates.
- **Human senses**: Can be used to detect many of the potential failures that process data or conditioning monitoring can detect. However, the downside of using the human senses is that at the point a human can detect them, there is little time before the equipment functionally fails.

Rolls-Royce has taken CBM to heart, with its business model of selling flight hours and not engines [2]. This incentivizes Rolls-Royce to utilize CBM, along with models to predict a failure and maximize the amount of time that the engines remain on the wing. This reduces the risk for the airline since they pay per flight hour, and if the engine is not operating, they are not paying for it. The condition monitoring data are put into various models, including artificial intelligence constructs, to better understand the failures. This has enabled Rolls-Royce to improve the uptime and availability of the engines.

9.3 Condition-based Maintenance vs. Time-based Maintenance

Moving away from traditional TBM brings numerous benefits to the organization. Often TBM was labor- and material-intensive, which would result in large costs as well as large amounts of planned downtime, which has a big impact on availability (see Chapter 15 for more on availability). In studying how equipment fails, it was discovered that the vast majority of failures are not related to time, therefore TBM is not the right approach, but CBM is. CBM is based on the premise that most potential failures, even random in nature, will exhibit signs prior to a functional failure. By monitoring theses signs, corrective actions can be taken to prevent the failure and its consequences. Monitoring these signs allows organizations to reduce planned downtime for preventative maintenance, while improving the reliability of the asset. Knowing when to use these two types of approaches to maintenance enables organizations to improve availability of the asset.

9.3.1 Time-based Maintenance

TBM is a periodic service wherein cleaning, replacement, or refurbishment of assets or their components will reduce the risk of unexpected failure. Traditionally, this has been the primary form of maintenance, which subjects assets to extended periods of planned downtime. This can reduce the capacity of the asset base, which requires organizations to either reduce their capacity or procure additional assets.

TBM should be used when there is a definite wear-out period or safe life of a component or asset. TBM should only be used when there are no signs of degradation occurring, or when components have a defined life. The defined life is often determined using a Weibull analysis, which may help drive the intervals at which the components should be replaced or refurbished. In order to properly utilize TBM, it is important to understand the different types of TBM.

9.3.2 Types of Time-based Maintenance

There are three primary types of TBM. These include:

- **Scheduled discard**: This is where a component is replaced regardless of condition at predefined intervals. The scheduled discard strategy is designed to replace the component just prior to failure based on statistical data. This approach may be used on components that cannot be serviced or those that cannot be serviced to a level achieving as good as new condition.
- **Scheduled restoration**: This is where a component is refurbished or overhauled regardless of condition at a predefined interval. This is often used for components that can be serviced and restored to a level that is as good as a new component.
- **Failure-finding task**: This is an inspection or testing task designed to determine whether a component has failed. This is often used for components that may not present signs of failure (hidden failures) or where the risk of failure is catastrophic. A common example would be a functional test on a pressure relief valve.

Knowing the different types of TBM is only part of the battle; the frequency of the tasks can determine how successful the program is and the operational availability of the asset.

9.3.3 Calculating Time-based Maintenance Intervals

Depending on the type of TBM activity to be performed, there are different techniques for determining when the task should take place. Ideally, the frequency of a scheduled task is governed by the age at which the item or component shows a rapid increase in the conditional probability of failure [3]. When establishing frequencies for TBM, it is required that the life be identified for the component based on data.

With time-based failures, a safe life and useful life exists. The safe life, which would only be used if the consequence of failure is either of safety or environmental significance, is when no failures occur before a defined date or time. The useful life (economic life limit) is when the cost of consequences of a failure starts to exceed the cost of the TBM activity. There is a trade-off at this point between the potential lost production and the cost of planned downtime, labor, and materials.

So how is the safe life or useful life established? They are established using failure data and history. This history can be analyzed statistically to determine the life of the component. Once that life is determined using a statistical analysis, the optimum cost-effective frequency must be established.

The optimum cost-effective frequency is the tradeoff point in which the optimum costs can be achieved by balancing the potential failure and its cost with the cost to perform the replacement. The optimum cost-effective frequency can be calculated using a cost model that helps to determine the optimum time to perform the TBM activity, specifically the scheduled discard and scheduled restoration tasks. The model uses research conducted, and is based on the minimum total costs, balancing the cost of failure with the cost of the maintenance actions [4]. The model is:

$$C_{\mathrm{T}} = \frac{C_{\mathrm{P}} \int_{T}^{\infty} f(t)dt + C_f \int_{0}^{T} f(t)dt}{T \int_{T}^{\infty} f(t)dt + \int_{0}^{T} tf(t)dt}$$

where C_{T} is the total cost per unit of time, C_{f} is the cost of failure, C_{P} is the cost of PM, and T is the time between PM actions.

A failure-finding task, which is also known as fault detection, is the process used when a system or component must be tested in order to verify proper operation. Often failure-finding tasks are associated with protective systems, with their failures not known to the operating crew under normal circumstances. For example, a tank may have a high switch, which is used to control the level in the tank. That same tank will also have a high high switch (or a safety high switch), which protects the tanks from overfilling if the high switch fails. The high switch is known as the protected system, which in a failed state may not be evident to the operating crew. The high high switch is a protective system, which protects the system in the event of the high switch failing. A failure-finding task may be to manually activate the high and high high switch to verify that they are operating as expected. Failure-finding task frequency has to be determined in a different way than the scheduled discard and restoration tasks, since it is designed to mitigate the risk on both the protected system and the protective system from failing at the same time. John Moubray [3] states that the failure-finding task can be determined by:

$$\text{Failure-finding Interval (FFI)} = (2 \times \text{MTIVE} \times \text{MTED})/\text{MMF}$$

where MTIVE is the Mean-Time-Between-Failure of the protective device or system, MTED is the Mean-Time-Between-Failure of the protected function, and MMF is the Mean Time Between Multiple Failures.

The failure-finding interval is the frequency at which the protective system must be checked to reduce the probability of both the protected and protective systems failing at the same time, resulting in severe consequences.

With the cost-effective and failure-finding frequencies of the TBM established, the organization can move forward, knowing that the right maintenance is being performed at the right time. However, this does not mean that those failure modes may never occur – just that the risk is reduced to a low enough level.

9.3.4 The P-F Curve

Because most failures are random in nature, it may be assumed that these failure cannot be predicted with any level of accuracy. However, even most random failure provides some evidence or warning that a potential failure is occurring, and will progress with time to a

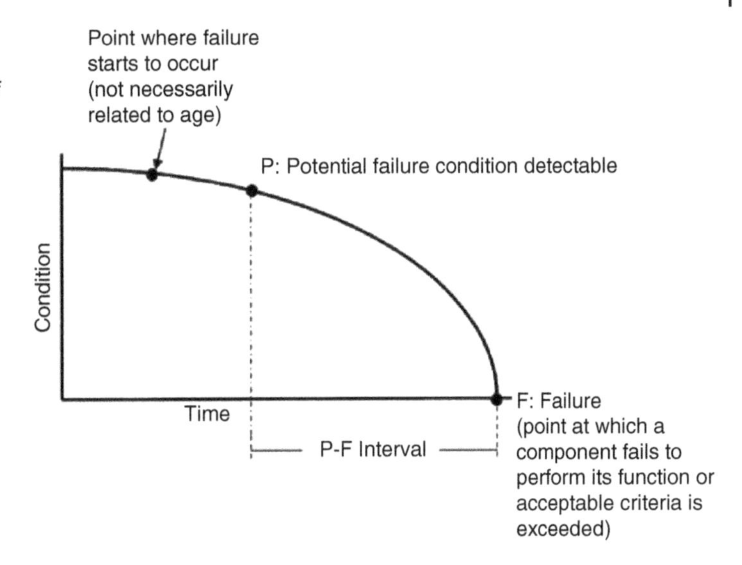

Figure 9.1 The P-F curve [5]. Source: US Department of Defense, *Condition Based Maintenance Plus DoD Guidebook*, Department of Defense, Washington, DC, 2008.

functionally failed state, known as a functional failure. If this evidence is detectable, then actions can be put in place to prevent the failure or the consequences of the failure [3]. The evidence may come from changes in the asset's temperature, vibration, current draw, and so on. Detecting this evidence enables organizations to take advantage of CBM.

The P-F curve (see Figure 9.1) illustrates that once a potential failure (point P) starts to occur, there will be a period of time, known as the P-F interval, before the asset fails (point F in Figure 9.1). A potential failure is defined as "an identifiable condition which indicates that a functional failure is either about to occur or is in the process of occurring" [3]. Closer to the potential failure point, technology is often used to detect minute changes in the asset operating characteristics. As the curve progresses closer to the functional failure point, changes detectable using the human senses may appear. This emphasizes the need for a quantitative method to detect these changes so that the organization has a longer period of time (the P-F interval) to react to or prepare for the impending failure.

The P-F interval is based entirely on the individual failure mode in the asset's present operating context. The P-F interval can vary from a few seconds to months, or even years, based on the specific failure mode and operating context. Take, for example, a large slow-moving bearing that starts to squeak. The bearing will likely operate for weeks or months before it functionally fails. However, a slow high-speed spindle bearing that starts to squeak will likely only last seconds to a few minutes. This interval, while not easily quantifiable, is vital as it helps to establish the appropriate inspection frequency using CBM. In its simplest form CBM tasks need to be done at an interval less than the P-F interval, while taking into account any logistical and administrative delays to ensure that the organization can detect and react to the potential failure before it becomes a functional failure. Further discussion on establishing intervals can be found in Sections 9.3.3 and 9.3.5.

Over the years, the P-F curve has been studied and expanded upon by numerous maintenance and reliability practitioners. One of these practitioners, Doug Plucknette, came up with the concept of the Design-Install-Potential Failure (D-I-P-F) Curve [6]. This version of the curve considers the impact that the design of the asset, as well as the installation/construction practices used, will have on the asset. Expanding on that a bit further, the D-S-I-P-F Curve (see Figure 9.2) also considers, in addition to the design and installation practices, the storage of the components before they have been installed. Why would storage matter? Consider false brinelling that occurs in bearings and motors while being transported and stored prior to installation. If false brinelling occurs, the time between the I-P portion of the curve will be greatly reduced, even if it is the right bearing and is installed correctly.

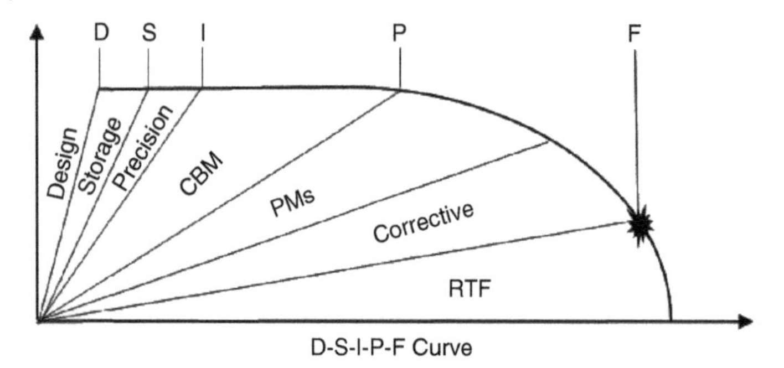

Figure 9.2 The D-S-I-P-F curve [6]. https://reliabilityweb.com/articles/entry/completing-the-curve

By using the D-S-I-P-F curve, organizations can understand the appropriate tasks to prevent potential failures from occurring at the various life-cycle stages of the asset. Proper design will ensure that the asset has a high level of inherent reliability while proper storage techniques will ensure that the components used are not contaminated or damaged prior to installation. Precision maintenance/construction techniques will ensure that assets are assembled and commissioned correctly. CBM will be used to monitor for potential failures further up the curve, while in some instances time-based visual inspections may occur further down the curve. Corrective maintenance may be used to correct the defective part prior to a functional failure, while Run-to-Failure (RTF) will be used to restore an asset after the functional failure has occurred.

9.3.5 Calculating Condition-based Maintenance Intervals

The first step to determining the inspection frequency for condition monitoring tasks is to construct the P-F curve and P-F interval. Constructing a P-F curve requires recording the results of the inspection and plotting the result versus the elapsed time. If enough measurements are taken, a wear-out curve can be developed for each failure mode, provided the curve is linear in nature and not exponential. Making sure that the data are gathered carefully and consistently will aid in increasing the quality of the P-F curve. However, if data are not readily available, experience and subject matter expertise can be used to develop the P-F interval. In addition, if the failure is exponential, as it is with vibration-related defects, then a curve can be constructed as well, but the rate of deterioration will increase dramatically as the failure gets closer to the functional failure point. With the P-F curve and P-F interval established, the frequency of condition monitoring inspections can be determined.

Thankfully it is not as complicated as establishing fixed-time maintenance frequencies. To determine the inspection frequency, the formula is either, "P-F interval/3" or "P-F interval/5", as shown:

- **Standard inspection**: The frequency of inspection for most equipment should be approximately 1/3 of the P-F interval (formula = P-F interval/3). For example, a failure mode with a P-F interval of 300 hours should be inspected every 100 hours.
- **Critical equipment inspection**: The frequency of inspection for critical equipment should be approximately 1/5 of the P-F interval (formula = P-F interval/5). For example, a failure mode on a critical piece of equipment with a P-F interval of 300 hours should be inspected every 60 hours.

Note that the above methods work well for linear P-F curves, but not for non-linear curves. How do you establish the frequency for the non-linear curves? You use the same approach as above for the initial inspection frequency. However, once a potential failure is detected, additional readings should be taken at progressively shorter intervals until a point is reached that a repair action must be taken. For example, the initial inspection frequency is every four weeks. Once a defect is detected, the next inspection will be at three weeks, then two weeks, and then every week.

These are only guidelines and should be adjusted based on the method used to track and trend data, the lead time of the repair parts (if not kept on site), how quickly the data will be analyzed, and the repair work planned. If your planning process is poor, the frequency should be more greater to allow for a high chance of detection sooner.

If the inspection frequency is relatively low, then online monitoring solutions may be required, or the monitoring may not be practical or cost-effective. Also, it is vital that the inspection frequency leaves enough time for a defect to be detected, and the corrective action to be planned, materials obtained, and the asset scheduled down for the repair, before a functional failure occurs. If there is not enough time for this to occur, then the frequency may need to be reduced or an online monitoring solution implemented.

9.4 Reduced Staffing Through CBM and Efficient TBM

The goal behind any organization is to ensure operational availability of their assets (see Chapter 15 for more information on availability). In order to ensure high levels of operational availability, it is vital that the time required for maintenance is balanced with the need for reliability. CBM enables organizations to reduce the need for planned downtime in a variety of ways.

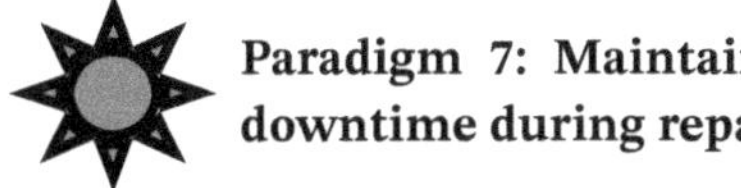

Paradigm 7: Maintainability predicts downtime during repairs

The use of CBM enables organizations to continuously monitor and trend parameters of the equipment (which generally must be taken as the equipment is operating), and to determine the optimum time to schedule downtime for the asset. This enables organizations to reduce the need for intrusive maintenance activities, and essentially only fix the asset when it is broken. By reducing the need for intrusive maintenance, which requires scheduled downtime, the asset has higher operational availability.

CBM can further be leveraged to reduce staffing and downtime by designing the condition monitoring devices into the equipment. In a traditional maintenance environment, a technician would be required to use a portable data collector and run routes to collect data points for various pieces of equipment. This type of CBM can present risk to the technician along with providing a degree of inaccuracy in the readings. In addition, readings can be missed or the data collector set up incorrectly. As people with technical skills become harder to recruit and retain, it is vital that asset designs include the monitoring devices in the equipment. By embedding these devices into the equipment, it enables remote monitoring of equipment, as well as real-time data collection and analysis, which enables organizations to react quicker and potentially reduce the consequences of the failure. In addition, by combining the real-time data collection with AI and other analytical models, fewer analysts and technicians may be required.

One additional means that can further reduce the staffing required to maintain the equipment is to make traditional TBM more efficient. Maintenance can be made more efficient with the use of visual factory. Visual factory is a method to communicate vital information in a workplace. The visual factory process applies a combination of signs, charts, and other visual representations of information that enables the quick dissemination of data to the workforce. The visual factory attempts to reduce the time and resources required to communicate the same information verbally or in written form. In terms of maintenance, it could mean adding red and green overlays to gauges which would allow operators to quickly monitor the operating parameters of the equipment and notify maintenance when action is required. In addition, various methods could be used to monitor chain slack, fluid levels, and so on. The use of visual factory allows organizations to transfer maintenance activities to operators, further reducing the need for maintenance. Lastly, designing the equipment to enable remote lubrication, quick change connectors, and so forth, can further reduce the time required to perform TBM.

If CBM is designed into the asset, the use of specialized tools and equipment can be reduced as well as labor-intensive data collection activities. More organizations are considering the labor shortage challenge and the need for increased availability of the assets. As such, more and more organizations are looking to improve CBM by integrating embedded sensors into the asset design, which not only enables labor savings, but also enables a more proactive approach. This approach is known as asset health monitoring and enables real-time evaluation of multiple data sources to quickly identify potential failures within the asset.

9.5 Integrated System Health Management

Integrated System Health Management (ISHM) or Integrated Vehicle Health Management (IVHM) is a system to assess the current and future state of an asset (e.g., electronic system, vehicle, or product) and to control the actions that are needed to sustain the asset and ensure its operational state. ISHM includes capabilities to monitor the health of the asset. This capability goes beyond normal fault detection from embedded diagnostic test capabilities. This ISHM capability also includes detection of off-nominal parameters that signal early warning of impending failures based on statistical data measurements. Health monitoring involves various types of sensors to detect early signs of an impending failure in an electronic system or product. These signs may take the form of off-nominal electronic circuit behaviors or anomalies, physical fatigue wear-out mechanisms, or indications of physical degradation that are not directly connected to test logic that triggers a functional failure mode. Strain gauges and pressure transducers are examples of two types of sensors that monitor the physical conditions of an assembly to determine parametric degradation in terms of pressure differentials. Piezo-resistive and piezo-electric devices are examples of pressure transducers that measure pressure differentials. Pressure differentials with constant pressure variation in the negative or positive directions mean that stressful conditions exist. If there are no pressure differentials, the assembly is considered to be stable and at equilibrium, ensuring the assembly will satisfy long-life expectations. The same is true for thermal and vibration conditions. Thermal and vibration fluctuations translate into unstable and stressful conditions to which the assembly is exposed. Constant thermal and vibration conditions mean stable conditions without stress on the assembly. Thermocouples are sensors that measure cold and hot temperature extremes to assess the health of an asset and compare its operating conditions with the design specifications for environmental exposure limitations. Accelerometers are sensors used to measure mechanical shock and vibration to uncover overstress conditions that limit the life of the asset. For electrical parameters, current, voltage, resistance, capacitance, inductance, frequency, and power sensors can be used to measure and ascertain the electrical health of a system. With the proper use of these different types of sensors, indicators of failure modes and failure mechanisms can be detected within an ISHM system to prevent mission-critical failures. Figure 9.3 illustrates an example of a process flow diagram for an ISHM and health monitoring system.

The focus of this process is on CBM and Prognostics and Health Management (PHM) data collection and process flow. Sensors embedded into a system design to monitor the health of a system of nodes enable prognostics through the use of prognostic targets. The prognostic targets are chosen based on the likelihood of failure symptom detection within a time-band of failure event occurrence. It is estimated that prognostic targets at those nodes where health monitoring sensors are placed have signals that change in response to degradation of devices, components, and assemblies. When these devices or circuits fail, there is a critical effect on the operation of the system so that it is unable to accomplish its mission. Health monitoring is the first step to avoid unexpected outages and prevent these failures. Referring back to Figure 9.3, an ISHM system needs to provide services for health management (HM), maintenance, and logistic support to schedule maintenance, locate and deliver parts and equipment, and to dispatch a maintenance team to restore operation and place the system back into service [7]. As seen in step 2 of the flow diagram, sensor data from step 1 are collected into a single data collection platform to analyze diagnostics and prognostics data. The data may take the form of real-time State of Health (SOH) or Remaining Useful Life (RUL) results. These data are reviewed by a human or an automated system for scheduling a certain type of action to improve the design or initiate a maintenance action. For the case of a design improvement, as shown in step 3, an Engineering Change Request (ECR) may be initiated to correct a pattern failure based on statistical trend analysis for a particular part with a suspected design flaw. The design improvement in step 3 involves actions performed by subsystem Original Equipment

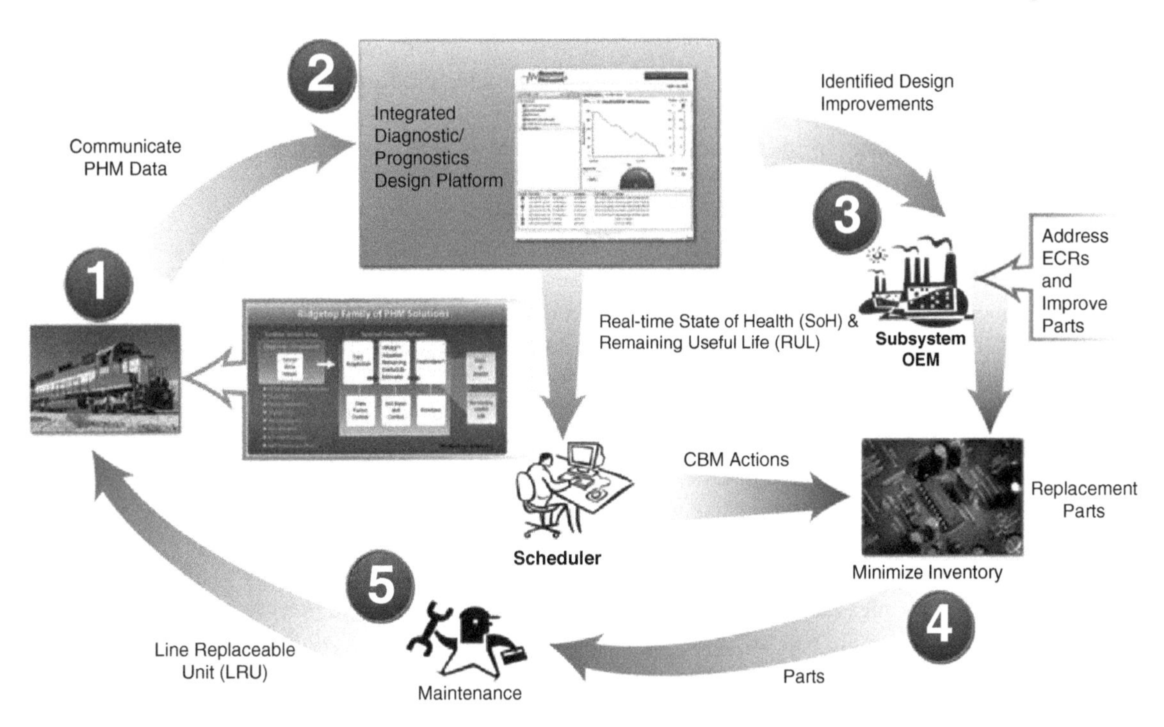

Figure 9.3 ISHM and health monitoring process flow diagram [7]. Source: Goodman, D., Hofmeister, J.P., and Szidarovszky, F., *Prognostics and Health Management: Practical Approach to Improving System Reliability Using Condition-Based Data*, First Edition, John Wiley and Sons, 2019. Reproduced with permission from John Wiley and Sons.

Manufacturers (OEMs) that are responsible for the designs integrated into a system where a part failure may cause a system failure. Step 4 involves the actions to replenish spare parts in a storage room inventory or replace parts that are defective with new parts based on approval of ECRs from the OEM. In step 4, a part replacement occurs through normal CBM processes from the maintenance scheduler. Maintenance occurs at the Line Replaceable Unit (LRU) level of repair with the actions in step 5, using the new parts sent through the design authority initiating a design change, or the performance of routine maintenance to simply remove and replace a part that was authorized in the original drawings without the need for an ECR. Following step 5, the LRU is sent to the system in the customer location for replacement.

The accuracy and completeness of the health monitoring data, with the reliability of the sensor devices and the data collection system, enable the development of predictive models for the purpose of prognostics and Condition-based Maintenance Plus (CBM+) capabilities.

Paradigm 2: Maintainability is directly proportional to testability and Prognostics and Health Monitoring (PHM)

9.6 Prognostics and CBM+

Prognostics are the prediction of future event outcome possibilities. Prognostics are the logical sequence of technology developments progressing beyond CBM methodologies. Prognostics involve the development of predictive analytical models based on the collection and analysis of CBM, parametric and environmental sensor, and health monitoring data. Once prognostic models are developed, they are implemented in a system to manage the prognostics capability and system health status. "Prognostics and health management is an approach to protect the integrity of equipment and avoid unanticipated operational problems leading to mission performance deficiencies, degradation, and adverse effects on mission safety. Researchers and application developers have developed a variety of approaches, methods, and tools that are useful for these purposes, but applications to real-world situations may be hindered by the lack of visibility into these tools, the lack of uniformity in the application of these tools, and the lack of consistency in their demonstrated results" [8].

IEEE Std 1856 provides information for the design implementation of PHM technologies and methodologies to enhance the capabilities of electronic systems and products. "This standard can be used by manufacturers and end users for planning the appropriate prognostics and health management methodology to implement the associated life cycle operations for the system of interest. This standard aims to provide practitioners with the information that will help them make business cases for prognostics and health management implementation and select the proper strategies and performance metrics to evaluate prognostics and health management results" [8].

IEEE Std 1856 provides guidance on how to measure the performance of a prognostic system to determine if the prognostic system goals are achieved, or to determine whether prognostic design improvements are needed. Prognostic system performance measures include the following categories:

1. Accuracy
2. Timeliness
3. Confidence
4. Effectiveness

Accuracy is a measure of the differences between predicted estimations of performance over a period of time and actual performance data and results collected over the same period of time. Examples of metrics used to assess accuracy are detection accuracy, isolation accuracy, and prognostic system accuracy. Detection accuracy is defined as the ratio of the number of times the prognostic system detected an actual failure occurrence divided by the number of actual failure occurrences over a period of time. Isolation accuracy is defined as the ratio of the number of times the prognostic system determined an actual failure occurrence was caused by a particular component or assembly, or a group of components or assemblies, divided by the number of actual failure occurrences over a period of time. Prognostic system accuracy is the difference between the predicted RUL and the actual End-of-Life (EOL) of the monitored product or system. RUL is defined as the estimated amount of time starting at the present time up to the time when a non-repairable product or system is no longer able to perform its functions or complete its intended mission as defined in the system requirements specification, and must be replaced.

Timeliness is a measure of the difference between the time when an actual failure event occurs and the time when the prognostic system produced an output to predict or detect the occurrence of the failure event. Examples of metrics used to assess timeliness are fault detection time, fault isolation time, diagnostic time, prognostic time, decision time, and response time.

Confidence is a measure of the probability that the prognostic system output is accurate and relevant based on an expected level of certainty and trustworthiness. Confidence may be expressed in terms of a confidence level. In general terms, confidence level is the probability that a parameter lies between two limits, or the parameter lies below an upper limit or above a lower limit. Confidence also means the probability that the true value of a parameter is within the preconceived boundaries established by the estimation of the parameter. Examples of metrics used to assess confidence are confidence level, confidence limit, confidence bounds, confidence interval, prediction uncertainty, prediction stability, robustness measures, and sensitivity analysis.

Effectiveness is a measure of the ability of the prognostic system to achieve its goals and objectives over a period of time while exposed to specified conditions. Typical effectiveness goals and objectives include cost, schedule, performance, reliability, maintainability, availability, security, and safety. Effectiveness is calculated as an aggregate of the previous three prognostic system performance measures (accuracy, timeliness, and confidence).

"A key objective of deploying a prognostic-enabled system is to monitor prognostic targets to provide advanced warning of failures in support of CBM. There are criteria associated with, for example, equipment availability and other metrics, test coverage and confidence levels. To meet criteria, the various sensing, signal processing, and computational (algorithms) routines in a Prognostic Health Monitoring (PHM) system need to be factored into the entire design. CBM methods and approaches, especially those using condition-based data (CBD) signatures that are ultimately transformed into functional-failure signature (FFS) data that are processed by a very good prognostic information program, provide significant advantages over (i) a system based on statistical or other methods that are applicable to populations rather than a specific instantiation of a population, and (ii) a system based on using CBD to detect damage without prognosing when such damage results in the system no longer operating within specifications [7]."

Figure 9.4 presents the critical timing parameters related to techniques for measuring a critical performance parameter of an assembly using a PHM system to monitor its health and predict future failure events. In the design of a PHM system, the designer will establish the upper and lower failure thresholds based on the assembly performance requirements. These assembly performance requirements are used to determine test requirements that decide when the assembly has passed or failed a test. The upper and lower off-nominal thresholds for the sensors in the PHM system may be calculated as sensor data are received based on statistical process control rules, or they may be established at the time when the designer establishes the upper and lower failure thresholds. t_0 starts when the assembly is turned on and begins to operate as described in its design performance requirements. A sensor is selected for the PHM system design to measure a critical performance parameter that is able to determine if the assembly is operating correctly. t_E is the time when the critical performance parameter exceeds the upper off-nominal threshold. t_D is the time when the sensor first detects that the critical performance parameter exceeds the upper off-nominal threshold. This t_D detection event could occur multiple times without the need for any maintenance action. The first t_D detection event triggers the PHM system to predict the time when a part or subsystem failure may occur. t_R is the time allotment for the PHM system to predict the failure time of the assembly. The PHM response time, $t_{R,}$ equals t_P minus t_D. t_P is the time when the PHM system predicts the assembly failure at the time t_F.

The most significant parameter in this figure is the prognostic distance. The effectiveness of the PHM

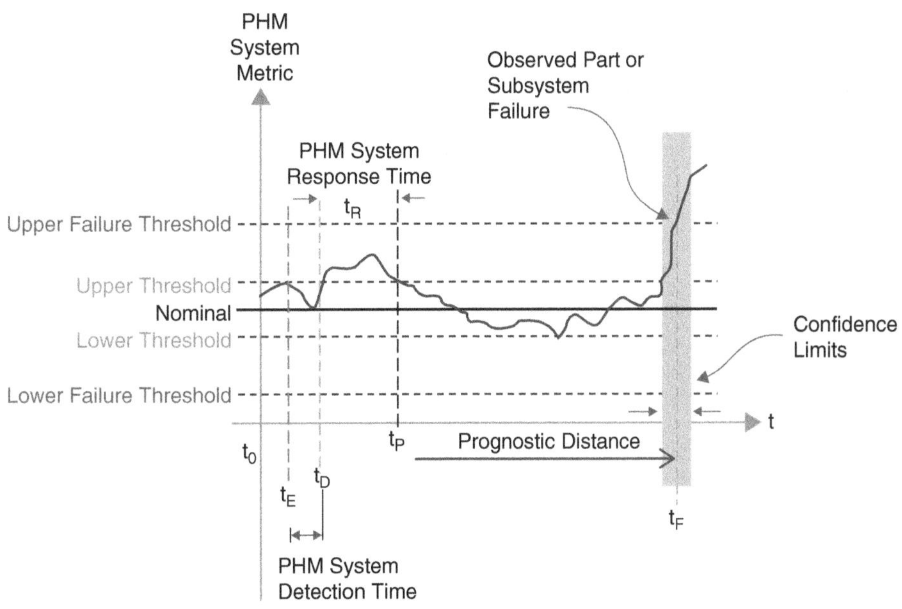

Figure 9.4 PHM control chart and timeline [8]. Source: *IEEE 1856-2017, IEEE Standard Framework for Prognostics and Health Management of Electronic Systems*, Institute of Electrical and Electronic Engineers (IEEE) Standards Association (SA), Piscataway, NJ. © 2017, IEEE.

system is directly correlated to the accuracy of and the speed with which the prognostic distance is measured with a high degree of confidence. Referring to Figure 9.4, prognostic distance is the difference between t_F and t_P. The longer the prognostic distance, the better the PHM system. When the sensor measures the value at the time the parameter exceeds the upper failure threshold, the assembly has failed. The challenge is to predict t_F with ample time to plan for and take action to prevent an operational failure.

Annex A of IEEE Std 1856 [8] provides guidance for implementation of PHM capabilities for complex electronic products and systems. When the term PHM is used in IEEE Std 1856, it is intended to refer to the sensor and health monitoring data, as well as the data processing functions used to manage and restore a product's or system's health. A PHM system consists of core capabilities, which are summarized in the PHM functional model diagram of Figure 9.5. As shown in Figure 9.5, there are five core PHM operational processes:

1. Sense
2. Acquire
3. Analyze
4. Advise
5. Act

These five core PHM operational processes are enabled by the PHM functions embedded within the system or distributed throughout the system, depending on the type of electronic system, such as a Real-time Operating System (RTOS) or a widely distributed computer network system. The core PHM functional elements include sensing (S), data acquisition (DA), data manipulation (DM), state detection (SD), health assessment (HA), prognostic assessment (PA), advisory generation (AG), and health management (HM).

"In this model, the lower three functional blocks provide low-level, application-specific functions. At the lowest level, a sensor(s) produces an output state that corresponds to a state in the object system or the object system's environment. At

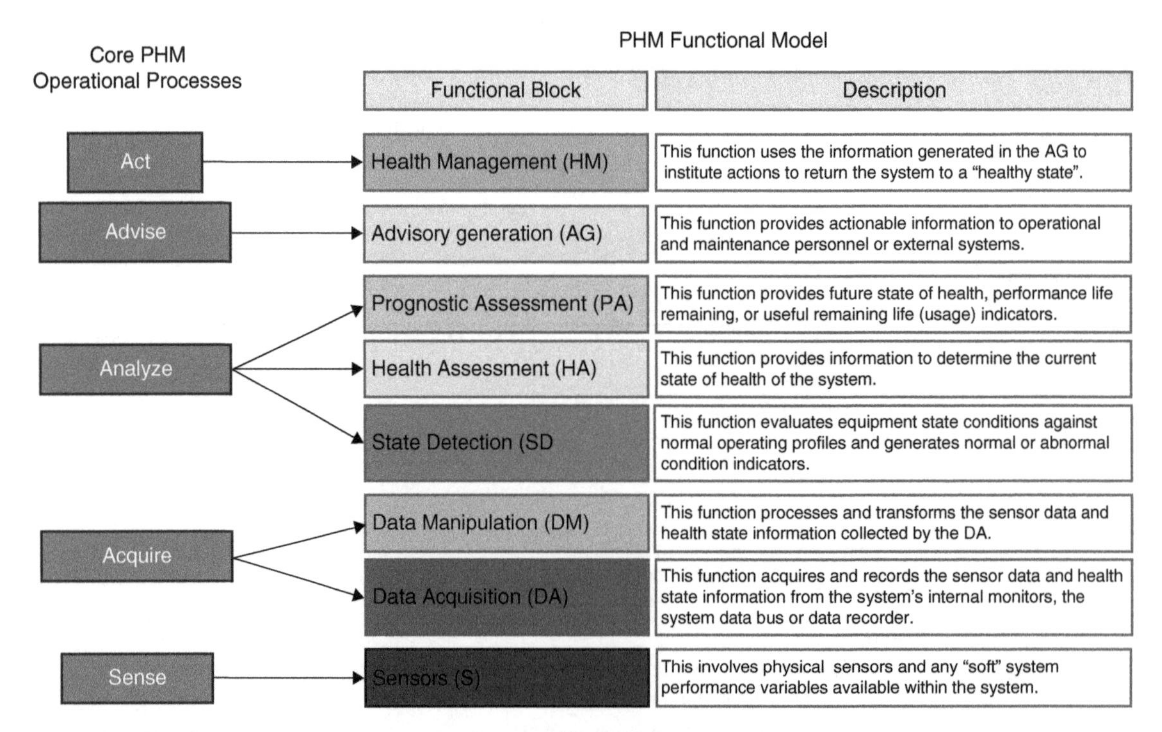

Figure 9.5 PHM functional model diagram [8]. Source: *IEEE 1856-2017, IEEE Standard Framework for Prognostics and Health Management of Electronic Systems*, Institute of Electrical and Electronic Engineers (IEEE) Standards Association (SA), Piscataway, NJ. © 2017, IEEE.

the next level, the data acquisition block converts the sensor output to digital data. The next level, the data manipulation block, implements low-level signal processing of raw measurements from the data acquisition block. The state detection function supports modeling of normal operation and the detection of operational abnormalities. The upper three functional blocks of the model provide decision support to operations and maintenance personnel based on the health of the target system. Within this group, the health assessment function provides fault diagnostics and health condition assessment. The prognostic assessment function forecasts the health condition based on a model, current data, and the projected usage loads, and computes performance RUL. The advisory generation function provides actionable information related to the health of the target system. Finally, the health management function converts the information from the advisory generation function into actions that will manage the health of the system to achieve overall system and mission objectives [8]."

Figure 9.6 provides an operational view of how the sensors, health monitoring, and assessment functions enable PHM capabilities using the core PHM operational processes. Starting the operational process on the left side of the diagram with the Sense block, the various physical sensors are designed into the PHM system. The Sense process accumulates the data from the physical sensor device outputs and any available "soft" system performance data that are inherent in the target system. The term "soft" refers to the capabilities of the target system that measure performance variables without the requirements placed on the system for the PHM capabilities. The Sense process transmits the physical sensor and "soft" variables data to the Acquire process. The Acquire process initiates the data acquisition (DA) and data manipulation (DM) functions. The Acquire process performs data capture, data processing, data storage, data management, and data communications. The Acquire process transmits data to the Analyze process. The Analyze process initiates the state detection (SD), health assessment (HA), and prognostic assessment (PA) functions. The Analyze process performs test diagnostic routines that include fault detection, fault isolation, and fault identification. The Analyze process also provides assessments of the system health state and estimations of the system performance life remaining and RUL metrics. When the Analyze process has completed, it sends its data to the Advise process. The Advise process initiates advisory generation (AG) functions that may be inherent in the target system or provided external to the target system. The Advise process

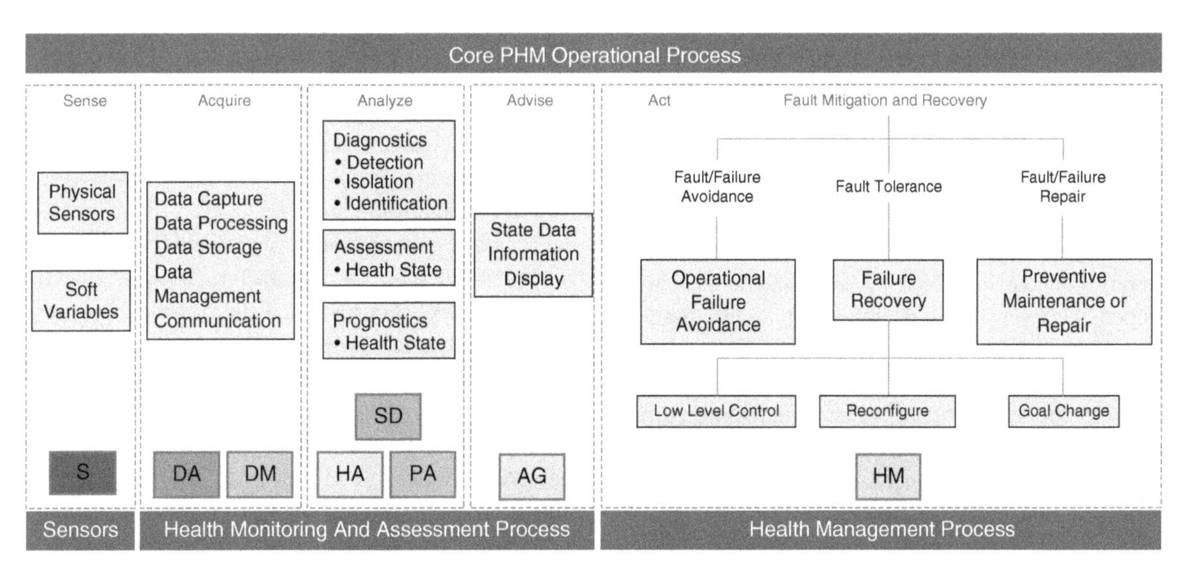

Figure 9.6 PHM operational process diagram [8]. Source: *IEEE 1856-2017, IEEE Standard Framework for Prognostics and Health Management of Electronic Systems*, Institute of Electrical and Electronic Engineers (IEEE) Standards Association (SA), Piscataway, NJ. © 2017, IEEE.

conducts all functions to interface with the system users, operators, and maintainers. The Advise process sends display advisories, such as messages, alerts, and warnings, to personnel who are responsible for taking action related to the continuous operation of the target system. The Advise process may also provide refined analytical charts and graphical health state data and prescriptive information. The last process in the PHM operational process diagram is the Act process. The Advise process transmit its data to the Act process, which conducts all health management (HM) functions. The Act process initiates all HM functions to restore the target system to a healthy state. The HM functions involving fault mitigation and fault recovery functions may be performed within the target system or provided external to the target system. The Act process provides both fault tolerance capabilities and fault recovery operations that are manually initiated, autonomous, and semi-autonomous. These fault tolerance capabilities and fault recovery operations are performed for failure avoidance, failure remediation, and preventive and corrective maintenance actions.

9.6.1 Essential Elements of CBM+

CBM+ is the application of processes, technologies, and expert knowledge to improve the reliability and maintenance effectiveness of systems, assemblies, and components. CBM+ is maintenance performed based on evidence of need provided by RCM analysis and other enabling processes and technologies. "CBM+ uses a systems engineering approach to collect data, enable analysis, and support the decision-making processes for system acquisition, sustainment, and operations" [5].

The CBM+ implementation strategy involves an operating concept in which systems and assets are equipped with sensors and embedded health management systems. These health management systems monitor the current health of the asset, predict future changes in the asset's health, and report status and problems for preventive and corrective maintenance actions. The embedded health management system uses information from the sensors and software integrated into the asset to measure and collect data, and store data to provide a detailed operating and maintenance history of the asset. The embedded health management system also uses automatic identification technologies on critical components and assemblies to sustain system configuration control over all hardware or software changes for any particular asset. The CBM+ implementation strategy includes communications for data transfers utilizing data networks and wireless connections to enterprise-level CBM+ data warehouses for remote data storage and data analysis. Figures 9.7 and 9.8 provide a CBM+ architecture overview, and CBM+ and the total system life cycle, respectively.

9.7 Digital Prescriptive Maintenance

As stated in Chapter 8, Machine Learning (ML) supports Design for Maintainability activities that facilitate PM and Digital Prescriptive Maintenance (DPM) along with PHM and CBM. One value-added benefit of ML, PM, and DPM when used with PHM and CBM is the potential to reduce or eliminate corrective maintenance.

ML with Prescriptive Maintenance and/or DPM leverages software for pattern recognition to allow automation to identify specific stress points and to explicitly diagnose root cause issues. Prescriptive maintenance with digitized decisions is intelligent, timely, and responsive when used with Internet of Things (IoT) technology. This IoT technology has been applied effectively in multiple industries, including automotive, energy, agriculture, mining, healthcare, defense and aerospace, and retail consumer products, such as home appliances, lighting, thermostats, and consumer wearables.

DPM allows manpower tasks related to preventive and predictive maintenance to become digital automated tasks. Maintenance digitization becomes prescriptive through Dynamic Case Management or Adaptive Case Management. DPM Dynamic Case Management is analogous to digitizing the logic from Failure Modes, Effects, and Criticality Analyses (FMECAs), Fault Tree Analyses (FTAs), and Probability Risk Assessments (PRAs) by embedding the functionality and logic into onboard processors for autonomous real-time decision-making. Some businesses are beginning to use ML and DPM programs leveraging existing predictive analytical software

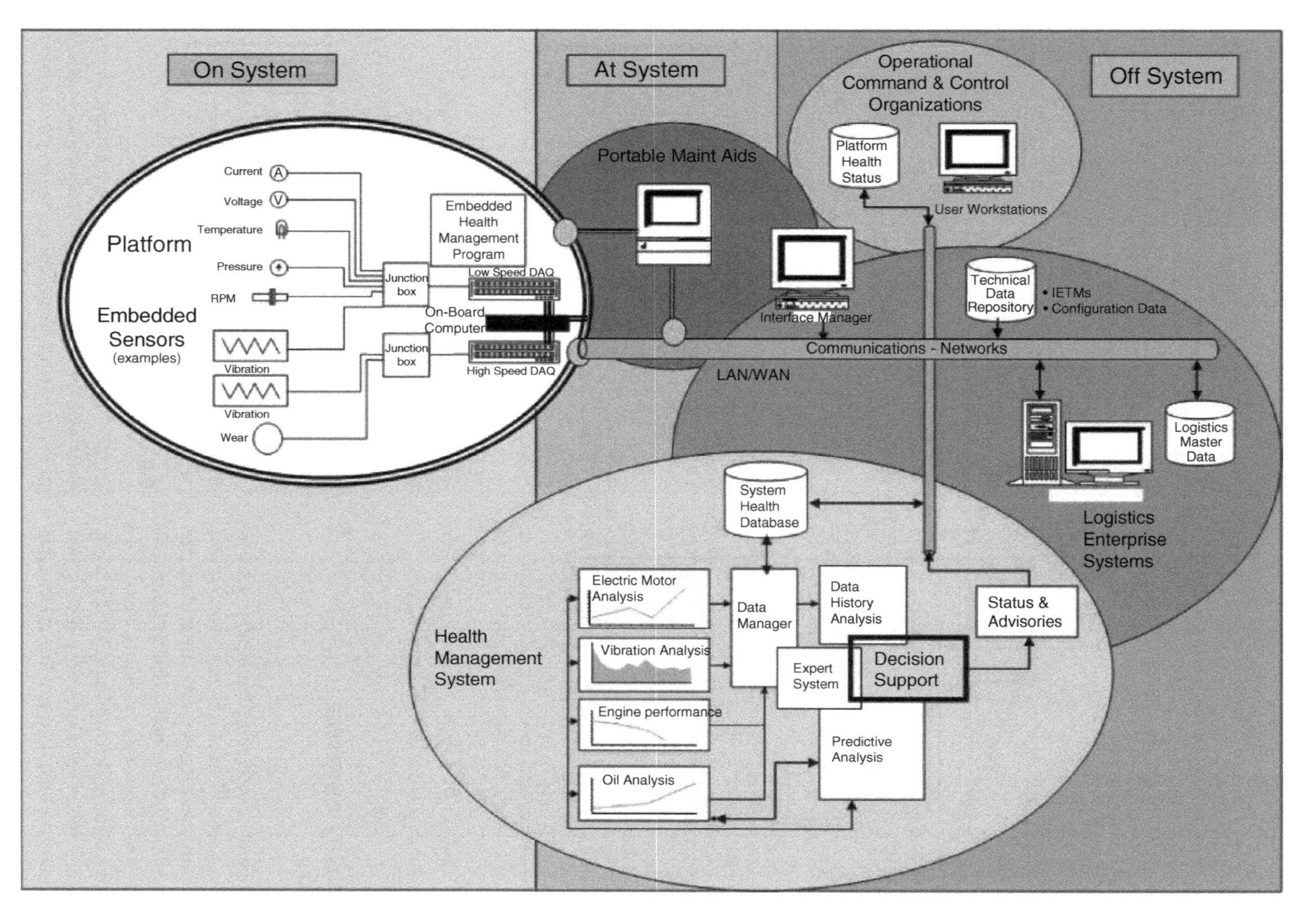

Figure 9.7 CBM+ architecture overview [5]. Source: US Department of Defense, *Condition Based Maintenance Plus DoD Guidebook*, Department of Defense, Washington, DC, 2008.

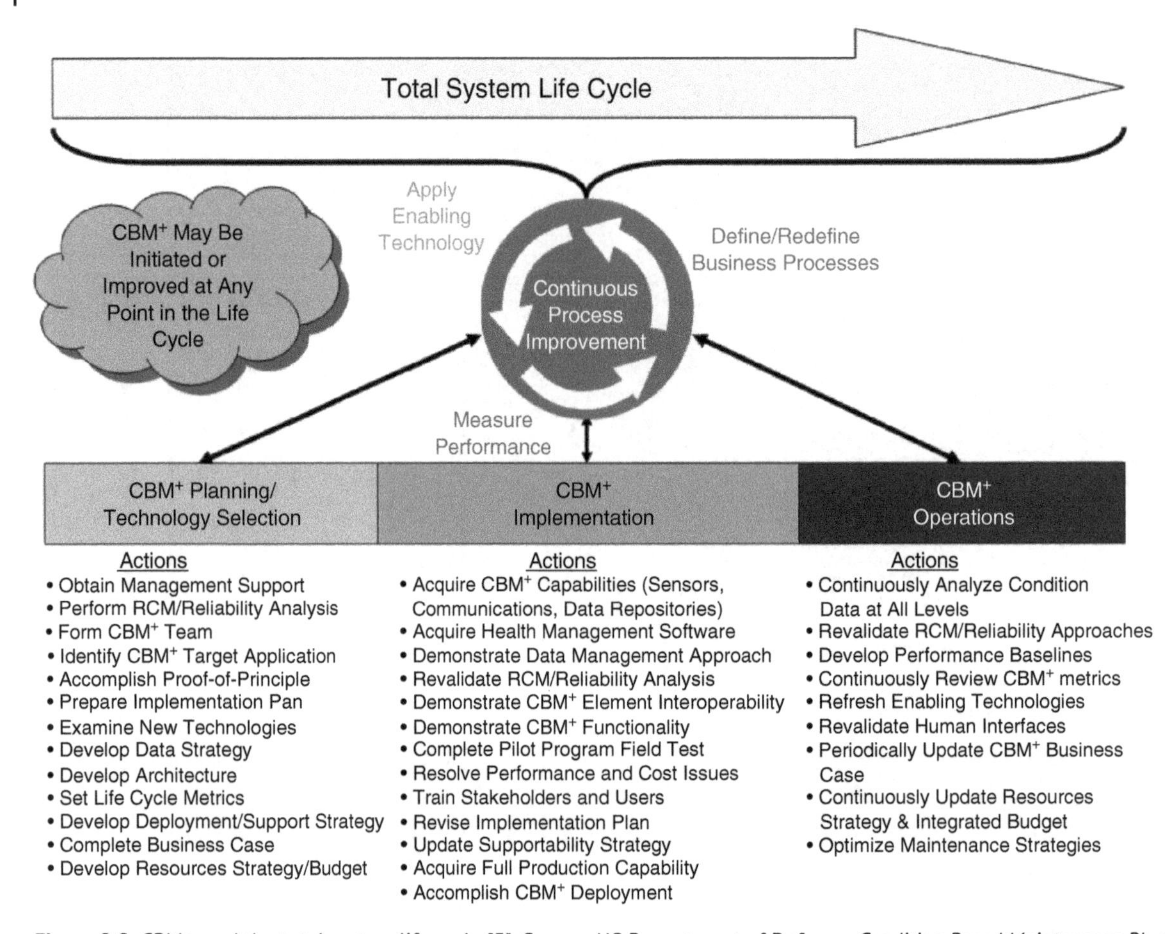

Figure 9.8 CBM+ and the total system life cycle [5]. Source: US Department of Defense, *Condition Based Maintenance Plus DoD Guidebook*, Department of Defense, Washington, DC, 2008.

to reduce or eliminate unplanned downtime and maximize profitability and equipment reliability.

9.8 Reliability-centered Maintenance

RCM is a structured process for developing plans to ensure that an asset will fulfill the needs of the end user in its present operating context. These plans are not just PM routines, but can include the development of procedures, or redesign of the systems, for example. RCM is a logic-based approach to ensure that the major consequences of a failure are addressed and mitigated in a cost-effective way that does not involve massive amounts of planned downtime. This has required a shift away from the traditional time-based overhauls and time-based replacement activities which have proved to be ineffective in most situations.

RCM has proven itself as a robust methodology to manage risks associated with the operation of assets, whether they are aircraft, nuclear submarines, nuclear reactors, or production equipment. Ideally, RCM is used initially during the design phase to ensure changes can be made easily, reducing the life cycle costs of the asset. But where did RCM come from?

9.8.1 History of RCM

In the 1960s the failure rate of jet aircraft was high even with the extensive maintenance programs that were put in place to prevent the failures. The programs required overhauls, rebuilds, and detailed inspections which required the various components to be disassembled, inspected, and rebuilt. All of these activities were based on an estimated safe life of the equipment, yet no matter how many of the activities were in place, the failure rate remained high. For example, the DC-8 had 339 components subjected to TBM activities [3].

In order to address this problem, the Federal Aviation Administration (FAA) ordered that the airlines investigate and determine the causes of the high failure rates being experienced even when extensive maintenance programs were in place. Under the guidance of the FAA, extensive engineering studies were conducted on all of the aircraft in service to determine the source of failures. United Airlines pioneered and published a report on the failures which turned the industry on its head. They concluded that only 11% of the failures were related to the age of the aircraft [3]. The remaining failures were random in nature or induced by the very maintenance work that was put in place to prevent them.

The output of the report from Stanley Nowlan and Harold Heap of United Airlines [9] was a technique to define maintenance activities titled Maintenance Steering Group 1 (MSG1), and was subsequently approved by the FAA for use in developing aircraft maintenance activities. As a result of these findings, many of the extensive maintenance programs were reduced, and the reliability of the aircraft went up! The next generation of McDonnel-Douglas aircraft, the DC-10, had the TBM activities reduced to just seven components. That is an astounding reduction of 98% [3].

In 1978, Nowlan and Heap published the research [9] on the use of MSG1 for the civil aviation industry, and released the first draft of RCM under contract from the Department of Defense. In 1980, MSG3 was published for use in designing and refining maintenance programs for all major types of civil aircraft. This methodology was further refined by the US Navy and its SSN submarine class in 1981 [10], while the nuclear power industry pilots the RCM at two sites under the guidance of the Electrical Power Research Institute (EPRI).

The late John Moubray published RCM2 in 1997 [3] as a methodology to apply RCM to the manufacturing industry. This book was quickly followed up in 1999 when the Society of Automotive Engineers (SAE) published SAE-JA-1011 [11] to standardize the use of RCM in the automotive industry.

9.8.2 What is RCM?

As seen above, RCM has proven itself through the test of time, and can deliver significant benefits to asset designers, owners, and operators through reduced maintenance costs, risk, and improved asset performance. But what is it that separates RCM from other methodologies? RCM is based on a set of principles that guide the entire process [3]:

- **Principle 1**: The primary objective of RCM is to preserve system function. This is often one of the most important but difficult principles for organizations to grasp. We may not care so much about the failure occurring, but ensuring that the critical functions will operate when a failure occurs.
- **Principle 2**: Identify failure modes that can defeat the functions. Since the objective of RCM is to preserve system function, the loss of the functions must be carefully considered. While the loss of functions is often summed up as too much, not enough, or not at all, it may be much more complicated than that. Often the loss of function, also known as a functional failure, should be based on specific performance considerations.
- **Principle 3**: Prioritize function needs. Not all functions of the asset or system are as important as others, therefore it is vital to prioritize system functions and failure modes in a systematic way. Consider, for example, an automobile, which has a function of being able to stop, and a function of passenger comfort. Obviously, the function of the braking system should be viewed as a higher priority than that of the air conditioning systems.
- **Principle 4**: Select appropriate and effective tasks that do one or more of the following:
  - Prevent or mitigate the failure
  - Detect onset of a failure
  - Discover a hidden failure

These tasks may be TBM, CBM, redesign of the asset, or even deciding to run-to-failure, if the consequences of the failure are not significant enough to warrant the maintenance tasks.

Without these principles, RCM would lead to extensive maintenance programs that do not necessarily add value to the organizations. But performing a RCM analysis is a resource-intensive activity, so outside of aviation and nuclear industries, why should an organization choose to utilize RCM?

9.8.3 Why RCM?

RCM is resource-intensive, but the results can provide organizations with numerous benefits. RCM is truly a system-driven approach to develop appropriate plans and strategies to ensure that the asset continues to function as defined. RCM also drives a culture change, in that both management and the maintainers are now required to do maintenance by number. This change requires organizational discipline, in the development of the PMs, the work management processes to ensure the work is completed, and the feedback loop to improve the maintenance program. But one of the most beneficial, yet often not considered, changes that organizations see from RCM is a change in technical knowledge of the asset from a maintainer's perspective and an operator's perspective. In addition, RCM provides very tangible improvements to organization in terms of:

- **Reliability**: By using the RCM methodology in the design phase, the inherent reliability can be improved. RCM can ensure that the equipment operates close to the inherent reliability levels. In addition, RCM creates a database of failure modes that can be shared with manufacturers and OEM for the next generation of assets.
- **Safety**: RCM prioritizes safety first. The preservation of human life and environmental risks are prioritized above all else. This thought process ensures that if there is a safety risk, it will be addressed and the risk reduced to a low enough level for the organizations.
- **Cost**: RCM prioritizes costs second, after safety. Using the RCM methodology, failure modes are prioritized based on the impact to the system function. From here, decisions are made based on cost effectiveness of any mitigation actions, against the cost of the consequence. Here, ineffective actions can be identified and not performed. In addition, since RCM considers the operating context, only actions required in the particular operating context are considered for implementation.
- **Supportability**: By understanding the PM requirements and the corrective actions that will take place with the asset, organizations can plan for and prepare for those activities. This may require organizations to gather the appropriate documentation in terms of drawings, manuals, procedures, checklists, and so on. In addition, spare parts needs can be estimated along with any specialized equipment or skills required to perform the tasks. With this understanding, organizations can decide which activities they will perform at the site or at a depot, and if a contractor or internal staff will perform the tasks.

So why not just use the OEM recommendation, and save the trouble of going through an RCM analysis? Well, most OEM recommendations (outside of aviation, defence, etc.) come out of past experience and are not specific to the operating context of the asset. They are more designed to ensure the asset makes it out of the warranty period in the most extreme operating context in which their clients operate. Knowing the benefits of RCM over other traditional methods, we can explore what the initial study from Nowlan and Heap [9] taught us, which was that failures are not predominately based on age, but rather on a variety of other factors which create random failure patterns. These must be addressed through CBM and not TBM.

9.8.4 What we Learned from RCM

The initial study from Nowlan and Heap [9] provided insights into how equipment fails, which previously up to that point in time, was thought to be driven mostly by time. This finding was one of the major drivers in the reduction of TBM activities, which made up most maintenance programs at that time. What Nowlan and Heap discovered was, in fact, that equipment fails in a variety of different ways.

In addition, Nowlan and Heap designed a concept to understand at what intervals TBM and CBM should be performed. This concept is known as the P-F curve, which is a visual representation of how equipment degrades once a failure has started. The P-F curve guides our decision-making on what can be done to

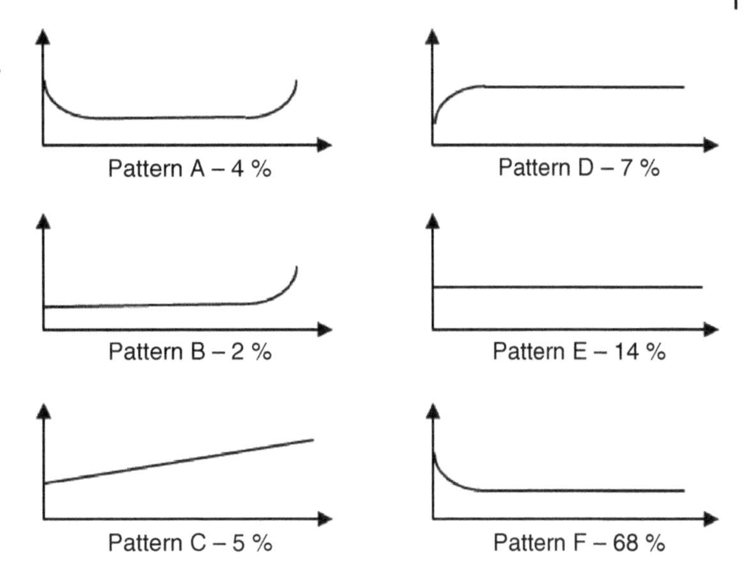

Figure 9.9 The six failure curves [3]. Source: Moubray, J., *RCM II* 2nd Edition, Industrial Press, Inc., New York, NY, 1997.

minimize the consequence of the failure, based on where a failure is in its degradation.

9.8.4.1 Failure Curves

The report from United Airlines highlighted six unique failure patterns (see Figure 9.9) [3]. Understanding these patterns illustrates why the reduction in maintenance could result in improved performance. The failure curves for Patterns A–F are described and shown as follows:

- Failure Pattern A is known as the bathtub curve and shows a high probability of failure when the equipment is new, followed by a low level of random failures, and followed by a sharp increase in failures at the end of its life. This pattern accounts for approximately 4% of failures.
- Failure Pattern B is known as the wear-out curve and consists of a low level of random failures, followed by a sharp increase in failures at the end of its life. The pattern accounts for approximately 2% of failures.
- Failure Pattern C is known as the fatigue curve and is characterized by a gradually increasing level of failures over the course of the equipment's life. This pattern accounts for approximately 5% of failures.
- Failure Pattern D is known as the initial break-in curve and starts off with a very low level of failure followed by a sharp rise to a constant level. This pattern accounts for approximately 7% of failures.
- Failure Pattern E is known as the random pattern and is a consistent level of random failures over the life of the equipment with no pronounced increases or decreased related to the age of the equipment. This pattern accounts for approximately 11% of failures.
- Failure Pattern F is known as the infant mortality curve and shows a high initial failure rate followed by a random level of failures. This pattern accounts for 68% of failures.

When looking at the failure patterns, the first three (A, B, and C) can be grouped together as the equipment having a defined life, in which the failure rates increase either with the age of the component, or once the component has reached a certain age. This age may be time or usage such as hours, widgets produced, and so on. The failures are usually related to wear, erosion, or corrosion, and are often caused by simple components which come into contact with the product. The total of these time-based failures only account for 11% of all failures.

The other patterns highlight the fact that during the initial start-up of the equipment is when the majority of failure will occur. This could be due to maintenance-induced failures or manufacturing defects in the components. Once the initial start-up period has passed, the failures are random. These patterns account for 89% of failures.

Now, these patterns show that the failures are random in nature, but that does not mean that the failures cannot be predicted or mitigated. It means that overhauls and replacements conducted at a specific frequency are only effective in 11% of the cases.

In the rest of the failures, the equipment can be monitored and the right time to conduct a repair, replacement, or overhaul can be identified based on the condition of the equipment. This is known as CBM, or Don't Fix It Unless It's Broke!

Looking at the different failure patterns, we can group the types of failures into three unique groups:

- Age-related failures – The term "life" is used to describe the point at which there is a rapid increase in the likelihood of failure. This is the point on the failure pattern before it curves up. Typically, these types of failures can be attributed to wear, erosion, or corrosion and involve simple components that are in contact with the product.
- Random failures – The term "life" cannot be used to describe the point of rapid increase in the likelihood of failure, as there is no specific point. These are the flat parts of the failure curve. These types of failures occur due to some introduced defect.
- Infant mortality – The term "life" cannot be used here either. Instead, there is a distinct point at which the likelihood of failure drops dramatically and transitions to a random level.

By understanding these unique differences, an effective maintenance strategy can be developed, which is what RCM is based on. In order to address the different types of failure patterns, different tactics can be taken which can reduce the probability of the failure occurring, or in some instances, even occurring at all. Managing the specific groups of failures can be accomplished using different techniques:

- Age-related – These types of failures can be addressed through TBM. TBM includes replacements, overhauls, and basic cleaning and lubrication. While cleaning and lubrication will not prevent the wear-out or corrosion, it can extend the "life" of the equipment.
- Random – These types of failures need to be detected using CBM as they are not predictable or based on a defined "life." The equipment must be monitored for specific indicators. These indicators may be changes in vibration, temperature, flow rates, and so on. These types of failures must be monitored using predictive or condition monitoring equipment. Cleaning and basic lubrication can prevent the defects from occurring in the first place if done properly. In addition, by using installation and commissioning standards, random failure can be prevented altogether by ensuring components are installed using precision maintenance techniques, thus ensuring some failure modes may not be introduced.
- Infant mortality – These types of failures cannot necessarily be addressed through fixed time, predictive, or CBM programs. Instead, the failures must be prevented through proper design and installation, repeatable work procedures, proper specifications, and quality assurance of parts.

Only when a maintenance program encompasses all of the above activities can asset performance improve.

But even with an extensive study on how equipment fails, many were non-believers, and did not believe that the majority of failures were not time-based. Numerous studies over the years have sought to prove or disprove the distribution amongst the failure curves, and these can be seen summarized in Table 9.1 [12].

Table 9.1 The six failure curves.

Failure curve	United Airlines Study 1968 [9]	Broberg 1973 [12]	MSP 1982 [12]	SUBMEPP 2001 [12][a]
A	4%	3%	3%	2%
B	2%	1%	17%	10%
C	5%	4%	3%	17%
Time-related total	*11%*	*8%*	*23%*	*29%*
D	7%	11%	6%	9%
E	14%	15%	42%	56%
F	68%	66%	29%	6%
Random-related total	*89%*	*92%*	*77%*	*71%*

Source: Allen, T. M., *US Navy Analysis of Submarine Maintenance Data and the Development of Age and Reliability Profiles*, Department of the Navy, Portsmouth 2001.

In order to understand the differences between the various studies, it is first vital to understand the context of each of them. We know that the initial study in 1968 was on civil aircraft, as was the 1973 Broberg study, which explains the similar results. The 1982 MSP is a US Navy study, while SUBMEPP was a study on US submarines. The dramatic increase in time-related failures in the MSP and the SUBMEPP is attributed to the corrosive environment within which the assets operate. In addition, the reduce of infant mortality in the MSP and SUBMEPP is minimized due to the extensive testing that takes place before the asset is put into service.

As the studies indicate, even in the corrosive environment, the majority of failures are still random in nature, and cannot be addressed through traditional TBM. Therefore, the principles learned from the initial Nowlan and Heap study still hold true. TBM is not the most effective approach to maintain assets, and can lead to decreased performance.

9.8.5 Applying RCM in Your Organization

Conducting an RCM analysis and implementing the findings is not a simple task. Many considerations are needed to ensure that the RCM analysis will be successful, as many RCM analyses end up on the shelf with little to no implementation. This results in a zero return on the analysis and leaves a bad taste in the mouth of those who participated. In addition, a poorly facilitated RCM analysis will result in having to drag people to the analysis, and not a lot of buy-in for the methodology. To ensure that the RCM analysis generates the expected results, organizations need to consider the following prior to the start of any RCM programs:

- **Facilitator**: The facilitator will have a tremendous impact on the success of the analysis. Experienced facilitators will keep the team moving, keep the analysis at the right level of detail (i.e. not too detailed, but not high-level), and ensure that when additional investigation is required, that it is performed outside of the analysis meeting. Facilitators are ideally not Subject Matter Experts (SMEs) on the product or system being analyzed, but they should be great at asking the right questions and keeping the team motivated.
- **Analysis team**: The analysis must be a team activity. The RCM analysis cannot be performed by a single person. The analysis team needs to be a cross-functional group with varying expertise and backgrounds. A typical group should be composed of 6–8 people and should include operations, maintenance, engineering, and various system experts.
- **Which RCM approach**: Determining which approach to follow can be dictated by industry, as civil aircraft require the use of MSG3 [13], while in automotive industries the SAE JA 1011 [11] standard will likely be used. In addition to those two approaches, one may choose to use the RCM2 [3] approach.
- **Training**: Prior to the start of any RCM analysis, the team members need to be trained in the particular variant of RCM to be used, including its principles and methodology. The terminology in RCM can be confusing for those not familiar with it, so training must be provided beforehand.
- **Meeting cadence and structure**: What will be the frequency of the analysis meetings, and how long will the meetings be? These are questions that will drive the length of time to complete the analysis and the level of engagement of the analysis team. Often, shorter, more frequent analysis meetings are better than fewer, extended sessions.

Once these considerations have been resolved, the actual RCM analysis may begin.

9.8.5.1 Inner Workings of RCM

RCM analysis seeks to answer a variety of questions to ensure a complete understanding of the asset and its function. The RCM analysis seeks to ensure an understanding of how the asset may fail, the consequences of failure, and what can be done to mitigate the consequences. As discussed previously, one of the major intangible benefits of RCM is a better understanding of the asset by all. In RCM2 [3], John Moubray defined the seven questions that RCM seeks to answer as:

1. What are the functions and associated performance standards of the asset in its present operating context?
2. In what ways does it fail to fulfill its functions?
3. What causes each functional failure?

4. What happens when each failure occurs?
5. In what way does each failure matter?
6. What can be done to predict or prevent each failure?
7. What should be done if a suitable proactive task cannot be found?

Imagine the power of truly understanding the answers to all of those questions for each asset in its unique operating context. However, one cannot simply answer these questions to arrive at a complete analysis. The RCM analysis is a methodology; therefore, it is a process with a series of steps. According to Ramesh Gulati, in order to complete a RCM analysis, the following should take place [10]:

1. **System selection and information collection**: The asset or system should be selected for analysis based on the criticality of the system, or whether the asset has experienced poor reliability in the past (a bad actor). In addition, new asset designs are often selected for analysis. It is also during this phase that the analysis team be selected, so that they may participate in the initial data collection. The data collected in this phase should include performance requirements of the asset, schematics, manuals, drawings, history, and maybe even photographs.
2. **System boundary definition**: The definition of the system boundary is critical for ensuring that the scope of the RCM analysis does not expand and result in a never-ending analysis. The scope of the analysis should include what are inputs to the systems, and what are outputs. For example, on a pump, 480 V 3-phase power is an input, therefore we would not go into the details of why power many not be available to the system. An output may be 100 gpm of water with a 25′ head. The boundary definition also includes defining the functional subsystems as well, to ensure there are no overlaps within the analysis.
3. **System description and functional block diagram**: The system description ensures there is a thorough understanding of the asset in its present operating context. This may include identifying whether the asset has any redundancy in the design, or work-arounds available, protection features such as pressure relief valves or alarms, and any key control features such as automatic or manual controls. The next step is to establish a Function Block Diagram (FBD), which is a top-level representation of the major subsystems and their interactions. Often components for each function block are then defined for the analysis.
4. **System functions and functional failures**: The functions of the assets need to be defined in a quantitative way. For example, a pump should have a primary function of "to pump water at a rate not less than 100 gpm, with a head of 25′." These functions can often be defined by looking at the FBD and identifying the outputs. There is also a need to define the secondary functions which consider other functions not completely associated with the primary function. These include protective, appearance, environmental, control, health and safety, efficiency, and superfluous functions. These may or may not be quantitative, but continuing with our pump example, an environmental function may

Table 9.2 Failure Modes and Effects Analysis example.

Function	Functional failure	Failure mode	Failure effect
Deliver 100 gpm of water at 25′ of head	Delivering less than 100 gpm	Impeller worn due to cavitation	Reduced output of pump, slow down of process. Two hours to repair.
		Seal leaking due to improper installation	Reduced output of pump, slow down of process. One hour to repair.
	Not delivering any water	Bearing seized due to lack of lubrication	Pump failed, process stopped, three hours to repair.
		Impeller jammed due to foreign material	Pump failed, process stopped, two hours to repair.

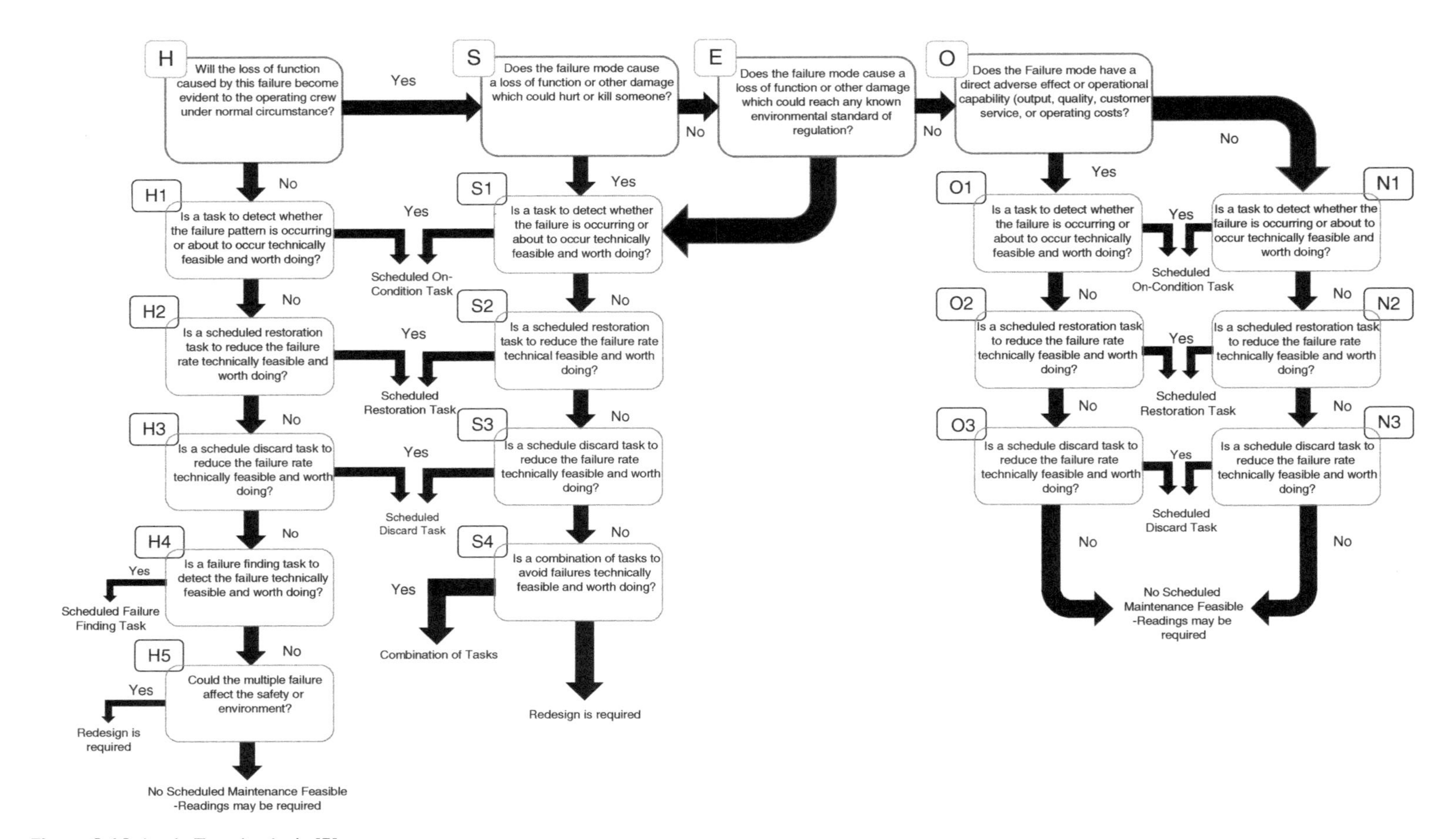

Figure 9.10 Logic Tree Analysis [3].

be not to leak. Once the primary and secondary functions have been defined, then the functional failures can be documented for each function. These may be total losses of the functions, partial losses, or excess of the defined function.

5. **Failure Modes and Effects Analysis (FMEA)**: The FMEA (see Table 9.2) is a common tool in reliability used to identify the failure modes. The FMEA defines what would cause each functional failure, as well as the effect that the individual failure mode will have on the system. For our pump, a function failure of the pump delivering <100 gpm may be an impeller eroded due to cavitation, which may have an impact on the entire system, and other dependent systems. It is important to note that the impeller could experience erosion for other causes, and those should be captured as well. If the system has redundancy, then that should be considered in the effects. The repair time is often included in the effect. These failure modes are related to human performance, management systems, and so on, and they should be captured for mitigation through redesign and procedure development. It is important to note that there are many different formats of FMEAs available, so be sure to choose one that is right for your organization.
6. **Logic Tree Analysis**: Logic Tree Analysis (LTA) (see Figure 9.10) [3] is designed to ensure that each failure mode is evaluated based on the consequences of the failure, and then actions implemented to either prevent the failure, or mitigate the consequences. LTA considers five types of consequences:

- **Hidden**: A failure mode that is not evident to the operating crew under normal circumstances.
- **Safety**: A failure mode that will result in an injury.
- **Environment**: A failure mode that will result in a breach of environmental regulations or standards.
- **Operational**: A failure mode that will affect the operational performance of the asset.
- **Non-operational**: A failure mode that does not impact safety, environmental, or operations, but results in the cost of repair only.

Using the Logic Tree Analysis, each failure mode is evaluated for its consequence, and the appropriate task arrived at to mitigate the risk associated with each.

7. **Selection of maintenance tasks**: Based on the outcome of the Logic Tree Analysis, the specific tasks need to be determined. For example, if there is an on-condition task defined, will the task be using vibration analysis, thermography, or similar? In addition, the frequency of the task needs to be defined. If RTF is the option, spare parts may need to be stocked. All of these outcomes need to be reviewed by the team so that a specific task or action for each failure mode is clearly defined.
8. **Task packaging and implementation**: Task packaging is the assembly of the maintenance tasks into job plans for entry into the Computerized Maintenance Management System (CMMS). This often involves the development of job plans. For non-maintenance tasks, redesign requests need to be submitted to engineers, spare parts set up in the storeroom, and so on. Often, this step is the most critical, as this is where the true benefits of the analysis will be realized.
9. **Continuous improvement:** A final review of the RCM analysis should be reviewed to ensure that any and all regulatory requirements have been met. In addition, CMMS should have the appropriate failure codes created to foster a feedback loop, which should drive an improvement in the RCM analysis. This loop allows failure data to be captured, compared with the RCM analysis, and improvements made. The RCM analysis should be a living program, and continuously updated.

RCM is a powerful process, which reduces risk, and drives improved performance of the asset, and cost-effectiveness. One of the key drivers of these benefits is the shift from TBM to CBM.

9.9 Conclusion

While CBM enables organizations to reduce the need for planned downtime, and improve operational availability, the organization must have the mechanisms in place to correct the defects found in a timely manner. Determining the right condition monitoring tasks requires a well thought-out approach like RCM, and can drastically reduce the staffing requirements to maintain the equipment. In addition, by considering the need for CBM in the design phase, asset designers can incorporate data collection devices into the asset, which further reduce the need for the hard-to-find maintenance technician.

Although RCM can enable organizations to reduce the need for planned downtime, it is vital that the organizations work to strike the appropriate balance between the need for proactive maintenance and the reliability of the asset. Without this balance, availability will fall, and the organization will not have the assets required to fulfill its mission. Also, with the addition of Artificial Intelligence and Machine Learning, maintenance can be moved past CBM to Prognostics and CBM+, further reducing planned and unplanned downtime.

References

1 Giffi, C., Wellener, P., Dollar, B. et al. (2018). *2018 Deloitte Skills Gap and Future of Work in Manufacturing Study*. London, UK: Deloitte Insights.
2 Vitasek, K. (2012). The Rolls-Royce of Effective Performance-based Collaboration. Efficient Plant. https://www.efficientplantmag.com/2012/06/the-rolls-royce-of-effective-performance-based-collaboration/ (accessed 20 September 2020).
3 Moubray, J. (1997). *RCM II*, 2e. New York, NY: Industrial Press, Inc.
4 Wortman, B. and Dovich, R. (2009). *CRE Primer*. Terre Haute, IN: Quality Council of Indiana.
5 US Department of Defense (2008). *Condition Based Maintenance Plus DoD Guidebook*. Washington, DC: Department of Defense.
6 Plucknette, D. (n.d.). *Completing the curve*. ReliabilityWeb.com. https://reliabilityweb.com/articles/entry/completing-the-curve (accessed 24 August 2020).
7 Goodman, D., Hofmeister, J.P., and Szidarovszky, F. (2019). *Prognostics and Health Management: Practical Approach to Improving System Reliability Using Condition-Based Data*, 1e. Hoboken, NJ: Wiley.
8 Institute of Electrical and Electronic Engineers (2017). *IEEE 1856-2017 – IEEE Standard Framework for Prognostics and Health Management of Electronic Systems*. Piscataway, NJ: Institute of Electrical and Electronic Engineers (IEEE) Standards Association (SA).
9 Nowlan, S. and Heap, H. (1978). *Reliability Centered Maintenance*. Washington, DC: Office of Assistant Secretary of Defense.
10 Gulati, R. (2019). *Maintenance & Reliability Best Practices*. New York, NY: Industrial Press Inc.
11 Society of Automotive Engineers (2009). *Evaluation Criteria for Reliability-Centered Maintenance (RCM) Processes JA1011_200908*. Warrendale, PA: Society of Automotive Engineers.
12 Allen, T.M. (2001). *US Navy Analysis of Submarine Maintenance Data and the Development of Age and Reliability Profiles*. Portsmouth: Department of the Navy.
13 Airlines for America (2018). *MSG-3: Operator/Manufacturer Scheduled Maintenance Development*, Fixed Wing Aircraft, vol. **1**. Washington, DC: Airlines for America.

Suggestion for Additional Reading

Mobley, K. (2002). *An Introduction to Predictive Maintenance*, 2e. Waltham, MA: Butterworth-Heinemann.

Nowlan, Stanley & Heap, Howard, *Reliability Centered Maintenance*, Office of Assistant Secretary of Defense, Washington, D.C., 1978.

Society of Automotive Engineers, *Evaluation Criteria for Reliability-Centered Maintenance (RCM) Processes JA1011_200908*, Society of Automotive Engineers, Warrendale, PA, 2009.

Airlines for America, *MSG-3: Operator/ Manufacturer Scheduled Maintenance Development,*

VOLUME 1 – FIXED WING AIRCRAFT, Airlines for America, Washington, DC, 2018.

Mobley, Keith, An Introduction to Predictive Maintenance, Second Edition, Butterworth-Heinemann, Waltham, Massachusetts, 2002.

Gulati, Ramesh, Maintenance & *Reliability Best Practices*, Industrial Press, Inc New York, N.Y., 2019.

Wortman, Bill, & Dovich, Robert. *CRE Primer*, Quality Council of Indiana, Terre Haute, IN 2009.

10

Safety and Human Factors Considerations in Maintainable Design

Jack Dixon

10.1 Introduction

The proactive incorporation of safety and human factors considerations in the design of products, equipment, and systems is important to ensure that they are maintainable by the user. Total system performance effectiveness recognizes that system performance is a function of both the equipment and the user. For a system or product to be at its most effective, it must be both safe to use and easy to use. Therefore, it is imperative that the human user be considered early and throughout the entire design and development process.

This chapter is divided into two sections: one section for safety considerations and a second section for human factors considerations. While these topics are treated separately, it is important for the reader to understand that they are also intimately related both to each other and to successful maintainable design.

10.2 Safety in Maintainable Design

Safety must be built into products and systems not only to make them safe to operate and use, but also to ensure that they are safe to maintain. It is always best practice to build safety in rather than requiring the maintainer to divert attention away from the maintenance task at hand in order to observe some ill-defined safety precaution that has been stuck on as an afterthought. In addition to ensuring the safety of the maintenance personnel, the designer must also work to ensure that the actions taken by the maintainers protect the safety of others and the equipment being maintained.

An example of how maintenance can affect the safety of personnel is illustrated by this excerpt from the *EHS Today*. "A gas-fired paint drying oven at a manufacturing plant shut down unexpectedly, and a crew was assigned to relight it. The crew was under pressure to troubleshoot and re-light the oven as soon as possible to minimize the unscheduled downtime. After a few attempts to light the pilot, gas had built up in the oven. One final attempt to light the pilot sparked an explosion in the oven that caused the death of a crew member. The fatal incident … was blamed partly on maintenance. Some critical parts of the oven were out of adjustment. Valves were out of alignment, not calibrated or nonfunctional. Important safety-related settings were incorrect. Further, similar equipment in the plant had been the center of some previous troubles that, if seen as warning signs, could have led to a detailed review of safety systems and devices. Tragically, the oven could likely have been much safer simply if a well-designed maintenance program had been in place" [1].

When designing a product or system, the designer must consider the safety of the maintainer and human behavior as design features being considered for incorporation. This aspect of good maintenance design is covered later in the human factors section of this chapter.

Design for Maintainability, First Edition. Edited by Louis J. Gullo and Jack Dixon.

10.2.1 Safety and its Relationship to Maintainability

Lack of maintenance or inadequate maintenance can lead to dangerous situations, accidents, or health problems. These conditions may be related to lack of, or poor maintenance of, vehicles, equipment, industrial machines, facilities, aircraft, military assets, or any other complex systems. Maintenance failures may contribute to large-scale disasters with damaging consequences for humans and the environment.

Maintenance can be risky. Maintenance-related accidents happen in all industries and in all countries. A sampling of accident statistics illustrates how risky maintenance is and how ubiquitous maintenance-related accidents are:

- The UK Health and Safety Executive reports that "Maintenance-related accidents are a serious cause of concern. For example, analysis of data from recent years indicates that 25–30% of manufacturing industry fatalities in Great Britain were related to maintenance activity" [2].
- The Joseph T. Nall Report, a comprehensive study on general aviation accidents, found that maintenance-related crashes represented 15% of all accidents for 2010 [3].
- "The present research determined that for the most recent 25-year period, accidents in which a maintenance deficiency was causal for, or contributory to, constituted 4.8% of all accidents in single engine piston aircraft operating under 14CFR Part 91 regulations" [4].
- "Between 2005 and 2015, flawed maintenance and inspection were causal factors in 14% to 21% of helicopter accidents in the U.S. civil fleet" [5].
- "…studies indicate specific safety challenges with stationary and mobile machinery – severe injuries involving these machines account for more than 40% of all severe accidents at mining operations in the United States. Most severe accidents are associated with the operation or maintenance of the machines" [6].
- "The Coast Guard has responded to an average of nearly one marine accident per day since the start of the commercial Dungeness crab season. Since the season started Jan. 15, the agency has responded to 28 accidents, … 'Most reported incidents are a result of equipment failures on vessels which are not ready for operation, poor maintenance and negligent operations while underway,' said Lt. Michael Tappan, chief of investigations division with the Marine Safety Unit Portland. 'These accidents endanger the crews onboard each of these commercial fishing vessels, other nearby vessels, and Coast Guard search and rescue personnel'" [7].
- "The data show that around 20% of all accidents in Belgium (in 2005–2006) were related to maintenance operations, as well as around 18–19% in Finland, 14–17% in Spain, and 10–14% in Italy (in 2003–2006). In addition, figures from a number of European countries indicate that around 10–20% of all fatal accidents in 2006 were related to maintenance operations" [8].
- "Improper maintenance contributes to approximately 1/3 of all infrastructure-related incidents and accidents linked to collisions and derailments on the Swedish railways" [9].

During maintenance activities, safety is important not only from the standpoint of protection of maintenance personnel, but also in terms of protecting the equipment being maintained. Improperly performing a maintenance task could result in damage to the particular piece of equipment being maintained, or it could result in an unsafe condition that could be the cause an accident during later equipment use. The improper maintenance could also induce failures in other items within a larger, complex system or possibly cause a failure of the entire system. This failure could result in an unsafe condition, or worse, an accident.

10.2.2 Safety Design Criteria

Over the years, various guidelines have been developed that product and system designers can use to improve safety in maintenance. Some general guidelines include:

- Develop designs and maintenance procedures such that maintenance errors are minimized
- Design for easy accessibility by ensuring that parts requiring maintenance are easy and safe to inspect, repair, and replace
- Incorporate effective fail-safe designs to prevent damage or injury in the event of a failure

- Eliminate or reduce the need to perform maintenance close to moving parts
- Incorporate early detection or prediction of failures so maintenance can be performed prior to actual failure, thus reducing risk
- Develop designs that minimize the probability that maintenance workers will be injured by electric shock, fire, radiation, and so on
- Minimize hazardous materials
- Consider human behavior when designing products
- Eliminate or reduce the need for special tools

Maintenance activities may expose the maintainer to many unique hazards. The design of equipment for ease and safety of maintenance must embody features for the protection of personnel from numerous hazards. Some of the types of hazards that must be considered include:

- Electrical hazards
- Mechanical hazards
  - Hot surfaces
  - Instability
- Fire
- Toxic and hazardous materials
- High noise levels

Design criteria to address these hazards should be considered during the design of products and systems. The following paragraphs provide a sampling of some basic design criteria designers can use to eliminate or control these hazards to an acceptable level of risk.

Electrical hazards present several risks during maintenance. The most obvious hazard is that of electric shock to personnel. There is also a risk of equipment damage and physical harm to personnel resulting from involuntary reactions to an electric shock, even if the shock is not sufficient by itself to cause injury or death.

The effects of electric shock depend upon the body's resistance, the current path through the body, the duration of the shock, the amount of current and voltage, the frequency of an alternating current, and the individual's physical condition. The most critical determinant of injuries is the amount of current conducted through the body. Besides the obvious risk of burns and injuries to the nervous system, electric shock can produce involuntary muscular reactions. All electrical systems of 30 V or more are potential shock hazards. Research indicates that most shock deaths result from contacts with electrical systems ranging from 70 to 500 V. Table 10.1 summarizes typical effects of various levels of electrical current. Table 10.2 shows the recommended exposure limits for personnel [10].

The danger to personnel from electric shock can be avoided by implementing suitable safeguards in the design of equipment. Safety labels, alarms, interlocks, proper grounding, enclosures, and protective devices are the main methods used for the prevention of

Table 10.1 Shock current intensities and their probable effects [10].

Current (mA)		
AC (60 Hz)	**DC**	**Effects**
0–1.0	0–4.0	Perception
1.0–4.0	4.0–15	Surprise
4.0–21	15–80	Reflex action
21–40	80–160	Muscular inhibition
41–100	160–300	Respiratory block
Over 100	Over 300	Usually fatal

Source: MIL-STD-1472G, Department of Defense Design Criteria Standard: Human Engineering, Washington, DC: US Department of Defense, Jan. 11, 2012.

Table 10.2 Electrical current exposure limits for all systems [10].

Frequency (Hz)	Maximum current (mA) (AC + DC components combined)
DC	40
15–2000	8.5
3000	13.5
4000	15
5000	16.5
6000	17.9
7000	19.4
8000	20.9
9000	22.5
>10 000	24.3

Source: MIL-STD-1472G, Department of Defense Design Criteria Standard: Human Engineering, Washington, DC: US Department of Defense, Jan. 11, 2012.

electrical shock to personnel. A few brief guidelines for safe electrical design include:

- Safety labels should be used to warn personnel of the electrical shock hazards within an enclosure.
- Alarms such as lights, bells, and horns can be used to warn personnel of potential danger.
- Interlocks are used to automatically turn power off when an access door, cover, or lid of the equipment enclosure is opened. These switches are ordinarily wired into the hot lead to the power supply and break the circuit whenever personnel enter the enclosure containing dangerous voltages.
- A main power switch that turns off all power to the item by opening all leads from the main power service connection should be provided for all equipment. This main power switch should be designed so that it can be locked out while the item is being maintained to prevent unexpected reactivation. Or, the item enclosure should be designed so that the enclosure cannot be opened when the main power switch is turned on.
- Capacitors can store lethal charges for relatively long periods of time. Bleed resistors should be incorporated in all circuits containing capacitors so that the voltage is bled off within 3.0 seconds.
- Equipment should be designed so that all external parts are at ground potential.

Mechanical hazards that have the potential to cause injury to maintenance personnel may also be present in equipment. Such hazards may include sharp edges and corners, protrusions, rotating or moving parts, hot surfaces, and unstable equipment. Design considerations for these hazards include:

- All edges and corners should be rounded using as large a rounding radius as practical. Designers shall avoid thin edges.
- Protrusions from equipment surfaces should be minimized. Flat-head screws should be used where possible. All exposed surfaces should be machined smooth, covered, or coated to reduce the risk of skin abrasion and cuts. Recessed mountings should be considered for any small projecting parts, such as toggle switches or knobs.
- Guards should be provided to cover all moving parts that may injure or entangle personnel. These include pulleys, belts, gears, and blades. All moving and rotating equipment should be disabled and unable to function as long as the guard is not permanently installed.
- Sufficient ventilation should be provided to keep parts and materials from getting so hot that they will be damaged or their useful life will be shortened, and to keep them from reaching temperatures that might endanger personnel.
- Equipment should be designed for maximum stability. Particular attention should be given to portable equipment such as maintenance stands, tables, benches, platforms, and ladders. Another important consideration is an item's center of gravity. The lower the center of gravity, the more stable the item will be. The center of gravity should be clearly marked.

Fire hazards should be minimized. Design considerations include:

- Equipment that presents a possible fire hazard should be enclosed by non-combustible material.
- Design equipment that will not emit flammable vapors during storage or operation.
- Provide switches that will cut off power if equipment overheats.
- Provide fire extinguishers where fire hazards exit.

Toxic fumes and hazardous materials should be eliminated from equipment design if possible, or minimized and controlled if present. Design guidelines include:

- Analyze all materials for suitability for use in the anticipated environments in which the equipment will operate.
- Follow guidelines for maximum allowable concentrations of any toxic materials.
- Provide adequate ventilation to protect personnel from toxic fumes.
- Ensure any connectors or tubing used in handling or controlling hazardous fluids are incompatible with the materials present.
- Provide automatic shutoff devices on fluid and fuel service equipment to prevent overflow and spillage.

High noise levels can damage the hearing of personnel. High levels of background noise can also interfere with communications. For hearing conservation purposes, continuous noise such as from machinery, vehicles, aircraft, and so on, is measured either in terms of the A-weighted sound level expressed in dBA, or in terms of the equivalent continuous sound level (8.0 hours), usually written as L_{eq} (8.0 hours) and also expressed in dBA. Guidelines for minimizing noise problems include:

- Design equipment that does not generate noise in excess of maximum allowable levels.
- Any areas where noise levels exceed 84 dBA are considered to be "high noise areas." Designers should consider relocating controls, displays, and workstations out of or away from these areas by implementing remote monitoring of equipment in separate, acoustically isolated spaces in order to reduce personnel exposure to high noise.
- Warning signs should be used to warn personnel of any noise issues.
- If noise levels cannot be reduced below allowable levels, then require personnel to use personal hearing protection devices.
- Where noise is present but not hazardous, and communication is required, workspace noise should be reduced to levels that permit necessary direct (person-to-person) or telephone communication.
- Use acoustic materials with high sound absorption coefficients as necessary in the construction of floors, walls, enclosures, and ceilings to provide the required sound control.

The use of these guidelines and those provided in Appendix A can help in equipment design.

While the use of design criteria, guidelines, and checklists can assist the designer to address hazards during maintenance, a more thorough analysis of the system relative to hazards presented is desirable. Using the techniques developed by the system safety engineering discipline is the best way for the design team to ensure that the design is safe to use and maintain. These techniques are described in the following sections.

10.2.3 Overview of System Safety Engineering

A system is a network or group of interdependent components and operational processes that work together to provide for a common purpose. It is important for a system to operate safely while meeting the goals and objectives of this common purpose.

MIL-STD-882 defines system safety as "The application of engineering and management principles, criteria, and techniques to achieve acceptable risk within the constraints of operational effectiveness and suitability, time, and cost throughout all phases of the system life cycle" [11]. More simply, system safety is the engineering discipline that works to prevent hazards and accidents in complex systems. System safety accomplishes this by applying a risk management approach focused on the identification of hazards, analysis of these hazards, and the application of design improvements, corrective actions, and risk mitigation. This system-based risk management approach requires the coordinated efforts of systems management, systems engineering, and system safety engineers having a diverse set of technical skills in hazard identification, in hazard analysis, and in the elimination or reduction of hazards throughout the system life cycle.

10.2.4 Risk Assessment and Risk Management

Risk assessment is an important aspect of system safety analysis. System safety requires that risk be evaluated and that the risk be accepted or rejected. "All risk management is about decision-making under uncertainty. It is a process wherein the risks are identified, ranked, assessed, documented, monitored, and mitigated" [12].

Risk is an expression of the possibility/impact of a mishap in terms of hazard severity and hazard probability. From the beginning of the design process, the goal is to design to eliminate hazards and minimize the risks. If an identified hazard cannot be eliminated, the risk associated with it must be reduced to an acceptable level.

Risk is the product of consequences of a particular outcome (or range of outcomes) and the probability of its (their) occurrence. The most common way of quantifying risk is as the product of consequences of a particular outcome (or range of outcomes) and the probability of its (their) occurrence [13]. This risk term is expressed mathematically in the following equation:

$$R = C \times P$$

where R is the risk priority, C the consequence severity as a result of the risk, and P the probability of the risk materializing or occurring.

The severity of the consequences of an undesirable event or outcome must be estimated. Severity is an assessment of the seriousness of the effects of the potential event if it does occur. The probability of occurrence is the likelihood that a particular cause or failure occurs [13].

The first step to controlling risk in maintenance operations is to identify those areas that might present a risk. The designer must attempt to figure out what can go wrong with the product or the operation being analyzed. This can be done using numerous tools such as Failure Modes, Effects, and Criticality Analysis (FMECA), in which all failure modes are identified, or a Preliminary Hazard Analysis (PHA) may be chosen to help determine what hazards might be present in a product or operation.

Risk estimations must then be made for each hazardous event to determine the risk it presents. This is done by combining the probability of occurrence of the undesired event with a measure of the severity of its consequences. This combination of severity and probability is the Risk Assessment Code (RAC). Usually an Initial Risk Assessment Code (IRAC) is made early as a worst-case assessment. As the design develops the RAC is updated to reflect the latest risk assessments. At the end of the development program, the Final Risk Assessment Code (FRAC) is established.

10.2.4.1 Probability

Initially a qualitative statement of estimated probability of occurrence as shown in Table 10.3 is used to make the initial risk estimate. As the design progresses, and time and budget permits, the probability of occurrence can be refined to become more quantitative. This can be accomplished by using reliability analysis techniques to estimate product or system failure rates, or other specialized analysis techniques such as Fault Tree Analysis (FTA) or event tree analysis.

Table 10.3 Hazard probability levels.

Probability level	Qualitative probability of occurrence	Applicability to item
A	Frequent	Likely to occur often in the life of an item
B	Probable	Will occur several times in the life of an item
C	Occasional	Likely to occur sometime in the life of an item
D	Remote	Unlikely, but possible to occur in the life of an item
E	Improbable	So unlikely, it can be assumed occurrence may not be experienced
F	Eliminated	Incapable of occurrence. This level is used when potential hazards are identified and later eliminated

10.2.4.2 Consequences

Usually the first step in assessing hazardous consequences is to make a qualitative ranking of the severity of the hazard is shown in Table 10.4. Much like estimating probability of occurrence, severity of consequences can be analyzed in greater depth as the design progresses using more quantitative techniques, such as FMECA or FTA.

Table 10.4 Hazard severity levels.

Severity	Category	Mishap definition
Catastrophic	1	Death, system loss, or severe environmental damage
Critical	2	Permanent partial disability, injuries, or occupational illness that may result in hospitalization of at least three personnel, reversible significant environmental impact
Marginal	3	Injury or occupational illness resulting in one or more lost work days, reversible moderate environmental impact
Negligible	4	Injury or occupational illness not resulting in a lost work day, minimal environmental impact

Table 10.5 Risk assessment matrix.

Frequency of occurrence	Hazard severity categories			
	1 Catastrophic	2 Critical	3 Marginal	4 Negligible
A – Frequent	High	High	Serious	Medium
B – Probable	High	High	Serious	Medium
C – Occasional	High	Serious	Medium	Low
D – Remote	Serious	Medium	Medium	Low
E – Improbable	Medium	Medium	Medium	Low
F – Eliminated	Eliminated			

10.2.4.3 Risk Evaluation

The next step is the evaluation of the risks to determine if further action is warranted or if the risk is acceptable.

Significance Risks must be evaluated to determine their significance and the urgency of their mitigation.

The RAC is used to determine whether further corrective action is required. The RAC is used to make this determination as shown in Table 10.5. The determination of RAC allows the risks typically to be categorized into five groups that represent their risk priority – high, serious, medium, low, or eliminated. The ranking of the risk into these groups determines whether further action is necessary and by what authority the hazard can be accepted and closed. Further action can then be taken on the most important risks first.

Risk acceptability The ranking of the hazards into groups by RAC determines whether further action is necessary or whether the hazard level is acceptable, and by what authority the hazard can be accepted and closed (see Table 10.6 for risk levels).

Cost In addition to the probability of occurrence and the severity of the consequences, it is often necessary to consider the cost of mitigation. There are almost always budget limitations, and these must be taken into account during the process of deciding how far to go in mitigating the risks and which ones to concentrate on. Obviously, the risks with the highest rank need the most attention, preferably to totally eliminate them or to reduce the consequences if they do occur. The lower ranking risks may have to be accepted.

Once the risk assessment has been made, the risk needs to be mitigated using design measures to reduce the probability of occurrence of the undesired event and/or by reducing the consequences that result if the undesired event occurs. The best solution is always to eliminate the risk by designing it out of the system or product. However, this is not always possible, so the designer must choose the most effective way of reducing the risk using the order of precedence.

The order of precedence for mitigating identified risks is provided in MIL-STD-882 [11]:

- Eliminate hazards through design selection. Ideally, the hazard should be eliminated by selecting a design or material alternative that removes the hazard altogether.
- Reduce risk through design alteration. If adopting an alternative design change or material to eliminate the hazard is not feasible, consider design changes that reduce the severity and/or the probability of the mishap potential caused by the hazard(s).
- Incorporate engineered features or devices. If mitigation of the risk through design alteration is not feasible, reduce the severity or the probability of the mishap potential caused by the hazard(s) using engineered features or devices. In general, engineered features actively interrupt the mishap sequence, and devices reduce the risk of a mishap.
- Provide warning devices. If engineered features and devices are not feasible or do not adequately lower the severity or probability of the mishap potential caused by the hazard, include detection and warning systems to alert personnel to the presence of a

Table 10.6 Risk levels.

Risk level	Risk assessment code	Guidance	Decision authority
High	1A, 1B, 1C, 2A, 2B	Unacceptable	Management
Serious	1D, 2C, 3A, 3B	Undesirable	Program manager
Medium	1E, 2D, 2E, 3C, 3D, 3E, 4A, 4B	Acceptable	Program manager or safety manager
Low	4C, 4D, 4E	Acceptable (without higher-level review)	Safety team

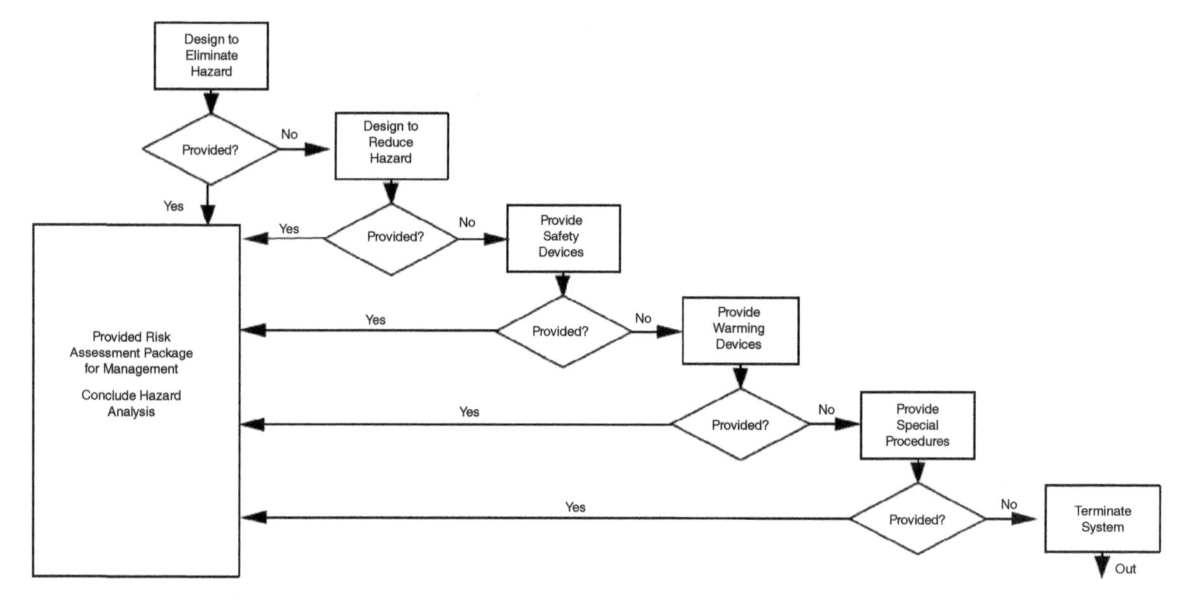

Figure 10.1 Hazard reduction precedence [14]. Source: NM 87117-5670 (2000) *Air Force System Safety Handbook*, Air Force Safety Agency, Kirtland Air Force.

hazardous condition or occurrence of a hazardous event.

- Incorporate signage, procedures, training, and Personal Protective Equipment (PPE). Where design alternatives, design changes, and engineered features and devices are not feasible and warning devices cannot adequately mitigate the severity or probability of the mishap potential caused by the hazard, incorporate signage, procedures, training, and PPE. Signage includes placards, labels, signs, and other visual graphics. Procedures and training should include appropriate warnings and cautions. Procedures may prescribe the use of PPE. For hazards assigned Catastrophic or Critical mishap severity categories, the use of signage, procedures, training, and PPE as the only risk reduction method should be avoided.

The hazard reduction precedence is shown in Figure 10.1.

10.2.5 System Safety Analysis

System safety analysis is a methodology for the assessment of risks associated with a product or system. The approach may be summarized simply as a process to determine:

- What can go wrong?
- What are the consequences if it happens?

- How likely is it to occur?
- What can be done to eliminate or reduce the risk?

The objective of safety analyses is to identify hazards associated with the product or operation and to assess the risk those hazards may present. The objective is to eliminate or control the risk to an acceptable level. Developing a safe system, process, or operation may require different types of system safety analyses. There are many types of system safety analyses. The approach to risk management and the various analyses planned for a particular development program should be delineated in a System Safety Program Plan (SSPP), and the various analyses should be scheduled at the appropriate time during the development process. There are many safety analysis techniques that are available for use. Table 10.7 lists a sampling of the most common types of safety analyses and a brief description of each.

While this is only a list of some the more commonly used analyses, there are many more. Covering all these techniques is beyond the scope of this book, and the reader is referred to *Design for Safety* [13] for an in-depth treatment of these and other techniques.

Hazard analysis is the cornerstone of safe systems development. Design for safety requires the elimination or mitigation of all hazards. Therefore, the first, and most important, step in eliminating hazards is to identify all the hazards. Once hazards are recognized, they can be evaluated and then eliminated or controlled to an acceptable level. The evaluation of hazards includes determining their causes and effects. Hazard analysis is also used to determine the risk presented by the hazard, which, in turn, provides a way to prioritize the hazards. This information is then used in the analysis to help determine design options that will eliminate or mitigate the risk of having an accident or mishap [13].

The most important and useful safety analysis techniques related to maintainability include hazard analysis in general, and more specifically Operating and Support Hazard Analysis (O&SHA).

10.2.5.1 Operating and Support Hazard Analysis

O&SHA is used to identify hazards that may occur during various modes of system operations including maintenance. O&SHA should be started as early as possible in the development process to allow problems to be fixed as early as possible. However, since the system design needs to be substantially complete and operating and maintenance procedures must be available, at least as drafts, the O&SHA is typically started late in the design phase or early in production.

The purpose of the O&SHA is to identify and evaluate the hazards associated with the operation, support, and maintenance of the system. Its emphasis is on procedures, training, human factors, and the human–machine interface. The O&SHA will also identify cautions and warnings that may be required.

The process of conducting an O&SHA will entail using design, operational, and support procedure information to identify hazards associated with all operational modes. The hazards are identified through the thorough analysis of every detailed procedure that is to be performed during system operation, maintenance, and support. Hazard-related checklists can also be used to stimulate the identification of hazards (see Appendix A).

Consideration should be given to the following when conducting an O&SHA:

- Facility/installation interfaces to the system
- Operation and support environments
- Tools or other equipment
- Support/test equipment
- Operating procedures
- Maintenance procedures
- Human factors and human errors
- Personnel requirements
- Workload
- Testing
- Installation
- Repair procedures
- Training
- Packaging, Handling, Storage, and Transportation (PHS&T)
- Disposal
- Emergency operations
- Personal Protective Equipment (PPE)
- Chemicals and hazardous materials used in production, operation, and maintenance

Most types of hazard analyses use some type of worksheet to help organize the analysis. For O&SHA the emphasis is on the tasks involved in the operation,

Table 10.7 Common types of system safety analysis.

Analysis technique	Description
Risk management	Design risk is the risk inherent in product design. Risk management attempts to control risk in product design by reducing the probability of having an accident and by reducing the consequences of an accident if an accident does occur. The purpose of risk management is to reduce risk to an acceptable level. The acceptable level of risk depends on many factors including the type of hazard, the risk tolerance of the user, and the particular industry involved. Accepting a risk ultimately boils down to a cost-benefit analysis. In a heavily regulated industry, an independent authority may determine whether to accept the risk.
Preliminary Hazard List (PHL)	The PHL is a compilation of potential hazards identified very early in development of the system and provides an initial assessment of the hazards. It helps to identify the hazards that may require special safety emphasis and areas where more in-depth analyses may be required.
Preliminary Hazard Analysis (PHA)	The PHA is typically an expansion of the PHL. It refines the analysis of the hazards in the PHL and identifies any new hazards that may arise as the system design is developed and more design details emerge. It is conducted in the early stages of development. The PHA identifies the causal factors, the consequences, and the risks associated with the preliminary design concept(s). Recommended hazard controls are also presented. The PHA helps to positively influence the design for safety early in the system development.
Subsystem Hazard Analysis (SSHA)	The SSHA provides a more detailed analysis than the PHA as the development proceeds. The SSHA only looks at a particular subsystem to identify hazards within that subsystem, and provides detailed analysis of each subsystem to determine hazards created by the subsystems to themselves, to related or nearby equipment, or to personnel.
Systems Hazard Analysis (SHA)	The SHA is a detailed analysis that focuses on the hazards at the system level and helps to develop the overall system risk. Its major focus is on the internal and external interfaces. The SHA evaluates hazards associated with system integration and involves evaluating all hazards across subsystem interfaces. The SHA can be viewed as an expansion of the SSHA in that it includes analysis of all subsystems and their interfaces, but does not include all the subsystem hazards identified in the SSHA.
Operating and Support Hazard Analysis (O&SHA)	The O&SHA is used to identify hazards that may occur during operation, support, and maintenance of the system. It is started as early as possible in the development process, but it requires the design to be substantially complete and operating and maintenance procedures must be available.
Health Hazard Analysis (HHA)	The HHA is used to identify hazards to human health presented by the system being developed. It is also used to evaluate any use of hazardous materials and processes that use hazardous materials. It typically proposes measures to eliminate or control the risk presented by the health hazards. The HHA describes the operational environment, including how and in what environment the equipment will be used and maintained.
Failure Modes, Effects, and Criticality Analysis (FMECA)	The FMECA is used to identify potential hazards that could occur during normal or worst-case system operation and includes quantitative criticality analysis. Worst-case system operation may be system missions performed in harsh environmental conditions or during high stress loading during peak usage scenarios. The FMECA is used to identify potential hazards that may result from failure symptoms that could develop from degradation, fatigue accumulation, or physical wear-out mechanisms. Besides failure symptoms, FMECAs are used to document and study failure modes, failure causes, and failure effects. The FMECA includes a quantitative criticality analysis which allows for the assessment of the severity of potential hazards caused by the failures in order to objectively rank hazard priorities for follow-up studies. FMECA is discussed in detail in Chapter 14.
Fault Tree Analysis (FTA)	FTA uses symbolic logic to create a graphical representation, in tree form, of the combination of failures, faults, and errors that can lead to the undesirable event being analyzed. FTA is a deductive (top-down) analysis technique that focuses on a particular undesirable event and is used to determine the root causes contributing to its occurrence. The process starts with identifying an undesirable event (top event) and working backwards through the system to determine the combinations of component failures that will cause the top event.

Operating & Support Hazard Analysis												
System: ______												
Preparer: ______												
Date: ______												
Identifier	Task	Operational Mode	Hazard	Cause	Effects	IRAC	Risk	Recommended Action	FRAC	Risk	Cmts	Status

Figure 10.2 Operating and Support Hazard Analysis format [13]. Source: Gullo, L. J. and Dixon, J., *Design for Safety*, John Wiley & Sons, Inc., Hoboken, NJ, 2018. Reproduced with permission from John Wiley & Sons.

maintenance, and support of the system. A typical worksheet for an O&SHA is shown in Figure 10.2.

Column descriptions

Header information: Self-explanatory. Any other generic information desired or required by the program could be included.

Identifier: This can simply be a number (e.g., 1, 2, 3… as hazards are listed), or it could be an identifier for a particular subsystem or piece of hardware in the system, with a series of numbers as multiple hazards are identified with that particular item (e.g., motor 1, motor 2, etc.).

Task: This column is used to identify the task being evaluated.

Operational Mode: This is a description of what mode the system is in when the hazard is experienced (e.g., operational, training, maintenance, etc.).

Hazard: This is the specific potential hazard that has been identified.

Cause: This column identifies conditions, events, or faults that could cause the hazard to exist and the events that can lead to a mishap.

Effects: This is a short description of the possible resulting mishap. What is the effect(s) resulting from the potential hazard – death, injury, damage, environmental damage, etc.? This would generally be the worst-case result.

IRAC: This is a preliminary, qualitative assessment of the risk.

Risk: This is a qualitative measure of the risk for the identified hazard, given that no mitigation techniques are applied.

Recommended Action: Provide recommended preventive measures to eliminate or mitigate the identified hazards. Hazard mitigation methods should follow the preferred order of precedence provided earlier in this chapter.

FRAC: This is a final, qualitative assessment of the risk, given that mitigation techniques and safety requirements identified in "Recommended Action" are applied to the hazard.

Comments: This column should include any assumptions made, recommended control, requirements, applicable standards, actions needed, and so on.

Status: States the current status of the hazard (open, monitored, closed).

While the O&SHA provides a focus on operational and procedural hazards, including maintenance, the analysis can become quite tedious if there are many tasks to be analyzed in a large system.

10.2.5.2 Health Hazard Analysis

Another system safety analysis technique that is important to maintainability is Health Hazard Analysis (HHA). The HHA is used to evaluate a system to

Health Hazard Analysis											
System: ______________											
Preparer: ______________											
Date: ____________											
Identifier	Hazard Type	Hazard	Cause	Effects	IRAC	Risk	Recommended Action	FRAC	Risk	Cmts	Status

Figure 10.3 Health Hazard Analysis format [13]. Source: Gullo, L. J. and Dixon, J., *Design for Safety*, John Wiley & Sons, Inc., Hoboken, NJ, 2018. Reproduced with permission from John Wiley & Sons.

identify hazards and to propose methods to eliminate or control the risk to human health presented by these hazards. It is also used to evaluate any use of hazardous materials and processes that use hazardous materials. The HHA describes the operational environment, including how and in what environment the equipment will be used and maintained.

As with other analysis techniques, the HHA uses design and operating information and knowledge of health hazards to identify the hazards involved with the system under development. The system is evaluated to identify potential sources of human health-related hazards. An assessment must then be made of the magnitude of the hazardous source and the exposure levels involved. Risk is assessed and mitigations are recommended for implementation. Other analysis, such as Preliminary Hazard List (PHL), Preliminary Hazard Analysis (PHA), and O&SHA can be sources of information for the HHA. Health hazard-related checklists can also be used to stimulate the identification of health hazards (see Appendix A). The HHA can be used to help identify the need for PPE.

Health hazard considerations include:

- Noise
- Chemicals
- Radiation (both ionizing and non-ionizing)
- Hazardous materials including carcinogens
- Shock and vibration
- Heat and cold
- Human–machine interfaces
- Extreme environments
- Stressors of the human operator or maintainer
- Biological hazards (e.g., bacteria, viruses, fungi, and mold)
- Ergonomic hazards (e.g., lifting, cognitive demands, long duration activity, etc.)

A worksheet similar to that used for the O&SHA can be used for the HHA. Actually, the two techniques are similar, but the O&SHA focuses on tasks performed during operation and maintenance, while the HHA is focused entirely on hazards to human health. The worksheet is shown in Figure 10.3.

Column descriptions

Header information: Self-explanatory. Any other generic information desired or required by the program could be included.

Identifier: This can simply be a number (e.g., 1, 2, 3… as hazards are listed), or it could be an identifier that identifies a particular subsystem or piece of hardware in the system with a series of numbers, as multiple hazards are identified with that particular item (e.g., motor 1, motor 2, etc.).

Hazard type: This column shows the type of health concern presented (e.g., noise, chemicals, radiation).

Hazard: This is the specific potential health hazard that has been identified.

Cause: This column identifies conditions, events, or faults that could cause the hazard to exist and the events that can lead to a mishap.

Effects: This is a short description of the possible resulting mishap – the effect(s) resulting from the potential hazard, such as death, injury, environmental damage, etc. This would generally be the worst-case result.

IRAC: This is a preliminary, qualitative assessment of the risk.

Risk: This is a qualitative measure of the risk for the identified hazard, given that no mitigation techniques are applied.

Recommended Action: Provides recommended preventive measures to eliminate or mitigate the identified hazards. Hazard mitigation methods should follow the preferred order of precedence provided earlier in this chapter.

FRAC: This is a final, qualitative assessment of the risk, given that mitigation techniques and safety requirements identified in "Recommended Action" are applied to the hazard.

Comments: This column should include any assumptions made, recommended controls, requirements, applicable standards, actions needed, and so on.

Status: States the current status of the hazard (open, monitored, closed).

While the matrix approach can provide for a thorough analysis and identification of health hazards presented by the system, often more detailed assessments of particular hazards may be required. These assessments are usually conducted by medical or health personnel and result in a detailed report being prepared.

10.3 Human Factors in Maintainable Design

The International Ergonomics Association defines Human Factors (or Ergonomics) as "...the scientific discipline concerned with the understanding of interactions among humans and other elements of a system, and the profession that applies theory, principles, data and methods to design in order to optimize human well-being and overall system performance" [15].

Human Factors Engineering (HFE) is focused on designing products for people. HFE attempts to strike a balance between the product design and the physical capabilities of people, their strengths and weaknesses, their mental abilities, and the environments in which they work. The goals of good user-centered design are to recognize people's limitations, to minimize negative consequences on them, and to not force them to adapt to the product but to adapt the product to the user.

Like the system safety engineering function described in the previous section, HFE is another important consideration in the systems engineering design process. While designing a product or system, in addition to considering the human user as related to operations, the product/system designers should always consider human factors when making decisions concerning maintainability.

10.3.1 Human Factors Engineering and its Relationship to Maintainability

During maintenance activities, what the maintenance personnel do and how they perform is important not only from the standpoint of efficient and cost-effective maintenance, but also for the protection of the equipment being maintained as well as the protection maintenance personal. Improperly performing a maintenance task could result in damage to the particular piece of equipment being maintained, or it could result in an unsafe condition that could be the cause an accident during the maintenance activities or later during equipment use. Improper maintenance due to poor human factors design could also induce failures in other items within a larger, complex system or possibly cause a failure of the entire system.

Paradigm 5: Consider the human as the maintainer

In addition to the most obvious human considerations, like physical constraints and capabilities, several additional human elements should be considered during the design of products or systems. These considerations may include some of the following.

Problem-solving and psychomotor skills are required. Maintenance requires skills that differ from those of an operator or user. The ability to work quickly and efficiently, sometimes in difficult environments, is an important human quality that maintenance personnel should possess. Equally important is their problem-solving ability, because maintenance can be a very complex task that may involve unanticipated and unfamiliar failures.

Job-oriented training is an important prerequisite for effective maintenance personnel. The proper prior education provides a strong background upon which an individual can build a successful maintenance career. There is a strong link between general education of the individual and training for the tasks presented in the workplace. Of course, selecting the right people is important, but so is the development of good maintenance training materials. The product or system design team must develop effective maintenance training materials to ensure effective maintenance. During new product development, one goal is to design the product such that the training requirements are minimized. The absence of effective human factors influences on the design during the development process may make the development of training materials more difficult, and the actual training of maintenance personnel challenging.

Human reliability is another key design consideration related to the human. As the saying goes, "To err is human" so the consideration of human reliability during the development of a system is imperative. Attention to human reliability and to human error is even more important if the system involves hazardous components or operations. The designer must be concerned with the reliability with which any maintenance task is performed.

10.3.2 Human Systems Integration

The goal of all good systems engineering design, as well as good HFE practice, is to reduce risk throughout the development cycle, which is accomplished by including all relevant disciplines. In the past, however, the risks associated with human systems integration were often been ignored. Many engineering risks are noticed at various times during the development due to implementation problems or cost overruns. However, the risks associated with human systems integration are often only noticed after a product or system is delivered to the customer. These problems may lead to customer dissatisfaction and rejection of the product, due to it being too difficult or inefficient to use and maintain, or, worse, they may lead to human error in the use of the product, which could lead to an accident having catastrophic consequences.

These operating risks can be traced back to failure to properly integrate the human needs, capabilities, and limitations at an early stage of the design process. Like all risk-reduction efforts, risk reduction in the area of human systems integration must be started early and continue throughout development of a product or system. This will ensure that requirements based on human factors are incorporated into the design. This will produce a high level of confidence that the product will be accepted, usable, maintainable, and safe for the customer to use.

Paradigm 8: Understand the maintenance requirements.

As has been emphasized earlier in this book, the key to design success rests in the development of requirements. Requirements involving human factors are just as important as other requirements. A thorough and complete effort to specify human factors requirements in the earliest stages of product or system development will reap large rewards later by resulting in a product or system that is usable, maintainable, and safe to use. As development progresses, other approaches to mitigating risk in the area of human systems integration include using task analysis to refine the requirements, and conducting trade studies, prototyping, simulations, and user evaluations [13].

10.3.3 Human Factors Design Criteria

There are numerous general guidelines that have been developed to help the product and system designers to improve the human–equipment integration as it is related to maintenance. Some of these general guidelines include:

- Simplify maintenance functions

- Develop designs and maintenance procedures such that human errors are minimized during maintenance operations
- Arrange and locate components to provide rapid access to those components that will require the most frequent maintenance
- Design for easy accessibility to all systems, equipment, and components requiring maintenance
- Design equipment to minimize skills and training requirements for maintenance personnel
- Develop designs that minimize the probability that maintenance workers will be exposed to hazards or injured
- Consider human behavior when designing products.

Maintenance activities take time, and time is money, therefore by incorporating good HFE into the design, costs can be minimized while at the same time making the work of the maintenance personnel as simple, easy, and effective as possible.

Design criteria that address the human capabilities and limitations should be considered during the design of all products and systems. A few of the major design criteria categories that must be considered to ensure good design for maintainability include:

- Accessibility
- Modularity
- Standardization
- Controls and displays
- Tools, test equipment, and test points
- Lifting

The following paragraphs provide a sampling of some basic design criteria designers can use to improve maintenance efforts.

Accessibility is a key criterion for ensuring effective maintenance of a product or system. All components that are required to be routinely inspected should be conveniently located. Likewise, components that are expected to fail most often or that will regularly need to be maintained should be located in such a way that allows maintenance personnel easy access and should not require removal of additional components in order to gain access. All connections, including electrical, mechanical, hydraulic, or pneumatic connections, should be of a quick connect/disconnect type and should be readily accessible. Any access panels should be removable with minimal effort.

Modularity provides for ease of maintenance. Standardized interfaces help to achieve maximum modularity. Using a modular approach within a system allows for quick replacement of any failed or defective modules with little or no readjustments.

Standardization of connectors, fasteners, tools, and test equipment helps to simplify the design and its maintenance requirements. Designers should consider standardizing the use of as many parts as possible, even down to the nuts and bolts that hold things together. While designers are creative people, they should resist the temptation to deviate from simplicity and from standard parts.

Controls and displays should be designed to facilitate human interactions with them. Efficient arrangement of maintenance workplaces, equipment, controls, and displays is of major importance.

The type of control selected and its location should ensure that the user and/or maintainer are properly accommodated so they can view, reach, operate, remove, and replace as necessary. Grouping of controls should be accomplished in such a way that task-driven sequence is respected, or if required to be operated together, they are grouped together along with their associated displays. The arrangement of functionally similar or identical controls should be consistent from panel to panel throughout the system or with similar equipment. Controls that are used solely for maintenance and adjustment should be covered during normal equipment operation, but readily accessible and visible when maintenance is required. Proper spacing between controls or between a control and any adjacent obstruction should be such that the maintainer can properly operate the control with bare hands or with gloves, or other protective equipment as dictated by system requirements.

Controls, displays, marking, coding, labeling, and arrangement schemes (e.g., equipment and panel layout) should be uniform for common functions across all equipment and systems. Control-display complexity and precision should not exceed the user's capability to discriminate details or manipulate controls. Manual dexterity, coordination, and reaction time must be considered along with the dynamic conditions and environments in which the human performance is expected to occur. Adjustable illumination should be provided for visual displays, including the display, controls, labels, and any critical markings, that must be read under darkened conditions. Display commonality is important to consider when multiple displays and multiple display formats are used. The symbology used on multiple displays should be consistent among all displays, and text or readout fields should be in standard locations on all displays. If a maintainer is required to use many controls and displays, they should be located and arranged to aid in identifying the controls that are associated with each display, the equipment component affected by each control, and the equipment component described by each display.

Tools, test equipment, and test points require special design considerations. Equipment should be designed to minimize the numbers, types, and complexity of tools required for maintenance. If possible, the designer should design the equipment so that only those tools normally found in the maintenance technician's toolkit will be needed. The design of the equipment should accommodate effective use of tools through their full range of motion. Minimize the use of special tools, and the variety and number of sizes of tools required. Special tools should only be required when common hand-tools cannot accomplish the job. A list of all tools needed for all maintenance tasks should accompany each product or system. The tools selected should be safe to use and ergonomically designed so they can be used with minimal risk of injury and component damage.

Sufficient space should be provided for the use of any required test equipment and other required tools without causing difficulty or presenting additional hazards to the maintenance personnel. Adequate storage space should be provided for the test equipment and within portable test equipment for any leads, probes, spares, and manuals required for operation. Instructions and manuals for operating test equipment should be available to the maintenance personnel. Calibrate test equipment should be made as simply as possible. Test equipment should be designed to be as rugged as conditions demand.

Test points should be readily accessible and visible during maintenance. Test and service points should be designed and located according to the frequency and duration of use. Test points used for adjustment should be located sufficiently close to the controls and displays needed to make the adjustment. Test points should be clearly labeled and the labeling should identify it in the maintenance instructions. Test points should be located for ease of access; they should be located on surfaces or behind access panels that can easily be reached when the equipment is fully assembled and installed. Test points should be provided with guards and shields to protect personnel; this is especially important if the equipment is to be tested while it is operating.

Lifting of equipment can present risks to the maintenance personnel and should be considered when designing equipment. Lifting limits of personnel involves consideration of the population of the personnel involved, the demand on the personnel to lift with one or both hands, the height to which the equipment must be lifted, the distance equipment must be carried, the distribution of the weight within the item to be lifted and/or carried, and the frequency of the lifting required. One of the best sources for analyzing lifting requirements is MIL-STD-1472 [10].

The use of these guidelines and those provided in Appendix A can help in equipment design.

10.3.4 Human Factors Engineering Analysis

Almost any technique used in systems analysis can be applied to address the human element. In addition, there are numerous human factors-specific analysis techniques from which a designer can choose. Many of the system safety analysis techniques described earlier will include the human element. Likewise, many Human Factors Analysis (HFA) techniques will consider safety.

The purpose of HFA is the inclusion of human considerations (e.g., man-in-the-loop, man–machine integration) in systems analyses to develop a reliable, usable, maintainable, and safe product or system. Various HFAs are conducted at different times in the development process and for different reasons.

As products and systems have become more complex, it has become imperative that the old approaches of totally ignoring human considerations or of making "educated guesses" based on intuition of how to accommodate the human be replaced by systematic analytical techniques to better match the human being and the machine. Although it is beyond the scope of this book to elaborate on all the various HFAs, Table 10.8 presents examples of the analyses that are available [13].

Of the techniques listed in Table 10.8, the most useful human factors analysis techniques related to maintainability are probably Task Analysis and Job Safety Analysis. While this list contains only some the more commonly used human factors-related analyses, there are many more. Covering all these techniques is beyond the scope of this book, and the reader can find more detailed coverage of these and many other techniques in the books of Raheja and Allocco [16] and Booher [17].

10.3.5 Maintainability Anthropometric Analysis

While many analyses are useful, probably one of the most important techniques available to the designer is Maintainability Anthropometric Analysis (MAA). The MAA technique helps to match the human to the equipment and the task at hand. It helps to ensure that the maintenance person can efficiently and safely perform the required maintenance. In order to achieve these goals, the designer/analyst must be familiar with the capabilities and limitations of the human body.

Anthropometry is the study of the measurements, strength, and the range of motion of the human body. These parameters are important considerations in designing products and systems for maintainability. The designed equipment ultimately must be operated and maintained by humans, and therefore it must accommodate operators and maintenance personnel of various sizes, shapes, capabilities, and limitations. Anthropometric data are found in numerous sources and are typically presented in ranges between upper and lower percentiles.

The best designs will always be able to accommodate the full range of people including the 5th through 95th percentiles of both males and females. Sometimes the user population will be a smaller subset of the 5th–95th percentile male and female; this usually occurs when the product is only planned to be used by a certain smaller group of the overall population.

Some of the most important anthropometric data related to good maintainability design include:

- Basic body dimensions
  - Stature
  - Eye height
  - Arm reach
  - Hand size
  - Strength
- Body mobility
- Dexterity
- Field of vision

There are numerous sources of anthropometric data, but probably the most widely used are defense-related documents since the military have led the effort to document body sizes and capabilities for many years. Table 10.9 lists some of the more common sources of these data.

The designer must first identify the user population involved and then consider the capabilities and limitations of the operators and maintainers. MAA will help to identify design requirements that will define the arrangement of the equipment so that the maintainer will be able to gain access to accomplish the needed maintenance, have sufficient space to perform the maintenance task, and have the strength and agility to accomplish the task. The analysis will also help to identify any shortcomings in the design that might hamper the performance of the maintenance task.

The MMA effort should start early in the development process. During the concept phase of development when few design details may exist, the early application of anthropometric data will help ensure success as the design matures. This early analysis can lead to the most feasible designs with adequate access, the efficient arrangements of components to facilitate ease of maintenance, and the timely development of procedures for performing maintenance tasks. Benefits of early consideration of anthropometric constraints include more efficient and cost-effective maintenance, enhanced supportability, improved safety, and reduced design changes. Continued focus on human capabilities and limitations throughout all phases of development is essential to designing a successful end product that is operable and maintainable by the user population.

Table 10.8 Human Factors Analysis tools [13].

Technique	Purpose	Description	Cost/difficulty	Pros	Cons
Prototyping	Addresses design and layout issues	Mock-up of user version	Cheap if done early; more expensive as design progresses	Quick way to show possible design before investing time and money on detailed development	Can be expensive for complex systems or if the prototypes have to be updated each time the product or system changes
Improved Performance Research Integration Tool (IMPRINT) (Reference "MANPRINT in Acquisition: A Handbook" [6])	IMPRINT is appropriate for use as both a system design and acquisition tool and a research tool. IMPRINT can be used to help set realistic system requirements; to identify user-driven constraints on system design; and to evaluate the capability of available manpower and personnel to operate effectively and maintain a system under environmental stressors. IMPRINT incorporates task analysis, workload modeling, performance shaping and degradation functions, and stressors, a personnel projection model and embedded personnel characteristics data.	IMPRINT, developed by the Human Research and Engineering Directorate of the US Army Research Laboratory, is a stochastic network modeling tool designed to help assess the interaction of soldier and system performance throughout the system life cycle – from concept and design through field testing and system upgrades. IMPRINT is the integrated, Windows follow-on to the Hardware vs. Manpower III (HARDMAN III) suite of nine separate tools.	Moderately time-consuming. Requires training. Inputs can be extensive.	Generates manpower estimates and can be used to estimate life cycle costs. Produces many types of reports.	Time-consuming and a lot of data are needed
Task analysis	Analyzing tasks and task flows that must be performed to complete a job	A job/task is broken down into increasingly detailed actions required to perform the job/task. Other data such as time, sequence, skills, etc., are included. This analysis is used in conjunction with other HFE analyses such as functional allocation, workload analysis, training needs analysis, etc.	Moderate, but can get expensive for large, complex systems	Relatively easy to learn and perform	Task analysis was first developed for factory assembly line jobs that were relatively simple, repetitive and physical, and easy to define and quantify. It is much more difficult to apply to complex or highly decision-based tasks.

Human Reliability Analysis (HRA)	Used to obtain an accurate assessment of product or system reliability, including the contribution of human error	Considers the factors that influence how humans perform various functions. It may include operators, maintainers, etc. The analysis is conducted using a framework of task analysis. First, the relevant tasks to be performed must be identified. Next, each task is broken down into subtasks and the interactions with the product or system are identified, and the possibility for errors is identified for each task, subtask, or operation. An assessment of the impact of the human actions identified is made. The next step is to quantify the analysis using historical data to assess the probability of success or failure of the various actions being taken. Also, any performance-shaping factors, such as training, stress, environment, etc., are taken into account. These factors may have an effect on the human error rate, which can be either positive or negative, but is usually negative. These might include: heat, noise, stress, distractions, vibration, motivation, fatigue, Boredom, etc.	Expensive and time-consuming on large systems	Thorough analysis of human errors and human–machine interactions	Requires extensive training and experience in several disciplines
Job safety analysis	Used to assess the various ways a task may be performed so that the most efficient and safest way to do that task may be selected	Each job or process is analyzed element-by-element to identify the hazards associated with each element. This is usually done by a team consisting of the worker, supervisor, and safety engineer. Expected hazards are identified and a matrix is created to analyze the controls that address each hazard.	Easy for simple jobs; harder for complex jobs	Good for structured jobs	Difficult if there is much variation in the job
Technique For Human Error Rate Prediction (THERP)	Used to provide a quantitative measure of human operator error in a process	THERP was initially developed during the 1960s for use in the nuclear industry for probabilistic risk assessment. It is a means of quantitatively estimating the probability of an accident being caused by a procedural error. The method involves defining the tasks, breaking them into steps, identifying the errors, estimating the probability of success/failure for each step, and calculating the probability of each task.	Can become expensive for processes with a lot of tasks	Can be very thorough	Obtaining good probability data

(Continued)

Table 10.8 (Continued)

Technique	Purpose	Description	Cost/difficulty	Pros	Cons
Link analysis	Used to evaluate the transmission of information between people and/or machines. It is focused on efficiency and is used to optimize workspace layouts and man–machine interfaces.	Links are identified between any elements of the system. The frequency of use of each link is determined. The importance of each link is then established. A link value is calculated based on frequency, time, and importance. System elements are then arranged so that the highest value links have the shortest length.	Moderate	Operations and training can be enhanced. Safety critical areas can be identified.	Can become cumbersome on large systems

Source: Gullo, L. J. and Dixon, J., *Design for Safety*, John Wiley & Sons, Inc., Hoboken, NJ, 2018. Reproduced with permission from John Wiley & Sons.

Table 10.9 Sources of anthropometric data.

Document	Source	Coverage	Availability
MIL-STD-1472, Human Engineering	US Department of Defense	Good overall coverage of male and female populations with some special subpopulations (e.g., pilots) and equipment design guidelines.	https://quicksearch.dla.mil/qsDocDetails.aspx?ident_number=36903
MIL-HDBK-759, Handbook for Human Engineering Design Guidelines	US Department of Defense	This handbook is a supplement to MIL-STD-1472. It has additional details on anthropometric measurements and equipment design guidelines.	https://quicksearch.dla.mil/qsDocDetails.aspx?ident_number=54086
NASA-STD-3000, Man-System Integration Standard NASA-STD-3001 Space Flight Human-System Standard Volume 1 (Crew Health) and Volume 2 (Human Factors, Habitability and Environmental Health)	National Aeronautics and Space Administration (NASA)	Male and female populations with additional information related to space travel.	https://msis.jsc.nasa.gov
HF-STD-001B	Federal Aviation Administration (FAA)	A compilation of human factors practices and principles related to the procurement, design, development, and testing of FAA systems, facilities, and equipment and includes anthropometric measurements.	https://hf.tc.faa.gov/hfds
Varies	Civilian American and European Surface Anthropometry Resource Project (CAESAR)	A database of human physical dimensions for men and women of various weights, between the ages of 18 and 65 in the United States and in Europe.	http://store.sae.org/caesar
Varies	Centers for Disease Control and Prevention	Sampling of numerous populations and subsets: • Firefighter hand anthropometry and structural glove sizing • Anthropometric Study of US Truck Drivers • Head-and-face shape variations of US civilian workers • Anthropometric criteria for the design of tractor cabs and protection frames • Anthropomorphic differences among Hispanic occupational groups	https://www.cdc.gov/niosh/topics/anthropometry/pubs.html

MAA is used to compare the user population anthropometric data to the workspace and equipment configuration. Some important considerations while designing the equipment include the strength of the personnel, the range of movement, illumination requirements, the proper display of important information, the ability to communicate, the environment in which personnel must function, and the hazards involved.

During the development process, engineering drawings evolve and are used by the designer/analyst to assess the compatibility of the equipment and the user. Sometimes the analysis can be done by making measurements of the equipment and comparing those dimensions to the anthropometric data. For example, the range of eye height for the population involved can be directly evaluated against the proposed location of information displays that must be observed (see Figure 10.4). At other times, the analysis may require more sophisticated models of human dimensions.

There are numerous approaches to determining the suitability of the design for human use. Different techniques can be used for different situations and/or at different times in the development cycle of the equipment.

Early in the design, often simple comparisons and calculations can be used, as the above example on eye height illustrates. Later, as drawings progress, two-dimensional manikins may be used to "fit" the human to the space. Of course, if a prototype of the equipment or workspace is available, a three-dimensional manikin or actual humans may be used. These full-scale prototypes can be used to simulate maintenance activities using people illustrative of the user population, to check the human–machine interfaces and interactions. Today, when most design work is accomplished using computer-aided design (CAD) applications, there are human models available that integrate with CAD programs and allow the designer/analyst to evaluate the design against various user population models. While the authors are not endorsing any particular products, a few examples of these human model add-ons are shown in Table 10.10 for the reader's convenience.

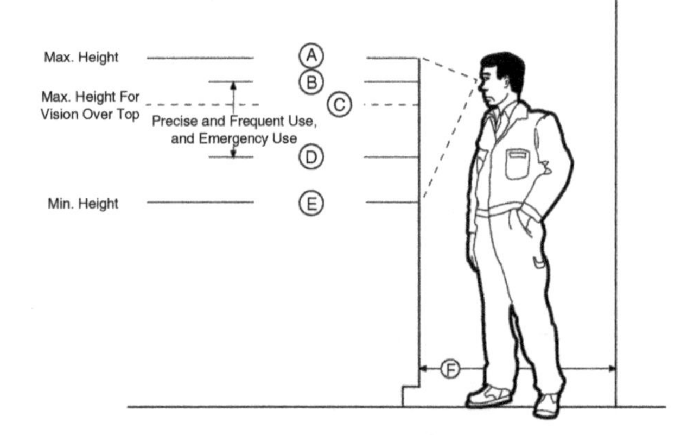

Dimension	Value
Maximum height (A)	177.8 cm (70 in)
Preferred max. height[1] (B)	165.1 cm (65 in)
Maximum lookover height[1]	(C) 150.1 cm (59 in)
Preferred min. height[1] (D)	139.7 cm (55 in)
Minimum height (E)	104 cm (41 in)
Preferred min. depth (F)	104 cm (41 in)
Minimum depth (F)	94 cm (37 in)
NOTES: (1) Preferred dimensions are for those controls that require precise, frequent, or emergency use.	

Figure 10.4 Display mounting height [10]. Source: *MIL-STD-1472G, Department of Defense Design Criteria Standard: Human Engineering*, Washington, DC: US Department of Defense, Jan. 11, 2012.

Table 10.10 Human model programs.

Human model program	Use	Company	Website
3D human model	Used by designers to optimize the physical human–product interaction. The models can be used in a 3D CAD environment.	SLIMDESIGN Veemkade 336 1019HD Amsterdam The Netherlands	http://www.3dhumanmodel.com
ManneQuinPRO	The ManneQuin is a series of human modeling programs.	NexGen Ergonomics Inc. 6600 Trans Canada Highway Suite 750 Pointe Claire, Quebec H9R 4S2 Canada	http://mannequinpro.software.informer.com/10.2
HumanCAD®	Creates digital humans in a three-dimensional environment in which a variety of ergonomic and human factor analysis can be performed. Aids users with the design of products and workplaces by determining what humans of different sizes can see, reach, or lift.	NexGen Ergonomics Inc. 6600 Trans Canada Highway Suite 750 Pointe Claire, Quebec H9R 4S2 Canada	http://www.nexgenergo.com/ergonomics/humancad.html
Tecnomatix Jack	Seamlessly integrates human factors and ergonomics into the planning, design, and validation stages of the product life cycle. Jack enables designer to size human models to match worker populations and test designs for injury risk, user comfort, reachability, energy expenditure, fatigue limits, and other important human parameters.	Production Modeling Corporation (PMC) World Headquarters 15 726 Michigan Ave Dearborn, MI 48126	https://www.pmcorp.com/Products/SiemensProducts/SiemensTecnomatix/Jack.aspx

Of course, the ultimate demonstration of the success of the design as it relates to maintenance is accomplished during the test phase. The test phase, and the Maintainability Demo in particular, provides the opportunity to validate the ability of actual users to perform the assigned maintainability tasks and verifies that the design accommodates the maintainer and provides adequate access and workspace to allow maintenance to be performed efficiently and safely.

10.4 Conclusion

Both system safety and HFA are important to all product and system developments. Safety and human factors should be addressed early in the development process, beginning in the concept formation stage. The depth of the analyses will vary with the phase of development and depending on the maturity of the design as it progresses from concept to fruition. This chapter has provided a sampling of the considerations needed to ensure equipment is designed to be safe and ergonomically effective to use. This chapter has also provided a range of analysis techniques that can be used by the designer to achieve cost-effective maintenance while protecting the maintenance personnel and making their job easier.

References

1 EHS Today (2001). *How Maintenance Contributes to Poor Safety Performance*. https://www.ehstoday.com/archive/article/21913215/how-maintenance-contributes-to-poor-safety-performance (accessed 26 August 2020).

2 UK Health and Safety Executive (n.d.). *Hazards during maintenance*. UK Health and Safety Executive. www.hse.gov.uk/safemaintenance/hazards.htm (accessed 26 August 2020).

3 Aviation Safety Institute (2012). *22nd Joseph T. Nall Report: General Aviation Accidents in 2010*. Frederick, MD: Aircraft Owners and Pilots Association.

4 Boyd, D. and Stolzer, A. (2015). Causes and trends in maintenance-related accidents in FAA-certified single engine piston aircraft. *Journal of Aviation Technology and Engineering* **5** (1): 17–24.

5 Saleh, J.H., Tikayat Ray, A., Zhang, K.S., and Churchwell, J.S. (2019). Maintenance and inspection as risk factors in helicopter accidents: analysis and recommendations. *PLoS ONE* **14** (2): e0211424. https://doi.org/10.1371/journal.pone.0211424.

6 Ruff, T., Coleman, P., and Martini, L. (2010). Machine-related injuries in the US mining industry and priorities for safety research. *International Journal of Injury Control and Safety Promotion* **18** (1): 11–20. https://doi.org/10.1080/17457300.2010.487154.

7 Heffernan, J. (2012). Coast Guard responds to nearly 30 accidents during crab season. *Daily Astorian*. https://www.dailyastorian.com/news/local/coast-guard-responds-to-nearly-accidents-during-crab-season/article_969f08d7-18ff-53e1-98e9-363d6543a513.html (accessed 26 August 2020).

8 EU-OSHA (2010). *Maintenance and Occupational Safety and Health – A Statistical Picture, TE-31-10-422-EN-N*. Luxembourg: European Agency for Safety and Health at Work.

9 Holmgren, M. (2006). Maintenance-related incidents and accidents: Aspects of hazard identification. Doctoral thesis, Luleå University of Technology, Department of Civil and Environmental Engineering, Luleå, Sweden.

10 US Department of Defense (2012). *Department of Defense Design Criteria Standard: Human Engineering*, MIL-STD-1472G. Washington, DC: US Department of Defense.

11 US Department of Defense (2012). *Department of Defense Standard Practice System Safety*, MIL-STD-882E. Washington, DC: US Department of Defense.

12 Raheja, D. and Gullo, L.J. (2012). *Design for Reliability*. Hoboken, NJ: Wiley.

13 Gullo, L.J. and Dixon, J. (2018). *Design for Safety*. Hoboken, NJ: Wiley.

14 Air Force Safety Agency (2000). *Air Force System Safety Handbook, NM 87117-5670*. Albuquerque, NM: Air Force Safety Agency, Kirtland Air Force Base.

15 International Ergonomics Association, https://iea.cc/what-is-ergonomics/ (accessed 20 September 2020).

16 Raheja, D. and Allocco, M. (2006). *Assurance Technologies Principles and Practices*. Hoboken, NJ: Wiley.

17 Booher, H.R. (2003). *Handbook of Human Systems Integration*. Hoboken, NJ: Wiley.

Suggestion for Additional Reading

Roland, H.E. and Moriarty, B. (1990). *Systems Safety Engineering and Management*. New York: Wiley.

Raheja, Dev and Allocco, M. (2006). *Assurance Technologies Principles and Practices, John Wiley & Sons*, Hoboken, N. J.

11

Design for Software Maintainability

Louis J. Gullo

11.1 Introduction

Usually, after a customer of a software product uses it for a while, becomes accustomed to the operation, and then the software fails, causing it to be inoperable, a person's first thought is how terrible the software is, regardless of how it performed when it was working. The next thought is how to fix the software so the person is able to resume operation and regain performance. The ease with which a person can fix this software product depends on the maintainability of the software design. Maintainable software typically means that it is easy for the user of the product to restore operation, and the user pays nothing or a small amount of money to fix the software when it fails.

If the software never fails, then it does not need to be maintained, right? Actually, the software may still need to be maintained even if it does not fail. Developers of software products will provide code updates or patches at intermittent times to their customers after initial product release to prevent potential failures. These code updates may require experienced members of a software development organization or trained employees of a maintenance service organization to perform the maintenance. This is true in the example of automobile manufacturer recalls for correction of software performing safety-critical functions. This type of software maintenance does not mean that the software is difficult to understand and impossible to correct by someone other than a member of the original software product development organization or trained member of a maintenance service organization. Software that fails will most probably be fixed by someone other than the original software designer or skilled software maintainer at some point during the software product's life cycle. For this reason, the software product should include simple-to-understand application interfaces, and user manuals for error code interpretation and software patch installation.

The majority of the life of any software system or product is spent in the Operations and Support (O&S) phase of the software life-cycle process where software maintenance occurs. This O&S phase is when the value of using the software system is realized by the customers. The majority of the cost of a system or product is spent in the O&S phase. The development and production costs of a program are a fraction of the total cost of ownership compared to the O&S costs. Therefore, it is imperative to perform software maintenance economically, effectively, and efficiently to reap the benefits from the software. Software maintenance should evolve as the state of the technology changes and as evidence of new software security threats in the software environment are discovered, leading to software functional enhancements in the system or product. The software maintenance process evolves based on improvements in the software maintainability, maintenance service models, maintenance strategies and concepts, and maintenance management planning.

Design for Maintainability, First Edition. Edited by Louis J. Gullo and Jack Dixon.

11.2 What is Software Maintainability?

Software maintainability is the composition of engineering activities and metrics within the software development process used to ensure that a software system or product is designed with the appropriate types and amounts of software features and capabilities so that it will be maintainable in the various user community applications, satisfying customer maintainability requirements. The *IEEE Standard Glossary of Software Engineering Terminology* states that software maintainability is the ease with which a software system or component can be modified to correct faults, improve performance or other attributes, or adapt to a changed environment [1]. Another source defines software maintainability as the probability that a maintenance activity can be carried out within a stated time interval [2].

Software developers responsible for maintaining source code that they did not originate appreciate knowing that the source code is maintainable. A software developer who is new to the code will take at least twice as long to understand and maintain the source code when the software's maintainability features are below average or nonexistent. If the software is not maintainable, any software design changes made to the code will risk the introduction of performance issues due to the probable creation of software bugs by the new software developer [3]. These performance issues are preventable with the use of certain software design practices.

The following is a summarized list of software design practice guidelines to aid a software developer in designing software that is maintainable to avoid the pitfalls for future developers who change the code later in the product life cycle and may experience performance issues as a result of their changes [3]:

1. Write clean code, correcting or documenting all bugs and providing adequate code comments
2. Write short units of code by limiting lengthy methods and functions
3. Write simple units of code by limiting branch points
4. Keep unit interfaces small by limiting parameters extracted into objects
5. Separate concerns in modules and limit the size of classes
6. Avoid redundant code by creating only one copy for each unit of code
7. Architecture components or binary executables should be coupled loosely while maintaining flexibility in architectural abstractions
8. Keep architecture components balanced in terms of number and size of the components
9. Ensure the codebase is as small as possible
10. Write code that is easily tested
11. Automate development pipeline and tests for the codebase

Many software products claim in their advertisements to be maintainable. The software developers of these products substantiate these claims by designing their products with fault-tolerant architecture and design features. Software fault tolerance is defined as the ability of a software system or product to withstand internal or external fault conditions without causing a system crash or critical failure effect. Fault-tolerant architecture usually leverages error detection and correction functions that enable a system or product to continue operating properly in the presence of one or more software or hardware fault conditions.

Further definitions of software maintainability are provided in ISO/IEC/IEEE 12207 [4] and ISO/IEC/IEEE 14764 [5]. In addition to the topic of software maintainability, ISO/IEC/IEEE 14764 and ISO/IEC/IEEE 12207 should be examined for assisting a person's understanding of the software development and software life-cycle processes that comprise a comprehensive software maintenance plan, describing all necessary software maintenance activities while gaining maximum benefit from software that was designed for maintainability.

11.3 Relevant Standards

The software standards that are most often used are:

ISO/IEC/IEEE 12207: 2017, Software Engineering – Software Life Cycle Processes defines all processes, activities, and tasks required for designing, developing, and maintaining software [4]. The software life cycle consists of five primary life-cycle process categories, which are Acquisition, Supply, Development, Operation, and Maintenance.

ISO/IEC/IEEE 14764, Software Engineering – Software Life Cycle Processes – Maintenance offers expanded guidance on the processes that comprise the Maintenance category defined in IEEE Std 12207. IEEE Std 14764 describes the importance of a software maintenance strategy, and provides guidance on how to formulate and compose a software maintenance strategy once a software maintenance concept is defined [5].

The software maintenance concept defines the high-level goals of the software maintenance effort in terms of the life-cycle program scope, personnel responsible for the software maintenance processes, estimated iterative software maintenance costs based on personnel and resources over the course of the software life cycle, and software maintenance plans that define the budget, tasks, controls, schedule, and standards. Further guidance on maintenance concept preparation is provided in Chapter 4.

Plans are underway to revise ISO/IEC/IEEE 14764 under a joint development project within the boundaries established by the Partner Standards Development Organization (PSDO) agreement between IEEE and ISO/IEC JTC 1. ISO/IEC/IEEE 14764 needs a revision to be consistent with the current best practices for software maintenance. The revision will also improve the alignment with the latest version of the ISO/IEC/IEEE 12207: 2017 standard.

11.4 Impact of Maintainability on Software Design

Software maintainability has the greatest impact on the software design when software maintainability activities are planned and performed early in the software development process. Software maintainability activities must be planned for during the initial software requirements generation phase of a development program, as well as after development, for future software modifications that will be necessary after the software is released to the customer. The necessity for this planning is based on the types and amount of software failures experienced in the field and the plans to add new features for technology insertion to refresh the software product and keep the customers delighted with benefits from technology advancements. In order to enable these future software modifications after initial product release, software should be designed up front for reuse. Designing the software for reuse achieves many of the objectives for designing the software for maintainability. Using a modular software design schema goes a long way to developing reusable code as well as maintainable code. Modular code design ensures that there is only one overall task to be accomplished for any particular software function. Using an object-oriented software design has advantages in the software development process in terms of development cycle time, and is also beneficial for software maintainability in terms of ease with which maintainers can understand the functions of the software and quickly restore the software to operational service. Following certain software coding practices and design rules, such as uniform naming conventions, commenting on each Line-of-Code (LOC), commonly used coding styles and templates, and standard documentation procedures, not only benefits the software design process but also enhances software maintainability.

Software design metrics derived from software maintainability impact the software design. Various software timing metrics are leveraged directly from software maintainability requirements. For instance, the Mean-Time-to-Repair (MTTR) maintainability metric was adapted for telecommunications system software applications to evolve into the metric referred to as Interruption of Service (IOS). IOS is considered a maintainability metric for software telecommunication systems, just as MTTR is considered a maintainability metric for hardware. IOS is used as a software maintainability metric predominantly in the telecommunications industry to monitor and track digital network system outages, to measure and analyze root causes of downtime occurrences, and to record and comparatively assess the recovery times associated with each downtime occurrence to develop network system design improvements. IOS metrics may be used to develop another metric in the telecommunications industry called Mean-Time-To-Restore-System (MTTRS) or Mean Time to Restore Service (MTRS). MTTRS or MTRS are used to calculate the average time delta from the time a data or voice network fails to the time when the network is fully restored to its normal system functionality. Further examples of software maintainability metrics are Mean-Time-to-Restore (MTTR) and Mean-Time-to-Recover-From-Abort (MTTRFA). MTTR and MTTRFA were developed considering embedded system software applications. And there are many other examples of maintainability metrics that impact the software design.

"Maintainable code is code that exhibits high cohesion and low coupling. Cohesion is a measure of how related, readable, and understandable code is. Coupling is a measure of how related code is, low coupling means that changing how A does something shouldn't affect B that uses A. Generally lower coupling means lower cohesion, and vice versa, so a developer needs to find an appropriate balance between the two [6]."

"Maintainability is itself a measure of the ease to modify code; higher maintainability means less time to make a change. Coding standards are a way to achieve high maintainability and are developed as a result of previous experiences; they aren't universal and are dependent on developer preferences [6]."

Lastly, use of common software design tools and software configuration management tools benefit the software design process as well as software maintainability.

11.5 How to Design Software that is Fault-tolerant and Requires Zero Maintenance

Designing software for maintainability means many things to many people. One of the things it means is fault tolerance in computer system applications. As previously defined in this chapter, fault tolerance is the ability of a software system or product to withstand exposure to fault conditions without causing a system crash or critical failure effect. Computer systems designed with fault-tolerant architecture are capable of detecting data errors and system faults without causing a detrimental effect to the data or the system performance. Fault-tolerant architecture is able to absorb the effects of defects, errors, or bugs that manifest as faults in the computer system, and allows the computer system to continue to work as intended in spite of the faults. Fault-tolerant architecture within a computer system may work in one of two ways:

1. The computer system allows the accumulation of errors or faults and takes no action.
2. The computer system assigns and schedules a corrective action to remedy the errors or faults when an error limit is reached or a high-priority error is flagged.

Paradigm 9: Support maintainability with data

These types of computing systems using fault-tolerant architecture are also called error-resilient computing, probabilistic computing, or stochastic processing. These types of fault-tolerant architectures may be considered the same or similar with only minor differences. For instance, error-resilient computing will most likely contain the capability to detect attempts from cyber threats that try to exploit vulnerabilities in the software design, resisting any attacks, and thereby fostering and bolstering an organization's cyber-security posture. For probabilistic computing, predictions of system performance are constructed to assess when or if a system is at risk of a crash and should be brought offline for maintenance. Probabilistic computing is different from deterministic computing, where data collected provides empirical evidence as proof that a system will crash at a certain time or during a certain condition, as opposed to a probability that a system will or will not crash. Whatever the name, the fault tolerance approach is not to immediately correct errors once they are identified, but rather assess them, or ignore them for the time being, or potentially ignore them forever. "Errors will abound in future processors … and that's okay" [7].

As semiconductor-based product quality variability worsens as circuit density increases, and hardware power consumption increases, causing concerns from the electronic products industry, probabilistic computing with error-resilient design features and architectures will become prolific in the marketplace, from supercomputers to smartphones.

Other types of fault-tolerant designs enable a system to continue operation at a somewhat reduced level of performance, rather than causing a system crash when the system detects a fault. In the event of a fault, this type of computer system is designed to continue, more or less fully functional, with perhaps a reduction in data or command throughput, or an increase in data latency or processing response time. This type of fault related to data latency has the effect of slowing down data transmission on a data bus, but the system user would not experience a change in system response times. It could also mean the system is robust and able to handle a single fault condition with a limited self-repair capability. For example, assume an automobile is designed to continue to be drivable if one of the tires is punctured. In this example, the automobile's sensors and software detect an air pressure leak in one of the tires, and initiates a command to repair the leak from inside the tire using a type of pressure leak sealant material.

Fault tolerance can be achieved by designing a system that anticipates every type of exceptional condition and aims for self-stabilization so that the system converges towards an error-free state. "Self-stabilization is a concept of fault tolerance in distributed computing. A distributed system that is self-stabilizing will end up in a correct state no matter what state it is initialized with. That correct state is reached after a finite number of execution steps" [8]. A self-stabilizing fault-tolerant system is one way to achieve zero maintenance. If a fault-tolerant system anticipates every fault condition and executes a process to mitigate the effects of the fault without alerting an operator or maintainer to perform a maintenance action, it is said that the system performs with zero maintenance. But this type of system may be very costly. If the cost of designing a fault-tolerant system is very high, and cannot be afforded, a better solution may be to use some form of duplication or redundancy [9].

Redundant hardware and software components allow for failures to occur in a particular system operation by leveraging spare capabilities or backup features. There are different types of configurations for multiple processors in fault-tolerant architecture. These processor configurations may be called master–master configuration and master–slave configuration. There are multiple types of backup features that may be employed using redundant components. These back-ups are hot backup, warm backup, and cold backup.

The thought process here is to let errors happen using redundant backup capabilities, without the need to conduct maintenance immediately, or potentially not at all. The number of redundant backup features for any particular function and application of a system are designed so that a certain number of errors can be tolerated over a period of time under specified conditions and use cases. Depending on the application and use case, error rates are tracked and kept under a pre-assigned performance threshold using algorithms or circuit techniques. "For many applications such as graphic processing or drawing inferences from huge amounts of data, errors in reasonable numbers do not materially impact the quality of the results. After all, your eye wouldn't even notice the presence of a single bad pixel in most images" [7].

For redundancy in a fault-tolerant system to be effective, the system must be able to revert back to a fail-safe mode, or last known-good state. This is similar to a rollback recovery. This function can be performed automatically without a human in the loop, or the function could possibly be performed as a human action if humans are available to perform the action in a timely and cost-effective manner.

Certain types of computer data storage devices or hard drives use a technology called Redundant Array of Independent Disks (RAID) to provide a fault-tolerance capability using software data redundancy with hardware redundancy. RAID is a type of fault-tolerant system called a lockstep machine. A lockstep machine uses replicated synchronous elements composed of hardware and software that operate in parallel as hot backups. "At any time, all the replications of each element should be in the same state. The same inputs are provided to each replication, and the same outputs are expected. The outputs of the replications are compared using a voting circuit. A machine with two replications of each element is termed Dual Modular Redundant (DMR). The voting circuit can then only detect a mismatch, and recovery relies on other methods. A machine with three replications of each element is termed Triple Modular Redundant (TMR). The voting circuit can determine which replication is in error when a two-to-one vote is observed. In this case, the voting circuit can output the correct result, and discard the erroneous version. After this, the internal state of the erroneous replication is assumed to be different from that of the other two, and the voting circuit can switch to a DMR mode. This model can be applied to any larger number of replications" [9].

"One variant of DMR is pair-and-spare. Two replicated elements operate in lockstep as a pair, with a voting circuit that detects any mismatch between their operations and outputs a signal indicating that there is an error. Another pair operates exactly the same way. A final circuit selects the output of the pair that does not proclaim that it is in error. Pair-and-spare requires four replicas rather than the three of TMR, but has been used commercially. [9]"

Another type of a lockstep design feature called lockstep memory is used for computer memory functions, as opposed to RAID, which is used for computer data storage functions. Lockstep memory uses Error-correcting Code (ECC) memory to describe a multichannel memory layout in which cache memory lines or blocks are distributed between two memory channels, so that half of the cache block is stored in a Dual In-line Memory Module (DIMM) on one channel, while the other half of the cache block is stored in a DIMM on a second channel. "By combining the single error correction and double error detection (SECDED) capabilities of two ECC-enabled DIMMs in a lockstep layout, their single-device data correction (SDDC) nature can be extended into double-device data correction (DDDC), providing protection against the failure of any single memory chip" [10].

Chipkill [11] is IBM's trademark for a form of advanced ECC memory technology that protects computer memory systems from any single memory chip failure, as well as multi-bit errors from any portion of a single memory chip. "Chipkill is frequently combined with dynamic bit-steering, so that if a chip fails (or has exceeded a threshold of bit errors), another, spare, memory chip is used to replace the failed chip. One simple scheme to perform this function scatters the bits of a Hamming code ECC word across multiple memory chips, such that the failure of any single memory chip will affect only one ECC bit per word. This allows memory contents to be reconstructed despite the complete failure of one chip" [11]. Hamming code is an ECC code used to detect and correct software errors that occur when data are transmitted from the originator or host to the intended data recipient or client, and the data is stored by the client. Hamming code uses redundant bits or extra parity bits to facilitate the identification of an error. This type of ECC technique was developed by Richard W. Hamming in 1950. Viterbi decoders use the Viterbi algorithm to decode a digital bitstream that was encoded using Forward Error Correction (FEC) code or channel code. FEC is a technique for controlling data errors in message transmissions over noisy communication channels. The Hamming distance is used as a metric for Viterbi decoders. Hamming distance measures the number of differences in bit values between two bit strings. More advanced codes may be used, such as a Bose–Chaudhuri–Hocquenghem (BCH) code that is a form of ECC code from a class of cyclic ECC codes, which can correct multiple bits with less computer processing overhead. BCH codes were invented in 1959 by Alexis Hocquenghem, and improved upon in 1960 by Raj Bose and D.K. Ray-Chaudhuri.

11.6 How to Design Software that is Self-aware of its Need for Maintenance

Electronic system self-awareness is becoming more and more prevalent throughout industry in a variety of technologies, such as artificial intelligence, machine learning, fully and semi-autonomous systems, robotics, communication networks, cyber security, and industrial automation controls. Recent developments in the Internet of Things (IoT) technology are an enabler for these system self-awareness capabilities. Robust system health, performance, and condition monitoring for self-awareness should be an integral part of any software fault-tolerant product design, especially where the mission is critical and catastrophic events may lead to safety-significant concerns. Self-awareness should be a key software design feature in any fully autonomous and semi-autonomous computing system incorporated into the Real-time Operating System (RTOS) processing elements that assess system health and conduct diagnostics with embedded test functions. A system is said to be self-aware when it detects a problem in its internal performance without the need for a human in the loop, and determines on its own that it needs some form of maintenance to prevent a failure or correct an error condition. One way to provide system health condition self-awareness features is through the use of a watchdog timer.

The watchdog timer is a counter that counts down from an initial value to zero. The watchdog timer function can be implemented in an application where a single loop controls the entire system. This watchdog timer function begins by initializing the timer to the preset limit stored in a buffer. The system controller begins to execute its instructions and functions in sequential order while the watchdog timer counts down. The instructions are stored in a stack from the top to the bottom of the stack. When the controller completes executing its instructions and functions, reaching the bottom of the stack, the controller loops back to the beginning of the stack. At this time, the watchdog timer is reset to the initial value. The watchdog timer should be reset before it counts down to zero. If the watchdog timer reaches zero, the watchdog timer times out and throws a flag to the controller. This condition may occur if the controller or a software function gets hung up, or spends too much time executing its instructions – for instance, when an Interrupt Service Routine (ISR) takes longer than expected to execute as a result of corrupted data caused by a line power anomaly, or an infinite loop causes a program lock-up. These types of ISR errors may cause the controller to execute an exception-handling routine that performs a software reboot and reinitializes the system, or the controller may be able to detect the ISR error, perform diagnostics to verify all functions are working, and gracefully resolve the problem without causing a software reboot or system shutdown.

The watchdog timer function can be implemented in another application called the check-twice method. This type of watchdog timer function tracks function completions using flag bits. A flag bit is assigned to each function. The flag bit indicates whether a function was completed or not completed. A flag bit that is set means the function was completed successfully. A flag bit that is cleared means the function was executed incorrectly or did not complete as instructed. During the course of a program cycle, the flag bits are monitored. If the controller sets all flag bits as complete, the watchdog timer resets, and the cycle starts over. The watchdog timer initiates a system reset if any of the flag bits are not set as complete. Following the system reset, all flags are cleared and the program cycle begins again.

"The watchdog timer should detect infinite loops and deadlocks as well as the failure to run lower-priority tasks. Infinite loops and deadlocks can occur within a task when two or more related tasks are so far out of synchronization that they lock up waiting for each other [12]."

As a third application, the watchdog timers and multi-tasking strategy may be used. In this application, watchdog timers should be high-priority functions run as separate tasks. The watchdog timer function should check the health status of the other functions at intermittent intervals. The watchdog timer should detect functions that halt or lag behind their timing requirements, providing the data to support a high-priority critical failure decision, so that a system restart may occur. If the watchdog timer detects that all functions in the system complete successfully, the system continues to operate as normal and the watchdog timer reinitializes. The watchdog timer is a critical function in managing the health of a system and a self-reliant recovery solution [12].

Further details about artificial intelligence and autonomous systems' self-awareness capabilities are discussed in Chapter 8.

11.7 How to Develop Maintainable Software that was Not Designed for Maintainability at the Start

Often, software is developed without consideration for maintainability. Software will be released and deployed in customer use applications without a strategy to correct code bugs or to provide enhanced design features in the aftermarket. The software product developers may realize over time that the software is broken or customers demand improvements, and the software product must be updated in an economic way for the customer and the product developer. Software warranties and software maintenance agreements are usually provided by the developer to the customers to establish the boundaries of the costs for software maintenance over the life of the software. These warranties and software maintenance agreements will state that the software product may be updated or patched at intermittent times during the life of the product.

Following initial release of a new software product, the software code is subject to change with incremental updates over time. These software updates are called patches or maintenance releases. The purpose of patches is to fix software bugs and correct software defects, such as memory leaks, while the customer is using the software product. Software patches are usually sent to customers as service pack downloads through the internet. Installation of service pack updates may occur at any time during the software life cycle when the software code developer feels responsible to service their software product. The service packs may not only contain software patches to fix bugs in the code, but they may also add new features to improve performance, thereby enhancing the software design functionality and delighting customers with improved and innovative user experiences. The software patches may even plug security holes and resolve cyber security weaknesses and vulnerabilities. These code updates are changes that are important to ensure a user's digital safety and privacy and avoid cyber attacks.

11.8 Software Field Support and Maintenance

The IEEE defines three basic types of software maintenance [5]. These methods are corrective maintenance, adaptive maintenance, and perfective maintenance. Corrective maintenance is the action taken to minimize product downtime following the occurrence of a software failure, to fix flaws in the original source code or software specifications, and to restore the product to operational service. Adaptive maintenance is the action taken to adjust software to comply with changes in the technological environment, including version upgrades, conversions, recompiles, and the reassembly and restructuring of code. Perfective maintenance is the actions taken to expand and improve the functionality of an existing software system.

Software maintenance is a primary process in the life cycle of a software product, as described in ISO/IEC/IEEE 12207 [4]. Maintenance is one of the five primary life-cycle processes that may be performed during the life cycle of software. These five primary processes are:

1. Planning
2. Analysis
3. Design
4. Implementation
5. Maintenance

The software maintenance process contains the activities and tasks necessary to modify an existing software product to implement design improvements and restore the product to its original condition, while preserving its integrity or enhancing the product to a condition that is better than the original intent of the product. ISO/IEC/IEEE 14764 [5] provides guidance on the software maintenance process. It describes an iterative process for managing and executing various software maintenance activities. The software maintenance process can be categorized and grouped into the following six major activities:

1. Process implementation
2. Problem and modification analysis
3. Modification implementation
4. Maintenance review and acceptance
5. Migration
6. Retirement

Each of these major activities is described in the following sections.

11.8.1 Software Maintenance Process Implementation

During process implementation, the maintainer establishes the plans and procedures which are to be executed during the maintenance process. The maintenance plan should be developed in parallel with the development plan.

The inputs for the process implementation include:

1. Relevant baselines
2. System documentation, if available
3. A Modification Request (MR) or Problem Report (PR), if applicable

In order to effectively implement the maintenance process, the maintainer should develop and document a strategy for performing the maintenance. To accomplish this effort, the maintainer should execute the following tasks:

1. Develop maintenance plans and procedures
2. Establish MR/PR data collection and handling procedures
3. Implement configuration management
4. Develop a configuration management plan

The outputs of this activity are:

1. The maintenance plan
2. Training plan
3. Maintenance procedures
4. Project management plan
5. Problem resolution procedures
6. Measurement plan
7. Maintenance manual
8. Plans for user feedback
9. The transition plan
10. Maintainability assessment
11. Configuration management plan

11.8.2 Software Problem Identification and Software Modification Analysis

This and subsequent activities are activated after the software transition and are called iteratively when the need for modification arises.

During the problem and modification analysis activity, the maintainer:

1. Analyzes MRs/PRs, if they exist
2. Replicates or verifies the problem
3. Develops options for implementing the modification
4. Documents the MR/PR, the results and execution options
5. Obtains approval for the selected modification option

11.8.3 Software Modification Implementation

During the modification implementation activity, the maintainer develops and tests the modification of the software product. Further discussion of software test activities appear later in this chapter.

11.8.4 Software Maintenance Review and Acceptance

This activity ensures that the modifications to the system are correct and that they were accomplished in accordance with the approved standards using the correct methodology. This review may occur during a software design review or as part of the software change control process. The software change control process is described later in this chapter.

11.8.5 Software Migration

During a system's life, software may need to be modified to correct or prevent causes of system downtime and to allow the system to be able to run in different applications and environments as technology evolves or new uses for the system are found. Software changes will most likely be required in order for the software to be migrated to a new environment or application. The software changes needed to enable the system migration path will involve software regression analysis and software development activities to some degree, depending on the percentage of new software that must be written.

11.8.6 Software Retirement

Once a software product has reached the end of its useful life, it must be retired. An analysis should be performed to assist in making the decision to retire a software product. Analysis should determine if it is cost-effective to do the following:

1. Retain outdated technology
2. Shift to new technology by developing a new software product
3. Develop a new software product to:

a. Achieve modularity and standardization
b. Facilitate maintenance and vendor independence

Paradigm 6: Modularity speeds repairs

11.8.7 Software Maintenance Maturity Model

A software maintenance maturity model [13] was created to assist organizations in launching and implementing a continuous improvement program tailored to software maintenance. This model helps organizations to benchmark their existing software maintenance process and compare it with an established software maintenance maturity model. This established Software Maintenance Maturity Model (S3m®) is a software process capability model leveraging the Carnegie Mellon University (CMU) Software Engineering Institute (SEI) Capability Maturity Model Integration (CMMI®) [14], ISO-9001 [15], and ISO/IEC 15504 [16]. There are some similarities between S3m and the SEI CMMI model, but there is one major difference. The SEI CMMI model defines five levels of capability, while the S3m model has two levels of capability. Other similar models are the Constructive Cost Model (COCOMO) Maintenance Model [17], the Adaptive Maintenance Effort Model (AMEffMo) [18], and an economic maintenance model based on the estimation of software rewriting and replacement times [19].

The S3m model helps software maintenance organizations identify their strengths and weaknesses in delivering software maintenance services to their customers. The software maintenance services are divided into three categories:

1. Internal Service Level Agreements (SLAs)
2. Maintenance service contracts (external service agreements)
3. Outsourcing contract (third-party agreements)

The S3m model allows organizations to uncover gaps in their software maintenance process. "With the identification of a gap, maintenance organizations can identify, through comparison with the model, what issues to address and how to address them, and by doing so, improve their software maintenance processes" [13].

For example, let's suppose a software development organization supports a software maintenance service organization within the same company and there is no SLA between them. Lack of an SLA will create confusion as to where resources should be assigned when unusual software services are needed, such as a software design change request from the software maintenance service organization to the software development organization to remedy a new software defect identified by a user application in an environment not previously tested, and what the priority of services should be to determine when resources will be assigned. The S3m model allows management to identify the weakness in the software support infrastructure and to create an agreement between the two organizations to establish rules and guidelines on how they will work together in the future to support their customers.

11.9 Software Changes and Configuration Management

As any software developer knows, software changes will occur after software is initially written, debugged, integrated, tested, and released. It is the prudent developer who accepts the fact that the software design must be locked down and tightly controlled for future changes after formal software release. Managing these software changes is among the most difficult parts of software product management. Software changes are expected during the software development phase of a program, but they may not be so obvious and planned for in the post-deployment and support phase of the program after the initial software release at the design baseline freeze point. Software design changes that affect a software product's maintainability performance are less obvious and require specialized engineering involvement to ensure program success in the O&S phase of the program. A prudent software program will include processes and procedures to guide the software design team in creating software change requests, preparing for software change orders, and adapting to software engineering design changes during the course of development and continuing into the O&S phase of the program to allow for software maintainability design improvements along the way based on user needs and the expected support environment. Software Change Control Management (SCCM) is a required function of any software development organization to keep up with the software changes during the life of the product.

Even if the software never fails, it should be easy to understand for future software design changes made by someone other than the original designer. For these cases when software is changed for reasons other than to repair a software defect, the software is said to be supportable as well as maintainable. Software supportability means the code is easy to change by adding new functionality and design features. Software maintainability and supportability engineers use SCCM to ensure the code is upgradeable and extensible in the future. Extensibility is the quality of a system design that allows for the addition of new capabilities or functionality. Extensibility of the software is important for supportability and maintainability. Extensibility allows for simple regression analysis and regression testing, ensuring that any single software Line-of-Code (LOC) change is not going to break the code in multiple places.

Documentation of software change requests and software change orders is important for software configuration management, as well as beneficial to the software design process and to software maintainability. For example, Microsoft's Developer Network (MSDN) Library includes useful guidelines on software project configuration management, which helps software maintainability. MSDN [20] is a branch of Microsoft that is responsible for managing their relationship with software developers and testers who are interested in changing the operating system (OS), such as MS Windows, and software developers developing on the various OS platforms or using the Application Programming Interface (API) or scripting languages of Microsoft's applications. The MSDN guidelines include: (i) the use of a few separate configuration files to reduce software complexity during software change control activities; (ii) assigning default values to optional configurable items; (iii) separating optional configurable items from required configurable items; and (iv) maintaining thorough and complete documentation, which describes all configurable setting relationships. The documentation should be stored in the formal software configuration control repository and its associated resources. This repository will help in the future when software changes are needed and documentation must be found to describe the last known software release.

11.10 Software Testing

Software testing is an activity that directly supports software maintainability in so many ways, too numerous to discuss within the confines of this chapter. At its best, software testing provides all those activities necessary to assess how a software product will perform when it is used under specified conditions, over a specified time, while exposed to various cases of user and environmental stresses. Software test results will be observed and recorded either autonomously, semi-autonomously, manually, or a combination thereof. As a conclusion of the testing, a decision will be made as to whether the software test results adequately met some or all of the test requirements. If software testing results in a failure, software design changes will be implemented using the established SCCM process to incorporate corrections in the software code design.

Software test activities seek to find software bugs, errors, defects, faults, and failures early in the software development process, and to determine how to handle these issues as the product develops and matures. Software test activities should strive to ensure that no critical software failures escape to the customer or are detected in any of the expected user application field environments that are possible. Software testing involves the development of multiple test cases. These test cases are documented instructions for a tester describing how a software function or combination of software functions shall be tested. Test coverage is the extent to which the test cases verify that a software function or combination of software functions met their requirements. Based on the comprehensiveness of the test cases to model the various expected customer use cases, and the high levels of test coverage to ensure software defects are detected, whether in software development test, production test, or customer field support test, software test capabilities will make or break software maintainability.

11.11 Summary

The majority of the life of any software system or product is spent in the software life cycle's maintenance and use phase, also known as the O&S phase. The customer should realize the ultimate value of the software system or product during this phase, with minimum downtime, substantiating the initial procurement or acquisition cost of the system or product. Therefore, it is imperative to design for software maintainability so that maintenance will be performed economically, effectively, and efficiently to unlock the benefits from the software continuously over the system or product life cycle. The software developers designing their products with fault-tolerant architecture and design features provide software maintainability with value-added benefits for their customers. Software maintainability design features and capabilities evolve as the state of the technology changes and as evidence of new software defects and cyber security threats in the software environment are discovered, leading to software functional enhancements in the system or product.

"Lehman (1980) states that "being unavoidable, changes force software applications to evolve, or else they progressively become less useful and obsolete" (cited in [13]). Maintenance is therefore considered inevitable for application software that is being used daily by employees everywhere in a changing organization."

References

1 IEEE (1990). *IEEE Standard Glossary of Software Engineering Terminology, Std 610.12-1990*. Piscataway, NJ: IEEE.

2 Pfleeger, S.L. (1998). *Software Engineering, Theory, and Practice*. Prentice-Hall, Inc.

3 Visser, J. (2016). *Building Maintainable Software, Ten Guidelines for Future Proof Code*. Software Improvement Group (SIG), O'Reilly.

4 ISO/IEC/IEEE (2017). *ISO/IEC/IEEE 12207, Software Engineering – Software Life Cycle Processes*. Geneva: ISO/IEC/IEEE.

5 ISO/IEC/IEEE (2006). *ISO/IEC/IEEE 14764, Software Engineering – Software Life Cycle Processes – Maintenance*. Geneva: ISO/IEC/IEEE.

6 Software Engineering Stack Exchange. What characteristics or features make code maintainable? https://softwareengineering.stackexchange.com/questions/134855/what-characteristics-or-features-make-code-maintainable (accessed 15 September 2020).

7 Lammers, D. (2010). The era of error-tolerant computing. *IEEE Spectrum* **47**: 15.

8 Wikipedia Self-stabilization. https://en.wikipedia.org/wiki/Self-stabilization (accessed 6 September 2020).

9 Wikipedia Fault tolerance. https://en.wikipedia.org/wiki/Fault_tolerance (accessed 6 September 2020).

10 Wikipedia Lockstep (computing). https://en.wikipedia.org/wiki/Lockstep_(computing) (accessed 6 September 2020).

11 Wikipedia Chipkill. https://en.wikipedia.org/wiki/Chipkill (accessed 6 September 2020).

12 Baker, B. (2003). Use a watchdog timer even with perfect code. EDN.com. https://m.eet.com/media/1136009/289962.pdf (accessed 6 September 2020).

13 April, A. and Abran, A. (2008). *Software Maintenance Management, Evaluation and Continuous Improvement*. Wiley.

14 Carnegie Mellon University (CMU) (2010). *Software Engineering Institute (SEI) Capability Maturity Model Integration (CMMI®)*. https://www.sei.cmu.edu/news-events/news/article.cfm?assetid=509086 (accessed 18 September 2020).

15 ISO (2015). *ISO-9001: 2015, Quality Management Systems*. Geneva: ISO.

16 ISO/IEC (2004). *ISO/IEC 15504, Information technology – Process assessment*. (Note: ISO/IEC 15504 has been replaced by ISO/IEC 33001:2015 *Information technology – Process assessment – Concepts and terminology* as of March 2015 and is no longer available at ISO.)

17 Boehm, B.W. (1981). *Software Engineering Economics*. Prentice-Hall.

18 Hayes, J.H., Patel, S.C., and Zhao, L. (2004). *A Metrics-Based Software Maintenance Effort Model*. Piscataway, NJ: IEEE.

19 Chan, T., Chung, S.L., and Ho, T.H. (1996). *An Economic Model to Estimate Software Rewriting and Replacement Times*. Piscataway, NJ: IEEE.

20 Wikipedia. Microsoft Developer Network. https://en.wikipedia.org/wiki/Microsoft_Developer_Network (accessed 6 September 2020).

12

Maintainability Testing and Demonstration

David E. Franck, CPL

12.1 Introduction

This chapter emphasizes the processes and tools involved in verifying the Design for Maintainability (DfMn) "goodness" of a system or product near the end of its design and development phase, called the demonstration and qualification test phase. The engineering effort that is performed during this phase determines whether or not the design actually meets the stated maintainability requirements. Testing for maintainability up to this point is performed to verify and validate the DfMn capabilities of the system or product design. The fundamental maintainability performance exhibited during the design's demonstration and qualification test phase is intrinsic to the attention, discipline, and rigor applied during the program's design phase.

Once the product transitions to the manufacturing and production phase, various types of inspections, analyses, and testing are performed to verify the production assembly, test, and workmanship processes. These types of inspections, analyses, and testing may be performed as part of the qualification test phase when a product line is initially started up, or it may be performed periodically as part of a lot-by-lot quality conformance inspection or sampling test during the entire production phase of a program. For certain customers, validation testing may continue when the product is delivered and deployed in the customer use application. It is worth noting that many product customers, particularly those from the military marketplace and large-scale system users, continue to track product maintainability performance while it is operated and maintained throughout its life cycle. Based on the results of this maintainability tracking and validation, customers may revise maintainability requirements and introduce incremental product improvement design changes to methodically and sequentially lower the cost to maintain a product over its life cycle. It must be considered, however, that later-in-life design changes are extremely costly when compared with design changes early in the product development process. Product design change costs increase throughout the life cycle from the concept and requirements phases, through the various design baselines and prototypes, through the production phase, and finally into the deployment and customer use phase. Thus, it is highly advisable to identify design deficiencies as early in the product design process as possible to minimize the cost of design changes that may occur later in the product development process or the product life cycle.

During the design phase, the process of testing the inherent maintainability characteristics of a product design involves a series of activities and events intended to analyze, measure, verify, and validate that a product's maintainability design does, in fact, achieve the stated design goals and documented specification performance requirements. The terms "verification" and "validation" are frequently used

Design for Maintainability, First Edition. Edited by Louis J. Gullo and Jack Dixon.

interchangeably, as if they are the same. These terms are not the same, but they are closely related. The different meanings between verification and validation are summarized as follows:

- Verification: Are you building the product right? This is a measure of properly designing and manufacturing the product as per the requirements and specifications. Verification includes design inspection, analysis, test and evaluation, and demonstration.
- Validation: Are you building the right product? This is a measure of whether the product meets the user needs. Validation includes usability tests, operational tests, and formal government testing and demonstrations involving the customer and representatives from the user community.

Early in the design process, requirements are established for the product's maintainability performance and a Maintainability Program Plan (MPP) is established. The MPP should address, as a minimum, the various roles and responsibilities of all participants in the product maintainability and design, the schedule of the plan activities aligned with program level milestones, the initial decomposition of the maintainability requirements to lower product design elements, and the tasks included in the test program. The MPP will be the guide for maintainability activities throughout the product's development, and needs to be updated as significant changes and progress are realized. Annual updates to the MPP are recommended.

The MPP will reference a Maintainability Test Plan as a separate document or an addendum to the MPP. The Maintainability Test Plan will document all inspections, analyses, and testing to be included in the maintainability testing process. This maintainability testing process starts early in the product design cycle with detailed planning, budgeting for costs of the tests, and scheduling milestone test events. The maintainability test process is conducted throughout the design evolution and continues into the operational use environment. More details of the Maintainability Test Plan (also known as the M-Demo Test Plan) are discussed later in this chapter, and more details about the overall MPP are discussed in Chapter 3 of this book.

12.2 When to Test

Testing is often thought of as an effort to be conducted after there is an actual product to test, or at least a prototype of the product. This is definitely not the case in the situation where the product's characteristics are as designed-in as maintainability. Maintainability testing should be an integrated part of a product design and development program from the very beginning of the process, and this should be reflected in the overall program schedule. Without an integrated approach to testing, certain risks and consequences can ensue, as shown in Table 12.1 [1].

An integrated, comprehensive maintainability testing program includes analysis and evaluation of the product requirements, the analyses and testing during the product design evolution, and the tools used to conduct the testing. Testing the design from the beginning provides early insight into whether the design in on the right track or not. It also provides early recognition of problems so that they can be corrected as

Table 12.1 Risks and consequences of a testing approach that is not integrated [1].

Risks	Consequences
Critical tests are omitted	Design shortcomings may appear after the customer assumes ownership of the product
Tests are duplicated	Development costs increase and schedules are affected
Test resources are inadequate	Tests are delayed, results are incomplete, results are inaccurate or invalid, faults are missed, and product performance suffers
Test schedules are not coordinated	Inadequate time for testing, tests occur in wrong sequence, tests compete for critical test equipment, test requirements are not met
Schedules are milestone-oriented	Test results seem to confirm progress but do not result in needed product design improvements

Source: US Department of Defense, *Designing and Developing Maintainable Products and Systems MIL-HDBK-470A*, Department of Defense, Washington, DC, 1997.

economically as possible. Recognizing problems early means they can be corrected for lower cost, compared with waiting until later in the development process.

A product is typically designed for a specified and bounded operational and maintenance environment which should be reflected in the Concept of Operations (CONOPS). To help the design and the operational support teams, the CONOPS should, among other items, define the expected maintenance and support infrastructure for the product. Early on, these terms are often less specific but can provide important insights for the design and support teams. Defining the maintenance environment, personnel, skill definitions, standard tool sets, available support equipment, and maintainer training provides a wealth of information that can be turned into requirements for the designers. Also, it would be helpful to know which country(s) will be using and supporting the product, whether the maintenance will be performed by a military customer or a contractor, and which branch(es) of the military will be performing the maintenance. Early in a product's life cycle, such general characteristics are usually known (or at least should be known) and shared with the potential design team. These initial considerations become important boundary markers for the design and the associated maintainability models. These boundary markers eventually become formal requirements and will be further refined and decomposed into more detailed requirements, further aiding the design process and the veracity of the model.

Product integration testing, where separate parts or subsystems of a product are combined, is an excellent time to test out planned or anticipated maintenance procedures and tooling. Naturally, there is a focus during such testing on design interfacing, performance, and interaction. However, maintenance is required to assemble and integrate the parts, which presents the occasion to either apply the intended maintenance tools and procedures or to observe the tools and procedures employed by the test team. If possible, it is advantageous to have these maintainability aspects written into the test plan, specifying the tools, procedures, and times to be used during assembly, disassembly, and maintenance. If the integration of maintenance processes with the testing processes is not possible, the maintainability engineer should try to be present during all testing events to collect all pertinent maintenance data and maintenance task times, as much as possible. This information can then be reviewed and merged into the maintainability model and into the tools and maintenance procedures analyses. Of course, any maintainability problems observed should be coordinated with the design team for resolution.

The maintainability engineer should attempt to be involved in or at least be available to witness all product testing of other team elements where maintainability may be a consideration. An experienced maintainability engineer is often able to see actual or potential maintainability issues that others don't see. Keen observations during the assembly and disassembly offer the chance to see the tools and procedures the test and engineering teams use, providing insight into potential issues for field maintenance. This opportunity is further enhanced as higher-level unit and subsystem integration testing is conducted as the product design matures. Additionally, product testing at all levels is known to experience problems and parts failures from time to time. These are important occasions to observe the processes, troubleshooting procedures, and the tools used for the correction of the problem experienced. An astute maintainability engineer might consider taking notes on the maintenance steps performed, noting the tools and the times required for the work. Some programs require a maintenance log to accompany all development products to capture and record all maintenance actions performed on the part, regardless of who performs it.

Generally, formal maintenance testing is conducted during a Maintainability Demonstration (M-Demo) which is a structured, formal testing of the maintenance characteristics of the product. The M-Demo is defined as "The joint contractor and procuring activity effort to determine whether specific maintainability contractual requirements have been achieved" [2]. The M-Demo is also a test of the infrastructure designed to support the product when in use, such as training, technical documentation, and tools. The M-Demo generally is conducted after there is a product to test that represents the final design, but it is sometimes appropriate to conduct the M-Demo earlier in the product development. This is especially appropriate in instances where there is a very long development cycle, when new technologies are incorporated in the design, or when there is a lack of confidence in the design and support structure.

The M-Demo is a major milestone for a product and should be taken seriously. The M-Demo is usually conducted near the end of the product development cycle and prior to production, Just as maintenance cannot be conducted without the appropriate support capabilities, the M-Demo also exercises and evaluates the support resources that enable product maintenance. These include support assets such as training materials, tools and test equipment, support and movement equipment, technical documentation, appropriate skills, and storage and transportation materials.

As the product approaches operational deployment and field maintenance, the last testing phase begins with operational maintenance. Operational maintenance, regardless of efforts to the contrary, invariably differs from that conducted in the sterile environs of a testing shop or the manufacturing area. Experienced product owners know this and anticipate some difference. However, until the product actually gets into the hands of the operational operators and maintainers, the differences and their impacts are unknown, undefined, and unquantified. Product owners typically have an established method of recording maintenance actions of the operators and of capturing issues encountered. This operational testing offers valuable insight to the product owners, operators, and developers alike as it provides a "real world" evaluation of maintainability performance. Operational experience can then be fed back to the developer's design team for consideration in future product developments. The operators also take this information and apply it to the operational support environment to improve their field performance.

12.3 Forms of Testing

The basic purpose of testing for maintainability is to verify that the product achieves the desired or specified maintainability performance requirements. Early maintainability test and evaluation efforts yield two benefits: (i) providing early indications of problems that require resolution, and (ii) providing confidence in design areas that are developing well. Both program management and the design team are further ahead with such knowledge. Valuable engineering efforts can be focused where known issues exist and additional resources are not expended tackling design details without focus. Furthermore, another benefit is that the program management and the design team are made aware of the problems as opposed to not knowing that there is a problem at all. Such a surprise is rarely received well and wastes program resources unnecessarily. Early and frequent testing in multiple forms reduces program risks and surprises.

There are several forms of testing or verification, each of which is applicable to one or more phases of development, the quality of the data to be verified, the characteristics of the data involved, and the requirement to be tested. Other factors in choosing a verification method may include the tools and workspace available, time constraints, funding, availability of equipment (actual equipment, mock-ups, prototypes), and safety considerations. Because such testing and evaluation requires forethought and planning, program resources should also be considered. The availability of test units, talent, test equipment, schedule opportunities, funding, customer participation, and other similar factors need to be considered.

Typical forms of maintainability testing or verification that may be applied at different stages of product design and development may include the following:

- Design process reviews
- Modeling or simulation
- Analysis of the design
- In-process testing
- Formal design reviews
- Formal maintainability demonstration

For complex systems, many of the test and review events will be carried out on lower-level parts of the product and then repeated as the parts come together as units or subsystems, and then again as completed systems. Naturally, the early reviews should help to identify any maintainability issues early in the design process, with higher-level testing to verify prior test findings. Caution is recommended for very large systems where there are likely to be several, or many, different design teams working on parts of the total product. It can be difficult for large, dispersed teams to focus on their assigned portion of the overall product and consider issues or features or characteristics that affect or are influenced by other teams. The engineering leadership must maintain vigilance over the diverse teams as well as help all teams maintain focus on the system-level requirements and goals. Effective requirements definition, decomposition, and allocation early in the design process will go a long way to minimizing team myopia.

12.3.1 Process Reviews

Engineering processes form the roadmap for successful design efforts. Proven and effective design processes do not guarantee excellence in a product, but they do go a long way to mitigating potential problems, keeping the design team on a path with processes that are known (proven) and trusted, and keeping the team focused on effective methods and specific design goals. The design methodologies and processes the maintainability design team plans to use should be reviewed and approved prior to beginning the design efforts (documented in the MPP), and should be reviewed periodically to make sure the team continues following the appropriate processes and transitions to new processes as development matures.

All design efforts should be guided by established design practices and processes that are systemic to and validated by the design organization. Maintainability design activities and guidance should be an integral part of these design processes. Such process and practices should be developed over time, incorporating the organization's best practices and lessons learned, providing a consistent and effective road map for design teams based on what is known, proven, and reduces risk from past experiences. The design processes should also include processes and tools for requirements management and should define the roles and responsibilities for design engineers and logistics engineers, especially maintainability and reliability, in addition to other technical specialties involved in the design process. The approved and recommended tools, models, features, methods, considerations, and other related factors should be identified by program management and the supporting organizational management for the design teams to use. These processes and evidence of their routine use should be inspected and included in process reviews to ensure they are being used correctly.

Process reviews are used to ensure that the appropriate processes are planned for and are put into place for the design and maintainability aspects of the program. Since processes are the tools and guidelines for the design efforts, using the wrong processes or misusing the right processes could lead to significant design issues. Ensuring that the design teams are appropriately applying the correct practices, and that they continue to use the correct processes throughout the design evolution, is a major step in design success. Process reviews also ensure that all appropriate processes are considered and an explanation provided if any are left out. Process reviews also consider that the processes are being applied properly.

The MPP will document the test, demonstration, and evaluation processes to be used in each product design project. The earliest process review should occur when these plans are created and reviewed prior to approval. This is the first time to evaluate and adjust the proposed design roadmap. If any of the planned processes are improperly or inadequately addressed or missing, this early review is the best opportunity to set the team on the right path, likely preventing potentially serious design missteps later.

While conducting process reviews during the planning phase helps to get the teams off on the right foot, periodic process reviews should be conducted throughout the development cycle to make sure that the processes continue to be applied as planned and as appropriate. As the design efforts progress, it is often the case that the applicable design processes change or their use is modified as the design matures. Frequent process reviews help the design team to see these changes coming and adjust in time to allow smooth transitions. They also help the design team stay focused on where they are and are going in their daily work.

12.3.2 Modeling or Simulation

Modeling and simulation are standard tools in the product design process for many design characteristics. These tools allow design work to proceed cost-effectively and efficiently by providing insights into the results and consequences of design possibilities and alternatives, without the expense of building, testing, and redesign that would otherwise be required. For maintainability purposes, early modeling of the basic, high-level design is used to assess whether the initial design can meet the maintainability objectives and where any design deficiencies exist. The early identification of areas that require special attention is a big benefit and advantage in the design evolution. Early model development also serves to pair the maintainability and design teams together for a common purpose, albeit one of many for the design team. A maintainability model, in conjunction with a reliability model of the design, is a common tool to evaluate predicted Mean-Time-to-Repair (MTTR) and is updated frequently as the design changes and matures.

There are many modeling tools available commercially to conduct maintainability, reliability, and Logistics Support Analysis (LSA) and simulations. LSA is formally defined in MIL-HDBK-1388 as "The selective application of scientific and engineering efforts undertaken during the acquisition process, as part of the system engineering and design process, to assist in complying with supportability and other Integrated Logistics Support (ILS) MIL-HDBK-1388 objectives" [3].

The LSA goal of optimized design for supportability and defined and optimized support resources is achieved through an iterative, multidisciplinary process having many system engineering and support interfaces. This process can be divided into two general parts: (i) analysis of supportability and (ii) assessment and verification of supportability.

Data collected in the LSA process is entered into a Logistics Support Analysis Record (LSAR) as part of an integrated LSA database for the product. An LSAR is a defined database record of many ILS-related data elements that define each product part. Taken together, the LSA database identifies which support elements are needed for a product and how much of each is needed. On a small product with rudimentary support considerations, such an LSA database can be created and managed using a spreadsheet program fairly easily. However, if a robust ILS program is part of the product development, using one of the many integrated ILS data collection and analysis tools is likely to be the best option. Caution is advised to ensure that capabilities of the product(s) being considered meet the full range of data needs for the project, as some products offer partial solutions for projects for which the full range of LSA data and analysis is not required or appropriate. Also, consider whether there is a need to export product LSAR data to other team members, to a prime contractor OEM, to a system integrator, or to the customer. Those data integration requirements may influence the LSAR product decision.

Paradigm 9: Support maintainability with data

This choice fits well with a design team that realizes that maintainability design cannot be properly accomplished without many other logistics elements as part of the design considerations and tradeoffs. This is the prime benefit of a logistics integrated maintainability analysis tool, where many types of alternatives can be assessed, where design possibilities can be modeled and the affects analyzed, and where various contributory logistics elements can be exercised.

It should be remembered, when using a maintainability model, that it represents a pristine and perfect maintenance environment. This is applicable when comparing the model output to typical requirements which also assume a perfect maintenance environment. For practical field operational planning, it is important to keep in mind that there are differences between the maintenance environment assumed by a model, and in requirements. Contrary to the real world, typical maintainability modeling assumes that all the needed tools are immediately available to perform whatever maintenance is conducted; that all maintenance personnel are properly trained and up-to-date on all technical manuals, design changes, and maintenance steps; that all technical documentation is correct and properly prepared; that all training provided to the maintainers is effective and accurate; that all needed test equipment is immediately available, in working order, and properly calibrated; and that any required spares are available with zero wait times. In short, the MTTR model is a perfect representation of an ideal maintenance environment. Thus it will not reflect the real world, but it does provide a defined and consistent measuring tool against the typical MTTR requirements, which also assume an ideal maintainability environment.

12.3.3 Analysis of the Design

As the design takes shape, it is appropriate to analyze the design to verify that the design is progressing appropriately towards achieving its maintainability goals. There are several forms of design analysis that can be used to verify the maintainability design characteristics and expected performance. Such analyses should be part of the design team's processes, both formal and informal. They should include the review of analyses and assumptions made, design metric allocations made, design decisions and the basis for those decisions, and review of the design tools used and their application. Basically, there are four levels of design analysis to consider. These are: self-review by the team members doing the design; peer reviews by informed engineers, either on the team or from other teams; oversight reviews by higher-level product design personnel and management; and formal reviews by outside or customer personnel.

The prime focus of design analyses is to verify that the team has been and continues to make sound and appropriate decisions, and that the design can achieve the desired maintainability performance. Thus, all elements that go into the design are subject to review. This would include assessments of the design models used (algorithmic, design, and maintenance), design and operational assumptions made, operational scenarios considered (such as "a day in the life of …"), mechanical connectors and supports (such as screws, bolts, electronic connectors, snaps, closing mechanisms), parts that require removal of other parts for access, safety considerations (such as reach, weights, locations of electrical cabling and power sources, slip considerations, PPE needs), tools required based on the design and comparison to the tools available to the anticipated maintainers, human modeling employed (anthropomorphic models and assumptions), MTTR [1, 4] prediction models and assumptions used, training and skills required, and so on.

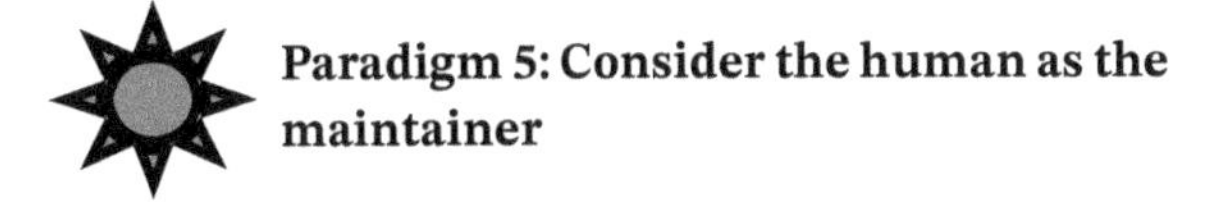
Paradigm 5: Consider the human as the maintainer

An excellent "best practice" is to ensure that in-team design analyses are conducted as a routine part of the development process. While such considerations are typically a normal part of the informal "give and take" that occurs within design teams, it behooves all involved that these reviews be conducted frequently to keep design problems transparent within the team(s) so that they can affect improvements and correct any deficiencies. Calling in peers from outside the team to review analysis decisions provides another layer of review and team-building, and functions as a "sanity check" by someone who is not intimately focused on the product. They are also important to ensure that the right tools are used for the right purpose and in the right way.

It is not uncommon that product development processes include design reviews by senior-level design personnel, usually acting as independent reviewers, to periodically review all of the design efforts, decisions, analyses, assumptions, and tool applications employed. Such a review is robust and thorough with the kind of pointed questions and insight one should expect from senior and experienced design engineers. Such design reviews are also very useful when a team is stuck on a problem, helping to get them moving again with minimal stagnant time working at a solution that isn't solving the problem.

Formal design analyses are generally conducted at specific points and milestones in a product's development cycle, and are events where the design team can share all of the work and design decisions made to date. With the customer often in attendance, formal design analysis reviews can be extensive and stressful events. Open communications and transparency are the best path forward.

12.3.4 In-process Testing

In-process testing is an informal test process whereby the design teams (outside engineers can be involved if deemed appropriate) test and evaluate their designs, processes, and products as an integral in-process part of the development process, and should be documented in the engineering schedules. Early confirmation of correctness in the design efforts and early identification of problem areas that need attention are valuable to prevent expensive issues arising later in the design process, or even more troublesome problems being discovered once the product is released for operational use. Such reviews should include evaluation of processes, models, simulations, testing, design parameters and analyses, and issue status and resolution as they relate to all aspects of maintainability.

In-process testing is intended to identify and resolve issues before final design decisions are established and to allow alternatives and impacts of options to be explored. Often seen as "sanity checks," these reviews encourage the exchange of information and ideas and bring together people with varying experiences, points of view, and skills. Because the maintainability characteristics of a product are established during the design process, maintainability engineers are integral to these reviews to focus appropriate attention on maintainability features. Additionally, testing the maintainability characteristic throughout the design process provides in-process verification of design features and their effectiveness without letting the design get too advanced, when design changes become painful to implement. Because maintainability of any product involves several other related technical support arenas, early verification that elements of the maintainability design are "good to go" frees up resources to continue other design activities and provides management with progress check points in their design and product development oversight roles.

12.3.5 Formal Design Reviews

As part of a comprehensive test and demonstration plan within a product development program, formal design reviews provide pause points where the design team, management, customers, users, and sustainment planners can gather together to view and evaluate design progress and assess forward plans. Formal design reviews should be considered as critical milestones in the design and contracting process at which the design is formally and thoroughly reviewed (including maintainability), usually by people from outside the design team and often including the customer and user community. Design reviews are milestones at which the design is formally declared to be at a given state of design completeness or readiness to move to the next stage of development and production. Informal design inspections, peer reviews, and dry-run reviews are conducted prior to the formal design reviews to make sure that they go smoothly.

Because of their criticality in the development and contracting processes, formal design reviews should be an integral part of the established design processes and with sufficient time and resources allocated in the design planning schedule for preparing for the review, the conduct of the review and the clearing up of action items resulting from the review.

As relates to maintainability, formal design reviews need to address all relevant aspects of maintainability to the same level of detail and rigor as all other engineering and design aspects of the product. This includes a thorough review of the maintainability requirements and the understanding of them, all requirements allocations and decompositions, the design and analysis tools being used and any LSA database analysis tools used, and the testing, verification and demonstration efforts planned.

Some of the common formal design reviews that may apply to a product include:

- Conceptual design review
- System Requirements Review (SRR)
- System Design Review (SDR)
- Preliminary Design Review (PDR)
- Critical Design Review (CDR)
- Test Readiness Review (TRR)
- Test Plan Review (TPR)
- Manufacturing Readiness Review (MRR)
- Production Readiness Review (PRR)

Details of the various design reviews can be found in Chapter 3 and in further detail in numerous DoD guidance documents, including MIL-HDBK-471A [2].

12.3.6 Maintainability Demonstration (M-Demo)

An M-Demo is perhaps the best known and, arguably, the most anxiety-inducing maintainability verification method in common practice. The M-Demo is a formal, hands-on joint OEM developer/producer and customer demonstration of actual production-ready (usually) hardware that evaluates "whether specific maintainability contractual requirements have been achieved" [2].

It should be stressed that the M-Demo is a big deal in the milestones of product development and contract execution. M-Demo is usually considered a formal test event and a critical program milestone with lots of customer visibility. Milestone approvals, contract payments, continued design progress, sustainability maturation, and positive business reputation are involved. There should be no doubt that the M-Demo should be taken seriously. This includes preparing a detailed M-Demo plan, coordination with all participants (including the customer and users), and practice and rehearsals prior to the actual demonstration.

Even though the M-Demo is a significant event that examines many aspects of the product's maintainability design and maintainability support, the distress often felt in preparing for the M-Demo is generally not justified. It is a major event with consequence, but all design and development projects include many significant events and the M-Demo is just one of them. Some aspects of the M-Demo are less controlled than is often desired, and sometimes results can be non-binary, leading to interpretation and sometimes difficult conversations. However, if the product design team has done their job well and produced a product that is compliant with the requirements, and if the test team is well prepared, there is nothing to cause apprehension. The M-Demo should produce no surprises.

Often, the M-Demo is thought of in terms of mechanical or electrical or other physical maintenance tasks, with software not being included. Modern systems and products are frequently software-intensive items. As such, much of the maintenance may well involve software-related activities, either on the system itself or in a backshop. Increasingly, software maintenance is taking large portions of maintenance efforts and resources and requires special equipment and skill sets. It is, therefore, considered increasingly prudent to consider including software maintenance in the planning for an M-Demo where appropriate. Consider the level of maintenance and what software tasks they may be required to perform. Some software tasks may include rebooting the system; reloading software; updating software; loading new software after equipment replacement; troubleshooting software; loading databases; updating databases; performing memory data dumps, purges, or erasures; updating or erasing storage media (e.g., disks, solid-state devices); and downloading data from the system.

The best guidance for preparing for and conducting the M-Demo continues to be MIL-STD-471A, *Maintainability Verification/Demonstration/Evaluation*. Even though MIL-STD-471A was canceled, it is still the standard-bearer for the guidance and conduct of an M-Demo. Other guidance for maintainability design and test standardization and verification has been developed since the cancelation of MIL-STD-471A, and all have incorporated MIL-STD-471A into them. Examples include the *DOD Guide for Achieving Reliability, Availability, and Maintainability* [5] and the *Applied R&M Manual for Defence Systems*, Chapter 11, Maintainability Demonstration Plans (UK) [6].

The best preparation for conducting an M-Demo is to become very familiar with the applicable portions of the development contract and the product requirements specifications early in the product design period. These requirements are considered mandatory reading to prepare for the M-Demo and in guiding the design team towards achieving the required maintainability performance. From this knowledge, in concert with contract guidance and coordination with the customer, the preparation of an M-Demo Test Plan is the next important step. Using contract guidance (such as any applicable Data Item Descriptions [DIDs]), a detailed M-Demo Test Plan will address all elements of planned maintainability test and verification activities. Naturally, it will also include the planned M-Demo.

12.3.6.1 M-Demo Test Plan

Typically, a separate M-Demo Test Plan will be called for to specifically address the conduct of the M-Demo. Preparation of the M-Demo Test Plan should be coordinated with the representatives of the numerous technical subject areas involved in the test, and with the customer. Depending on specific contract guidance, the plan should include the test planning activities and test execution. Typical parts of the plan include the plan and M-Demo purpose, scope, directing and reference documents, and goals of the testing. In addition, it will include the roles and responsibilities of the participants, the maintenance concept, and the schedule. Then, the test plan will address the specifics of the planned testing, providing step-by-step actions to be taken, by whom, and including the desired results. The test plan should also include space for, or other accommodations to record the attendees, the actual results from each test, and any action items that result from the testing.

Participants and resources included in a typical M-Demo may include: test management, the logistics team, the design team, production, quality assurance, training development, technical documentation, reliability engineering, facility management, tools management, calibration lab, safety shop, maintenance personnel (developer or procuring agency or user), program management, and customer/user representatives.

Select topics to include in the M-Demo test plan also include:

- The formal, detailed maintenance concept – including all levels of maintenance, on-equipment, and off-equipment
- Clear, detailed descriptions of the expected maintenance skills supporting the product
- All technical documentation to be used by maintenance personnel
- The skill definitions (i.e., Military Occupational Specialty code (MOS code), Air Force Specialty Codes (AFSC), Navy ratings or Navy Enlisted Classification (NEC)) for each maintainer participating in the training
- The training each maintainer is to have (or equivalent) prior to testing
- The tools, test and support equipment available to, provided to, or required by the maintainers at each maintenance level
- The shop capabilities and resources associated with each test
- PPE or other safety precautions or considerations needed or available for maintenance
- Clothing and other gear that the maintainer may wear that would hinder maintenance, such as cold weather gear, rain gear, flight gear, or Mission Oriented Protective Posture (MOPP) suits
- Shipping materials and containers associated with the product
- Any special handling or disposal instructions, processes, or materials
- The facility requirements and environmental conditions in which the maintenance tests will be conducted
- Fault insertion methods and responsibilities
- The anthropomorphic characteristics of the maintainers, for example, 90% maximum/10% minimum male and/or female, or European vs US maintainers (yes, it does make a difference).

12.3.6.2 M-Demo Maintenance Task Sample Selection

Many years ago, it was realized that maintainability, especially the MTTR parameter, could not be sufficiently defined or bounded if the maintainability and design communities tried to define and measure all possible circumstances, conditions, environments, and situations associated with product maintenance. Indeed, for most products, it is not practical to test every possible corrective, preventive, and scheduled maintenance action under different circumstances and varying skill levels. As evident: by the development of military standards of many countries and similar commercial and civilian design guidelines, maintainability performance, especially MTTR, is recognized as a parametric feature rather than an empirical characteristic (such as volts, pounds mass, frequency, etc.). It was further realized that maintainability could best be understood and measured as a statistical parameter with variability.

As a result, MTTR was defined as the average time required for a particular maintenance action (or group of maintenance actions) to be performed by properly trained maintainers possessing the appropriate skills, who are familiar with the maintenance task, using proper facilities, with all specified and proper tools and test equipment immediately available, in perfect weather, with all necessary technical documentation on hand – essentially, maintenance conducted under perfect conditions. Because there is no real-world condition as a perfect maintenance environment, and because there is no realistic way to define all possible maintenance conditions, the perfect mean (MTTR) has prevailed as the basis for measuring maintainability and for assessing product suitability. Logistics engineers and managers know this and adjust the term MTTR as appropriate when planning support resource needs and timing.

Through the early years when the statistical science of reliability and maintainability were evolving, it was further determined that general maintainability could be characterized statistically using the MTTR parameter. Of course, this did not address any and all situations, but it was recognized to be sufficiently representative of the product to be useful in developing military and complex mechanical systems. Exceptions, it was decided, would be defined as separate, specifically defined maintainability requirements as the situation warranted.

Also, once defined in a given condition, maintainability characteristics could be reasonably extrapolated and effectively used in other maintenance conditions. As a statistic, it was recognized that MTTR, which in itself is an averaging metric, could provide accurate insight into a product's actual maintainability performance based on statistical sampling, and still provide sufficient statistical confidence to approve the product based on that sampling. Today, the statistical nature of MTTR (and its related cousin parameters) is sufficient to allow a product that satisfies the test requirement to move forward.

This leads to an aspect of M-Demo test planning that involves careful consideration and some statistical acumen: choosing the proper quantity of maintenance actions to test and appropriate types of maintenance actions. There is abundant technical guidance, especially in MIL-HDBK-470A, to lead one to understand and choose the best solution for their situation. It should be understood that choosing the appropriate maintenance task sampling method and choosing the task samples can be technically detailed and heavily involved in statistics. It is recommended that expertise in statistics and test methodology be involved in this portion of the test definition.

During the M-Demo, a statistically significant number of "representative" maintenance actions will be performed which will determine whether the product does or does not pass the maintainability test. It is relevant for the design team and test leader to be aware of considerations in selecting the candidate maintenance actions for the demonstration. Carefully review the test sample requirements for the product under test in advance. Sample selection involves dividing up the totality of all heterogeneous maintenance tasks into similar, or homogenous, subgroups to ensure all representative types of maintenance are included in the testing. Subgrouping is intended to allow testing of similar tasks, thus allowing statistically "fair" grouping of results. For example, preventive maintenance tasks should not be mixed with corrective maintenance tasks, and mechanically intensive tasks should not be mixed with electronic go/no-go checks. Again, existing available guidance should be consulted for details.

It should be kept in mind that one of the purposes of the M-Demo is to determine whether the product meets the maintainability requirements and the impact on real-world operations. As such, part of the test sample selection evaluation is the failure rate of each of the components or assemblies replaced within maintenance tasks under test in the M-Demo. The failure rates provide the probabilities of occurrence of each maintenance task that the maintainers are likely to experience. Failure rates are provided as part of the maintainability predictions [4]. Obviously, one does not want to select test tasks solely at random for complex products, because there is no way to ensure that tasks which are truly representative of the maintenance responsibilities expected in the field will be selected. Frequent, maintenance-intensive tasks or tasks which are going to be experienced infrequently will inappropriately skew the results. Therefore, maintenance task selection criteria have been developed using frequency of occurrence and type of maintenance.

Stratification is a key practice in the maintenance task sample selection process. Stratification is the arrangement of data into different groups. The objective of stratification is to divide a large maintenance task data-set into smaller homogenous data subsets. A maintenance task sample will be extracted from each data subset. The selection of a maintenance task sample from a homogenous data subset will result in a representative sample of that data subset. The sum of samples from all data subsets represents the total population of the large maintenance task data-set.

There are two basic types of tests that may be used for statistical maintainability data selection: sequential and non-sequential. MIL-HDBK-471A [2] describes fixed sample size test methods for proportional stratified sampling that could be used for selection of corrective maintenance tasks to be demonstrated in the M-Demo. In sequential testing, testing continues until a decision to accept or reject the hypothesis under consideration can be made. One drawback of sequential testing is that the length of the test cannot be determined in advance. However, sequential testing will accept very low MTTRs or reject very high MTTRs very quickly. A non-sequential, or fixed sample size is best when the maintainability must be demonstrated with a given confidence level. Sequential methods must involve simple random sampling techniques. Preventive maintenance or servicing tasks should not be combined with corrective maintenance tasks during the stratification process. For system-level M-Demos, the stratification process would be applied to each separate subsystem or equipment to demonstrate and verify the maintainability requirements for each.

The first step in the stratification process is to choose the data grouping criteria. This step involves the designation of a data characteristic by which to group the data, the number of homogenous data subsets to group, and the boundaries defining the data subsets. To accomplish this, the maintenance tasks within each group should require approximately the same amount of maintenance time for the same type of repairable assembly or component. Repairing an electronic assembly within a system may take approximately the same amount of time as repairing a motor within the same system; however, the differences between the two types of maintenance actions would make it inappropriate to place the two types of equipment in the same data subset grouping. The analyst should ensure that there are similarities among the tasks assigned to a data subset grouping so that each data subset grouping is homogenous. The next steps are to fill in additional data about each task, such as failure rates (frequency of occurrence) and types of maintenance tasks. The final step is the test sample selection. The selection process is different depending on the type of test that is used for statistical maintainability data selection: sequential or non-sequential. Examples with further details of the test sample selection process are contained in MIL-HDBK-471A [2].

Once the test samples are selected and agreed upon and all test resources are available, testing can commence. The testing of maintenance tasks may involve simple tasks such as the reconnection of connectors or pressing a switch to interconnect certain functions. Some tasks may require invasive electrical operations so that known faulty parts are inserted into the product prior to the start of testing. During the testing, each of the selected maintenance tasks is performed on the unit under test in a properly equipped facility that is free from distraction. For each task, the assigned maintainers are informed of the maintenance to perform; they make sure they have all the specified tools and test equipment (including software) for that task at hand, and any replacement parts or materials are made available. The maintainers are generally allowed to familiarize themselves with the maintenance task by reviewing the authorized technical documentation prior to beginning the task.

As each maintenance task is started, a timekeeper records the time used for completion of each step in the task. Special note should be taken in the records of times when maintainers stop the maintenance actions to request a tool or part that was not provided with the test setup. Time would also stop if the maintainer takes any action that is deemed not directly part of the maintenance task itself, such as asking a question of the test team. The intent is to record only direct maintenance times. Delay times that are not directly part of the maintenance task will be disallowed.

The test plan should make clear what maintenance activities are to be included or excluded in the time for the record. Common examples of test times of interest include: fault verification time, troubleshooting time, fault isolation time, disassembly and reassembly times, part removal time, system shutdown or reboot times, time for curing of sealants and adhesives, alignment time, part insertion time, and Built-In-Test (BIT) run time.

This is a common method, especially for electronic and complex systems, that exercises the product's ability to identify that a fault exists, possibly isolate the fault and inform the maintainers, and verify that the product is returned to proper operations post maintenance. Inserting known faulty parts into a product can be expensive. Inserting a specific fault into expensive parts means that those faulted parts can no longer be used in the product unless approved refurbishment processes are authorized. Care should be used when selecting when and how to use faulted parts in the M-Demo. The expense needs to be weighed against the importance of the task and the value of the information to be gained. Other fault insertion possibilities are offered in software-intensive products in which software "errors" can be created or simulated without corrupting the baseline software. These possibilities should be considered where practical.

One additional caution is warranted. By the nature of the M-Demo, there exists the possibility that the unit under test may sustain damage. This is a possibility with certain types of maintenance actions, so the possibility of this should be considered in the test planning, and repair responsibility determined in advance. For example, spare parts should be made available for items that are subject to damage from dropping, mishandling, being stepped on, removal efforts, stripped screw heads, incorrect placement and orientation, and so on. As another example, experience dictates that if a connector can be inserted incorrectly, someone will try to force it, causing damage to the connector. If a part can be installed in the wrong position or location, testing may find this. Designing for keyed connectors is one way to prevent connector damage. One should be prepared for the unexpected mistakes during any maintenance task.

12.3.6.3 M-Demo Test Report

At the conclusion of the M-Demo, it is appropriate, and sometimes required, that an M-Demo Test Report be generated in accordance with company or contract requirements. The test report is a detailed description of the conduct of the M-Demo, including the test results, with the purpose of formally documenting the test activity, test findings, test results, and any recommendations that result from the test activity. There are many ways that an M-Demo Test Report can be formatted, with variations in content and format depending on company or contractual guidance. The responsible author(s) should research and verify the report contents, format, distribution, and schedule while preparing the M-Demo Test Plan. Reasonable guidance can be gleaned from MIL-HDBK-470A [1]. The responsibility for creating the test report may also vary from organization to organization, with the engineering, testing, and logistics organizations (in no particular order) often being the primary responsible parties.

With no single established guidance for an M-Demo Test Report, the following suggested list of topics to be addressed in the report is offered. This list is intended to serve as a generic guide for preparing a typical M-Demo report. The order of some topics may be changed, and some topics may even be deleted as desired. The purpose of this list is to provide ideas to consider in generating the report.

- Introduction – short discussion of the purpose for the test report.
- Background – short discussion of the history and background information that provides context for the test report and the equipment under test.
- Scope of test – description of the intent and limitations of the M-Demo testing, providing context and definition to the rest of the report.
- Documents and guidance – provide a list of relevant documents and guidance that is applicable to the conduct of the M-Demo. Include the M-Demo Test Plan, applicable contract reference, contract guidance, applicable DIDs, MIL-STDs, or MIL-HDBKs, company policy or guidance documentation, master test plans, and any other appropriate guiding documents.
- Key personnel – identify the key personnel for the M-Demo, including the test team.
- Equipment description – provide a description of the equipment under test, with a list of significant parts. Short descriptions of the purpose of the equipment along with applicable illustrations are appropriate to give the reader an understanding and context of the testing.
- Maintenance concept – describe the overall maintenance concept for the equipment under test and focus on the elements of the maintenance concept that apply to the M-Demo.
- Test objectives – delineate the objectives for the M-Demo, including the maintainability requirements, and identify any non-required, but desired, objectives.
- Test setup – describe the test setup, to include facilities layout, working area, equipment under test location and the locations of technical documents, spare parts, tools and test equipment, test personnel, and observers.
- Conditions of test – describe the physical and environmental conditions of the M-Demo.
- Maintenance Task Sample Selection – document the maintenance tasks chosen for the M-Demo including the method used to select the task

samples and the underlying rationale in choosing the method. Identify the applicable reference documentation that guided the selection.

- Test conduct description – describe how the M-Demo was conducted. Include descriptions of resources used, test documentation followed, test fault insertions or simulators used, data logging used, training provided, and the steps followed in the conduct of the testing.
- Maintenance tasks performed – identify the maintenance tasks performed during the M-Demo, identifying the tools and resources used for them (a table is applicable to present much of this information).
- Observations – identify relevant observations made during the testing that could skew the results or present issues with the confidence in the test data or could present opportunities for further study or improvement.
- Results – identify the results of the M-Demo, overall and for each test. Identify the data analysis conducted to reach the results, and compare the results against the requirements or goals. Include information on the status of each test as to whether it is chargeable or non-chargeable or non-test. Address all test issues or dissenting opinions, special situations, retesting required, and whether any test procedure or technical documentation errors or deficiencies were observed.
- Discuss any discrepancies that occurred during the conduct of testing, any dissenting opinions about the analysis or results.
- Summary or conclusions – provide a summation of the conduct of the M-Demo and a summation of the test results, including a summary of the results of the testing and any issues encountered or suggestions resulting from the testing.

12.3.6.4 AN/UGC-144 M-Demo Example

The AN/UGC-144 Communications Terminal was a ruggedized computer developed by the US Army Signal Corps and the US Army Communications-Electronics Command (CECOM) in the late 1980s, to replace the aging AN/UGC-74 terminal. AN/UGC-144 was a digital computer with a dedicated communications processor for automated messaging and data storage. A picture of the AN/UGC-144 Communications Terminal is provided in Figure 12.1.

Figure 12.1 AN/UGC-144 communications terminal. Source: FAS.org Military Analysis Network, https://fas.org/man/dod-101/sys/ac/equip/an-ugc144.jpg. Public domain.

The AN/UGC-144 made significant design improvements on its predecessor, the AN/UGC-74. The AN/UGC-74 terminal, also known as the 74, had various mechanical parts that failed and required maintenance very frequently, which plagued the soldiers in the field. The AN/UGC-144, also known as the 144, had fewer moving parts than the 74 and was deemed to be more reliable and easier to maintain. The 144 was a costly item to develop, but the estimated reduction in total life cycle costs for the 144 compared to the 74 was considered a big Return on Investment (ROI) to offset the development costs. The US Army customer considered the 144 a big technological step forward to automate many processes that were done manually on the 74, such as the storage of messages and the acknowledgment of receipt of messages transmitted from a distant terminal, but it was not sure that the life cycle ROI that was advertised by the OEM developer was possible.

In order to verify that the 144's life cycle maintenance cost reduction analysis would be able to live up to the hype, the customer required the OEM to conduct an M-Demo. The M-Demo requirement was split into three tasks: preparing an M-Demo plan, executing the M-Demo, and writing an M-Demo report. The M-Demo requirement cited MIL-STD-470 and MIL-STD-471 as the standards to conduct the M-Demo. The M-Demo was described as one of the tasks in the MPP. The 144 M-Demo program created an M-Demo plan as a separate document from the MPP. The M-Demo plan described all the detailed planning and procedural tasks that would be performed during the M-Demo. The OEM and customer approved the MPP and the M-Demo Plan prior to releasing the documents and starting the M-Demo. The customer wanted to ensure representatives of the user community were involved in the M-Demo exercise. Customer involvement during the M-Demo was a key part of the planning to ensure the results were accurate.

The 144 M-Demo execution began with an analysis of the electronic parts and circuits of the assemblies in the 144. A list of thousands of failure modes were prepared from this analysis, based on the results of a Failure Modes and Effects Analysis (FMEA), which was performed as a task required in the Reliability Program Plan. From this list of potential failure modes, a filtering process narrowed down the list of thousands of failure modes to 120 failure modes that could be selected and inserted into the 144 for evaluating the maintainability requirements during the M-Demo. This filtering process involved analysis of each of the 120 failure modes of the electronic parts and electrical circuit interconnects to make sure that a failure mode inserted into the 144 would not cause a catastrophic failure effect, such as a high-voltage short-circuit that could start a fire or electrically shock an individual performing the test. Failure modes judged to potentially cause personal harm to M-Demo participants or physical damage to the 144 were removed from the list. Mechanical failure modes were also added to the list.

Once the list was narrowed down to the top 120 failure modes as a representative sample of failure modes from all assemblies in the 144, the failure mode stratification process began. The 120 failure modes were documented with the actions required to insert the faults into the 144, the estimated time it would take to detect, isolate and repair the failure mode, and the actions required by the user to perform a test to checkout the 144 to verify the repair action was done properly. A random sample of the 120 failure modes was conducted to narrow down the list to 30 failure modes that could be performed during the M-Demo. The list of the 120 failure modes, and the 30 failure modes randomly selected from the 120, were provided to the customer for approval prior to the start of the M-Demo. The customer made a few changes to the list of 30, substituting out failure modes that were felt to be redundant or ambiguous with other failure modes.

When the list of 30 failure modes was agreed, the final preparation for the M-Demo began. The 30 failure modes were prepared for the fault insertion process. This process involved the creation of faulty components or rewiring techniques to demonstrate the faulty conditions as described by each of the failure modes. The faults were inserted into the 144 in a sequential order. The 144 was operated and tested one at a time by the OEM to determine the expected fault symptoms would be found by the test. If there were no symptoms that the test discovered, additional instructions were provided in the maintenance manual to allow the user/maintainer to conduct troubleshooting procedures to detect and isolate the fault. When the OEM completed the verification of the 30 failure modes, the official M-Demo began. Customer provided several individuals to be the test participants. Each test participant was given training on the operation and maintenance of the 144. This training included a familiarity with the 144 test programs and the maintenance manual.

Following training of the test participants such that they were comfortable with the operation and maintenance of the 144, the M-Demo started officially. Each fault was inserted into the 144 M-Demo unit one at a time, in a lab that was hidden from the test participants. After insertion of each fault, the 144 unit was brought out of the lab and placed in front of the test participant in the M-Demo area. Then the time clock started to capture that particular maintenance task time. The test participants used the training they received to operate the 144 and detect a failure mode, then run a test or perform a troubleshooting procedure to determine what had failed and how to repair the failure. When a test participant completed a repair action and ran the checkout procedure to determine that the fault was corrected, the time clock was stopped and the time recorded for each failure mode on an M-Demo test data sheet. At the end of the official M-Demo, the test data and test results were collected. Approximately one month after the formal M-Demo, the M-Demo report was prepared and sent to the customer to gain approval of the M-Demo test requirement and to verify that the maintainability requirements were met.

12.3.7 Operational Maintainability Testing

Following a formal M-Demo and other maintainability-related developmental testing, it is common for the customer to conduct some form of operational testing that includes evaluating the maintainability characteristics and suitability of a product as an Independent Verification and Validation (IV&V). The purpose of operational testing and IV&V, generally, is to gather insight and data about the suitability of the product for use under operational conditions by actual users and maintainers without involvement from the OEM. Such tests also provide useful data and insight into problems that may require resolution, and the adequacy of the planned supportability infrastructure.

There are different forms and methods of operational testing, ranging from very formal user-only testing to general use operations with verbal feedback from maintainers. Commercial customers tend to be less inclined to invoke additional, separate operational and maintainability testing, often accepting the results of development testing by the producer, thus foregoing the additional expenses and time involved. In lieu of formal testing, the commercial user can be expected to insist on having their own personnel and/or experts observe the developer testing. Commercial customers may require testing and certification by certain types of regulatory labs, such as Underwriters Laboratories (UL).

The most rigorous and formal operational testing tends to occur with Government customers where maintainability is a major construct. The military has formal operational testing guides and organizations that define when, where, and how to conduct operational testing. For the United States, excellent reference sources are the Defense Acquisition Guidebook (DAG) [7] and the DoD Test and Evaluation Management Guide [8]. Non-US militaries have similar guidance documents. Some significant guidance can be found in UK standards DEF STAN 00-60 [9], DEF STAN 00-40 [10], and DEF STAN 00-42 [11].

Within the US military, the testing organizations are:

- Air Force Operational Test and Evaluation Center (AFOTEC)
- Navy Commander Operational Test and Evaluation Force
- Army Test and Evaluation Command
- Marine Corps Operational Test and Evaluation Activity

There are several types of operational testing in the US Department of Defense (DoD) that include maintainability testing as part of the validation process. The most applicable among these are:

- Operational Test and Evaluation (OT&E): field test, under realistic combat conditions
- Initial Operational Test and Evaluation (IOT&E) or Operational Evaluation – OPEVAL: conducted on a production or production-representative system using typical operational personnel in a realistic combat scenario
- Follow-on Operational Test and Evaluation (FOT&E): conducted during fielding/deployment and operational support to assess system training and logistics status not evaluated previously

Operational testing, regardless of the customer or rigor employed, presents a product developer with feedback from the user community concerning the usability and suitability of the product in operational situations. Maintainability experiences under such conditions offer insights that are valuable lessons learned that can be driven back into the design and support resources within the product design teams.

12.4 Data Collection

Data collection for all maintainability testing activities is necessary to effectively manage product design and development efforts. It is an opportunity to discover and document the maintainability performance of a product and to use that information to benefit the product. Without collecting and disseminating test data, the maintainability program is essentially blind. It is often difficult to remember that testing is conducted to learn about the design as much as it is to verify the design. If there are no test data, then there is nothing to be learned about the product. As a corollary, if test data are collected but not used or shared with those who could use it, there is no effective benefit to be gained.

Throughout all maintainability test and verification activities, it is important to carefully collect data about the activity that occurred. The data that are recorded during these activities may include the identification of the types of tools and equipment employed, skill sets of the personnel involved, any product-specific training provided, technical documentation used, elapsed times of the steps taken, and any problems encountered. Test data provide a valuable record of past activities that is useful for identifying problem areas that have been or are yet to be corrected, and a record of progress and trends that provides insight into design maturation.

The importance of maintainability-related test data is underscored by the development (by military and industry) of formalized test data collection and analysis tools and processes. Two acronyms for data collection systems in common use are DCACAS (Data Collection, Analysis, and Corrective Action System) and FRACAS (Failure Reporting, Analysis, and Corrective Action System). The US and UK militaries, among others, have developed minimum-required data-sets for DCACAS and FRACAS tools, and they often mandate their use during formal testing. Additionally, military development contracts often require a product developer to employ a data collection system that can be in their own in-house design. Similarly, commercial product development may require a data collection process during formal testing, and customers are likely to require access to the data for oversight and verification purposes. Careful reviews of contract terms and customer desires are appropriate.

There is no single, universal format or established data element set required of a DCACAS or FRACAS. Each implementation of either of these systems is tailored to the particular application, type of testing involved, and program contract requirements. Certain data elements may be considered desirable, but not critical or required, so these may be tailored out of the data collection system requirements depending on cost and schedule constraints. However, useful test data collection systems have certain mandatory data elements in common regardless of the application. Among these mandatory data elements are:

- Date, time of occurrence
- Location of the test
- Test conditions
- Purpose of the test
- Identification of the unit under test, serial number, and part number
- Identity and roles of the participants and their qualifications
- Test setup, including tools and test and support equipment
- Name and description of testing
- Test procedures included in the test, including test pass/fail criteria
- Test results, including expected outcomes and actual outcomes
- Test failure symptoms
- Test failure analysis
- Recommended corrective actions
- Issues identified that may not be related to a test failure

One characteristic of a successful data collection system is a feature that is normally referred to as a closed-loop process. A closed-loop process is one by which test results are fed back into the design process to provide lessons learned and results to the engineering team. The closed-loop nature of a FRACAS ensures that test data are evaluated, causation is determined, and recommended corrective actions are fed back into the design process. However, as logical as it may seem, such feedback is so often forgotten. To prevent this from happening, FRACAS is a required feature of most formal testing data collection systems. Feedback from testing is considered a necessity for successful engineering design efforts because engineering staff are not always involved in or privy to the results of testing.

Design teams require feedback on the results of testing their designs in order to improve the product. This is especially relevant where deficiencies or issues are uncovered during testing. Test results, whether good or not, that are not provided back to the engineering teams cannot be addressed at the times when they should be addressed. Successful testing helps prevent spending unnecessary time and resources on product designs that work. Alternatively, problems not identified early cannot be corrected in a timely manner and may linger through development, only to be rediscovered later in the development process.

The closed-loop feature for data collection can be created using processes that ensure that test results are provided to the appropriate design teams. A better implementation involves using an automated data gathering and collection system, such as a common database, to capture the test data in electronic form and then automatically transmit any relevant data to other data collection and analysis tools. For example, if a single database were used throughout the development of a product, it could be used to aggregate data into specific groupings or categories and then apply data filters for detailed analysis using all relevant testing data. Data collected from the closed loop data collection system could be fed into that database where the pedigree, frequency, and accuracy of the data are maintained without losing any data or causing erroneous data errors. The data extracted from the database would prove to be useful for more engineers than it was originally intended for. Enlightened engineering teams understand that testing in one area, like reliability, has relevance to other design areas, like maintainability, for example. The common database has the potential to provide huge returns on the investment in reducing the number of data repositories maintained by different engineering teams.

To ensure that test data are efficiently put to effective use, the closed-loop data collection process, commonly referred to as a FRACAS, was developed and implemented. The analysis of the collected test data will indicate whether the design is working, is effective, meets requirements, and is fit for use. The FRACAS process, and the information so collected, is very useful in determining where additional design efforts and resources may be required. In today's engineering world, failure to collect, assess, and feed test failure data back to design teams is not an acceptable situation. There is no possibility for lessons to be learned if the experiences observed during the testing are not shared with the design teams. Also, there can be no product improvement if design issues are not known and addressed. Thus it is critical to achieve performance improvement and to correct identified design issues that the testing FRACAS data include a feedback loop and corrective action process.

Figure 12.2 is a depiction of a 14-step FRACAS process in which unit test data are collected and evaluated for units tested over a period of time for a particular product type; root cause analysis occurs until conclusions are reached, corrective actions are identified and recommended, and design changes are implemented. Formalizing this FRACAS process is a major benefit to a product design and development process.

The 14 steps in Figure 12.2 are briefly discussed here. Note that some of these FRACAS-type process steps may differ slightly based on customer or industry preferences, but the basic steps will remain the same.

Step 1: Failure Observation
The test team conducts unit testing and initially observes an indication of a failure or an anomaly, which may or may not be a failure. It is preferred that the test team identifies all anomalous occurrences because at this early point in time, it may not be known whether the event observed is actually a failure or the origination of a performance parameter degradation condition, which ultimately becomes the cause of a failure.

Step 2: Failure Documentation
The test team collects all observable data and prior documented information on related FRACAS reports and test failure documentation from historical records, using all types of formats and methods, to explain similar failure or anomalous events. This documentation may include all automated electronic data collected from test and measurement equipment, and the manually written observations of test observers documented in lab or test station notebooks or logs.

Step 3: Failure Verification
The test team, along with supporting engineers and specialists, review the test event, the unit under test, and test data to assess whether the event was indeed a failure in accordance with the test requirements and procedure established for the testing. The test may be repeated several times on the unit that failed, to verify the failure is correctly recorded against a particular unit. Note that it is not uncommon for unit testing to experience test events that are not directly attributed to the test unit or to the definition of failure. Examples of non-failure events might include the failure of test equipment, the failure of other equipment than the unit under test, running out of fuel or power source, stoppage for preventive maintenance, an event cause by personnel errors or mishandling test units, and incorrect documentation.

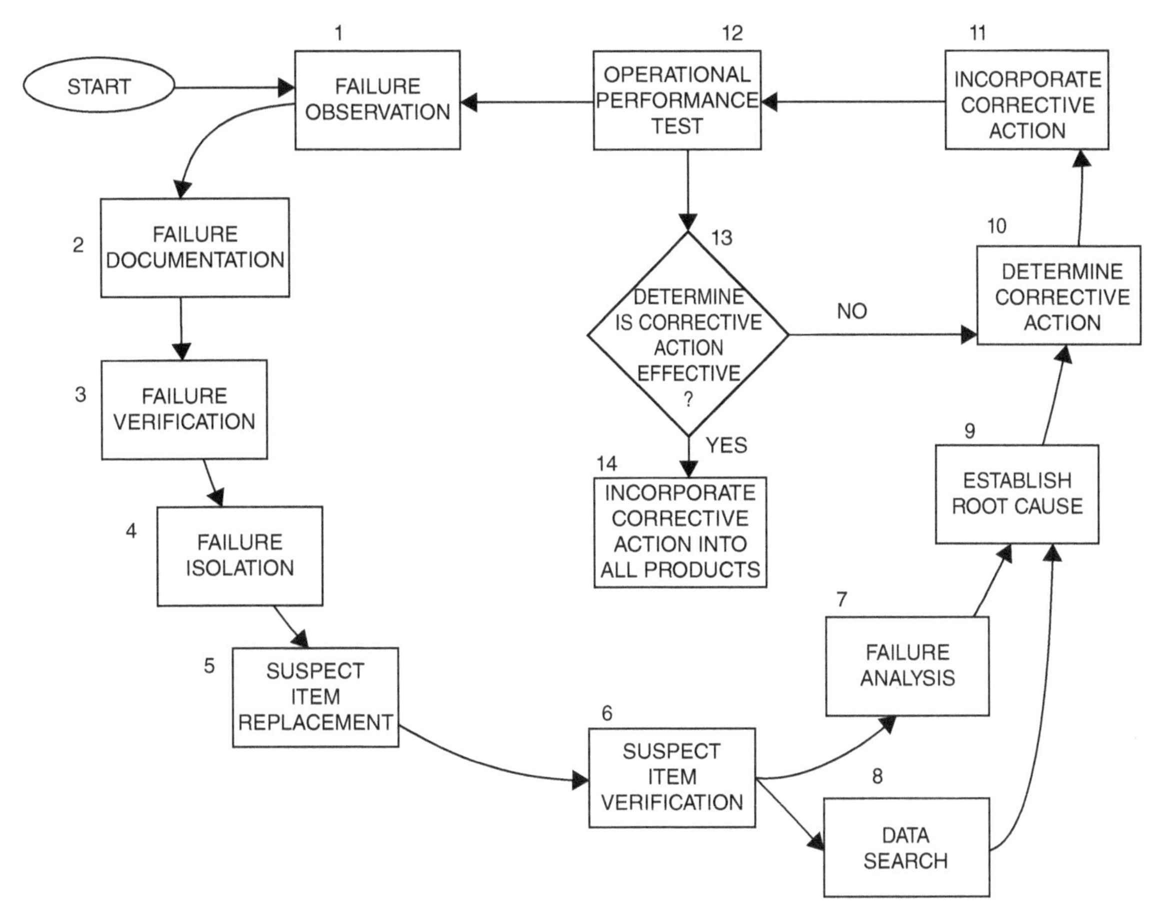

Figure 12.2 Closed loop failure reporting, analysis and corrective actions system [12]. Source: US Department of Defense, *Electronic Reliability Design Handbook MIL-HDBK-338B*, Department of Defense, Washington, DC, 1998.

Step 4: Failure Isolation

The test team conducts fault isolation procedures to determine the source of the failure. It is the goal of this step to determine the component or assembly within the unit that failed. The isolation or troubleshooting procedures may result in the identification of multiple components or assemblies that failed. Note that it is preferred to identify the fault to a single replaceable component part or assembly (such as a Line Replaceable Unit (LRU) or a Shop Replaceable Unit (SRU)) within the unit that failed. If more than one component or assembly is identified as potential causes of the unit failure, they should all be tested and verified to be failing a specified parameter. These isolated components or assemblies could be members of an ambiguity group that could be replaced in troubleshooting to speed up the failure isolation process. The documentation of this ambiguity group could prove to be valuable because considerable resources may be expended later to explore and investigate the fault in other occurrences, causing a waste of time and energy. It is best to have the investigative team spend the time and resources on the initial failure occurrence rather than on other test teams constantly repeating the search for a fault source in the future occurrences of this same failure or anomaly.

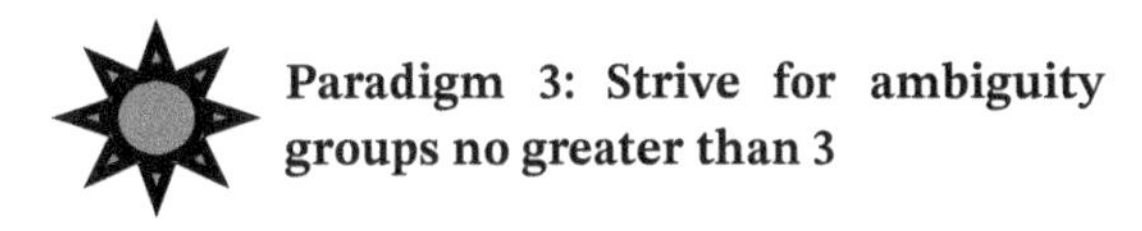

Paradigm 3: Strive for ambiguity groups no greater than 3

Step 5: Suspect Item Replacement

Once the test team has isolated the fault to a replaceable part or assembly, a replacement part or assembly can be installed, in accordance with the established test procedures. Attention should be paid to replacement parts that require special attention prior to being ready for use, such as calibration, burn in, data or software loading, alignment, tuning, fueling charging, or lubrication.

Step 6: Suspect Item Verification

The test support team conducts fault verification using the same test setup as used on the initial failure occurrence. During this step, the team uses the data collected during the testing and observations of the fault event to attempt to confirm that the fault actually happened and to document the conditions and events occurred surrounding the event, which may include all test parameters and actions. One goal of this step is to be able to replicate the fault event using multiple test units, and ensure that a failed component that is removed from a failed unit could be installed on another unit and cause a second "known-good" unit to fail. This means that the failure follows the item that caused the unit failure. This step could be repeated on several units until the test support team has confidence that the failure is verified.

Step 7: Failure Analysis

The test support team conducts detailed analysis of the failed unit under test and the test conditions to identify and document how the fault occurred, what the contributors were that led to the fault, and how the fault manifested itself. This step is important to support the root cause analysis to determine any and all failure mechanisms associated with the component, components, assembly, or multiple assembly failures, and to develop the appropriate corrective action.

Step 8: Data Search

The test support team is expected to conduct data searches for any information that can assist in understanding and supporting the failure analysis and root cause efforts. Such information may include failure analysis lab case studies, operational histories of parts similar to the one that failed, and units or systems similar to that to which the failed part belonged. This research of data may include component/assembly supplier vendor design and test data. It may also include the large customer data repositories of testing and analysis for similar parts and units with similar functionality. On some cases, Subject Matter Experts (SMEs) in industry or academia may be sought out for assistance.

Step 9: Establish Root Cause

The test support team uses all available test and analysis data and product design data to assess and verify the root cause of the unit test failure. If appropriate and possible, the team's goal is to identify a single root cause that, once corrected, will prevent the identified failure mode from recurring. Any contributing factors or secondary failures associated with the event should also be identified so that they can also be addressed.

Step 10: Determine Corrective Action

The test support team identifies and recommends design changes, manufacturing process changes, and any actions which may prevent recurrence of the observed test failure. It is not uncommon for multiple possible corrective actions to be identified during this step. It often requires engineering design research and analysis, which may involve multiple iterations of design, analysis, and testing before a single corrective action or a set of corrective actions is identified, verified, and agreed upon. It is possible that a set of corrective actions does not just involve redesigned parts or assemblies, but may include changes to training, operational procedures, test requirements and documentation, usage limits, and so on, to ensure a failure occurrence does not resurface.

Step 11: Incorporate Corrective Action

The test and product teams create the materials and processes required to implement the identified corrective actions that may affect one or more test units, depending on how many are required by the customer to be built. As part of implementing the corrective actions, all aspects of the unit design and the unit build processes that are, or may be, affected by the changes should be reviewed and updated to reflect the changes. Updating design requirement specifications, maintenance and troubleshooting procedures, training documentation, technical publications, operating manuals, engineering drawings and parts lists, manufacturing assembly instructions and processes, and sustainment planning are some of the ancillary product elements that should be addressed. These updated supporting materials should be used in the next step to test and evaluate all aspects of the corrective actions.

Step 12: Operational Performance Test
The test team determines the appropriate process for testing the updated unit and conducts the appropriate testing. The testing may be a continuation of the prior testing in which the initial unit failure was observed. It may be a subset of the prior testing, or it may be specific isolated testing, or it may be some other form of testing. The objective is to convince the product management team with high confidence that the design changes, along with any other changes identified, will achieve the requirements of the unit. Determining the appropriate testing to achieve high confidence in the corrective actions and changes can be controversial, but must be thorough to ensure effectiveness of the corrective action, which is the next step in this process.

Step 13: Determine – Is Corrective Action Effective?
Note that this step is a critical decision step (diamond-shaped object) in the middle of the FRACAS flow diagram, which is provided to designate where the corrective action test results are analyzed in the FRACAS process flow to determine the effectiveness of the change. Retesting is usually required to focus down on the minimum set of high-value corrective actions to verify that the changes are the right changes. The test and product management teams review all the associated test data from testing the changes, such as a redesigned part, to determine whether the results demonstrate that the design changes are effective. Consideration includes the long-term costs and impacts on production and sustainment, as well as any impacts on operational effectiveness. Evaluating whether or not the redesigned part prevents the observed failure mode from repeating is not the only question to answer. Such related questions might include the cost-effectiveness of the changes on the planned production processes, materials and costs, or the cost of changing parts that are already fielded, or the cost of changing technical documentation and training, or the impacts on support planning and maintenance resources. Another factor might also include the timing of the changes and the potential costs associated with any contract changes needed.

Step 14: Incorporate Corrective Action into All Products
The product management team leads the efforts to create a change management plan to implement the approved changes in all units fielded, in storage, and in the production process. This plan may be incorporated into other plans for implementing additional design changes, such as with an Engineering Change Notice (ECN) or Engineering Change Proposal (ECP), or as part of a major upgrade of the unit in products that use them, or part of a revision rollout upgrade or technology refresh/transition plan. Regardless, the plan to incorporate the design changes will require consideration and involvement from numerous organizations and skill representatives to ensure that all aspects of the unit, its use, and its sustainment are included in the plan. Once the incorporation plan is developed and approved, it is put into action with oversight by the product management team and the user community.

12.5 Summary

Testing and demonstrating maintainability is not a singular on–off event. A system or product's maintainability characteristics are design characteristics, which are purposeful and not considered accidental fallout from the design process. With this in mind, one can readily see how waiting until late in the development process, or the end of the process, to discover whether the product meets its maintainability requirements is not an effective strategy. Like reliability, maintainability performance can only be changed by changing the design in some way. As we know, the later in the design and development process that design changes are made, the more costly those changes are. Those costs are not just costs against the design engineering budget. The extra costs include extra time against the schedule to do the redesigns, to fabricate prototypes, to conduct informal and formal tests for compliance, and to re-baseline the design. The costs also include finding and qualifying new parts and suppliers, updating logistics plans and analyses, updating design documentation, and repeat customer and user evaluations.

In this chapter, we have emphasized a maintainability test and demonstration philosophy that embraces the totality of the design and development spectrum, while still including the more traditional test and demonstration events and expectations. It is recognized that the design process is often lengthy and full of conflicting requirements that need compromise. Further, the longer one waits to evaluate the design process, including the tools used and assumptions applied, the longer it will take to find out if the design meets requirements and the harder it will be to make needed changes. It is suggested that all aspects of the design process be evaluated for applicability to the product design, and that the design decisions and considerations be subjected to frequent review and oversight. Such rigor in the design team is intended to instill within the teams an understanding that testing and questioning design options and decisions is an integral part of the design process. Also, it is intended to provide engineering and program management with insight and confidence that the design process is working well and is applying appropriate design rigor and discipline to yield satisfactory results.

Early and continuing reviews of designs and design processes provide an excellent path for continuous discovery and correction of maintainability design performance. All relevant outputs from design tools should be included in these reviews, such as: requirements capture status; requirement assumptions; design tradeoff analyses and their underlying considerations; concerns from design analysis tool results involving various runs from design model simulation tools; definitions of operational and maintenance environments; definitions of maintainer skills and resources; the conduct of design peer reviews; and maintainability and logistics analysis models. If a maintainability issue is caught early, it can be corrected early (and less expensively). A design team that double-checks all aspects of its processes, tools, and decisions is more likely to discover problems earlier rather than later. Formal maintainability testing events are not the moments to discover problems. In fact, they are the worst time to discover problems that could, or should, have been avoided or discovered earlier by the design team. There is nothing positive about maintainability issues discovered late in the product development process in formal test events. The goal is to test early, test often, and test everything.

References

1 US Department of Defense (1997). *Designing and Developing Maintainable Products and System, MIL-HDBK-470A*. Washington, DC: Department of Defense.

2 US Department of Defense (1997). *Maintainability Verification/Demonstration/Evaluation, MIL-HDBK-471A*. Washington, DC: Department of Defense.

3 US Department of Defense (1996). *Logistics Support Analysis, MIL-HDBK-1388*. Washington, DC: Department of Defense.

4 US Department of Defense (1984). *Maintainability Prediction, MIL-HDBK-472*. Washington, DC: Department of Defense.

5 US Department of Defense (2005). *DOD Guide for Achieving Reliability, Availability, and Maintainability*. Washington, DC: Department of Defense.

6 UK Ministry of Defence (2009). *MDES JSC TLS POL REL, Applied R&M Manual, for Defence Systems (GR-77 Issue 2009), Maintainability Demonstration Plan*, Chapter 11. Bristol, UK: MoD Abbey Wood South.

7 Defense Acquisition University (2017). *Defense Acquisition Guidebook*. Fort Belvoir, VA: Department of Defense.

8 US Department of Defense (2012). *DoD Test and Evaluation Management Guide*. Washington, DC: Department of Defense.

9 UK Ministry of Defence (2010). *Integrated Logistic Support, Defence Standard 00-60*. Glasgow, UK: Directorate of Standardization.

10 UK Ministry of Defence (2012). *Reliability and Maintainability, Defence Standard 00-40*. Glasgow, UK: Directorate of Standardization.

11 UK Ministry of Defence (2016). *Reliability and Maintainability Assurance Guides, Defence Standard 00-42*. Glasgow, UK: Directorate of Standardization.

12 US Department of Defense, Electronic Reliability Design Handbook, MIL-HDBK-338B, Department of Defense, Washington, DC, 1998.

Suggestions for Additional Reading

North Atlantic Treaty Organization (NATO) (2014). *Guidance for Dependability Management, STANREC 4174*. Brussels, Belgium: NATO.

Raheja, D. and Allocco, M. (2006). *Assurance Technologies Principles and Practices*. Hoboken, NJ: Wiley.

Raheja, D. and Gullo, L.J. (2012). *Design for Reliability*. Hoboken, NJ: Wiley.

US Department of Defense (1996). *DOD Preferred Methods for Acceptance of Product, MIL-STD-1916*. Washington, DC: Department of Defense.

US Department of Defense (1998). *Maintainability Design Techniques, DOD-HDBK-791*. Washington, DC: Department of Defense.

US Department of Defense (2001). *Configuration Management Guidance, MIL-HDBK-61A*. Washington, DC: Department of Defense.

US Department of Defense. *The Defense Acquisition System, DODD-5000.01*. Washington, DC, 2003: Department of Defense.

US Department of Defense (2017). *Best Practices for Using Systems Engineering Standards (ISO/IEC/IEEE 15288, IEEE 15288.1, and IEEE 15288.2) on Contracts for Department of Defense Acquisition Programs*. Washington, DC: Department of Defense.

US Department of Defense (2019). *DoD Test and Evaluation Management Guide*. Washington, DC: Department of Defense.

13

Design for Test and Testability

Anne Meixner and Louis J. Gullo

13.1 Introduction

This chapter defines the engineering role of designing for maintainability concurrent with Design for Test and Design for Testability. Depending upon the industry sector, engineers split up the definition of Design for Test (DfT) into Design for Test and Design for Testability. Design for Test considers designing test capability as applied during development, production, and sustainment. Design for Testability refers to the analysis of the test capability applied during development, production, and sustainment. Design for Test refers to changes in the test setup for an item or the item's circuit topology to make it testable with a particular test setup. Design for Testability refers to changes in the circuit design to increase test coverage, Probability of Fault Detection (PFD), and/or Probability of Fault Isolation (PFI). Some circuit techniques accomplish both – make a circuit testable and increase fault isolation. For the purposes of discussion in this chapter, DfT encompasses both meanings, Design for Test and Design for Testability.

DfT usually begins early in the system and product development process. The culmination of the DfT effort in development results in the Design Qualification Test (DQT) and First Article Test (FAT) for the product or system. DQT and FAT are performed to ensure the design meets its specifications. The DfT effort continues into the system and product manufacturing process, leveraging the DQT and FAT processes created in the development phase. The DfT process occurs at all levels of assembly: system or product, circuit board, electronic device or integrated circuit (IC) component.

Manufacturing tests focus on detecting defects or faults that would adversely impact the end customer's usage of the final product. Like DQT and FAT, electronic manufacturing tests adopt DfT techniques to keep the test costs low and increase test coverage.

This chapter discusses DfT techniques for three distinct categories of hardware levels:

- DfT at system or product level
- DfT at electronic circuit board level
- DfT at electronic device or component level

DFT techniques used for development and manufacturing test activities at these three hardware levels turn out to be very handy for maintainability activities.

13.2 What is Testability?

The *Testability Handbook for Systems and Equipment*, MIL-HDBK-2165, dated 31 July 1995 [1], defines testability as a design characteristic, which allows the functional status of an item to be determined and the isolation of faults within the item to be performed.

Design for Maintainability, First Edition. Edited by Louis J. Gullo and Jack Dixon.

Paradigm 2: Maintainability is directly proportional to testability and Prognostics and Health Monitoring (PHM)

Testability design features ensure that previously known and new failure modes, gaps in the testing methodology, test result ambiguities, and test tolerance difficulties are recognized and defined. Testability analysis verifies testability metrics and determines whether improvements are needed in the electronic design or test processes to meet the system/product requirements. Testability engineering provides an approach for the analysis of production and acceptance test procedures, and evaluation of test results. The analysis determines a Built-In-Test (BIT) solution and Automated Test Equipment (ATE) solution that will satisfy production test and testability requirements. This analysis also considers the maintainability needs to service the product/system throughout its life cycle.

From the "8 Factors" detailed in Chapter 1 of this book, the following two factors are relevant to how DfT supports Design for Maintainability (DfMn):

- Fault detection (Factor 6)
- Fault isolation (Factor 7)

Fault detection and fault isolation capabilities of a test are measured using DfT techniques. These DfT techniques are important to differentiate the types of tests that can be performed and determine how thorough these tests are.

DfT techniques enable improved fault detection and fault isolation in the electronic systems, products, or electronic devices that are tested. DfT engineers use metrics to comprehend quality and thoroughness of a test for specific detection, isolation, and maintenance needs. The subsequent paragraphs describe these metrics.

Fault coverage, or fault detection percentage, is the ratio of failure modes found using a test program or test procedure compared with the total population of potential failure modes of an item at whatever level a test is designed for.

Test coverage is the percentage of defects or faults detected for all tests applied to an item at a level that a test is designed for, such as system level, board level, or component level. As test coverage increases, test effectiveness typically increases as well. Digital circuits lend themselves to easier test coverage calculations compared with analog circuits. Digital circuit analysis only considers logic values and signal path delays. In contrast, analog circuit analysis needs to consider a wide range of continuous values to measure.

Test effectiveness is a measure of a whole test solution's quality level, and it can be assessed at each stage of a continuous-flow manufacturing process as well as at a customer's system. Basically, it is the test escape rate: the ratio of devices that fail over the total number of devices tested. Quantification of test effectiveness is measured in parts per million (PPM), often described as defects per million (DPM). For microprocessors used in a mobile phone, an escape rate of 300 DPM is typical. In automobiles, which have a product life cycle of 15 years, the test effectiveness requirement is of the order of 10 DPM related to product recalls, and defects not related to normal preventive maintenance items.

Fault resolution percentage is a measure of the percentage of faulty items within a set of systems or products that are tested, faults detected, items replaced, repair actions accomplished, and the set of systems or products retested and failing for the same reason; compared with the number of times the tests are run and the system or product passes following the replacements and repair actions. If the test does not pass, then another item is replaced, and the test rerun. This cycle continues until the replacement and retest iteration is successful in correcting the cause of a failure. The more a test fails after a replacement, the lower the fault resolution percentage.

Probability of Fault Detection (PFD) is the probability that a fault will be detected by the test. This measure of testability is directly correlated to test coverage, fault coverage, fault resolution percentage, and test effectiveness.

Probability of Fault Isolation (PFI) is the probability that a fault will be isolated to a single item or group of items by the test. If a group of items is isolated by a test for replacement, this is called an ambiguity group. A common design criterion is to select a group of items to replace that is no larger than 3. Sometimes, the ambiguity group will include a sequential order in which to replace each item one at a time, with a test in between each replacement. When the ambiguity group does not provide a sequential order of replacement, the ambiguity group is considered a shotgun approach, since a maintainer is expected to remove and replace all items in the grouping at one time in the hope that the faulty item is included in this group and the replacement time is shortened.

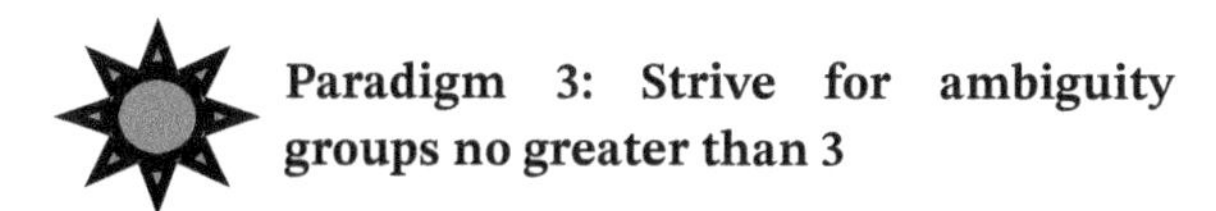

Paradigm 3: Strive for ambiguity groups no greater than 3

False alarm rate is a measure signifying the number of times an indication by BIT or other test circuitry incorrectly reports the status of a healthy or failed circuit or component. False alarms may be classified as false positives or false negatives. A false positive is an indication that a fault was detected, but an actual fault did not exist. A false negative is an indication that a circuit or component was healthy and functioning, but actually, a fault existed.

Overall test effectiveness is a measure of the test's combined fault coverage percentages, fault resolution percentages, fault detection times, fault isolation times, and false alarm rates.

The above metrics will each have an impact on the decisions made regarding DfT techniques. For the system design, engineers need to consider all metrics for each operational step considered for development, manufacturing and maintenance.

13.3 DfT Considerations for Electronic Test at All Levels

Engineers use DfT techniques for development and manufacturing tests at all hardware levels of a system. These DfT techniques consider the existing test capabilities available in development test laboratories in which similar predecessor systems, assemblies, or components have been tested. Likewise, these DfT techniques consider the manufacturing and production factories where the products, boards, and components will be built, assembled, and integrated into a system. Often, these same DfT techniques are used at a customer's site in the field with external support test equipment, to verify product performance before deciding to remove and replace an item from a system. Maintainability engineers need to consider the test equipment capabilities available at each location. These capabilities may require additional features in the electronics or the test setup to make use of new DfT features provided in a new product upgrade. Maintainability engineers should gather the documentation from the test design team that details how to activate all pertinent test modes. This documentation should also provide guidelines for deciding if or when an electronic device needs to be placed into a particular test mode to satisfy certain testability objectives or requirements. With this documentation, maintainability engineers ensure that faults at a device level will be detected at the system/product level using the established system tests. Often the maintainability engineer will perform a Maintainability Demonstration (M-Demo) to verify that critical device failure modes are detected at the system level.

13.3.1 What is Electronic Test?

Electronic test refers to the ability to determine the state of an item. Following a broad generalized categorization, test ultimately determines one of two states for an item being tested: operational state or a failed state. These two states may be further simplified by referring to the "Go" or "No-Go" states. Sometimes a more detailed test is created to determine three states for an item under test: an operational state, a degraded state, or a failed state.

Electronic test consists of a set of test equipment or test apparatus, environmental or test bench conditions, and a sequence of applied tests. The electronic device could be a tested as a bare silicon wafer, a bare silicon die, a packaged die(s), an electronic circuit board, an electronic box or chassis composed of multiple circuit boards, or within an electronic product or system.

A manufacturing test has a defined flow of test modules. Each test module has a specific manufacturing test objective, test process, test measurements, environmental conditions, and test equipment associated with it. Test measurements include checking for digital logic values, timing properties, frequency response, memory functionality, and analog circuit responses. At the end of the test flow the product needs to meet the electrical requirements specified. Fundamentally, if it works in the customer system/application for the defined product life cycle, then the goal has been met. Test equipment can vary depending on the item being tested. The test equipment could be a simple rack of test modules, a fully automated test set, or a diagnostic system that reads the output codes from a device. A test system needs the following:

- Input stimulus
- Output observation capability
- A fixture to hold the Device Under Test (DUT), often called a load board
- Power supplies
- The ability to maintain a controlled temperature

In a maintainability scenario, a test system capability applied in a customer field application often differs from the manufacturing test system for which DfT was originally applied and designed. This could result in significant differences in test coverage, fault coverage, PFD, and PFI between the manufacturing test system and the field test system capabilities. So, engineers need to be aware of these differences when considering DfT usage outside of manufacturing test.

Before proceeding we need to consider the types of testing done on electronics and the nature of electronic circuits. The nature of the electronic circuit depends on the type of electronic devices used in designing the circuit to perform a particular function. There are many types of electronic circuit functions, but this chapter does not attempt to list them. For the purposes of explaining a few examples of electronic circuits with their corresponding electronic devices, three broad categories of electronic circuits and components are described. These three broad categories of electronic circuits with components and their corresponding section numbers are:

- Logic/digital testing (Section 13.6.3)
- Memory testing (Section 13.6.4)
- Analog and mixed-signal testing (Section 13.6.5)

Test types that are applied to these three broad categories of electronic circuits with components include:

- Functional/mission system or product level test: Run application software (e.g., Flight Software [FSW], system maintenance diagnostics, etc.) in a customer use environment.
- Structural circuit board or component level test: Check for correct manufacturing (e.g., wires connected, components present, devices installed backwards, etc.) in a production factory.
- Parametric circuit board or component level test: Measure timing, voltage, currents, and leakage in an engineering laboratory, engineering test bench, or ATE for board or device-level testing.

13.3.2 Test Coverage and Effectiveness

The three types of tests for the three categories of electronic circuits are analyzed for their test coverage and test effectiveness. Test coverage and test effectiveness may be defined interchangeably or may be viewed as different sides of the same coin. One axiom that is used is: "The higher the test coverage, the better the test effectiveness." For the DUT, tradeoffs can drive the DfT capabilities in a positive or negative direction. If the design costs are being cut back, the options and decision criteria between design constraints, test time, and DfT features may result in lower test coverage, and hence, lower test effectiveness. If the program design budget and schedule allow for improvements in DfT, test designers will be making choices in design tradeoffs that enhance and/or optimize test coverage and enhance test effectiveness.

DfT techniques focus on screening all manufacturing-related failures before sending the system, board or component to the customer. Defects at the component level that are not screened out at the component factory are delivered to the circuit card assembly factory. If tests are not adequate to detect the component defect at the board test, then the defective component is shipped to the system factory. If the system test is not adequate to test for the defective component, then the component is shipped in the system to the customer, where it ultimately fails, possibly as a mission-critical or safety-critical failure with catastrophic effects.

For manufacturing test, no matter what the level of hardware is (system, board or component), a test engineer's goal is to use the least amount of time to meet a device's test coverage and effectiveness targets. This time increases as the signals propagate through the board circuitry and the system circuitry. The time to detect a component failure is much longer at the board level than at the component level, and is much longer time to detect that same component at the system level compared to the board or component levels. The same is true for fault isolation times comparing board-level and system-level tests. System test leverages board-level tests and integrated circuit (IC) device-level tests, each with varying degrees of test detection time, test isolation time, test coverage, and effectiveness to consider.

Fault isolation times vary widely due to the complexity of the sequence of fault isolation activities for any particular component applied in a system. An example sequence of these fault isolation activities follows:

- What system fails or is near failure if not maintained?
- Which electronic board/module fails?
- In that module in which an electronic part fails, what is the device that failed and where is it located? Is the device that caused the board/module failure a discrete electronic component or a packaged silicon/IC electronic device?
- If the failed device is a packaged IC, what is inside that package part? Is it another IC and how did the IC fail?

Consider next the complexity of fault isolation activities with each of these items from a test coverage perspective:

- Circuit board assembly electronics: Percentage of connections tested at slow speed and at functional speed.
- System in Package (SiP) and Multi-Chip Module (MCM): Percentage of connections tested.
- Very Large Scale Integration (VLSI) device (packaged or bare die) that may use one or more combinations of the following functions with their corresponding approaches to test coverage:
  - o Digital/logic: Stuck at Fault coverage (e.g., stuck at "1" or stuck at "0").
  - o Memory: Number of memory test patterns applied from the test pattern universe to uncover stuck bits or sporadically changing bits.
  - o Analog and mixed-signal: Number of performance specifications checked related to voltage or current levels.

Section 13.6 discusses in more depth the test coverage of the three electronic device types: logic, memory, and analog/mixed-signal.

A manufacturing test cannot reasonably test for all possible faults. As the percentage of test coverage approaches 100%, the cost exponentially increases and the time to design a test for each fault increases. Because of the results of design tradeoffs with established limitations on cost, schedule and performance, a certain amount of test coverage is deemed acceptable for a certain type of customer that signs a contract agreement for delivery of certain systems, products, circuit boards, or electronic components. The acceptance of a test coverage number may be documented in an explicit customer contractual agreement, which guarantees the escape rate that the customer will accept. Obviously, such terms in a contract agreement may not exist for products sold to the general public on the open market.

13.3.3 Accessibility Design Criteria Related to Testability

A key testability requirement is accessibility. To meet test coverage goals, the tester and/or the test setup (manual or automated) needs access to the signals to apply the required tests and to observe the test results. Engineers talk about this access in terms of controllability and observability. Some DfT techniques that enhance access include:

- IC scan-based design [2, 3].
- IC pins that provide access to embedded memory.
- Test access points added on a circuit board.
- External test connectors added to a system chassis or frame to perform system-critical functional diagnostics and collect diagnostic data for maintenance actions.

More information on design criteria that improves testability and maintainability is discussed in Chapter 5.

13.4 DfT at System or Product Level

In the end customer solution, the ability of a system or product to sustain operation in the event of failure varies. In an enterprise data center computer system, if a part stops functioning, the system must be fault-tolerant to avoid system crashes. An enterprise data center computer differs significantly from a desktop computer in terms of system reliability, availability, maintainability, and serviceability. An enterprise computer system needs the ability to keep running, while flagging a part for repair later. This repair can occur either when the system is powered down for a scheduled maintenance cycle, or when a replacement circuit card is available for a hot swap. The fault-tolerant capabilities of the system should be able to sustain acceptable system operation when a failed part on a circuit board is detected, and flag the failed part on a circuit board using a maintenance data log for access by a data center maintainer in a future repair action. The fault-tolerant system might also have the ability to maintain operation while a fault board is removed and replaced, and the system repaired. This is the motivation behind "hot" plug-in spares, to permit the replacement of a faulty electronic board with a functioning board without disrupting the system performance. Because of this later case, test designers should be motivated to use hot plug-in spares whenever possible to avoid system downtime while a part is replaced during a scheduled or unscheduled maintenance cycle.

For discussion of the DfT techniques using a system-level example in an application that more readers may be familiar with, let's consider the electronics in today's commercially available automobiles that are privately owned transportation vehicles. Automobiles require maintenance and the ability to diagnose faulty modules that cause the check-engine or check-battery light in an automobile's dashboard to turn on. With the increasing safety features to assist drivers (a precursor to driver capability), ISO 26262-11, a standard for road vehicle functional safety [4], has a set of requirements for integrated circuits (ICs). One specification requires in-field testing to be done on a periodic basis throughout the life of the IC. Mechanics use diagnostic computer systems, which take the signals coming from an automobile's computer, and diagnose where the failure is to dictate what part must be replaced.

Continuing with the automobile example, automobile maintenance embodies diagnosis of multiple signals and performance factors, which can cause a check-engine light warning. The diagnostic system isolates the fault, which leads to knowing which part to replace. An automobile mechanic uses a diagnostic test system to analyze the fault data and isolate the faulty module related to causing the indicator light to become lit. The diagnostic system applies tests and observes resultant signals from the automobile's electromechanical modules.

Consider the Engine Control Unit (ECU) module in an automobile, and how it is designed. Assume the ECU consists of two ICs, one memory device and one digital logic device, and assorted passive components. When the car's check-engine light comes on, the DfT circuitry detects a fault originating from a failed component in the ECU. The concerned owner drives the car to their local automobile mechanic for servicing the problem. To isolate the fault to the ECU, the local automotive mechanic uses their diagnostic equipment to interface with the automobile's embedded processor to read an on-board diagnostic code. This code indicates that the ECU needs to be replaced. Thus, the automobile's DfT circuitry has isolated the fault to the ECU. At an organizational level of maintenance, the goal for the car has been satisfied by using the car's DfT capabilities in combination with the mechanic's test device. However, this might not be the end of the situation. It is possible that another level of maintenance, such as an intermediate maintenance level, is warranted.

Assume that the automobile manufacturer requests all faulty ECU's to be sent back to them. This request would be reasonable if a product safety recall was issued by the manufacturer to all its service centers. The manufacturer's customer service center facilities, considered intermediate-level maintenance facilities, have additional diagnostic and repair equipment that can be used to do further fault detection and isolation. With appropriate DfT features that have been incorporated into the ECU design, the automobile manufacturer can potentially use them to learn more about location of the failure inside the ECU and the nature or cause of the failure. The manufacturer might be able to determine whether the cause of the ECU failure is one of the two ICs in the module, or a passive component, or an interconnection between the components. If it is one of the ICs, the manufacturer can use the embedded memory tests and logic scan chains to determine whether the cause of the ECU failure is a logic failure or a memory failure. If the manufacturer's service center determines that a defective memory device or logic device caused the ECU failure, and a design change has been incorporated into replacement parts, then a third level of maintenance will be involved. This is the depot level of maintenance where detailed component and board-level testing and failure analysis occurs. The depot for the automobile manufacturer provides all the data collected from the ECU failure analysis to ensure that the manufacturer's ECU designers make an effective ECU design change, thus ensuring no future ECU failures occur during the life of the vehicle model once new replacements are incorporated during the product recall period.

For these DfT features to exist in the design of the ECU, they would have been implemented during the ECU's design phase for that particular automobile model and type. DfT engineers concerned with the detection of failures need to be involved in the early stage of product development for good reasons. The greatest impact on DfT occurs at this stage. Historical examples of what happens when DfT features are not incorporated can be used to convince designers and management of the wisdom of incorporating them into the design early.

13.4.1 Power-On Self-Test and On-Line Testing

To explain what Power-On Self-Test (POST) is, let's refer to the automobile example. When a person starts their automobile, the alert lights click on and off on the dashboard as the car goes through a sequence of electromechanical steps. Once the engine has started, the lights go off. If the "battery light" or "check-engine light" remain on, that is an indication of a possible issue, and an automobile mechanic is needed to understand and resolve this issue. This is an example of a POST. Similar test sequences occur in household appliances, military equipment such as a guided missile system, telecommunication data center equipment racks composed of blade servers and blade switches, and ordinary desktop personal computers. Checking that any electronic device can perform some basic functions at the point of system or product power-up is the goal of POST.

POST tests are important and useful DfMn techniques. They inform the user about the functionality of basic electronic element pathways. Depending upon the sophistication of the POST implementation, the level of diagnostic indication may include such things as "this module is bad" or "this particular component within the module is bad." The test sequences in POST may use the boundary scan or Built-In Self-Test (BIST) on an integrated circuit. For reliability purposes in mission-critical systems, avionic electronics, and the soon-to-be autonomous automobiles, these same tests can be run on a periodic basis and are then called on-line testing. On-line testing can also be more extensive than POST testing. With the increased safety features to assist drivers, the latest ISO 26262-11 standard [4] has extended a set of requirements to ICs, requiring periodic on-line testing throughout the life of the car. If the on-line test fails, then ISO 26262-11 prescribes several options for responding to that failure to ensure safe operation of the car.

13.5 DfT at Electronic Circuit Board Level

This section describes the electronic circuit board assembly level of testability. Referencing the automobile dashboard indicator, consider this example. Your mechanic has diagnosed it to a faulty ECU module. Inside that module is a small printed circuit board (PCB) with several components (ICs and passive elements). Prior to assembly, the manufacturer has two test methods to use: in-circuit test (ICT) and DfT methods. These test methods assume all components have passed their respective manufacturing test. An ICT board test checks for defects in the PCB and component bonding or soldering. Test coverage considers checking all wires or interconnects in the

PCB and all component signal pins. The power supply interconnects and component pins are included in DfT techniques involving unpowered circuit boards to perform continuity tests. Once continuity tests pass, more powered tests occur on the circuit to ensure voltages and critical circuit nodes are correct.

Recent ICT research [5] describes automatic test generation that provides a coverage assessment. This process can guide adding extra test probe access points to improve coverage. As a non-powered test, ICT uses a bed-of-nails tester to test for opens and shorts between test points and verify circuit connections. Figure 13.1 shows the connectivity diagram for ICT test points between the Device Under Test (DUT) and the ICT tester.

Accommodating more VLSI components increases board interconnect density. The engineering community took notice and developed the IEEE-Std-1149.1 to use VLSI devices to check board connectivity between themselves [6]. Figure 13.2 illustrates testing for shorts and opens (common interconnect faults) enabled by this standard. Device 1 sends a test pattern to Device 2 to check for interconnect faults on the board. As shown in Figure 13.2, specific patterns detect shorts and opens. To enable this test in a design, the standard requires adding flip-flops (1-bit storage circuits) to all input, output, and Input/Output (I/O) circuitry, located around the boundary of a VLSI device on the circuit board. Hence, engineers refer to IEEE-Std-1149.1 types of board testing as Boundary Scan.

For the implementation of boundary scan functions on a single IC device, Figure 13.3 shows added circuitry in a single VLSI device and the connectivity of the DfT signals between the four devices on a board. Figure 13.3a shows the added circuitry for a single VLSI device. In addition to the flip-flop cells at every pin, the DfT implementation requires a Test Access

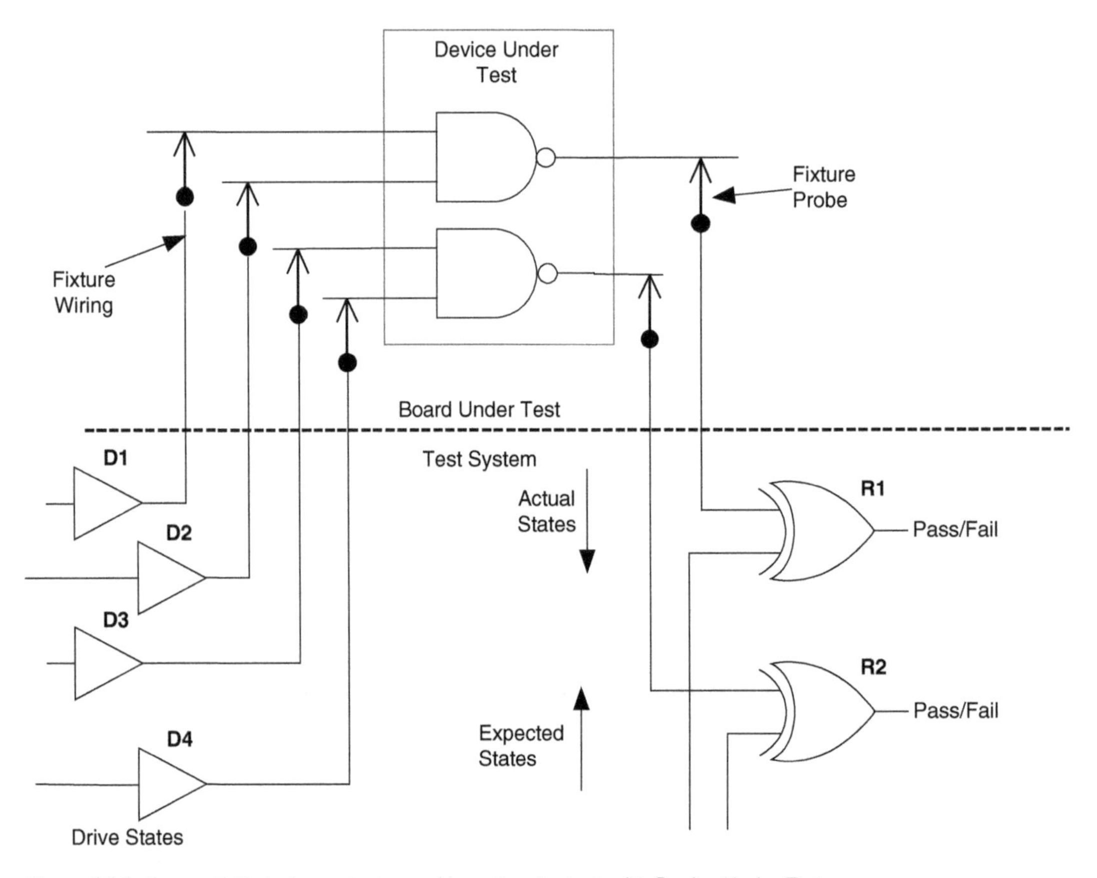

Figure 13.1 Connectivity between tester and board under test with Device Under Test.

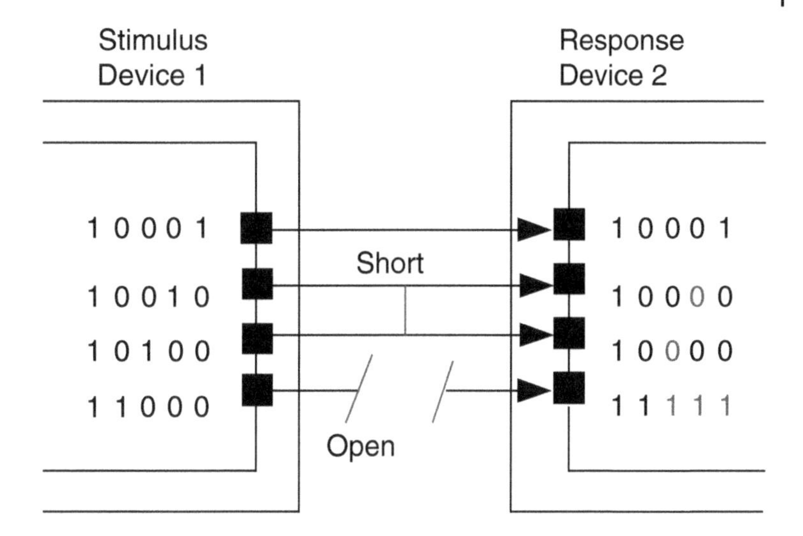

Figure 13.2 Interconnect test using boundary scan DfT.

Port (TAP) controller, an instruction register, a miscellaneous register, and a bypass register. The standard requires a minimal number of pins on electronic circuit boards:

- Data pins: TDI (Test Data In), TDO (Test Data Out)
- Control pins: TCK (Test Clock), TMS (Test Mode Select), TRST (Test Reset, optional)

The connectivity between all four parts on the board use these required signals as illustrated by Figure 13.3b, in which the board has four devices (same as Figure 13.3a) connected together. TDI and TDO are serially connected and the remaining signals are connected in parallel.

Boundary Scan has been co-opted by DfT and test engineers to perform tests on their VLSI devices. Using the TAP, a test setup can send stimuli and observe the responses. Note that it is a slow interface (e.g., 100 KHz to 10 MHz). Tests engineers use the TAP for:

- Executing Built-In Self-Tests (BIST on any component)
- Running logic scan tests on a component

With a PCB the first objective is to detect a fault and isolate it to a part on the board. Engineers can use DfT features described in this section to assist with fault detection and fault isolation. If the failing part is an integrated circuit, engineers may have access to that part to run some of the DfT features to further isolate the failure while on the board.

13.6 DfT at Electronic Component Level

This section describes the electronic component or device level of testability. In this section, five device types are explained. These device types are SiP or multi-chip packaged devices, Very Large Scale Integrated Circuits (VLSICs), logic devices, memory devices, and analog devices or mixed-signal devices.

13.6.1 System in Package/Multi-chip Package Test and DfT Techniques

This section focuses on the DfT strategy used with multiple die on a package. Some of the names for electronic devices with multiple die or chips in a single package are SiP and MCMs.

SiP parts consist of two or more die bonded to a substrate. These packages may also include passive components. Die may be connected on the same substrate or they may be stacked vertically. Package technology options vary from a ceramic substrate to a silicon substrate. The latter significantly increases the interconnect density. The IEEE Electronics Packaging Society has developed a roadmap for packaging technologies [7]. The advanced packaging technology permits mixing IC die (chiplets) from a multitude of fabrication and design sources. These chiplets may include logic devices, radio-frequency components, and Micro-Electro-Mechanical System (MEMS) devices.

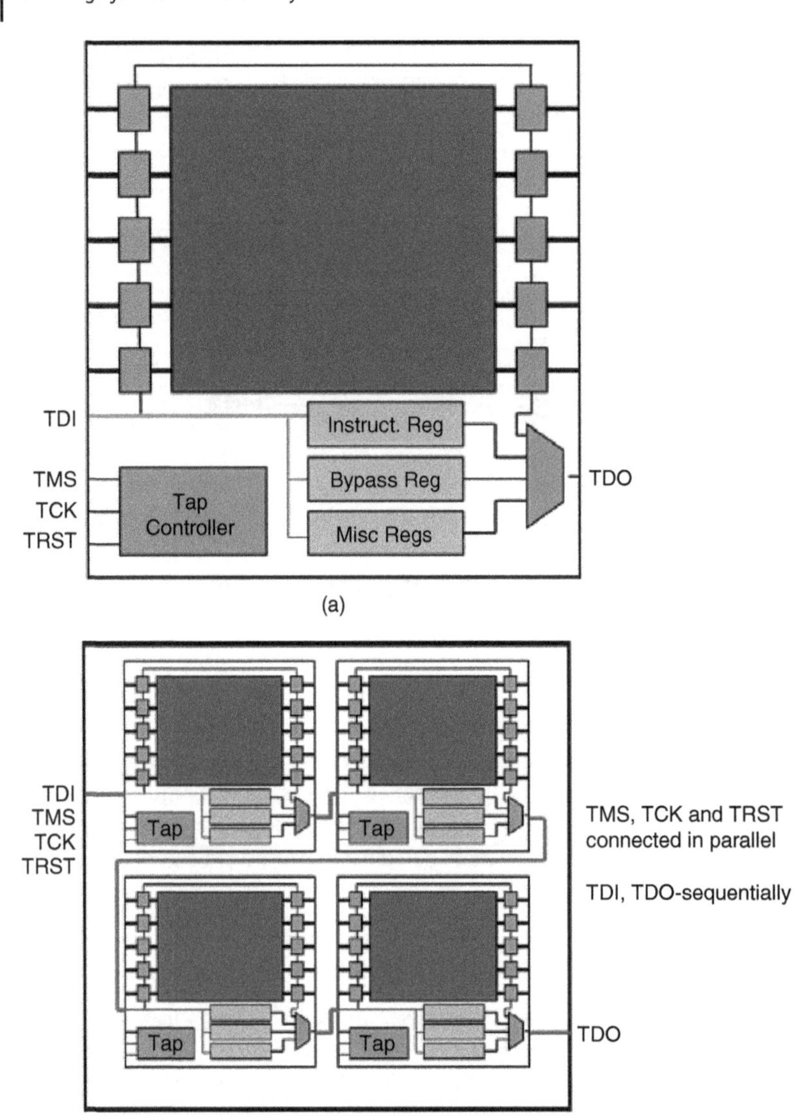

Figure 13.3 (a) Boundary Scan implementation – single device; (b) Boundary Scan implementation on a board with four identical devices. Source: From Alex Gnusin, *Introduction to DFT*, http://sparkeda.com/technology/, 2002.

The SIPs are tested the same as a single packaged die, and the test process must meet the quality standards for the end customer's system requirements. For test coverage one considers connectivity within the package, bonding between die and package, and die-level functionality. SIPs share both board-level and VLSI device coverage metrics. Often, to arrive at higher test coverage, engineers implement a DfT strategy. DfT strategies for SiPs borrow from a Boundary Scan approach used in boards. Often called a wrapper, this strategy permits using a minimum number of signals for a test access. In [8], the authors' wrapper solution requires three signals.

Fundamentally, SIPs have a test challenge related to wafer test, guaranteeing high die/chiplet quality: Known Good Die (KGD). The impact of a faulty die could mean throwing away the whole device or replacing a faulty die, if the packaging technology permits replacement. Engineers call the latter option "rework." This places a high test coverage requirement on the incoming die. This impact fosters debate amongst engineers and engineering managers on the tradeoff in higher test coverage for KGD vs. impact on the final product's package and test cost.

13.6.2 VLSI and DfT Techniques

This section provides knowledge on common DfT techniques used in VLSI circuitry. VLSI is a complex design of ICs. Without DfT circuitry, the development and manufacturing of complex semiconductor products would not be possible. DfT techniques evolved from providing access between the Automated Test Equipment (ATE) and the DUT to adding test circuitry on the DUT for BIT/BIST.

Often a DfT technique is expanded upon (i.e., more features added to use it for the specific intention of silicon validation, debugging, and failure analysis) to make improvements in DfT capabilities during the test flow of the silicon during manufacturing and during post-silicon manufacturing. Manufacturing test is not the only post-silicon (after chip-level testing) usage of DfT capabilities. Post-silicon usage includes silicon debug along with engineering test and maintenance activities at board level and system level. Post-silicon usage of DfT capabilities also occurs after system or product delivery to the customer in field applications during maintenance tasks performed on the product or system throughout its entire useful life.

The next subsections cover common DfT approaches that engineers use in logic, memory, analog, and mixed-signal circuitry.

13.6.3 Logic Test and Design For Test

Logic design can be purely combinational digital circuitry (no feedback paths, no storage elements). A logic design can also include digital storage bits (latches or flip-flops), feedback paths, and clock signals. Test coverage metrics consider the digital state faults and at-speed faults. Common failure modes include:

- Stuck at fault: Inputs and outputs of logic gates stuck at a 1 or a 0 (S@0, S@1).
- Transition/delay fault: Changing states from a 1 to a 0 or 0 to a 1 takes longer than expected.
- Bridge fault: Two signals connected together.
- Cell-aware fault: Only faults based upon a layout analysis of the standard cell under test.

ATE applies the test stimulus and observes the outputs. Automatic Test Pattern Generators (ATPGs) analyze a logic design with a specific fault model and determine the stimulus/output pairs [3]. The oldest coverage metric is "stuck at fault" coverage. As transistor size decreased and clock rates increased, DfT engineers began to assess transition and path delay fault coverage. With the shift of Complementary Metal Oxide Semiconductor (CMOS) integrated circuits, bridge fault coverage has been added to the engineer's test assessment metrics. With advanced process nodes (less than 45 nm), DfT test engineers became interested in defect models (i.e., at the physical implementation layer; these have been dubbed cell-aware faults) [9].

Often the test coverage targets for stuck at fault coverage range from 92% to 98%. Achieving a 100% stuck at fault coverage should result in zero escapes, which is the ideal test effectiveness metric. To reduce escapes, engineers test at higher clock data rates [4] and measure delay faults coverage; a lower target of 80% for this approach is not uncommon. It is important to understand that coverage metrics are not perfect and that the fault models are imperfect. To improve test effectiveness, engineers apply test methods that target multiple fault models. In the mid-1990s, the electronics industry sponsored a detailed empirical study comparing functional, scan, IDDQ (common bridge fault test), and delay test methods on an IBM Application Specific Integrated Circuit (ASIC) CMOS product [10]. IDDQ is the IEEE symbol for CMOS quiescent power supply current. The study showed that each applied test method had unique failures. Hence, all these test methods contributed to decreasing potential escapes to a customer.

To improve test coverage, DfT engineers apply digital design techniques that modify the design. For instance, to maximize test coverage an engineer wants the ability to set or observe a signal on every node within the circuit under test. In assessing an integrated circuit's testability, engineers consider the concepts of controllability and observability [3]. These two terms are defined as follows:

- *Controllability*: Ability to establish a specific signal value at each node in a circuit from setting values at the circuit's inputs.

- *Observability*: Ability to determine the signal value at any node in a circuit by controlling the circuit's inputs and observing its outputs.

A circuit's design impacts its controllability and observability. These attributes impact the design of an algorithm used for generating tests. Test generation costs less for circuits with better controllability and observability. DfT techniques improve these attributes, which lead to increased fault coverage, decreased test generation costs, and often decreased test time.

Most digital circuit blocks found in VLSI designs are composed of combinational and sequential logic circuits. Such blocks can be designed as synchronous or asynchronous implementations. This section addresses the more prevalent synchronous designs (i.e., a clocked design). The 1-bit storage elements, flip-flops, form the basic building block in sequential circuitry, which presents controllability and observability challenges. Fortunately, the very nature of flip-flops can be exploited to address these challenges. Flip-flops can be modified to increase controllability and observability. Basically, the design engineer connects all the flip-flops together into a shift register, which is called a scan chain. During test this enables access to internal states of the digital circuitry which can be set and observed. It is analogous to PCB test point insertion as it enables independent control of flip-flop states, enabling scanning of their internal states. Engineers often refer to this DfT technique as "scan design."

Figure 13.4 provides a simplistic overview of scan design as it assumes dedicated pins for scan test inputs and scan test outputs. One is more likely to come across designs in which scan and functional pins are shared. During scan tests a mode pin is used to multiplex the signals from the functional/mission mode paths to the internal paths.

Figure 13.5 depicts the scan test process. The sequence is as follows:

- Pattern is applied to the scan inputs by shifting in the values through the flip-flops in the scan chains.
- Pattern is applied to the functional inputs.
- Responses at the functional outputs are observed against expected values.
- Values captured in the scan chain may then be observed by shifting out the flip-flop contents to the scan outputs and comparing with expected values.

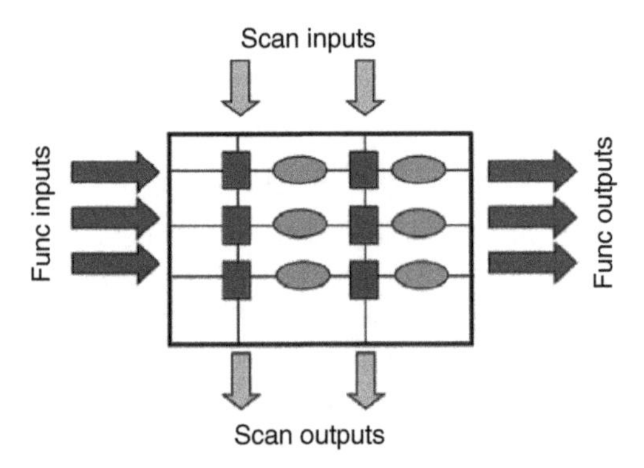

Figure 13.4 Conceptual diagram of scan insertion. Source: From Alex Gnusin, *Introduction to DFT*, http://sparkeda.com/technology/, 2002.

- The outputs with red bit values indicate a value that is different from the expected value.

For DfMn purposes, engineers need to inquire: Is this a full or partial scan design? Scan design has an implementation range described by the following terms:

- Full-scan design: All flip-flops connected.
- Partial-scan design: Selected flip-flops connected.

This difference in implementation impacts the maintainability attributes of fault detection and fault isolation.

Logic testing can be done at slow speed or at normal speed. Typically, it is done at a slower speed first, then at a speed to distinguish between the severities of the impact to circuit performance that a defect has on the digital testing.

From a DfMn perspective the following characteristics of a logic DfT implementation matter:

- Full vs partial scan design: While requiring more silicon area, full scan designs are easier to test and provide better diagnosis.
- Test coverage metrics for stuck at fault, delay test, bridge test, and path delay testing.
- Dedicated test pins or mission/test pins multiplexed. If the latter, be sure the instructions of how to get into the test mode are readily available.

13.6.4 Memory Test and Design for Test

Semiconductor memory comes in many forms, such as Dynamic Random-Access Memory (DRAM), Static

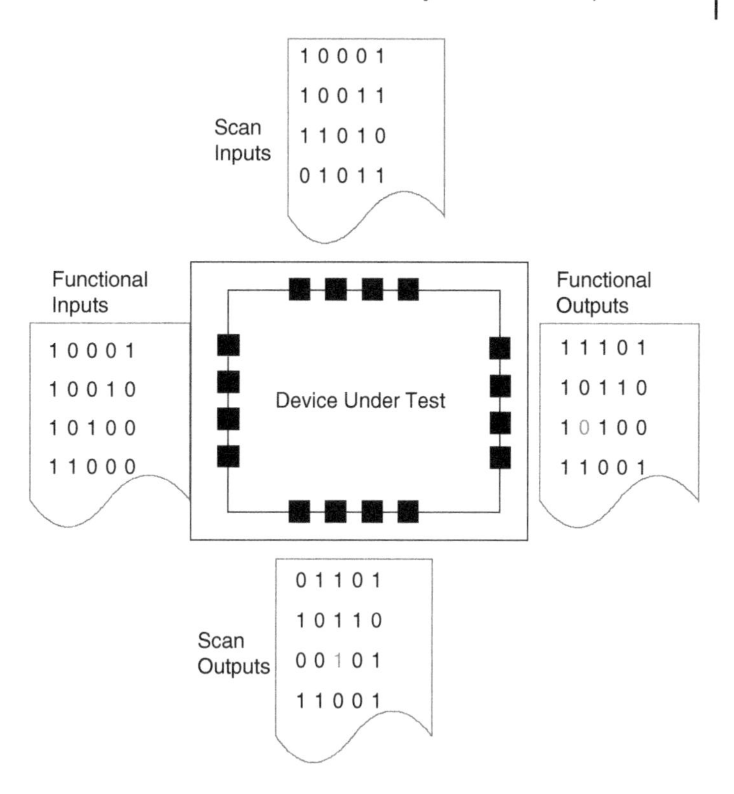

Figure 13.5 Application of scan test.

Random-Access Memory (SRAM), and NAND/NOR memory (also known as Flash memory). Memory circuits can be stand-alone or embedded into a larger integrated circuit. Memory module sizes range from 512 MB (megabytes) up to 16 GB (gigabytes) for a standard personal computer or laptop, and up to 1 TB (terabyte) or 1024 GB for network servers and larger complex computers. This subsection uses SRAMs to explain memory test basics and common DfT techniques. The general concepts can be applied to other memories, and the reader can learn more by consulting references listed at the end of this chapter.

Memory is organized in blocks, and each block has a grid system of columns and rows of bit cells. Figure 13.6 illustrates a SRAM's memory block organization and shows a six-transistor bit cell. Address circuitry connects a row of bits (often organized in bytes: 8 bits) to the read and write circuitry via each cell's bit lines. A row connects to a word line to activate that row, and a column connects to bit lines (bit, bit_b). The bit lines feed into a differential read amplifier circuit. Registers hold the bit values to be written or read.

Memory tests must cover storage registers, address circuitry, the timing of read/write circuitry and the bit cells. Often, memory test coverage is assessed by the number of memory test patterns that are applied. Due to the density of DRAM and SRAM circuitry and the architecture of word-lines and bit-lines, the data patterns written can become quite complex if the desired coverage goal is all possible data patterns. For small embedded memories, often 5–10 pattern types are applied. For large embedded memories, all patterns are initially applied (often called "the kitchen sink") and then during the manufacturing test start-up engineering data collection is used to reduce the number of patterns applied. DfT engineers make tradeoff decisions in test coverage vs. test effectiveness based upon the accumulated results for the tested devices.

Testing of memory needs to cover the following circuitry:

- Address circuitry: Did it write to the correct location?
- Bit cell: Is the bit faulty? Is the bit cell's state influenced by neighboring cells?

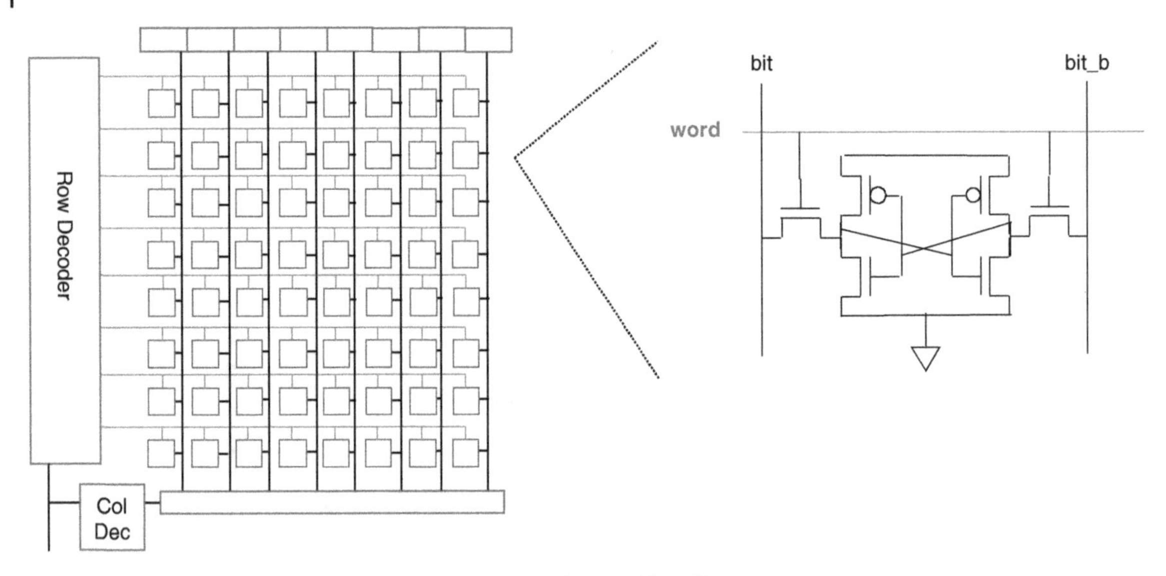

Figure 13.6 SRAM block organization and six-transistor bit cell (6 T-Cell).

- Read circuitry: Did it read correctly within time allowed?
- Write circuitry: Did it write the correct value within the time allowed?
- Storage buffers/registers: Is it faulty there?

To fully test a memory, there exists a set of structurally based tests that can be performed based upon knowing the memory's physical organization. The regular structure of a memory block combined with one bit cell's close proximity to another provides a key design characteristic that guides structurally based tests. Their physical implementation of memory cells leads to a form of "neighborhood pattern sensitivity" for faults or defects. The regular structure of memory design lends itself to algorithmically derived test patterns. Some algorithmic tests have been developed from past experiences of experts in the field who discovered defects that became silicon escapes. Names like the Hammer test and Pause test reference very specific customer functional failure types. Van de Goor's book on memory testing provides a list of pattern test types [11]. Van de Goor classified memory fault types as follows:

- Stuck at fault: The logic value of a cell or a line is always 0 or 1.
- Transition fault: A cell or a line that fails to undergo a 0-to-1 or a 1-to-0 transition.
- Coupling fault: A write operation to one cell changes the content of a second cell.
- Neighborhood pattern sensitive fault: The content of a cell, or the ability to change its content, is influenced by the contents of some other cells in the memory.
- Address decoder faults: Any fault that affects address decoding operation. Some examples include:
    - With a certain address, no cell will be accessed.
    - A certain cell is never accessed.
    - With a certain address, multiple cells are accessed simultaneously.
    - A certain cell can be accessed by multiple addresses.

The memory test patterns described can be extensive, yet they are easily described in a memory architecture-based language. The regular structure of memory architecture lends itself to on-the-fly-pattern generation at ATE tester pins or within an on-die pattern generator.

Engineers apply DfT circuitry-based techniques to do one of the following:

- Permit direct access from the ATE or other test equipment to an embedded memory block.
- Speed up test applications by providing on-die pattern and evaluation circuitry.

- Provide very specific defect/faulty behavioral coverage.

Addressing specific defect/faulty behavior coverage typically requires in-depth knowledge of the memory architecture and circuit design. For instance, if an engineer wanted to cover "word line" defects ranging from stuck at fault behavior to transitional fault behavior, knowing the actual circuitry informs a DfT circuitry strategy to target actual behavior. SRAM cell stability testing provides examples of DfT implementations, which require such knowledge:

- Application of a "weak write" on the bit lines to an opposite state [12].
- Application of a "weak write" on the word line to an opposite state [13].

Stand-alone memory products have dedicated ATE in their factories, but this type of test equipment is not available in the field at a customer site. DfT on the memory product permits a minimum set of tests to be applied for triage. The type of test equipment in the field may be able to detect microprocessor faults, but not memory faults. If memory is embedded into a microprocessor (i.e., a System on a Chip), then the field test application will not be able to thoroughly test the microprocessor. To connect the new System on a Chip to an ATE dedicated to memory increases the costs for microprocessor silicon design (in the performance and silicon area), manufacturing test flow (requiring two unique ATEs), and customer return test equipment (requiring two unique ATEs). DfT can be added to the microprocessor (e.g., BIST, as described in Section 13.7.1) and the manufacturing test flow goes down to one-pass testing. Such DfT decisions are made by trading-off the component cost versus the product's cost of ownership.

13.6.5 Analog and Mixed-Signal Test and DfT

Analog circuits have continuous waveforms at their input and outputs. A mixed-signal circuit combines analog and digital circuitry, and their input and output signals can be digital only, analog only, or analog and digital. Analog circuits have properties of impedance, 3 decibels (dB) roll-off points in the frequency domain, and transform functions. Analog circuits either produce or receive signals composed of non-periodic or continuous waveforms at respective outputs and inputs.

As stand-alone device engineers focus on testing a mixed-signal device, they refer to the function requirements contained in the device's specifications and count the total number of functions for use in their test coverage calculations. Therefore, the test coverage is measured in terms of the percentage of specified functions tested out of all the functional specification requirements. Checking the specifications requires test conditions, which can differ from one specification to another, and requires measurement equipment that generates and analyzes the waveforms. Test equipment can be specialized depending on the test situation. For example, waveform generators produce a variety of waveforms such as saw-tooth waves, periodic voltage ramps, or sine waves.

Within a larger VLSI design, test engineers often face limits in access to analog and mixed-signal circuit and device inputs and outputs. In turn, these limits challenge the DfT engineer's ability to directly assess all of a circuit's specifications. This has generated interest in alternative test methods, which take a defect-based test strategy. Such methods focus on checking a circuit for presence of a defect causing poor or degraded circuit performance, but not necessarily the presence of a fault. These approaches assume that correct manufacturing results in a circuit that won't impact a customer system.

During the circuit design process, defects are described, modeled, and simulated at the circuit schematic or netlist level during the circuit design process. In defect simulation, applied tests are assessed for which defects may be detected [14]. Engineers perform test coverage or fault coverage calculations based upon their defect circuit modeling and simulation strategy. Defect coverage numbers may be challenging to accept, since not all netlist defects result in faulty behavior. In addition, engineers simulate multiple defect types, such as shorts, opens, and parametric drift. They may simulate netlist defects for a range of process parameter values, but they probably won't be able to cover them all. The right percentage of coverage depends on each situation. For example, a 70% defect or fault coverage may well meet the test effectiveness goal for a marketplace expectation. For some customers, who require high reliability and safety-critical systems, fault coverage of less than 95% is not acceptable. The impact to a customer's system provides the ultimate measure of test effectiveness.

DfT facilitates testing mixed-signal circuitry residing in a larger VLSI design with one of the following options:

- Test access to internal signals from external pins:
  - Digital pins that need to provide access to analog signals.
  - Analog pins that need to provide access to analog signals.
  - Analog pins that need to provide access to digital signals [15].
- On-die instruments (e.g., waveform generators, D/A converters, and A/D converters).
- DfT circuitry to support a defect-based test approach.
- Test station load-board modifications to support a test function [16].

DfT test approaches support either a specification test approach or a defect-based test approach. Two common defect-based test strategies involve outlier detection or transformation. In outlier detection, a defect's impact on a measurement shows a significant difference from that measurement in the normal population. Transforming a measurement from one domain (current) to another (time) can facilitate measuring something on a tester or a die. Two examples of DfT defect-based test methods for I/O circuits are provided. One example is a test performed for I/O pin leakage, and the other is a test for I/O timing.

I/O pin leakage is a time-consuming yet necessary test. Typically, this is done on an ATE. The traditional measurement for I/O pin leakage starts by making contact with and placing the single I/O pin in tri-state mode. Next, the ATE forces this pin to the voltage supply rail, using the existing voltage source supplied for the circuit by the ATE. Finally, the ATE measures the sourcing/sinking current for that pin.

Pin leakage needs to be tested when the ATE doesn't touch the pins, for instance, a reduced pin-count ATE. This has motivated innovative DfT solutions. An alternative approach transforms current into the time domain by noting that leakage and the parasitic pin capacitance represent a resistive-capacitive (RC) current [17]. A voltage stored on the capacitor changes over time, and one simply needs to sample the voltage at a specified time. Another approach transforms a current measurement into a voltage measurement [18].

I/O timings represent an analog measurement in the time domain. An I/O pin will have output and input timings relative to a clock source. Traditional explicit measurement of these timing relationships measures each one separately. Presuming a single pin and single defective timing, engineers have used DfT circuitry to perform a timing margin test [19, 20]. By making two timing margining measurements, these defect-based test methods cover four timing measurements and use on-die delay circuitry in increments of picoseconds. Based upon the statistics from good parts, engineers set pass/fail limits.

For the DfMn design goals of fault detection and fault isolation, analog/mixed-signal DfT circuitry adequately provides fault detection. Fault isolation within a VLSI device for analog devices may be limited to a circuit block. The ability to probe further will be determined by DfT control of signals in a diagnostic test scenario and the capabilities of the diagnostic test software.

13.6.6 Design and Test Tradeoffs

Constructive tension plays into any discussion on design and test tradeoffs. Engineers in the semiconductor industry often focus this discussion on changes to the design to ease test and diagnostic burden. Alternatively, the discussion can focus on modifying test equipment or using a specific test method (e.g., IDDQ testing for CMOS circuits). Manufacturing test remains cost-sensitive, yet product quality goals need to be met. Cost and quality concerns drive innovation in the test space, whether it is focused on design modifications to test equipment or DfT techniques. The following examples from the semiconductor industry illustrate two related innovations:

- SRAM's had long required checks on data retention, and the test literally was as follows: write a 1/0, pause for 500 ms, and then read. Innovation came with DfT circuitry. The test time was reduced from 1 second to 100 ms [12].
- High-speed differential interfaces (e.g., USB 3.0 [21]) possess stringent specifications called Bit

Error Rate (BER). Loopback testing on the load board, a common DfT technique, cannot detect all faulty behaviors. By inserting a passive filter in the loopback interconnect, engineers enhance its capability to cover jitter testing [16].

13.7 Leveraging DfT for Maintainability and Sustainment

Both manufacturing tests and sustainment tests used for maintainability require the ability to detect a failure. They typically differ in the availability of test equipment. A customer system will have less capability than a manufacturing test system. A customer return facility may or may not have the same capability as a manufacturing facility. These facilities also differ in their goals. Manufacturing test sorts good parts from bad parts. Maintainability requirements go beyond detecting the failing part by attempting to repair the product containing the failed part. In a customer's system, DfT techniques can support maintainability objectives as follows:

- Identifying the failing subsystem
- Flagging that there is a failure
- Conducting in-field diagnosis to guide repair steps
- Conducting failure analysis on the failing parts.

In the field, there may be limited equipment to use. In these situations, BIST capabilities come in handy. For mission mode field/mission mode test and diagnosis, BIT and BIST provide excellent tools to the DfMn engineer. When considering DfMn, engineers can leverage BIST solutions for in-field testing and for diagnosing internal to a VLSI chip or a larger system board. As BIST solutions can have capability limitations, designing in direct test access to critical blocks (e.g., functional safety) will more fully support DfMn goals.

13.7.1 Built-In-Test/Built-In Self-Test

BIT, BIST, and embedded test are common names for circuitry that requires minimal interaction from test equipment. Let's start with a discussion of BIST, which is embedded test at the component or silicon level. BIST facilitates shorter test times and increased test coverage. BIST requires the following capabilities:

- A control mechanism to start and stop the on-die test
- A generator for inputs and analyzer for outputs
- Communication to the tester, which can be a simple pass/fail indication or message

DfT engineers design BIST solutions for specific blocks of circuitry. In the mid-1980s, IC designers began using BIST circuitry for Logic Built-In Self-Test (LBIST) and Memory Built-In Self-Test (MBIST). By the mid-1990s LBIST and MBIST became more prevalent. Motivations for using these DfT solutions include testing multiple types of logic and memory blocks in a large system on a chip (e.g., a microprocessor), providing high test coverage in a limited test equipment setup, and enabling on-line testing throughout the product life cycle.

As mentioned previously, memory devices easily lend themselves to automatic pattern generation on testers due to their structured design. Thus, implementing a MBIST solution can be straightforward except for one caveat: how many pattern types should be supported? For small memories, 5–10 pattern types can provide the test effectiveness required. For much larger memories, MBIST solutions need to cover more pattern types. One way of doing this is to design a programmable BIST with a user interface, which requires some interaction with the tester to determine the best mix of patterns to achieve the desired test effectiveness while considering test time.

BIT testing is embedded diagnostics testing at all hardware levels above the component level, which leverages the component BIST capabilities of the components. BIST capabilities prove to be very valuable in any post-silicon engineering activity, especially for board-level manufacturing test and maintenance testing in the customer use applications. BIT testing is designed for use at the circuit board and product/system levels of hardware. BIT at the board level usually involves boundary scan testing. At the product or system level, there may be many types of BIT. Several examples of BIT include: PBIT (Power-up BIT) or POST (also see Section 13.4.1); I-BIT (Intermediate Maintenance BIT); C-BIT (Continuous BIT performed in the background while a system is operational); M-BIT (Maintenance BIT that requires manual connections to external test systems such as Built-In-Test Equipment [BITE]); or D-BIT (Depot Level Maintenance BIT). All these forms of board, product, and system BIT will use different combinations of component-level BIST at different times for different purposes.

13.8 BITE and External Support Equipment

Engineers may design Built-In-Test Equipment (BITE) to diagnose and manage faults during system operation as well as to support system maintenance during routine maintenance. The use of BITE integrated with systems or products is pervasive across multiple industries and marketplaces. BITE or external support equipment interfaces with the embedded test diagnostics of the product BIT or component BIST inside each system or product. The BITE equipment can include a wide range of instrumentation such as frequency generators, plug-in to TAPs, multimeters, and oscilloscopes.

Diagnostic computer systems that interpret internal signals may be part of the BITE equipment. Referring back to the automobile example, an automobile mechanic uses a diagnostic computer system to troubleshoot the cause of a check-engine light turning on. This diagnostic computer system plugs into the car's data interface bus, which connects to every electronically controlled module in the automobile. This example of BITE is applicable for DfT and DfMn for many types of system or product applications. This automobile diagnostic computer system is a mobile, transportable form of BITE. BITE system sizes vary greatly from a mobile handheld device to equipment filling up large rooms or several decks on a naval vessel.

13.9 Summary

Design engineers focus on the system from a system performance perspective, and rarely do they consider the needs of manufacturing the system or the maintenance of the system throughout its expected life. Maintainability engineers should be conversant in the approaches to DfT and DfT capabilities to assist the design team in test planning and in incorporating DfT into the product or system design.

It is important for the maintainability engineer to understand the full range of test capabilities within a system. Whatever test capability exists for the smallest part of the system, that capability could be leveraged and brought up to the level of test for the final deliverable product or system.

To arrive at a properly designed DfT feature, the test objectives or needs must be fully specified in released test requirements documentation, including test programs and test procedures.

This chapter has provided a broad overview, as well as a few specific examples, of what DfT means and how it is implemented at the three levels of hardware indenture: system/product, board, and device levels. This chapter provided some key definitions of test and design for test methods for electronic circuit boards and devices. Additional detailed information on DfT and test may be found in the cited references and in the recommended reading sections.

References

1 US Department of Defense (1995). *Testability Handbook for Systems and Equipment*, MIL-HDBK-2165. Washington, DC: Department of Defense.

2 Semiconductor Engineering (n.d.). Scan Test. https://semiengineering.com/knowledge_centers/test/scan-test-2/ (accessed 10 September 2020).

3 Abramovici, M., Breuer, M., and Friedman, A.D. (1990). *Digital Systems: Testing and Testable Design*. New York, NY: Computer Science Press.

4 International Standards Organization (2018). *ISO 26262-11:2018, Road Vehicles – Functional Safety – Part 11: Guidelines on application of ISO 26262 to semiconductors*. Geneva, Switzerland: International Standards Organization.

5 van Schaaijk, H., Spierings, M., and Marinissen, E.J. (2018). *Automatic generation of in-circuit tests for board assembly defects*. In: *IEEE International Test Conference in Asia (ITC-Asia)*, 13–18. Harbin:

Institute of Electrical and Electronics Engineers https://doi.org/10.1109/ITC-Asia.2018.00013.
6 IEEE (2013). *1149.1-2013 – IEEE Standard for Test Access Port and Boundary-Scan Architecture (Revision of IEEE Std 1149.1-2001)*. Piscataway, NJ: Institute of Electrical and Electronics Engineers.
7 IEEE Electronics Packaging Society (2019). Heterogeneous integration roadmap 2019 edition. https://eps.ieee.org/technology/heterogeneous-integration-roadmap/2019-edition.html (accessed 10 September 2020).
8 Appello, D., Bernardi, P., Grosso, M., and Reorda, M.S. (2006). System-in-package testing: problems and solutions. *IEEE Design and Test of Computers* **23** (3): 203–211. https://doi.org/10.1109/MDT.2006.79.
9 Hapke, F., Krenz-Baath, R., Glowatz, A. et al. (2009). Defect-oriented cell-aware ATPG and fault simulation for industrial cell libraries and designs. In: *IEEE Proceedings of 2009 International Test Conference*, Austin, TX. doi: https://doi.org/10.1109/TEST.2009.5355741.
10 Nigh, P., Needham, W., Butler, K. et al. (1997). An experimental study comparing the relative effectiveness of functional, scan, IDDq and delay-fault testing. In: *Proceedings of the 15th IEEE VLSI Test Symposium*, Monterey, CA, USA, 459–464. https://doi.org/10.1109/VTEST.1997.600334.
11 van de Goor, A.J. (1996). *Testing Semiconductor Memories*. Chichester, UK: Wiley.
12 Meixner, A. and Banik, J. (1997). Weak Write Test Mode: an SRAM cell stability design for test technique. In: *Proceedings of International Test Conference 1997*, Washington, DC, USA. doi:https://doi.org/10.1109/TEST.1997.639732.
13 Ney, A., Dilillo, L., Girard, P. et al. (2009). A New Design-For-Test Technique for SRAM Core Cell Stability Faults. In: *2009 Design, Automation and Test in Europe Conference*, Nice, France.
14 Kruseman, B., Tasic, B., Hora, C. et al. (2012). Defect oriented testing for analog/mixed-signal designs. *IEEE Design and Test of Computers* **29** (5): 72–80. https://doi.org/10.1109/MDT.2012.2210852.
15 Laisne, M., von Staudt, H.M., Bhalerao, S., and Eason, M. (2017). Single-pin Test control for Big A, Little D Devices. In: *Proceedings of IEEE International Test Conference 2017*, Fort Worth, TX.
16 Laquai, B. and Cai, Y. (2001). Testing gigabit multilane SerDes interfaces with passive jitter injection filters. In: *Proceedings of International Test Conference 2001*, Baltimore, MD, USA. doi:https://doi.org/10.1109/TEST.2001.966645.
17 Rahal-Arabi, T. and Taylor, G. (2001). A JTAG based AC leakage self-test. *2001 Symposium on VLSI Circuits*, Digest of Technical Papers, Kyoto, Japan. doi: https://doi.org/10.1109/VLSIC.2001.934239
18 Muhtaroglu, A., Provost, B., Rahal-Arabi, T., and Taylor, G. (2004). I/O self-leakage test. *Proceedings of 2004 International Conference on Test*, Charlotte, NC, USA.
19 Tripp, M., Mak, T.M., and Meixner, A. (2003). Elimination of traditional functional testing of interface timings at Intel. In: *Proceedings of International Test Conference 2003*, Charlotte, NC, USA. doi:https://doi.org/10.1109/TEST.2003.1271089.
20 Robertson, I., Hetherington, G., Leslie, T. et al. (2005). Testing high-speed, large scale implementation of SerDes I/Os on chips used in throughput computing systems. In: *Proceedings of International Conference on Test*, Austin, TX. 10.1109/TEST.2005.1584065.
21 Wikipedia USB 3.0. https://en.wikipedia.org/wiki/USB_3.0 (accessed 12 September 2020).

Suggestions for Additional Reading

Arnold, R. et al., (1998). Test methods used to produce highly reliable Known Good Die (KGD). *Proceedings International Conference on Multichip Modules and High Density Packaging*, Denver, CO, USA.

Bateson, J.T. (1985). *In-Circuit Testing*. Springer.

Burns, M., Roberts, G.W., and Taenzler, F. (2001). *An Introduction to Mixed-Signal IC Test and Measurement*. Oxford University Press.

Bushnell, M. and Agrawal, V. (2004). *Essentials of Electronic Testing for Digital, Memory and Mixed-signal VLSI Circuits*. Springer.

Gnusin, A. (2002). Introduction to DFT. http://sparkeda.com/technology/ (accessed 12 September 2020).

The IEEE journal, *IEEE Design & Test of Computers*, 1984 to present.

IEEE (2005). *1500-2005 – IEEE Standard Testability Method for Embedded Core-based Integrated Circuits*. Piscataway, NJ: Institute of Electrical and Electronics Engineers.

IEEE (2014). *1687-2014 – IEEE Standard for Access and Control of Instrumentation Embedded within a Semiconductor Device*. Piscataway, NJ: Institute of Electrical and Electronics Engineers.

Park, E.S., Mercer, M.R., and Williams, T.W. (1989). A statistical model for delay-fault testing. *IEEE Design and Test of Computers* **6** (1): 45–55.

Parker, K.P. (2016). *Boundary-Scan Handbook*, 4e. Springer.

Patil, S. and Reddy, S.M. (1989). A test generation system for path delay faults. In: *Proceedings IEEE International Conference on Computer Design: VLSI in Computers and Processors*, Cambridge, MA, USA.

Zorian, Y. (1993). A distributed BIST control scheme for complex VLSI devices. *Digest of Papers, Eleventh Annual IEEE VLSI Test Symposium*, Atlantic City, NJ, USA.

Zorian, Y., Marinissen, E.J., and Dey, S. (1998). Testing embedded core-based system chips. In: *IEEE Proceedings of 2009 International Test Conference*, 130–143.

14

Reliability Analyses

Jack Dixon

14.1 Introduction

"Reliability engineering is a sub-discipline of systems engineering that emphasizes dependability in the life cycle management of a product. Reliability describes the ability of a system or component to function under stated conditions for a specified period of time" [1].

"Reliability is the probability of an item to perform a required function under stated conditions for a specified period of time" [2].

A reliability program is essential for achieving the high levels of reliability expected in today's products. Reliability efforts support other desirable system characteristics, such as testability, maintainability, and availability. The reliability effort is best begun early in system development and refined over the system's life cycle.

While the focus of this book is designing for maintainability, this chapter focuses on reliability. In spite of this, the reader must not lose sight of the fact that the two system characteristics are closely interrelated. One can focus on a reliability strategy to produce a certain level of availability, or the focus could be on a maintainability strategy to achieve the desired level of availability. Improving maintainability is generally easier than improving reliability. Maintainability estimates (repair rates) are also generally more accurate. When reliability is not under control, more complicated issues may arise, like manpower (maintainers/customer service capability) shortages, spare part availability, logistic delays, lack of repair facilities, extensive retro-fit and complex configuration management costs, and others. The problem of unreliability may also be increased due to the "domino effect" of maintenance-induced failures after repairs of the initial failure are completed. One failure leads to a repair action that causes another failure that also needs to be repaired.

Focusing only on maintainability, therefore, is not enough. If failures are prevented or eliminated, none of the other issues related to maintainability are of any importance. Therefore, reliability is generally regarded as the most important part of availability. Reliability needs to be evaluated and improved in order to improve availability and to reduce the Total Cost of Ownership (TCO) due to cost of spare parts, maintenance man-hours, transport costs, storage cost, part obsolescence risks, and so on. Often a design tradeoff is needed, and a decision needs to be made between designing for reliability and designing for maintainability. Testability of a system should also be addressed since it serves as a link between reliability and maintainability. The maintenance and testability strategy can influence the reliability of a system (e.g., by preventive and/or predictive maintenance), helping the system to achieve its inherent design reliability, but the maintenance and testability strategy can never increase the reliability above the inherent design reliability as established through extensive design reliability analysis [3].

Design for Maintainability, First Edition. Edited by Louis J. Gullo and Jack Dixon.

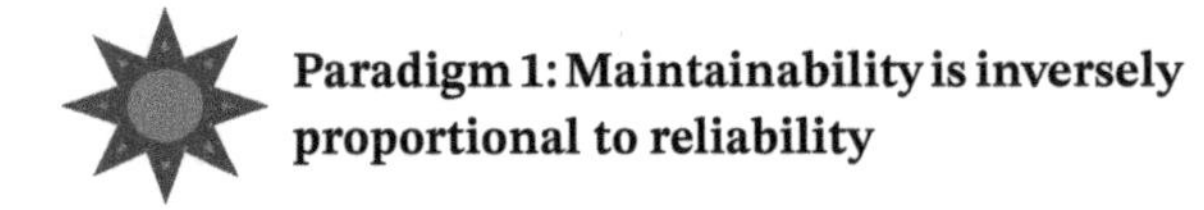
Paradigm 1: Maintainability is inversely proportional to reliability

All organizations, governments, or private companies want to acquire and/or produce quality products that satisfy user or customer needs in a timely manner and at a reasonable cost or price. Reliability, along with maintainability and availability, are essential to meeting these goals. While reliability, maintainability, and availability are inextricably linked, this chapter will focus on reliability, and in particular some of the most common reliability analyses and modeling techniques.

14.2 Reliability Analysis and Modeling

A simple definition of modeling appears in the *Oxford Advanced Learners Dictionary* as "the work of making a simple description of a system or a process that can be used to explain it" [4]. The *Merriam-Webster Collegiate Dictionary* provides a more technical definition of modeling as "a system of postulates, data, and inferences presented as a mathematical description of an entity or state of affairs; also a computer simulation based on such a system" [5].

Reliability modeling typically uses graphic and mathematical representations supported by data to represent systems or products, and uses these representations for performing various types of analyses. Reliability modeling is used to understand or predict the reliability of a component or system prior to its implementation. Reliability modeling is an important tool for:

- Assessing the performance of product designs.
- Quantifying system effectiveness.
- Uncovering design problems that might cause the system to not meet its requirements.
- Performing trade studies between alternative design solutions.
- Performing tradeoffs between design characteristics such as performance, reliability, and maintainability.
- Performing sensitivity analyses to optimize performance and cost-effectiveness.
- Determining the required redundancy needed in fault-tolerant or non-repairable systems.
- Determining and evaluating warranty provisions.
- Providing inputs to logistic support and life cycle cost functions.

During all phases of a system development project, reliability models are used to facilitate reliability allocations, evaluate alternative design options, conduct reliability predictions, and perform stress–strength analyses.

The most common type of reliability models are:

- Reliability Block Diagrams
- Reliability allocation
- Reliability mathematical model
- Reliability prediction
- Fault Tree Analysis (FTA)
- Failure Modes, Effects, and Criticality Analysis (FMECA)

These models will be covered in the following sections.

14.3 Reliability Block Diagrams

Initial reliability models are developed as Reliability Block Diagrams (RBDs), which typically evolve from the systems engineering functional analysis effort. The RBD describes the component or functional level series and parallel relationships (see Figure 14.1) to designate the presence of redundancies in a fault-tolerant architecture. The purpose of the RBD is to develop, very early in the life cycle, a reasonable facsimile (i.e., model) of the system elements that must function for the operation of the system to be successful.

The first step in the construction of a system RBD is to create the top-level RBD representing the overall

Figure 14.1 Examples of series and parallel component configurations.

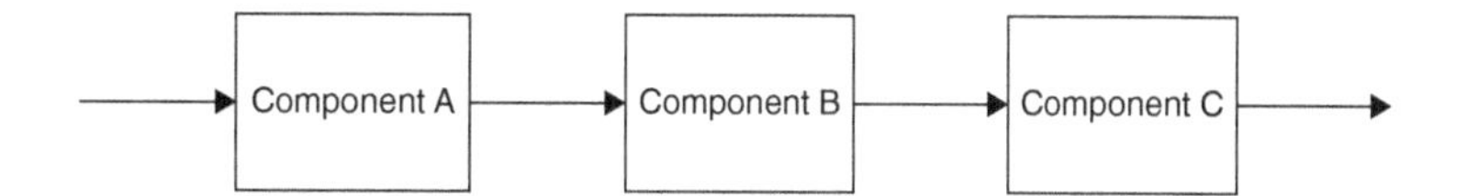

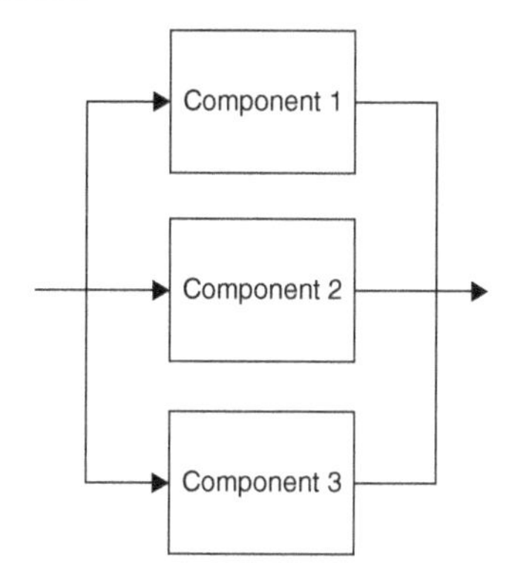

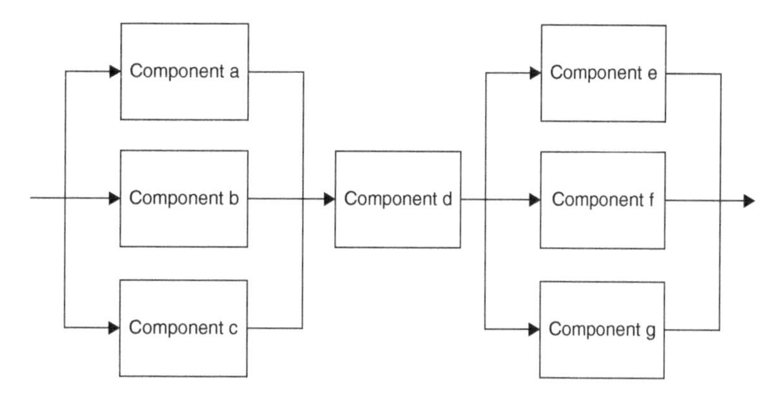

system (Level 1). Once the top-level RBD for the system has been created using the various series-parallel components, the RBD can be expanded to lower and lower levels, ultimately describing the entire system as shown in Figure 14.2.

In this example, component B is expanded to show the single string components 1 through 4 in a series circuit. This means that any one of these components in Level 2 will cause component B to fail. Component 3 is expanded to the serial-parallel diagram at Level 3 for components a through f. Component d at Level 3 is expanded to show components U though Z at level 4 in their series-parallel configuration. This figure shows an expansion only for component block B at Level 1; similar expansions must be made for Level 1 blocks A, C, and D to represent the breakdown of the system components that make up those blocks at Levels 2 through 4.

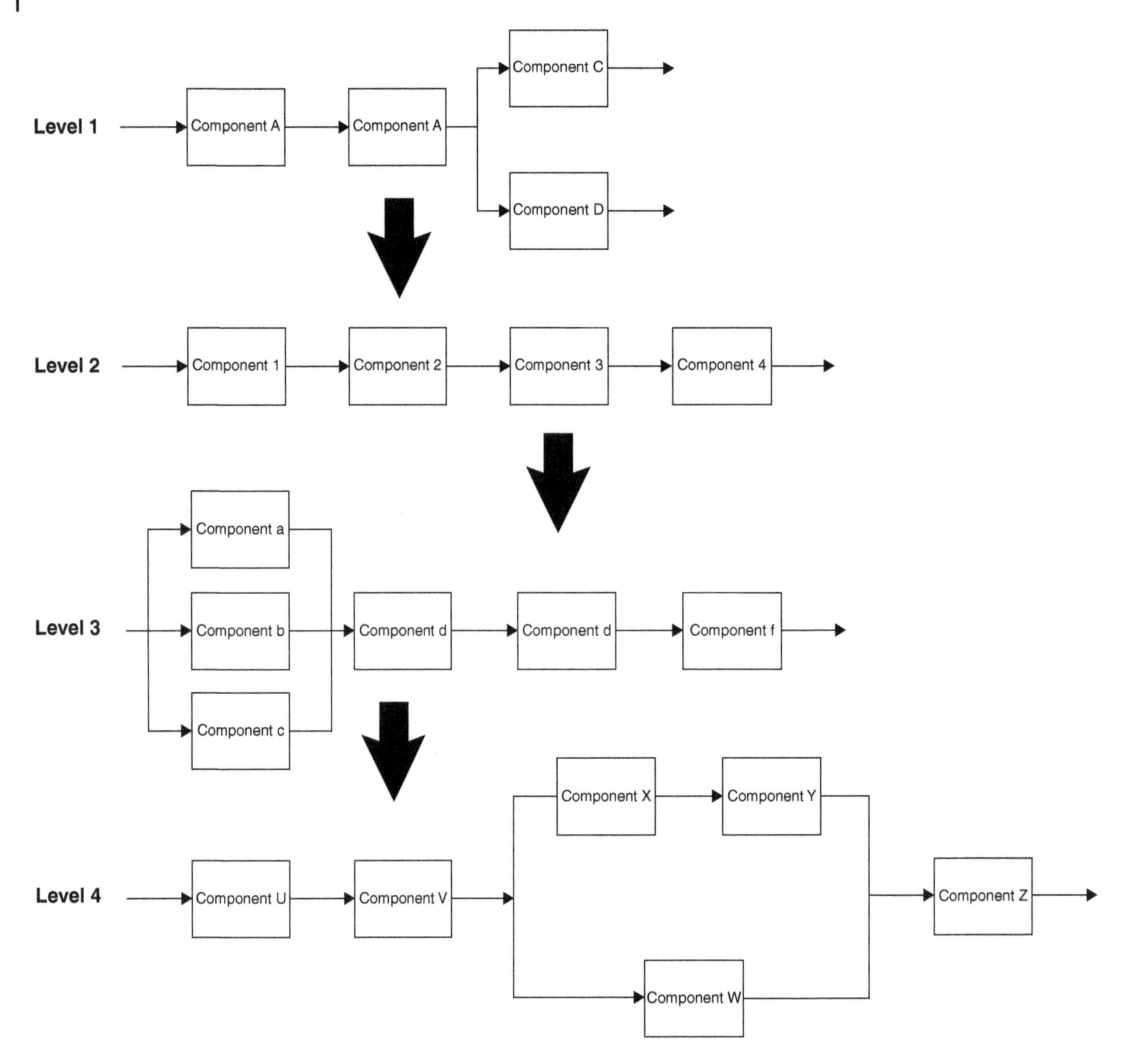

Figure 14.2 Reliability block diagram.

14.4 Reliability Allocation

After the overall system RBD has been developed, the next step is to use it to do reliability allocation. The top-level reliability requirement is typically provided as part of the customer requirements. This top-level reliability requirement must then be allocated among the lower-level subsystems, assemblies, components, and so on. Referring back to Figure 14.2, the system-level reliability requirement applies to Level 1. An allocated reliability requirement is subsequently specified for each block, A, B, C, and D, in the system Level 1 block diagram, creating lower-level block diagrams. This process is continued until the lowest level of components is reached. When these values are combined at Level 1, they will represent the reliability of the system. Each of these requirements is then subsequently broken down into allocated requirements for every lower level of the RBD until each block has been assigned an allocated reliability requirement.

14.5 Reliability Mathematical Model

Once the RBD is completed and the reliability allocation has been done, a reliability mathematical model can then be developed. This procedure calls for developing a mathematical representation of the reliability of each portion of the RBD, then combining these and reducing the complex result to a single probability statement consisting of an equivalent serial expression.

Referring back to Figure 14.1, the following formulae are used to describe the reliability of each configuration.

For series components:

$$R_{System} = R_A \times R_B \times R_C \times \ldots R_N$$

For parallel components:

$$R_{System} = 1 - (1 - R_1) \times (1 - R_2) \times (1 - R_3) \times \ldots (1 - R_n)$$

These formulae can be combined to represent any combination of series and parallel components. The RBD establishes a model that can be used for performing reliability prediction.

14.6 Reliability Prediction

Reliability prediction is an indispensable technique used to evaluate a system throughout its life cycle from conception, through development, and into production. Prediction helps to provide a basis for making design decisions, choosing between alternative concepts, and selecting quality levels for components. A good reliability prediction is invaluable to those making decisions regarding the feasibility and adequacy of a particular design approach.

Reliability prediction is usually considered to be the best estimate of the reliability of a particular design within the limits of data and the accuracy of the system definition. While systems and reliability engineers are good at system definition, the data that goes into reliability predictions is usually going to be the limiting factor. The input for the models and for reliability predictions typically comes from numerous sources including testing, experience, field data, and data handbooks. Regardless of where the data come from, all data must be used with caution, because the predictions will only be as good as the data used. The analyst must ensure that the data, particularly data based on similarity, are valid to use for the product under development. Since reliability predictions are often based on data from similar items, or their components, that are used in the same or similar manner, caution must be exercised in establishing the similarity with other items and the degree of similarity in the conditions of their use.

Paradigm 9: Support maintainability with data

Reliability modeling and reliability prediction are typically started in the early stages of the development cycle to aid in the evaluation of the concept design and to provide a basis for reliability allocation. Reliability models and predictions should be updated throughout the system development as design details emerge, when there is a significant change in the design or in the system use, or when additional data such as environmental requirements, stress data, or failure rate data become available. Reliability prediction updates are usually provided at all design reviews and other major program milestones.

MIL-STD-756 [6] is commonly used as a guide for the performance of reliability predictions. The steps below are adapted from MIL-STD-756 and define the procedure for developing a reliability model and performing a reliability prediction.

Step 1: Define the item for which the prediction is applicable.
Step 2: Define the service use (life cycle) for which item reliability will be modeled and predicted.
Step 3: Define the item Reliability Block Diagrams.
Step 4: Define the mathematical models for computing item reliability.
Step 5: Define the parts of the item.
Step 6: Define the environmental data.
Step 7: Define the stress data.
Step 8: Define the failure distribution.
Step 9: Define the failure rates.
Step 10: Compute the item reliability.

In order to calculate the item reliability in the last step, the analyst must use the RBD developed earlier and apply the rules for calculating series and parallel components as described previously. This approach results in a value that is commonly referred to as the Mission Reliability. For a system of any significant size or complexity, doing these calculations by hand very quickly becomes unwieldy. There are a number of commercial software products on the market that can streamline the calculations.

A second, simpler approach is called the Basic Reliability Prediction. This approach uses a series model for calculating the reliability. All elements of the item, even those provided for redundancy or alternate modes of operation, are included in the series. Then the formula for calculating reliability of a series as shown previously is used to determine the reliability. This approach is used for estimating the demand for maintenance and logistic support caused by an item and its component parts.

14.7 Fault Tree Analysis

"A Fault Tree Analysis (FTA) is a systematic, deductive methodology for defining a single specific undesirable event and determining all possible reasons (failures) that could cause that event to occur. The undesired event constitutes the top event in the fault tree diagram, and generally represents a complete or catastrophic failure of the product. The FTA focuses on a select subset of all possible system failures, specifically those that can cause a catastrophic top event [2]."

Since FTA was invented at Bell Laboratories by H. A. Watson in 1961 [7], it has become a very popular and powerful analysis technique. FTA was initially applied, by Watson, to the study of the Minuteman Launch Control System. Dave Hassl, of Boeing, later expanded its use to the entire Minuteman Missile System. He presented an historic paper on FTA at the first System Safety Conference in Seattle in 1965 [8].

The initial use of FTA was predominately as a system safety tool. FTA is a root-cause analysis tool used to determine the underlying causes of an undesirable event that has been identified by a hazard analysis technique. It is an important technique often used by system safety engineers. However, FTA has also become a favorite tool of reliability practitioners.

FTA can be used for all of the following [2]:

- Functional analysis of highly complex systems.
- Observation of combined effects of simultaneous, non-critical events on the top event.
- Evaluation of safety requirements and specifications.
- Evaluation of system reliability.
- Evaluation of human interfaces.
- Evaluation of software interfaces.
- Identification of potential design defects and safety hazards.
- Evaluation of potential corrective actions.
- Simplifying maintenance and troubleshooting.
- Logical elimination of causes for an observed failure.

From a reliability perspective, the FTA can estimate whether a product will or will not meet performance reliability requirements.

This section is meant only to serve as a brief overview of FTA. It is not intended to be an exhaustive, detailed treatment. The reader is invited to review the books recommended in the suggested reading section at the end of the chapter for more in-depth treatment of FTA.

14.7.1 What is a Fault Tree?

A fault tree is a representation in tree form of the combination of causes (failures, faults, errors, etc.) contributing to a particular undesirable event. It uses symbolic logic to create a graphical representation of the combination of failures, faults, and errors that can lead to the undesirable event being analyzed. The purpose of FTA is to identify the combinations of failures and errors that can result in the undesirable event. The fault tree allows the analyst to focus resources on the most likely and most important basic causes of the top event.

FTA is a deductive (top-down) analysis technique that focuses on a particular undesirable event and is used to determine the root causes contributing to the occurrence of the undesirable event. The process starts with identifying an undesirable event (top event) and working backwards through the system to determine the combinations of component failures that will cause the top event.

Using Boolean algebra, the fault tree can be "solved" for all the combinations of basic events that can cause the top event. This results in a qualitative analysis of the fault tree. The fault tree can also be used as a quantitative analysis tool. In this case, failure rates or probability of occurrence values are assigned to the basic events and the probability of the occurrence of the top level, undesirable event can be calculated. [9]

While the qualitative analysis is heavily relied upon in safety analysis, the quantitative analysis is almost always used in reliability analysis.

14.7.2 Gates and Events

Logical gates are used, along with basic events, to create the fault tree. The standard symbols used in the construction of fault trees and the descriptions of these symbols are shown in Figure 14.3.

14.7.3 Definitions

Some basic definitions are needed to be able to understand FTA.

Cut set: A cut set is a combination of hardware and software component failures that will cause system failure.

Minimal cut set: A minimal cut set is a smallest combination of hardware and software component failures which, if they all occur, will cause the top event to occur [10].

Failure: A failure is a basic abnormal occurrence in a hardware or software (HW/SW) component. Failure is the inability of a function to meet its requirement specification. For example, a relay fails closed when it should be open under normal operation, according to its specification.

Fault: A fault is an undesired state of a function composed of hardware and software components, which occurs at an undesirable time. For example, a relay closes when it is supposed to be open, but not because the relay failed closed, but because it was "commanded" to close at the wrong time.

Primary fault: A primary fault is any fault of a HW/SW component that occurs in an environment for which the component is qualified. For example, a pressure tank, designed to withstand pressures up to and including a pressure p_0, ruptures at some pressure $p \leq p_0$ because of a defective weld [10].

Secondary fault: A secondary fault is any fault of a (HW/SW) component that occurs in an environment for which it has not been qualified. In other words, the component fails in a situation that exceeds the conditions for which it was designed. For example, a pressure tank, designed to withstand pressure up to and including a pressure p_0, ruptures under a pressure $p > p_0$ [10].

Command fault: A command fault involves the proper operation of a HW/SW component but at the wrong time or in the wrong place. For example, an arming device in a warhead train closes too soon because of a premature or otherwise erroneous signal origination from some upstream device [10].

Exposure time: The exposure time is the time the system is operated, thus exposing the components of the system to failure. The longer the exposure time period, the higher the probability of failure.

Undesirable event: The undesirable event is the top level event. It is the subject of the FTA. In safety engineering it is typically an event that has been identified by hazard analysis. For example, in a weapon system, the undesirable event may be "Inadvertent Detonation of Warhead." In reliability engineering the event may be some undesirable system failure. For example, it might be "Solar panel fails to deploy on satellite."

14.7.4 Methodology

FTA begins by identifying the undesirable event that is to be analyzed. Then the immediate causes of that event are identified. This process continues until the most basic causes have been identified (e.g., resistor fails open). The completed fault tree model then represents the relationships between all of the basic causes and the undesirable event. There may be multiple ways the undesirable event can occur, each of which is represented by a combination of basic causes. These

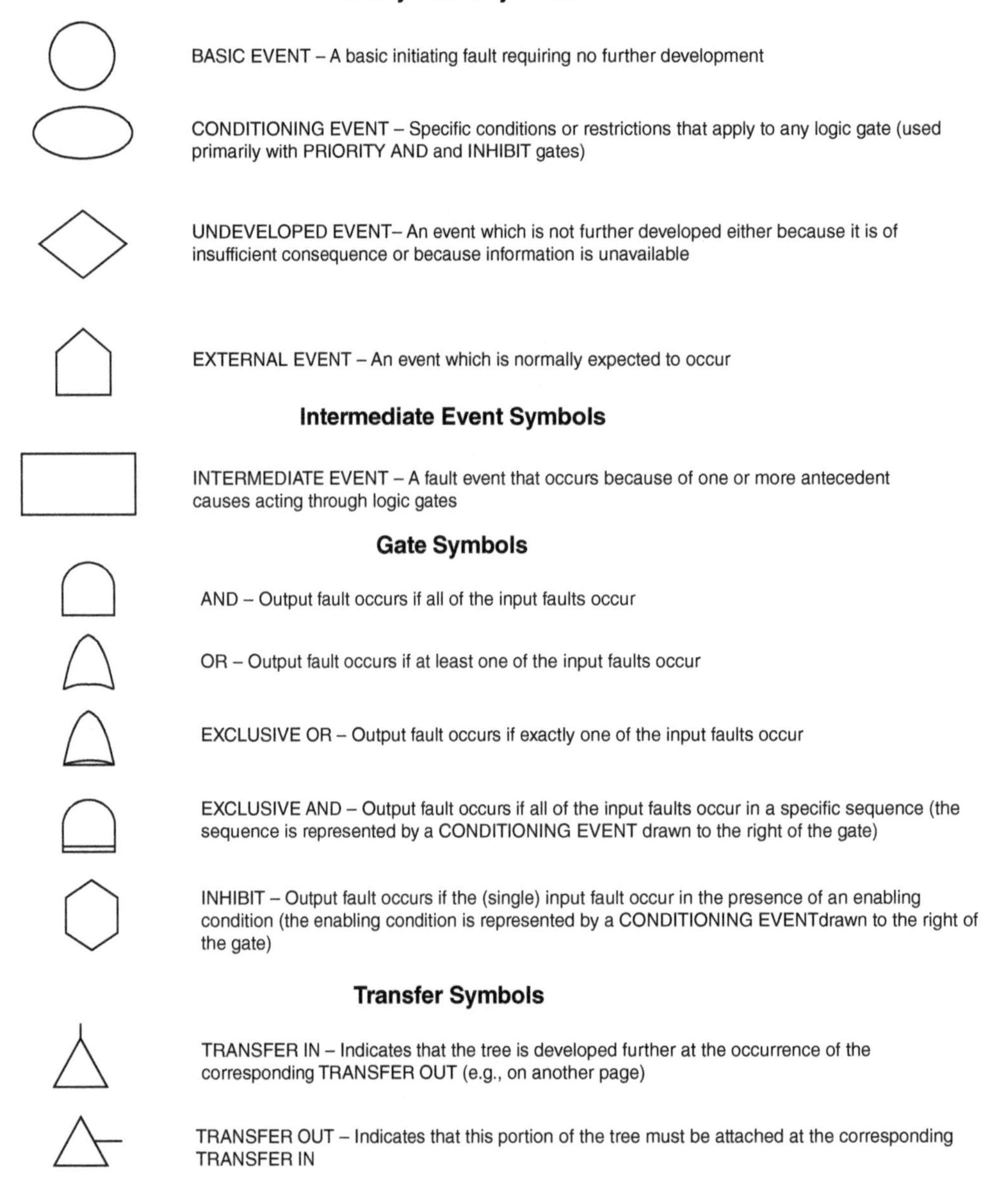

Primary Event Symbols
BASIC EVENT – A basic initiating fault requiring no further development
CONDITIONING EVENT – Specific conditions or restrictions that apply to any logic gate (used primarily with PRIORITY AND and INHIBIT gates)
UNDEVELOPED EVENT– An event which is not further developed either because it is of insufficient consequence or because information is unavailable
EXTERNAL EVENT – An event which is normally expected to occur
Intermediate Event Symbols
INTERMEDIATE EVENT – A fault event that occurs because of one or more antecedent causes acting through logic gates
Gate Symbols
AND – Output fault occurs if all of the input faults occur
OR – Output fault occurs if at least one of the input faults occur
EXCLUSIVE OR – Output fault occurs if exactly one of the input faults occur
EXCLUSIVE AND – Output fault occurs if all of the input faults occur in a specific sequence (the sequence is represented by a CONDITIONING EVENT drawn to the right of the gate)
INHIBIT – Output fault occurs if the (single) input fault occur in the presence of an enabling condition (the enabling condition is represented by a CONDITIONING EVENTdrawn to the right of the gate)
Transfer Symbols
TRANSFER IN – Indicates that the tree is developed further at the occurrence of the corresponding TRANSFER OUT (e.g., on another page)
TRANSFER OUT – Indicates that this portion of the tree must be attached at the corresponding TRANSFER IN

Figure 14.3 Fault tree symbols. Source: From Gullo, L. J. and Dixon, J., *Design for Safety*, John Wiley & Sons, Inc., Hoboken, NJ, 2018. Reproduced with permission from John Wiley & Sons.

combinations are known as cut sets. A minimal cut set is the smallest combination of basic causes that cause the top, undesirable event to occur.

Fault trees consist of gates and events connected together like branches on a tree. AND and OR gates are the two most commonly used gates in FTA. Figure 14.4 shows two simple fault trees consisting of two input events that can lead to the output or top event. If either input event can cause the output event, these input events are connected using an OR gate. If both input events are required to cause the output event then they are connected using an AND gate.

Construction and analysis of fault trees involves following the process as shown in Figure 14.5.

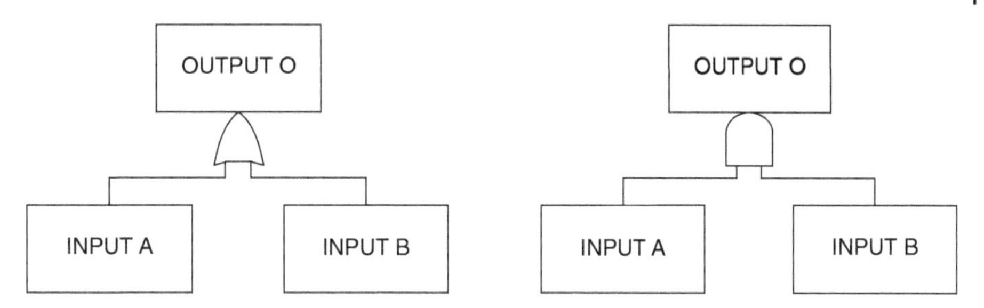

Figure 14.4 Simple fault trees. Source: From Gullo, L. J. and Dixon, J., *Design for Safety*, John Wiley & Sons, Inc., Hoboken, NJ, 2018. Reproduced with permission from John Wiley & Sons.

Successfully drawing fault trees requires following a set of basic rules. These rules are adapted from NUREG-0492 [10].

- **Ground Rule 1.** Describe exactly each event by writing statements that are entered in the event boxes as faults; state precisely what the fault is and when it occurs (e.g., motor fails to start when power is applied).
- **Ground Rule 2**. If the answer to the question "Can this fault consist of a component failure?" is "Yes," classify the event as a "state-of-component fault." If the answer is "No," classify the event as a "state-of-system fault."
- **Ground Rule 3**. No Miracles Rule. If the normal functioning of a component propagates a fault sequence, then it is assumed that the component functions normally.
- **Ground Rule 4**. Complete-the-Gate Rule. All inputs to a particular gate should be completely defined before further analysis of any one of them is undertaken.
- **Ground Rule 5**. No Gate-to-Gate Rule. Gate inputs should be properly defined fault events, and gates should not be directly connected to other gates.

14.7.5 Cut Sets

After the fault tree has been constructed, it must be solved in order to determine the cut sets. The following examples are adapted from Gullo and Dixon [9]. For very simple trees, as in Figure 14.6, cut sets can be determined by sight.

The cut sets are:

A
B, D
C, D

As fault trees become more complex, the reduction of the tree to cut sets becomes more complicated. Boolean algebra must be used. It is assumed that the reader is familiar with the rules of Boolean algebra. The following illustrates the calculation of cut sets and minimal cut sets in a slightly more complicated tree (Figure 14.7).The cut sets are:

$(A + B)$
$(A + C)$
$(D + B)$
$(D + C)$

From these, we must determine the minimal cut sets. Recalling the definition from an earlier section, a minimal cut set is a smallest combination of HW/SW component failures which, if they all occur, will cause the top event to occur [10]. So the Boolean expression for the cut sets for the tree in Figure 14.7 is:

$$(A + B) \cdot (A + C) \cdot (D + B) \cdot (D + C)$$

Using the rules of Boolean algebra, we reduce this expression to its minimal cut sets as follows:

$$[A + (B \cdot C)] \cdot [D + (B \cdot C)]$$

$$(B \cdot C) + (A \cdot D)$$

Therefore the minimal cut sets are:

$$(B \cdot C)$$

$$(A \cdot D)$$

Even in this simple example, solving fault trees manually using Boolean algebra can get cumbersome very quickly. Because of this, numerous algorithms and computer programs have been developed to assist in the generation of cut sets and minimal cut sets.

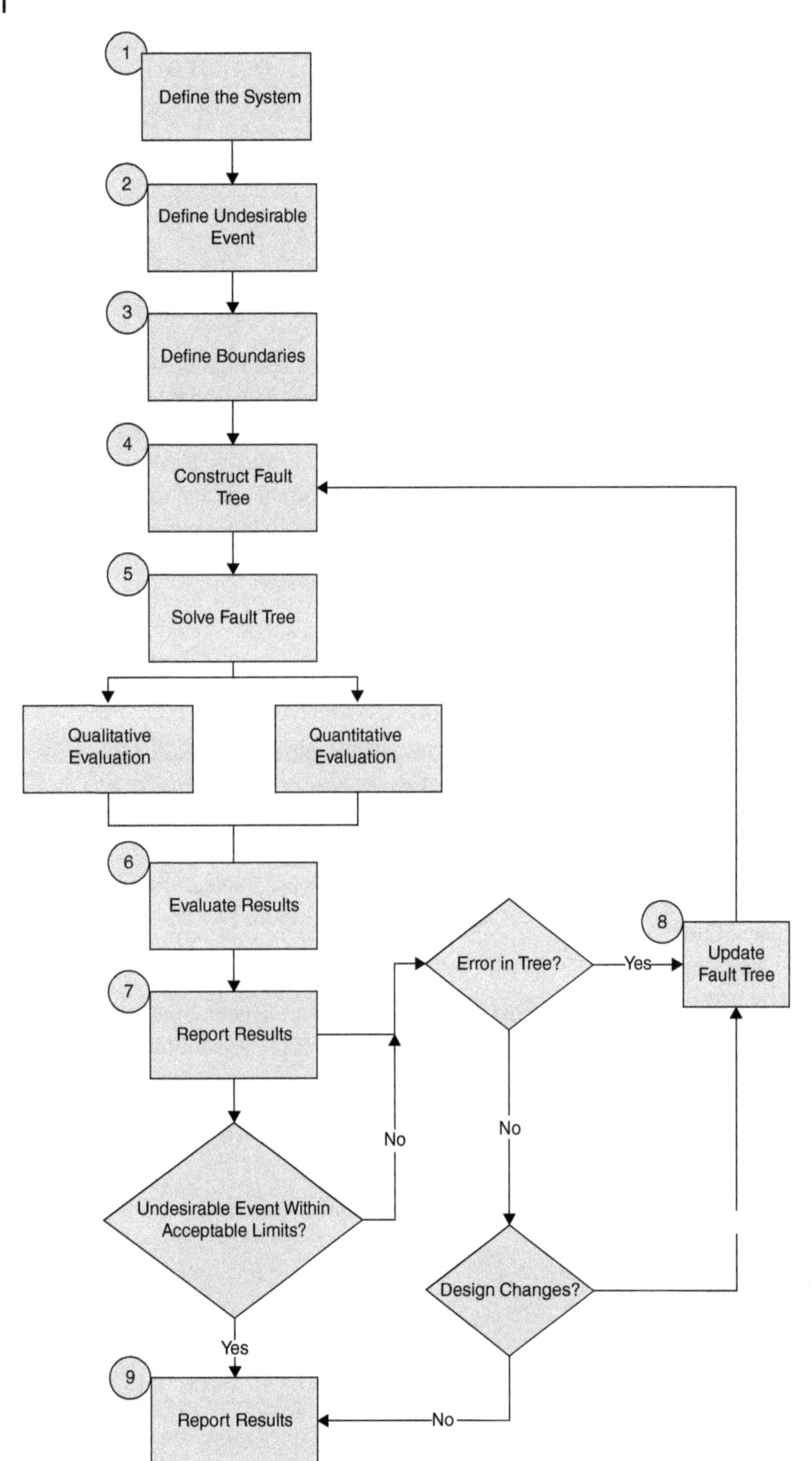

Figure 14.5 Fault tree construction process. Source: From Gullo, L. J. and Dixon, J., *Design for Safety*, John Wiley & Sons, Inc., Hoboken, NJ, 2018. Reproduced with permission from John Wiley & Sons.

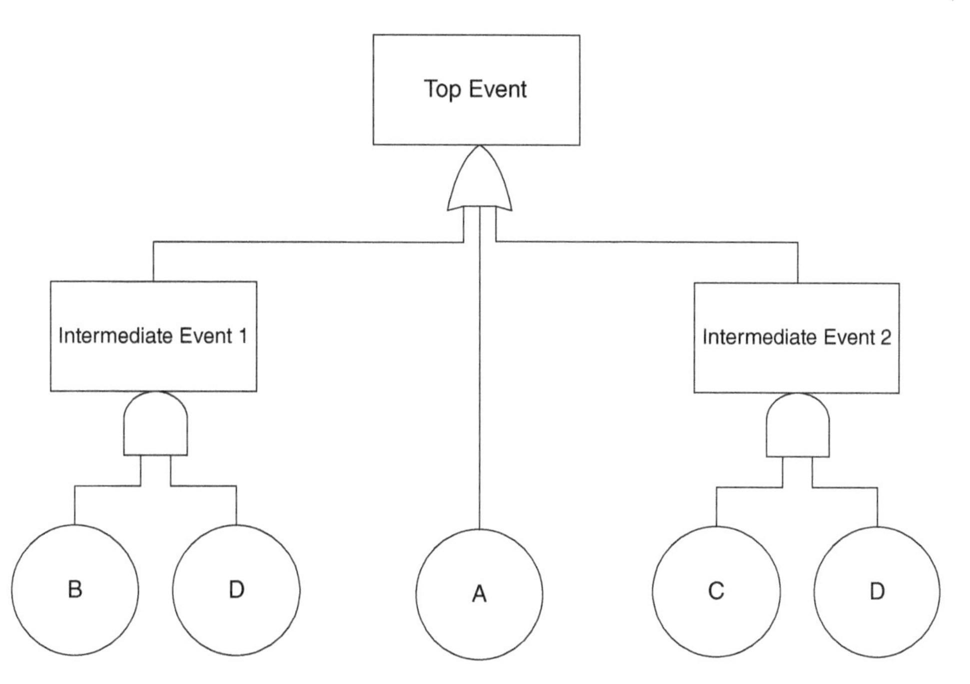

Figure 14.6 Simple cut set determination. Source: From Gullo, L. J. and Dixon, J., *Design for Safety*, John Wiley & Sons, Inc., Hoboken, NJ, 2018. Reproduced with permission from John Wiley & Sons.

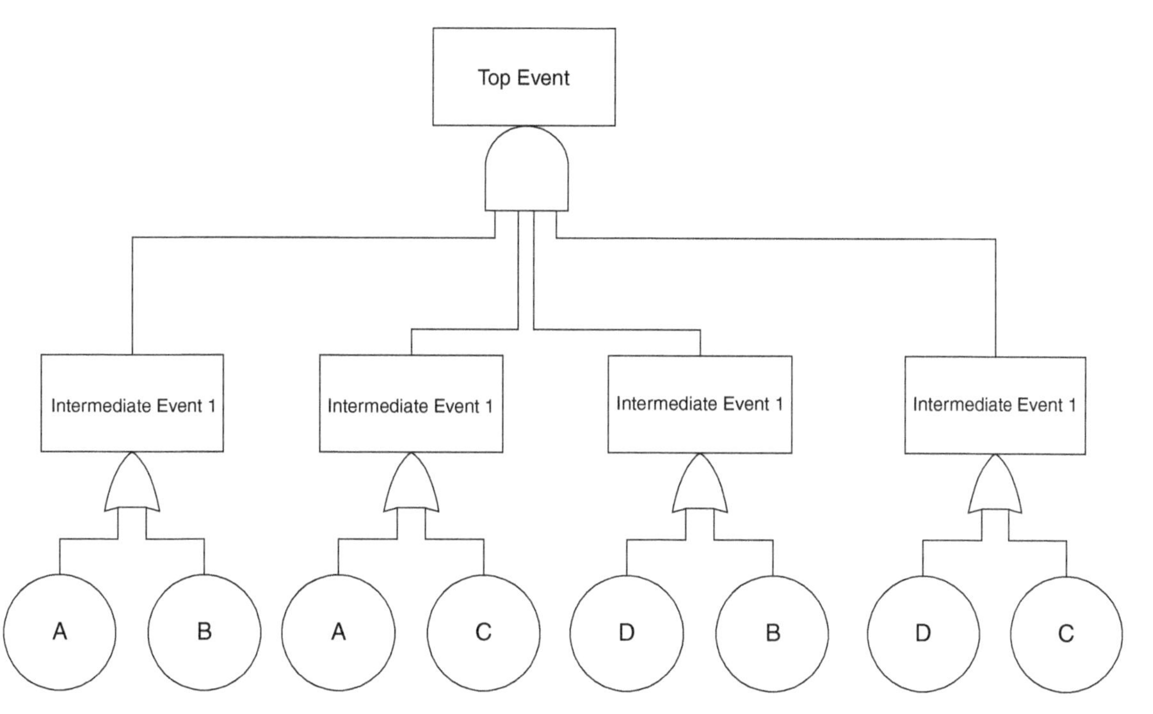

Figure 14.7 Minimal cut set determination. Source: From Gullo, L. J. and Dixon, J., *Design for Safety*, John Wiley & Sons, Inc., Hoboken, NJ, 2018. Reproduced with permission from John Wiley & Sons.

14.7.6 Quantitative Analysis of Fault Trees

There are two methods used to calculate the probability of occurrence of the top, undesirable, event. In both methods, the failure rates for all of the basic events must be determined. This is done by using reliability calculations and/or by obtaining failure rate data using a reliability handbook. One of the most commonly used references is *Military Handbook: Reliability Prediction of Electronic Equipment*, MIL-HDBK-217F [11].

The formula used to calculate the reliability of a HW/SW component is:

$$R = e^{-\lambda t}$$

where R is reliability, λ is the HW/SW component failure rate, and t is the exposure time.

The probability of failure is:

$$P_f = 1 - R = 1 - e^{-\lambda t}$$

Once all the failure rates have been determined, the probability of occurrence of the top event can be calculated.

First Method – Summing Probabilities: To calculate the probability of occurrence of the top event using this method, the analyst starts at the bottom of the tree and calculates the probabilities for each gate moving up the tree to the top event. To do this, the following rules apply to the common AND and OR gates:

For an AND gate the probability of failure is:

$$P_f = P_A \times P_B$$

For an OR gate the probability of failure is:

$$P_f = P_A + P_B - (P_A \times P_B)$$

This method does not work, however, if there are multiple occurrences of events or branches within the tree. If these are present, then a cut set method must be used to obtain accurate results.

Second Method – Cut Set Method: For the cut set method of probability calculation, once the minimal cut sets are obtained, if desired, the analyst can perform quantitative analysis of the fault tree by evaluating probabilities. The quantitative evaluations are most easily performed in a sequential manner by first determining the component failure probabilities, then determining the minimal cut set probabilities, and finally determining the probability of the top event occurring by summing up the probabilities of each of the minimal cut sets using the appropriate gate equations for determining the cut set probability [9].

These calculations get very unwieldy and difficult for anything but the smallest fault trees, and thus automated methods are typically applied using readily available commercial software.

14.7.7 Advantages and Disadvantages

FTA is a powerful analytical technique that has been used in many different industries for many years. It can be implemented at any stage of design from conceptual to very detailed. It can provide valuable insights into reliability problems and a valuable complement to tradeoff studies. FTA has the advantage over some other analysis techniques in that it provides a very visible presentation of how the particular undesirable event happens. FTA provides a practical approach to test planning. By viewing the FTA, a test engineer can readily see key points in the system network where testing would be effective and improve test fault coverage. Automated FTA software tools are readily available and cost-effective for executing complex fault tree designs.

The biggest disadvantage of FTA is that it can be very time-consuming and expensive, if performed manually for large complex systems.

14.8 Failure Modes, Effects, and Criticality Analysis

The FMECA is a very popular and useful tool used by reliability engineers. The process of conducting a FMECA is carried out in the reverse order of that of FTA. The FTA is a top-down approach starting with an undesirable top event and working down to component failure level. The FMECA, on the other hand, starts at the bottom level of the hardware (i.e., piece part component) looking for failure modes and failure causes, and flows the analysis up to determine the failure effects at the local assembly level, the next higher assembly level, up to the system-level effects.

There are actually two types of related analyses: Failure Modes and Effects Analysis (FMEA) and the FMECA. The FMEA is the process of reviewing the components, assemblies, and subsystems to identify potential failure modes and their causes and effects. Using a worksheet format, the failure modes and their effects on the system under investigation are recorded. The FMEA can be a qualitative or quantitative analysis. The FMEA is often the first step in a reliability study of a system.

The FMEA is usually extended to become a FMECA to include criticality analysis. The FMECA is an analysis procedure which documents all probable failures in a system, determines by failure mode analysis the effect of each failure on system operation, identifies single failure points, and ranks each failure according to a severity classification of the failure effect. "While the objective of an FMECA is to identify all modes of failure within a system design, its first purpose is the early identification of all catastrophic and critical failure possibilities so they can be eliminated or minimized through design correction at the earliest possible time. Therefore, the FMECA should be initiated as soon as preliminary design information is available at the higher system levels and extended to the lower levels as more information becomes available on the items in question" [12].

FMECA is essentially a reliability task, but it also provides information for other purposes. FMECA is also sometimes used in maintainability analysis, system safety analysis, logistics support analysis, maintenance plan analysis, and survivability/vulnerability studies. Because of these multiple uses, it is prudent to carefully plan the FMECA effort to minimize the duplication of efforts within the development program.

There are several types of FMECA, including Design FMECA, Process FMECA, and Software FMECA, and their use depends on the focus of the analysis:

- The **Design FMECA** analyzes a system's performance to determine what happens when or if a failure occurs. This type of FMECA is performed by examining assembly drawings, part datasheets, electrical schematics, and specifications.
- The **Process FMECA** is focused on the manufacturing and test processes required to build a reliable product. To perform this type of FMECA the analyst examines the materials, parts, manufacturing processes, tools, equipment, inspection methods, human errors, and documentation to identify possible risks of failures in process steps during production, maintenance, or use.
- The **Software FMECA** may be applied at different points in the software development life cycle. These various applications of FMECA to software products may occur at any point in the software development cycle, such as during development of software requirements, creation of software architecture, high-level software design, low-level software design, coding, or software testing.

While there are similarities between the various FMECA analyses, this chapter focuses on the Design FMECA to illustrate the technique. This section is meant only to serve as a brief overview of FMECA and is not intended to be an exhaustive, detailed treatment of the subject. The reader is invited to review the books listed in the suggested reading section at the end of the chapter for more in-depth treatment of FMECA and the three types of FMECA listed above.

The FMECA is conducted using a worksheet. While there are many variations to the worksheet, Figure 14.8 will serve as a general example.

Column Descriptions

Header information: Self-explanatory. Any other generic information desired or required by the program could be included.

Identifier: This can simply be an identifying number that identifies a particular subsystem, component, etc.

Function: This is a description of the function being performed by the system, subsystem, etc.

Failure Mode: This is a description of all possible failure modes for the identified item. This information may be obtained from historical data, manufacturer's data, experience, similar items, or testing. Items may have multiple failure modes; therefore, each failure mode should be recorded and analyzed for its effect on the system.

Effects of Failure: This is a short description of the possible effect each failure mode has on the system, subsystem, etc.

Causes of Failure: This is a brief description of what can cause the specific failure mode. Causes might include things such as physical failure, wear-out, temperature stress, vibration stress, and so on.

Failure Modes, Effects, and Criticality Analysis									
System: ________									
Preparer: ________									
Date: ________									
Identifier	**Function**	**Failure Mode**	**Effects of Failure**	**Causes of Failure**	**P**	**S**	**D**	**RPN**	**Recommended Action & Status**

Figure 14.8 Failure Modes, Effects, and Criticality Analysis worksheet.

All conditions that might affect an item should be recorded, including any special periods of operation, stress, or combinations of events that could increase the probability of failure or severity of damage.

P: The probability of failure in the particular mode.
S: The severity of the failure; its consequence.
D: The probability of detection of the failure.
RPN: This column contains the Risk Priority Number (RPN). The RPN is a function of and is determined by combining P, S, and D.
Recommended Action and Status: Provide recommended preventive measures to eliminate or mitigate the effects of the potential failure mode. This column should also include current status of the action.

Different methods are used to calculate the RPN. One common method uses the multiplication of the three factors, P, S, and D as follows:

$$\text{RPN} = (\text{probability of failure}) \times (\text{severity of the failure}) \times (\text{probability of detection of the failure})$$

$$\text{RPN} = \text{P} \times \text{S} \times \text{D}$$

Ratings from 1 to 10 for each term are established. Tables 14.1, 14.2, and 14.3 provide simplified versions of each of these factors.

Table 14.1 Probability of failure rating.

Probability of failure rating	Failure probability (*P*)
10	1 in 2
9	1 in 8
8	1 in 20
7	1 in 40
6	1 in 80
5	1 in 400
4	1 in 1000
3	1 in 4000
2	1 in 20 000
1	1 in 1 000 000

Obviously, some judgment by the analyst is necessary to determine the ratings for the three factors, P, S, and D, After the analyst establishes the ratings they are multiplied together to obtain the RPN, which reflects the failure mode criticality.

Since the three ratings vary from 1 to 10, their product will vary from 1 to 1000. Once the RPN is determined, it is a simple matter to separate the important from the unimportant. The RPN values can then be sorted by highest to lowest value, providing insight into the most, and least, important issues. However, this approach cannot be used blindly; it requires some caution on the part of the analyst because potentially

Table 14.2 Severity of failure rating.

Severity of failure rating	Severity of failure
9–10	Very high. Failure affects mission or safety (ranging from injury to catastrophic) or causes noncompliance with government regulations.
7–8	High. High customer dissatisfaction. Inoperable system, but does not violate safety or government regulations.
4–6	Moderate. Customer dissatisfaction, some degradation in performance.
2–3	Low. Only slight deterioration of performance. Annoyance.
1	Minor. The minor effect will not have any substantial effect on the system.

Table 14.3 Detection of failure rating.

Probability of detection of the failure rating	Probability of detection of the failure
10	Absolute certainty of non-detection. No means of detection. No way to find problem in time to affect an outcome.
9	Very low. Failure will probably not be detected by controls.
7–8	Low. Failure not likely to be detected by controls.
5–6	Moderate. Failure may be detected by controls.
3–4	High. Controls have a good chance of detecting failure.
1–2	Very high. Controls will always or almost always detect the failure.

catastrophic failures can be masked by low ratings for probability of failure and probability of detection. For example, if $\text{RPN} = \text{P} \times \text{S} \times \text{D} = 2 \times 10 \times 2 = 40$. This RPN of 40 may appear to be low and unimportant when listed with others having a higher RPNs. However, since an $\text{S} = 10$ is catastrophic, it may be more important than it seems. The analyst must ensure that every failure with a severity rating of 9 or 10 is assessed for safety or mission failure, no matter what the overall rating is.

14.9 Complementary Reliability Analyses and Models

While there are many additional techniques for performing reliability analyses and creating models that have been described in the literature and used for various projects, we have focused on the most common in this chapter.

Some of the other popular but less often used techniques include:

- Success Tree Analysis (STA)
- Event trees
- Markov models
- Monte Carlo simulation
- Petri nets
- Physics of Failure (POF) models
- Strength–Stress models
- Reliability growth models

Among these many analysis and modeling options, many can be used to either complement or supplement the more common ones discussed in this chapter. These additional techniques are a selection of analysis and modeling techniques that provide a different, and usually more specific, focus than the all-encompassing ones presented in detail here. The techniques listed above are typically used to investigate particular types of problems in greater detail than the more general techniques discussed previously.

The "Suggestions for Additional Reading" should be consulted for more detailed discussions of these and additional, less often used, analysis techniques.

14.10 Conclusions

This chapter has focused on several of the more common types of reliability analyses and models. Besides the techniques discussed here, there are many others too numerous to mention. Analysis and modeling during the development of any system or product is essential to ensuring that the design will meet the specified performance requirements. Analysis and modeling are used iteratively to guide the design toward the end goal of meeting the specified requirements, thus ensuring program success and customer satisfaction. The key to the value of any model rests with the fact that it must be useful in the decision-making processes during the development of a system. The creation of any model and associated analyses must be done to support the design team and recommend design improvements. It should help them measure system effectiveness and performance, evaluate design trade study alternatives, make design decisions, and evaluate risks.

References

1 IEEE (1990). *IEEE Standard Computer Dictionary: A Compilation of IEEE Standard Computer Glossaries*. Piscataway, NJ: Institute of Electrical and Electronics Engineers.
2 US Department of Defense (2005). *DOD Guide for Achieving Reliability, Availability, and Maintainability*. Washington, DC: Department of Defense.
3 Wikipedia Reliability engineering. https://en.wikipedia.org/wiki/Reliability_engineering (accessed 12 September 2020).
4 Oxford Advanced Learners Dictionary. Modelling. https://www.oxfordlearnersdictionaries.com/us/definition/english/modelling?q=modelling (accessed 12 September 2020).
5 Merriam-Webster Collegiate Dictionary. "Model." https://www.merriam-webster.com/dictionary/modeling (accessed 12 September 2020).
6 US Department of Defense (1981). *Reliability Modeling and Prediction, MIL-STD-756B*. Washington, DC: Department of Defense.
7 Watson, H.A. (1961). *Launch Control Safety Study*, Section VII, vol. **1**. Bell Laboratories.
8 Hassl, D. (1965). Advanced concepts in fault tree analysis. Presented at the System Safety Symposium (8–9 June).
9 Gullo, L.J. and Dixon, J. (2018). *Design for Safety*. Hoboken, NJ: Wiley.
10 US Nuclear Regulatory Commission (1981). *Fault Tree Handbook, NUREG-0492*. Washington, DC: US Nuclear Regulatory Commission. https://www.nrc.gov/reading-rm/doc-collections/nuregs/staff/sr0492/ (accessed 14 September 2020).
11 US Department of Defense (1991). *Military Handbook: Reliability Prediction of Electronic Equipment, MIL-HDBK-217F*. Washington, DC: Department of Defense.
12 US Department of Defense (1980). *Procedures for Performing a Failure Mode, Effects and Criticality Analysis, MIL-STD-1629A*. Washington, DC: Department of Defense.

Suggestions for Additional Reading

Gullo, L.J. and Dixon, J. (2018). *Design for Safety*. Hoboken, NJ: Wiley.

Raheja, D. and Allocco, M. (2006). *Assurance Technologies Principles and Practices*. Hoboken, NJ: Wiley.

Raheja, D. and Gullo, L.J. (2012). *Design for Reliability*. Hoboken, NJ: Wiley.

15

Design for Availability

James Kovacevic

15.1 Introduction

Would you like to know if your assets are in an operational state when they are needed and can be used at any given point in time? If so, then you are interested in availability. Availability is a measure of the degree to which an item is in an operable state and can be committed at the start of a mission when the mission is called for at an unknown (random) point in time [1]. Availability considers the time difference between the times when an asset is expected to be operating compared with the time when an asset is not able to be operated. In addition, availability may also be defined as the percentage of time that the asset is in an operable state. Typically, availability is measured at the asset level, which may be the system level, the platform level, the equipment level, or the product level. The definition and configuration of the asset must be clearly understood before considering availability as a requirement. Availability is a covariance function of how reliable the asset is and how easy the asset is to maintain. Reliability is based on how often failures of the asset occur. Maintainability is not only how quickly the asset can be restored to a functional state, but also considers how often preventive maintenance (PM) is performed, and how long the asset is down for the planned and unplanned maintenance. One form of availability considers any logistical delays in restoring the asset to a functional state or while performing preventive maintenance. Occasionally, individuals may confuse availability with reliability. To ensure the difference is clear, here is a simple way to remember the differences:

- Availability focuses on time utilization
- Maintenance focuses on process restoration
- Reliability focuses on failure elimination

15.2 What is Availability?

Availability comes in many different variations, each with its own use. The various types of availability will be covered later in the chapter, but essentially, availability considers the total time (TT) available (e.g., a 30-day period, or 720 hours), and what amount of time the asset is able to perform its intended function (e.g., 672 hours). In the example shown in Figure 15.1, the asset would be available for three periods of time: 320 hours, 210 hours, and 142 hours. This totals 672 hours out of the 30-day (720 hours) period, which would result in an availability of 93.3% (672/720).

Availability is a powerful metric that allows asset owners to understand how ready their assets are to perform the intended mission. Once understood, availability enables asset owners to determine whether the lack of availability is a result of poor maintainability, poor reliability, or logistical delays. Availability also allows asset designers to perform a delicate balance and perform tradeoffs in reliability and maintainability to deliver the most cost-effective solution for the

Design for Maintainability, First Edition. Edited by Louis J. Gullo and Jack Dixon.

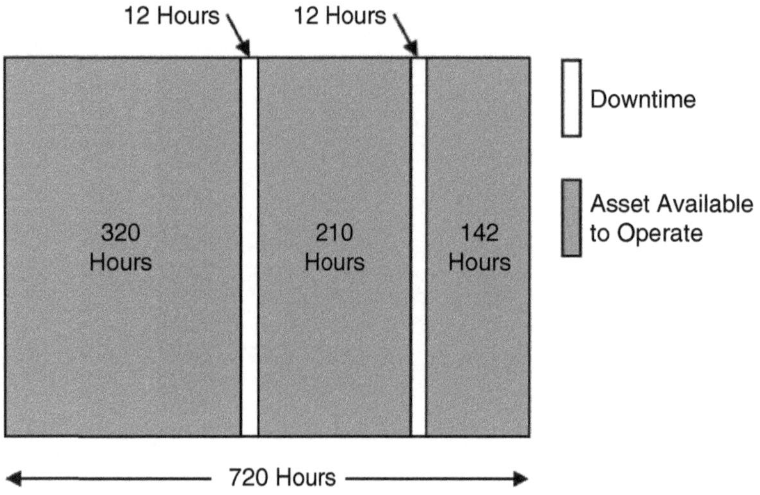

Figure 15.1 Availability in its simplest form (see text).

asset owner. Lastly, availability enables asset owners to determine how many assets may be needed to ensure the achievement of the organization's objectives. For example, if an organization requires a certain amount of production capacity for a manufacturing asset, or flight hours per year for an airborne platform, and they know the availability of the asset, the organization can procure the correct amount of assets to achieve the desired production throughput or the total number of flight hours for airworthy assets.

The example below (see Table 15.1) shows what availability actually means in terms of lost time over the course of a one-year period. Assume that our system operates 365 days per year, 24 hours per day, which is 8760 hours per year. For how long will that asset be unable to operate over the course of the year?

Table 15.1 Availability and downtime correlation.

Availability (%)	Downtime (h)
50	4380
55	3942
60	3504
65	3066
70	2628
75	2190
80	1752
85	1314
90	876
95	438
96	350
97	263
98	175
99	88
99.9	9
99.99	0.88

Problem: Assume an organization requires 13 200 hours of aircraft flight time over the course of a year (using 8760 hours/year), and the availability of an individual asset is 65%. From this example, determine how many airborne assets the organization requires to ensure it can meet the flight time requirements in a year?

Solution: With an availability of 65%, each aircraft asset would potentially have 5634 hours available for flight time per year. Using Table 15.1, given 65% availability, this equates to 3066 downtime hours per aircraft asset. With 8760 hours in a year and 65% availability that is 5694 hours available for each asset flight time per year (8760 – 3066 = 5694). Therefore, the organization would need three assets to meet the flight requirements over the course of the year. The three assets are based on 13 200 hours of flight time required per year and 5694 hours available per year per aircraft (13 200 hours/5694 available hours per asset = 2.32, rounded up to 3).

As with other metrics in reliability and maintainability, availability is also dependent upon the operating

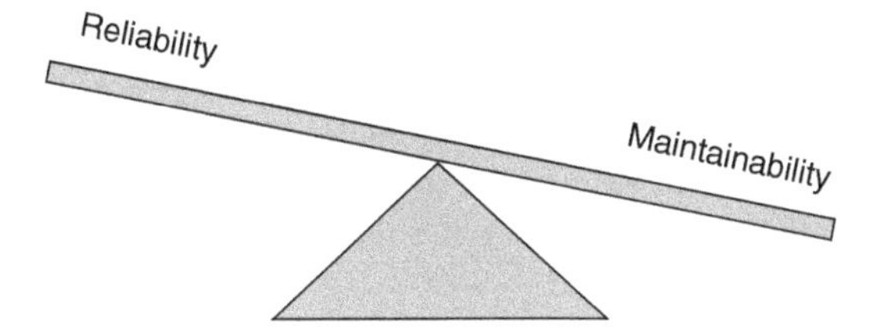

Figure 15.2 The availability balance.

context. The operating context is a combination of the specific influences on the asset, such as environmental conditions, staff competencies, operating parameters, shift patterns, maintenance strategy, and so on. Availability for one specific asset at one location may be drastically different for the same asset at another location, due to spare part lead times, environmental extremes, or skill levels of the maintainers.

Availability is a key metric that asset owners, designers, or operators need to track to ensure that the organization can meet their objectives. It is critical to ensure that the right availability metric with the right operating context is used for the specific situation and needs of the organization.

15.3 Concepts of Availability

Availability in its simplest form is a function of reliability and maintainability. Consider reliability to be represented by Mean-Time-Between-Failure (MTBF), while maintainability is represented by Mean-Time-to-Repair (MTTR). If focus is only put into MTBF, a consequence may be more preventive maintenance routines, which may drive overall availability lower. Conversely, if focus is put only into MTTR, and reducing the downtime, then a large amount of failures may still occur and drive availability lower. Therefore, it is absolutely vital that organizations take a balanced approach (see Figure 15.2) to achieving availability goals.

The balance between reliability and maintainability needs to be understood so that organizations can focus on the right improvement activities, whether the activities are in the design or operating phase of the asset. Without this understanding, organizations often focus on reliability, to the detriment of maintainability, and they fail to achieve their availability goals. Table 15.2 helps to demonstrate the relationship between reliability and maintainability.

Table 15.2 Impact of MTBF and MTTR on availability.

MTBF	MTTR	Availability
Increases	Remains constant	Increases
Increases	Increases	Decreases/ Increases/ Remains constant
Increases	Decreases	Increases
Decreases	Remains constant	Decreases
Decreases	Increases	Decreases
Decreases	Decreases	Decreases/ Increases/ Remains constant
Remains constant	Increases	Decreases
Remains constant	Decreases	Increases

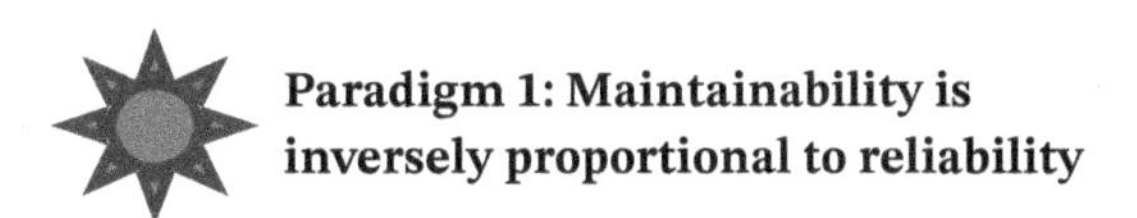

Paradigm 1: Maintainability is inversely proportional to reliability

In another scenario, the MTBF decreases while MTTR decreases, which again results in a constant availability. This tradeoff may occur when the cost of failure is non-consequential, but the cost of performing the preventive maintenance routines is expensive. For example, in an epoxy injection machine, the cost of having a clogged mixing nozzle is non-consequential, as the operator can quickly change the nozzle when needed. However, the cost of proactively changing out the mixing nozzle is expensive, as there are no time data to support a change, and it will result in ineffective maintenance. In this situation, we may experience a minor decrease in MTBF, and reduce the MTTR, by making it an operator activity and not relying on maintenance staff. As a result, the availability would stay the same. There is a cost benefit to the organization as well, which is realized in tradeoffs in MTBF and MTTR. While this is a general rule, there are many exceptions to it. To assist with understanding the relationship between MTBF and MTTR, Table 15.3 shows some examples of how changes to MTBF and MTTR impact availability.

Table 15.3 MTBF, MTTR, and availability.

MTBF (h)	MTTR (h)	Availability
1	1	0.5
1	2	0.333 333
1	0.5	0.666 667
1	0.1	0.909 091
2	1	0.666 667
2	2	0.5
2	3	0.4
2	4	0.333 3333
3	1	0.75
0.9	1	0.473 684
0.1	1	0.090 909
0.01	1	0.009 901
100	100	0.5
90	90	0.5
80	80	0.5
100	25	0.8
90	22.5	0.8
80	20	0.8
70	17.5	0.8
60	15	0.8

It is vital to note that these tradeoffs are often made at the failure mode level in the form of detailed electrical or mechanical design tradeoffs. The failure mode describes how the asset may fail, either electrically or mechanically, or both. The design tradeoffs consider the ways failure modes occur, such as physical fatigue degradation and root cause failure mechanisms, and how the design could be changed to prevent the failure mode occurrence or minimize the effect of its occurrence. These types of tradeoffs are part of Design for Reliability (DfR) and Design for Maintainability (DfMn). As a result of the design tradeoffs, optimization of preventive and corrective actions can be determined for the specific failure mode. During design tradeoff analysis for a particular failure mode, the cost of failure events with their corresponding corrective maintenance (CM) actions and the cost of preventive maintenance actions to deter failure events are determined and balanced. It is important to note that the total cost of failure with corrective maintenance is not just the cost of repair (i.e., parts and labor), but should also include any additional costs such as staff availability and training costs, transportation and supply chain logistics costs, lost opportunity costs of the failed asset, fines, and penalties.

Often, it is helpful to visualize the impact that MTBF and MTTR can have on availability. Figures 15.3–15.6 are four scenarios that show how changes in either MTBF or MTTR can influence availability. Figure 15.3 demonstrates an asset that has a relatively short MTTR, compared with the MTBF. Figure 15.4 demonstrates a longer MTBF than Figure 15.3, with the same MTTR. Figure 15.5 demonstrates a shorter MTBF with a slightly shorter MTTR. Lastly, Figure 15.6 demonstrates an asset with a MTBF that is relatively short, compared with a long MTTR. As a result, the availability is much lower than the other examples. The terms short or long are relative, and one should expect to see absolute values in the equipment requirements. The values used in the examples for short or long are not guidelines for what short and long MTBF values are. The impacts on availability are summarized in Table 15.4.

Figure 15.3 represents a long MTBF combined with a short MTTR, which will result in an acceptable availability (83.3%). In this instance, the organization would need to perform an analysis to determine whether the MTBF or MTTR should be targeted for improvement. Comparing this to Figure 15.4, where the MTBF is doubled, with the MTTR remaining constant, the doubling of MTBF only provides a minor improvement in availability (90.9%). In this instance, the organization may benefit most by targeting the MTTR for improvement to improve availability.

Figure 15.5 represents a frequently failing asset with short MTBF, with a short MTTR. This often represents organizations that do well at reactive maintenance and perfect component replacements. The system has an availability of 66.7%. Based on the information provided in the graphic, the organization would benefit from improving the MTBF rather than the MTTR.

Figure 15.6 represents a frequently failing asset with short MTBF, with a long MTTR. The system has an availability of 54.5%. Based on the information provided in the graphic, the organization would benefit from improving either the MTBF or the MTTR.

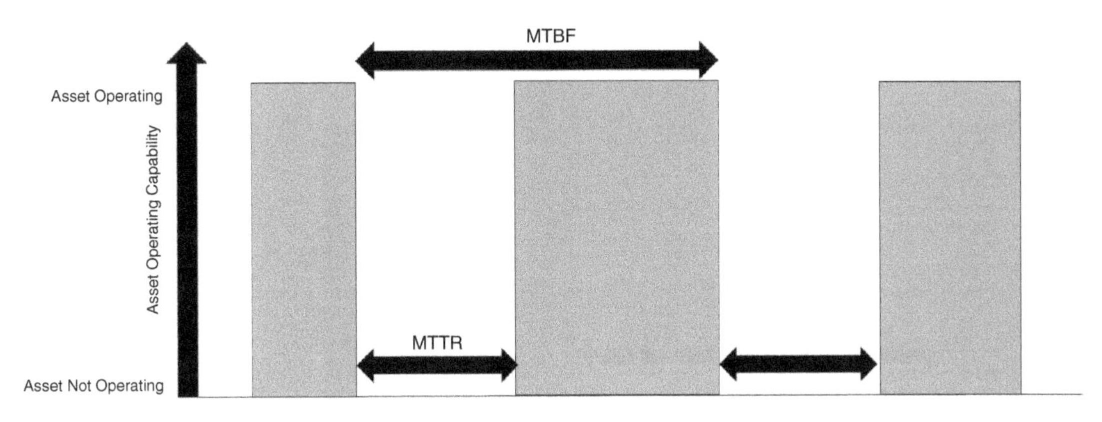

Figure 15.3 Availability with long MTBF and short MTTR.

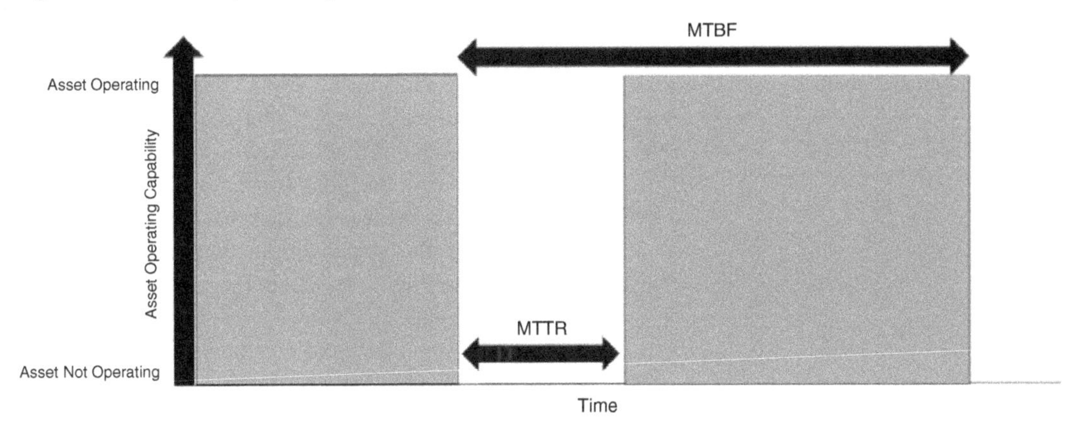

Figure 15.4 Availability with long MTBF and short MTTR.

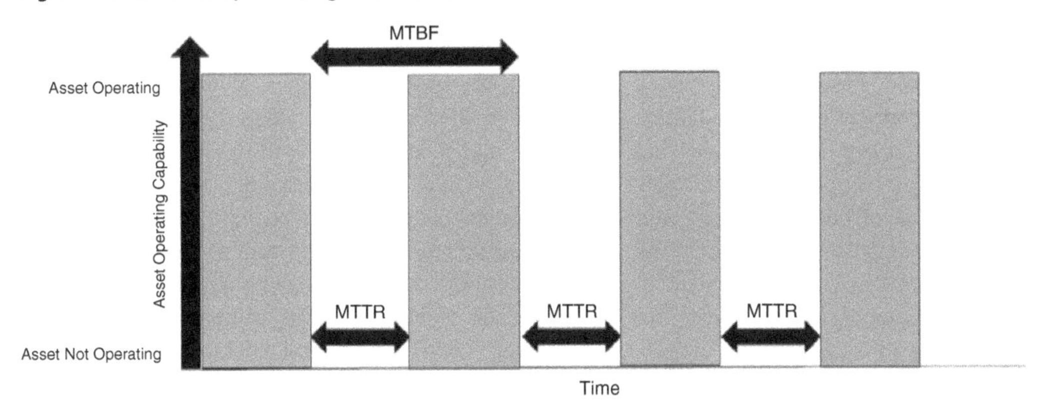

Figure 15.5 Availability with short MTBF and short MTTR.

As one can see in Figures 15.3–15.6, availability can be improved or reduced by incorporating design changes in either the reliability or maintainability of the asset. This is why it is absolutely vital that all asset designs incorporate a balanced mix of DfR and DfMn features.

15.3.1 Elements of Availability

There is more than one type of availability parameter, and more than one set of elements that are input into an availability equation. To properly understand availability, it is vital to understand all of the elements

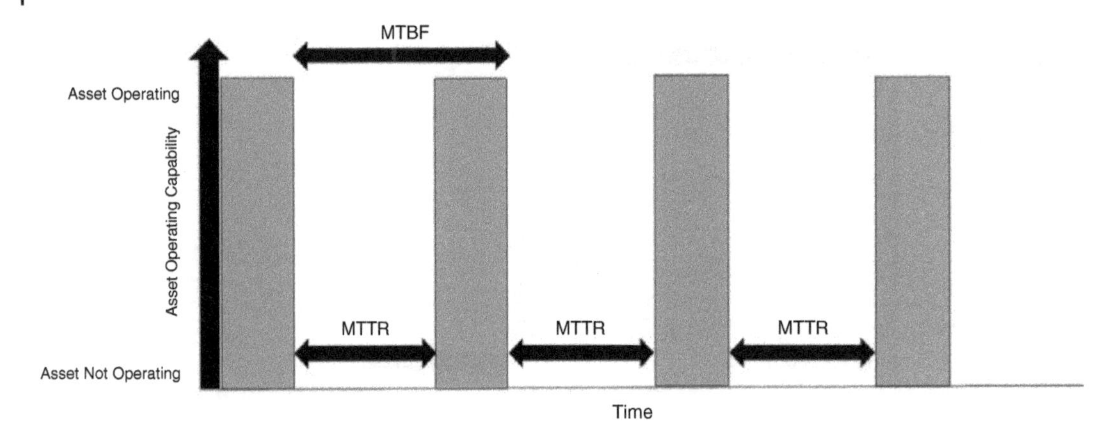

Figure 15.6 Availability with short MTBF and long MTTR.

Table 15.4 MTBF, MTTR, and availability.

Scenario	MTBF (h)	MTTR (h)	Availability (%)
1 (Figure 15.3)	50	10	83.3
2 (Figure 15.4)	100	10	90.9
3 (Figure 15.5)	10	5	66.7
4 (Figure 15.6)	30	25	54.5

that go into the various availability calculations. Since availability is time-related, the elements associated with availability, such as MTBF and MTTR, are time-based. There are other elements that will be discussed. All of these time-based elements play a role in calculating the appropriate type of availability. Since an asset is designed for availability considering when the asset is required to be operational and able to sustain operation over a given time period and under given stress conditions, any time when the asset is not required to be operational is not included in the availability calculations. Availability does not consider planned asset non-utilization periods of time. A planned non-utilization period of time may be the time when an asset is shut down due to lack of demand for its use and will not be utilized, or when the asset is in storage, or when it is being transported.

15.3.1.1 Time-related Elements

Availability can be broken down into many time-related elements, as shown in Figure 15.7, starting first with the Total Time (TT). The TT is defined as the intended utilization period, such as 720 hours, and does not include the planned non-utilization period. TT can be further broken down into two main elements: uptime and downtime. Uptime is defined as the time that the asset is ready to operate and meet the needs of the organization, while downtime is the non-operating time when the asset is required to be operational, but is unable to meet the needs of the organization. Downtime includes all asset time scheduled for planned preventive maintenance, or unscheduled corrective maintenance [2].

Uptime can be further broken down into Operating Time (OT) and Standby Time (ST). Operating Time can be defined as the time when the asset is operating and meeting the needs of the organization, while Standby Time is defined as the time when the asset is ready to operate and meet the needs of the organization if it is called upon to do so. Operating Time and Standby Time are described with an example of a water utility pumping system that uses a backup pump. The two pump systems require one main pump to operate at all times to handle the water volume and flow rates to properly distribute water to customers. The second pump is a backup to the main pump. While the main pump is operating, it is the primary source of water meeting the needs of the water utility organization and its customers. As long as the main pump is operational, the redundant backup pump is not operating. If the main pump were to fail, the backup pump would start up, take over operation from the main pump, and meet the needs of the water utility organization and its customers.

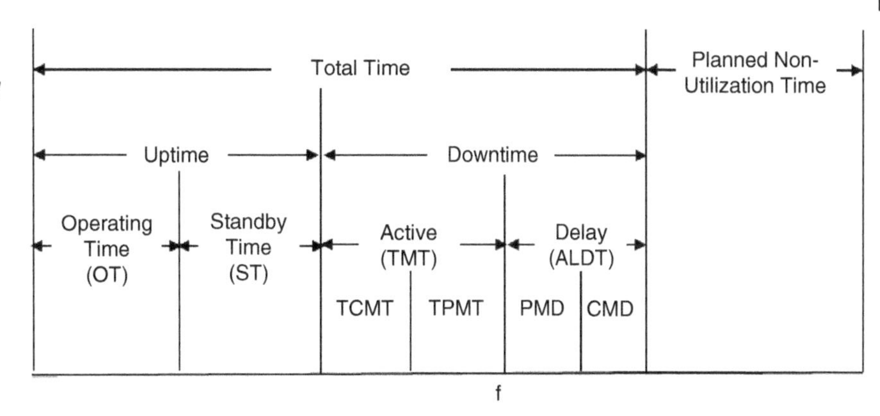

Figure 15.7 Elements of availability. Source: From US Department of Defense, *Test and Evaluation of System Reliability, Availability, and Maintainability, DOD 3235.1-H*, Department of Defense, Washington, DC, 1982.

Downtime can be broken down into many smaller elements such as Total Maintenance Time (TMT) and Administrative Logistics Delay Time (ALDT). TMT is the period of time in which work is being performed on the asset, whether it is preventive maintenance or corrective maintenance. Total Preventive Maintenance Time (TPMT) is the period of time when preventive maintenance activities are being performed on the asset to ensure operational readiness. Total Corrective Maintenance Time (TCMT) is the period of time for which work is being performed to restore the function of the asset to an operational readiness state following a failure event. Both TCMT and TPMT should not include any time associated with administrative or logistical delays. This exclusion of ALDT data in the downtime equation is done to ensure that the inherent availability of the design can be determined, which guides improvements to the design of the asset, if needed. The ALDT data is used for the operational availability calculations, which should guide improvements to the maintenance and support processes, along with the design of the asset.

ALDT should capture all delays, whether logistical, (such as lack of parts, skills, or equipment), or administrative (such as scheduling and coordination issues) for both Corrective Maintenance Delays (CMDs) and Preventive Maintenance Delays (PMDs). CMD typically includes delays such as expediting a part due to an unexpected need for the part, which is out of stock in the parts inventory. PMD typically includes delays due to hanger or shop space limitations, lack of specialized test equipment capacity, or scheduling issues. It is vital that these delays are tracked, so appropriate actions can be taken to increase operational availability closer to the level of the inherent availability of the asset.

15.3.1.2 Mean Metrics

In addition to the individual time elements required for availability calculations, availability can be calculated using a variety of "mean metrics." A "mean metric" is a metric that is used to monitor reliability and maintainability by using the average or mean time between specific events, or the mean of specific events. That being said, many of the mean metrics can be classified as either reliability-centered or maintainability-centered [3].

Reliability-centered mean metrics include:

- Mean-Time-Between-Failure (MTBF), which is used for repairable assets and components. MTBF represents the average time between failures. This time may represent running time, calendar time, cycles, units, etc. MTBF can be calculated by measuring total time that the system is operational, divided by the number of failure events for the total time period.
- Mean-Time-to-Failure (MTTF), which is used for non-repairable assets and components. MTTF assumes the asset or component is thrown away and discarded after the occurrence of a failure. Since availability is typically not used for non-repairable assets, this metric should not be seen in any availability calculations.

Maintainability-centered mean metrics typically include:

- Mean-Time-to-Repair (MTTR), which is used for repairable and non-repairable assets and components. MTTR represents the average time taken to perform corrective maintenance on an asset after a failure has occurred. MTTR typically only considers the active repair time and does not include delays associated with the corrective work. MTTR can be calculated by measuring the total corrective maintenance downtime accumulated during a specific period, divided by the total number of corrective maintenance actions completed during the same period. MTTR does not include any preventive maintenance activities.
- Mean Time Between Maintenance (MTBM), which is used to evaluate maintenance processes. While MTBF is used for maintenance time associated with failures only, MTBM includes all maintenance actions such as failures, preventive maintenance (PM) activities, cleaning, and inspections. MTBM is the mean of the distribution of the time intervals between corrective maintenance actions, preventive maintenance actions or all maintenance actions. MTBM can be calculated by measuring total time that the system is operational, divided by the number of maintenance actions taken place for the total time period. If there are no PM activities, only corrective maintenance actions, then MTBF = MTBM.
- Mean Downtime (MDT), which is used to evaluate the total downtime for maintenance actions, for both preventive and corrective action including any delays. MDT can be calculated by measuring the total downtime for an asset, divided by the number of maintenance actions that have taken place. MDT should include all delays associated with any maintenance actions.
- Mean Downtime for Maintenance (MDTM), which is used to evaluate the total downtime for maintenance activities, for both corrective and preventive actions. However, unlike MDT, MDTM does not include logistical and administrative delays.

All activities leading up to the restoration of the asset should be included in the MDT calculations, while MTTR only includes active maintenance time – see Figure 15.8 for an example. As discussed previously, any delay times associated with the repair or failure prevention activities need to be measured, as they will adversely affect the availability of the asset. The contributors to excessive delay times need to be evaluated and corrected to improve availability. Contributors to excessive delays include:

- Time to notify and dispatch the maintainer
- Travel time to the asset
- Time required to receive the appropriate permits to be allowed to access the asset
- Time to shut down power to an asset
- Time to gain access to the asset's location and to the space surrounding the asset
- Troubleshooting activities to diagnose the cause of the failure

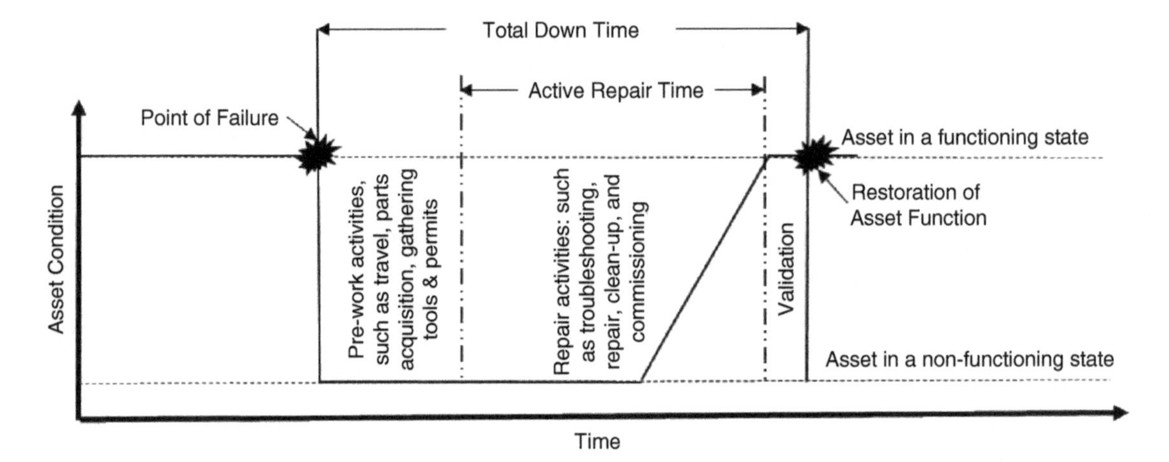

Figure 15.8 Active vs. total downtime.

- Time to procure and/or receive the parts required to restore the asset from a failed state
- Time to perform Removal and Replacement (R&R) of parts or assemblies during the active maintenance repair
- Time to perform calibrations and adjustments following replacement of parts or assemblies
- Clean-up and restoration of safeguards and safety-related systems
- Start-up and commissioning of the asset, which includes time to bring the asset back to its operating specification
- Validation that the asset is functioning within defined operating parameters [4].

With all of the different elements, it can be hard to keep them all straight, but it is vital to capture the elements correctly and consistently to ensure proper availability calculation, and more importantly, the proper evaluation of availability. The various elements will be used for the different types of availability. Each type of availability serves a distinct purpose. Now, with an understanding of the elements, the different types of availability can be explored.

15.4 Types of Availability

The availability metric is one that can serve many different purposes, and which purpose it serves often depends on how it was calculated. Yes, availability can be calculated in many different ways, and therefore it is vital that when the term availability is used, it is used with the correct preface. By defining which type of availability is being discussed, users can truly understand what it means. There are many types of availability in use, with the three most common being Inherent Availability, Achieved Availability, and Operational Availability. There are many more types of availability that are less commonly used variations of these three main types of availability.

15.4.1 Inherent Availability

Inherent Availability (A_i) is focused on the impact that functional failures will have on operations, whether or not the asset is running or shut down. As such, Inherent Availability looks at the availability of an asset from a design perspective. A_i is a measure of the ideal performance of the asset assuming there are no losses due to delays or preventive maintenance. A_i is a measure of the capability of the design. On many government programs, A_i is a Key Performance Parameter (KPP) or a Key System Attribute (KSA), which is a requirement that the asset developer or Original Equipment Manufacturer (OEM) would be solely responsible for. The customer would not share in this A_i requirement responsibility, but would share the responsibility if Operational Availability was the KPP or KSA. A_i is used to evaluate the design and perform tradeoffs between performance, reliability, and maintainability in design. Both reliability and maintainability need to be considered and measured to ensure an available asset.

Inherent Availability should result in the largest value of the availability calculations, as it is design focused, and it assumes the asset is operated in ideal conditions with ideal corrective maintenance. As seen from the definition, Inherent Availability does not really provide a true representation of what an organization should expect from the asset, since all preventive maintenance and delays are excluded from the calculation. A_i is usually used in an equation or expression with "Capability" to evaluate overall system effectiveness. Capability is used to measure the productive output of the asset compared with the inherent productive output. Capability enables organizations to understand not only how available the asset is, but whether it is efficient and meeting the output requirements.

To calculate A_i, one only needs the design MTBF and MTTR values. This relatively simple calculation is defined as:

$$A_i = \mathrm{MTBF}/(\mathrm{MTBF} + \mathrm{MTTR})$$

An example of an A_i calculation can be shown for a pump, which under ideal conditions, will operate for 6750 hours between failures and have an MTTR of eight hours. In this situation the pump would have an A_i of:

$$\begin{aligned} A_i &= \mathrm{MTBF}/(\mathrm{MTBF} + \mathrm{MTTR}) \\ &= 6750/(6750 + 8) \\ &= 0.9988 \\ &= 99.88\% \end{aligned}$$

This A_i is not bad for a pump, but in reality most organizations would never achieve this level of availability due to the operating context for the pump, the need for preventive maintenance, and associated delays with maintenance activities.

15.4.2 Achieved Availability

Achieved Availability (A_a) is used to evaluate the overall impact on operations but assumes that the asset operates under ideal support requirements, which does not include any logistical or administrative delays. Therefore, A_a only measures the active maintenance time, which includes both corrective and preventive maintenance. A_a is used to evaluate the steady-state availability with maintenance being performed.

Achieved Availability should result in an availability of less than the Inherent Availability, but would usually be larger than the Operational Availability. Achieved Availability is the goal that organizations should set for themselves for the asset, as it represents a perfect delivery of corrective and preventive maintenance. A_a sets the benchmark for what could be provided if the organization has ideal support conditions in place. The difference between Achieved Availability and Operational Availability is the gap that organizations need to close to drive availability and improve the performance of their organization.

To calculate A_a, one only needs two elements: MTBM and MDTM. Achieved availability can be defined as:

$$A_a = \text{MTBM}/(\text{MTBM} + \text{MDTM})$$

To continue using our pump example from the Inherent Availability calculation, that same pump will have a PM action performed every 2000 hours which will take 4 hours of active maintenance time to perform. In addition, the pump is expected to have one failure over the period of 6750 hours, which will take 8 hours to repair. Assuming we are measuring achieved availability for a period of 8000 hours, the achieved availability will be:

$$A_a = \text{MTBM}/(\text{MTBM} + \text{MDTM})$$

where MTBM = 8000 hours/5 downtime events = 1600 hours

[5 downtime events = 4 preventive maintenance actions (one every 2000 hours) + 1 corrective action at 6750 hours]

and MDTM = 24 hours/5 events = 4.8 hours

[24 hours = (4 preventive maintenance actions × 4 hours each = 16) + (1 corrective maintenance action × 8 hours)]

$$\begin{aligned} A_a &= (8000\ \text{hours}/5\ \text{events})/ \\ &((8000\ \text{hours}/5\ \text{events}) + (24\ \text{hours}/5\ \text{events})) \\ &= 1600/(1600 + 4.8) \\ &= 0.9970 \\ &= 99.7\% \end{aligned}$$

There is one variation of Achieved Availability, based on the elements defined within Figure 15.7. This variation of achieved availability is defined as:

$$A_a = (\text{OT})/(\text{OT} + \text{TPMT} + \text{TCMT})$$

Using the values above we can see that variation of A_a will result in an availability of 99.7%, which is the same result as the more traditional calculation:

$$\begin{aligned} A_a &= (\text{OT})/(\text{OT} + \text{TPMT} + \text{TCMT}) \\ &= (8000\ \text{hours})/(8000\ \text{hours} + \\ &((8000\ \text{hours}/2000\ \text{hours}) \times 4\ \text{hours}) + 8) \\ &= (8000\ \text{hours})/(8000\ \text{hours} + \\ &(4\ \text{events} \times 4\ \text{hours}) + 8) \\ &= (8000)/(8000 + 16 + 8) \\ &= 0.9970 \\ &= 99.7\% \end{aligned}$$

As can be seen in the examples, availability is still quite high, but slightly less than the Inherent Availability since it considers the time the asset is down for preventive maintenance activities. It is in the best interests of organizations to reduce the impact from PM activities as much as possible through techniques such as condition-based monitoring and Lean techniques such as Single Minute Exchange of Dies (SMED) (see Chapter 17 for more on SMED). However, since achieved availability assumes the asset is supported under ideal support conditions, it does not reflect the true impact to operations. This is where Operational Availability comes in. However, it is using Achieved Availability that an organization can design the right support systems, as covered in Chapter 16, Design for Supportability.

15.4.3 Operational Availability

Operational Availability (A_o) is used to evaluate the overall impact on operations, but takes into account the actual operating environment that the asset operates in, including any and all logistical or administrative delays. Operational Availability includes all downtime of the asset, whether for preventive or corrective maintenance, plus all delays including scheduling, spare parts availability, lack of skills, and so on. Operational Availability allows organizations to truly understand the impact that a support program has on the asset.

Paradigm 7: Maintainability predicts downtime during repairs

In an ideal world, Operational and Achieved Availability would be the same, but that rarely happens in real life. Generally, A_o is the lowest availability percentage; however, it enables organizations to drive not only reliability and maintainability improvements to the asset, but also improvements to the support programs behind the asset. Any gaps between A_a and A_o can be improved through better supportability, which is often in the hands of the operating organizations.

There are four methods to calculate A_o. These methods are described in this section with numerical examples to show how to calculate A_o using each method.

15.4.3.1 A_o Method 1

The first method to calculate A_o involves two elements: MTBM and MDT. Operational availability can be defined as:

$$A_o = \text{MTBM}/(\text{MTBM} + \text{MDT})$$

Continuing with the previous pump example, MTBF is 6750 hours. Let's assume PM activity occurs every 2000 hours for a 4-hour period; however, all delays will be included in the calculation. Table 15.5 represents the true downtime experienced for each of the events:

The total delay time for PM activities is 10.5 hours. The total PM downtime associated with maintenance activities and logistics delays is 27.5 hours. With the one delay for corrective maintenance measured at 48 hours, the total delay time for all PM and CM maintenance is 58.5 hours. The total downtime for both active maintenance time and the delays is 84.5 hours. To calculate the MDT, divide the total delay time by 5 maintenance activities, 4 for PM and 1 for CM. This results in an MDT = 16.9 hours. The 1600 hours comes from the five downtime events (4 preventive maintenance actions (one every 2000 hours) + 1 corrective action at 6750 hours). With these delays identified and MDT calculated, Operational Availability can be calculated:

Table 15.5 Mean Downtime (MDT) examples.

Activity	Active maintenance time (h)	Administrative delay time (h)	Administrative delay reason	Mean downtime (h)
PM activity #1	4	1.5	Scheduling	5.5
PM activity #2	4.5	5	Can't find the part	9.5
PM activity #3	3.5	0		3.5
PM activity #4	5	4	Verification	9
Corrective activity #1	9	48	Part not available	57
PM time totals	17	10.5		27.5
Corrective time totals	9	48		57
Totals	26	58.5		84.5

$$A_o = \text{MTBM}/(\text{MTBM} + \text{MDT})$$
$$= 1600/(1600 + 16.9)$$
$$= 0.9895$$
$$= 98.95\%$$

Building on the examples for Achieved Availability and Operational Availability, the lost availability due to logistical delays can be calculated. Using the Achieved Availability above of 99.7%, and the Operational Availability 98.95%, the loses can be calculated by subtracting the Operational Availability from the Achieved Availability:

$$\text{Logistical losses (Availability)} = A_a - A_o$$
$$= 99.7\% - 98.95\%$$
$$= 0.75\%$$

As shown in the example, the impact on Operational Availability due to the delays is approximately 0.75% availability. While that may not seem like much, in a demanding operating environment this could result in significant lost production or lost capabilities of the asset or organization. To determine the total number of hours lost, one can simply take the percentage of the 8000 hours the asset is expected to operate:

$$\text{Logistical Losses (Hours)} = \text{Expected runtime}$$
$$\times \text{Logistical losses (Availability)}$$
$$= 8000 \times 0.75\%$$
$$= 60 \text{ hours}$$

It can also be assumed that if this asset is losing as much as 60 hours due to poor support systems, other assets are likely experiencing the same issue.

15.4.3.2 A_o Method 2

Operational Availability can be calculated in another manner, which is often used in separating the A_i and A_o requirements that are the responsibilities of the OEM designer and the customer of the system.

As a design requirement, A_i is strictly the responsibility of the OEM. The OEM designs the test architecture for the system using a combination of embedded diagnostics and external support test equipment. A_i can be determined to be compliant based on the results from Built-In-Test (BIT) and external testing of the system, with maintenance task time data collected in the field applications. A_o, on the other hand, is partly the responsibility of the OEM since the MTBF and MTTR inputs to the A_i are also required in the A_o calculation. A third input to the A_o calculation is Mean Logistics Downtime (MLDT), which is derived from the ALDT data. MLDT is the responsibility of the customer since the OEM does not control the customer's logistics infrastructure in the field application.

$$A_o = \text{MTBF}/(\text{MTBF} + \text{MTTR} + \text{MLDT})$$

Using the previous pump example, MTBF is 6750 hours and MTTR is 8 hours. Now let's assume MLDT was calculated from an ALDT data-set associated with corrective maintenance on pumps in the field application during a five-year period, and was determined to be 79 hours. With the mean logistics delay time identified, Operational Availability can be calculated:

$$A_o = \text{MTBF}/(\text{MTBF} + \text{MTTR} + \text{MLDT})$$
$$= 6750/(6750 + 8 + 79)$$
$$= 0.9873$$
$$= 98.73\%$$

15.4.3.3 A_o Method 3

Interestingly, Operational Availability can be calculated in another manner, which is often used in Overall Equipment Effectiveness (OEE), and Total Effective Equipment Performance (TEEP) measures. While the concept is still the same, which is to determine the overall impact on operations, the formula is slightly different. This variation of A_o can be defined by:

$$A_o = \text{Uptime}/(\text{Uptime} + \text{Downtime})$$

Using the values from the previous example (A_o Method 1), the variation of Operational Availability can be calculated as:

$$A_o = \text{Uptime}/(\text{Uptime} + \text{Downtime})$$
$$= (8000 \text{ hours} - 84.5 \text{ hours})/$$
$$((8000 \text{ hours} - 84.5 \text{ hours}) + 84.5 \text{ hours})$$
$$= 7915.5/(7915.5 + 84.5)$$
$$= 0.9894$$
$$= 98.94\%$$

As you can see, the result of all three variations of Operational Availability arrive at almost the exact same value, so it is not very important which variation is to be used, as long as the variation selected is used and measured consistently. Also, remember when calculating the A_0 = Uptime/(Uptime + Downtime) variation, do not include the planned non-utilization time in the downtime.

15.4.3.4 A_0 Method 4

Lastly, the final variation of Operational Availability can be calculated using the elements of availability found in Figure 15.7. This variation of Operational Availability can be defined as:

$$A_0 = (\text{OT} + \text{ST}) / (\text{OT} + \text{ST} + \text{TPMT} + \text{TCMT} + \text{ALDT}) \quad [2]$$

This calculation is the same as the A_0 = Uptime/(Uptime + Downtime) calculation, except this method breaks down further the elements of uptime and downtime.

The planned utilization time of the asset can play a large role in any of the above variations. The less an asset is utilized in a given period, the higher the operational availability will be. Therefore, it is beneficial to the operator of the asset to exclude lengthy periods of planned non-utilization time taking place, to ensure that the calculation accurately reflects the reality.

In summary, Operational Availability is the most useful of the availability measures, as it enables organizations to fully understand the current performance of the asset, within its present operating context. Operational availability enables organizations to plan for and determine the number of assets required to fulfill its given mission or organizational objectives in the operating context. Operational Availability also enables organizations to identify all forms of inefficiencies with the design and supportability of an asset, which would allow it to make systematic changes and drive improved performance of all assets. A_0 is an all-encompassing metric to understand the asset, the support system, and the operating environment.

Table 15.6 summarized the three types of availability described in this section.

Table 15.6 Types of availability summary.

Type of availability	Availability formula	Exclusions	When to use
Inherent Availability	A_i = MTBF/(MTBF + MTTR).	Preventive maintenance, logistics, and administrative delays	During design of the asset to balance MTBF and MTTR
Achieved Availability	A_a = MTBM/(MTBM + MDTM) or A_a = (OT)/(OT + TPMT + TCMT)	Logistical and administrative time delays	To understand the operational impact to availability, under ideal support conditions. This can be used to identify PM activities to be optimized through condition-based maintenance and SMED to get closer to inherent availability levels.
Operational Availability	A_0 = MTBM/(MTBM + MDT) or A_0 = MTBF/(MTBF + MTTR + MLDT) or A_0 = Uptime/(Uptime + Downtime) or A_0 = (OT + ST)/(OT + ST + TPMT + TCMT + ALDT)	N/A	To understand the actual availability of the asset under current support conditions. This can be used to correct issues in the support environment, and bring operational availability closer to achieved availability.

15.5 Availability Prediction

The ability of an organization to predict the availability of its assets is vital and can make the difference between success and failure. If an organization chooses to purchase more assets than required, then the organization may be in a situation where assets are not used, a large outlay of capital was unnecessary, and the assets may not be generating value for the organization. On the other hand, if an organization does not procure enough assets, then the organization may not meet the market demand or mission objectives, and customers will experience long periods of downtime waiting for completion of maintenance of failed assets.

In addition, by predicting the availability of the individual assets, or the fleet, organizations can then begin to plan properly for operational impacts and develop the appropriate supportability plans. As seen earlier in this chapter, supportability plays a large role in ensuring Operational Availability. Therefore, an availability prediction should be completed once the initial design of the asset has been developed. This initial availability prediction will enable the design team to make revisions to the initial design to bring availability up to the requirements of the end user. Often this prediction is an iterative process, involving each design being evaluated and compared to optimize the total life cycle costs of the asset. Keep in mind that often the same design will be used in multiple operating contexts, such as in a desert situation or in an arctic environment, so predictions models should be made for different operating scenarios to ensure the asset meets the needs wherever it may be used.

Availability prediction can be performed at many different levels, including the asset and fleet levels. At the most basic level, the availability prediction is based upon the Reliability Block Diagram (RBD) (see example in Figure 15.9), which is a visual representation of a system and its components showing how the individual reliability characteristics of components contribute to the reliability of the system. Since assets are generally made up of several systems, each with their own reliability and maintainability considerations, it is vital to consider the interfaces, as they can have a profound impact on the availability of the asset.

For example, consider a redundant hydraulic system in an aircraft analyzed using the RBD in Figure 15.9.

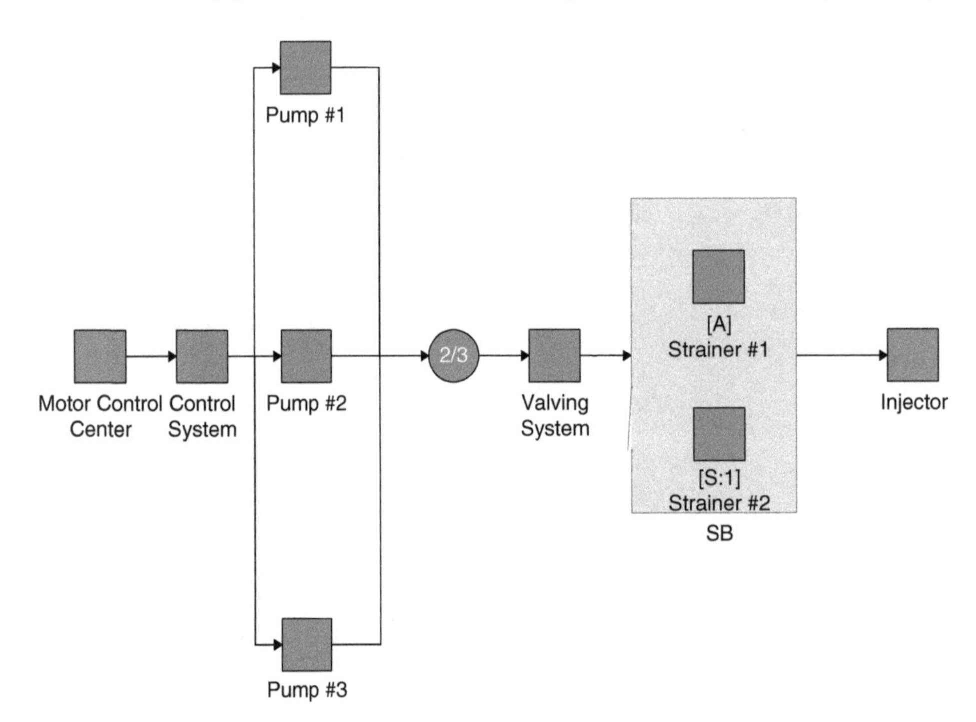

Figure 15.9 Reliability block diagram.

A subsystem could fail which would result in a corrective maintenance action after the mission, but will not impact the ability of that aircraft to complete its mission. The same approach can be used for a fleet of aircraft, when a k-out-of-N function is used. The k-out-of-N approach is used when a certain number of systems (or assets within a fleet) are required to meet the objectives. The k-out-of-N approach allows for redundancy as a fault-tolerant capability so that the system is able to meet its objectives with failures. An example of where k-out-of-N is used is in commercial aviation. An aircraft that is to operate further than 60 minutes from an alternate airport must demonstrate that the aircraft can reliably operate with an engine in a failed state for a given period of time. This ability and certification are known as Extended Operations (ETOPS). The ETOPS certification is based on the number of minutes that the aircraft can successfully operate reliably with a reduction in engines. For twin-engine aircraft, that means being able to operate at cruising speed with a single engine until the aircraft can land at an alternate airport within the ETOPS rating such as 120 minutes at any given point on the flight. For larger aircraft with four engines, that may mean being able to land at an alternate airport with one or two engines in a failed state within the approved ETOPS rating of 180 minutes. By using k-out-of-N, organizations can understand how many backup systems or assets may be required to achieve the defined level of availability.

While availability prediction can be extremely powerful for organizations, the accuracy of the predictions is always cause for great concern. The availability predictions are accurate based on the quantity and quality of the data used to calculate the elements of availability. The predictions require the right data to be successful and valuable to the organization. When the data are not right, then the confidence in the availability predictions suffers. You know what they say: "Garbage in equals garbage out."

15.5.1 Data for Availability Prediction

In order to start the availability prediction, one first needs to understand the design of the asset as well as the operating context in which the asset is to be used. Understanding the operating context involves understanding the users' expected operating environments, their ability to support the asset, the demand requirements, and so on. Once the end user requirements are established and understood by the asset designers, the specific reliability and maintainability data of the asset are needed in order to calculate the different time-related elements and mean metrics. These data are used to prepare the availability prediction and provide confidence that the results are accurate and correct.

So, where do the data for availability prediction come from? Well, the data come from the reliability and maintainability characteristics for each subsystem in the RBD. In a simple availability prediction model, the MTBF, MTTR, and associated data around preventive maintenance can be defined for each of the blocks. This often comes from the design Failure Modes and Effects Analysis (FMEA), as the FMEA establishes what failure modes are applicable to the subsystem, as well as the effects, which would drive the preventive and corrective maintenance activities.

However, depending on the level of maturity of the organization, various statistical distributions can be used to better define the true reliability and maintainability of each block. These distributions can include two-parameter Weibull, or three-parameter Weibull, log-normal, exponentials, and others. It is imperative to collect all available data for all assets in the field and fleet applications, as much as possible within the cost and schedule constraints, to determine these distribution functions. It is vital to collect the data from both internal and external sources. Much of the data may be collected from sources internal to the organization, such as Accelerated Life Testing (ALT). The data may also come from the various organizations that make up the asset's supply chain. The data may be supplied as part of the supplier requirements, which contain all relevant reliability and maintainability data.

In order for the reliability and maintainability data to be useful in building an accurate availability prediction model, there must be an understanding of where the data originated from, the pedigree of the data, the time period of the data-sets, when the data were collected, and the operating context in which it was gathered. The operating context involving the way the asset is used and maintained in the field and fleet is important as that may skew the results of the availability prediction. Ideally, the reliability data include the type of test from which the data were gathered (e.g., ALT), the environment data (e.g., temperature, vibration, etc.), the length of the test, and the specific failure mode(s) observed [5]. In addition, to truly understand how the asset may operate in real life, there needs to be consideration for wear-out failures, random failures, and infant mortality failures. Lastly, the impact (or consequence) of each failure mode needs to be documented to understand whether the failure results in partial output, intermittent failures, or complete failure.

Reliability and maintainability data may also come from industry data sources such as the IEEE, military or industry handbooks, engineering analysis, public data (Government–Industry Data Exchange Program [GIDEP]), or subject matter expertise. In some cases, data may not be readily available, and in that case, any and all assumptions made must be documented in the availability prediction report. By documenting these assumptions, the asset designer and operator can either make corrections or take into account the assumptions and derive an appropriate availability for their unique operating context.

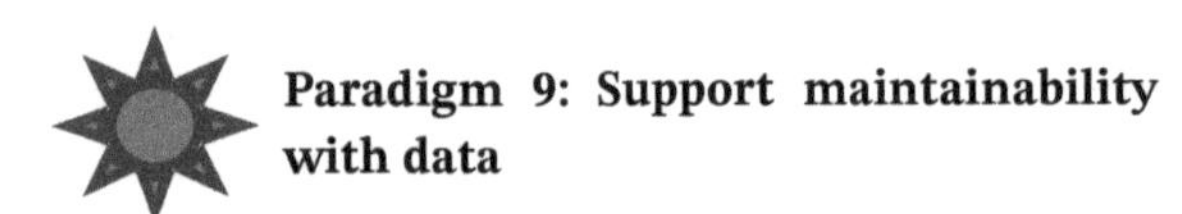
Paradigm 9: Support maintainability with data

15.5.2 Calculating Availability

As covered previously, there are various ways to calculate availability for a single asset or system, whether it is for Inherent, Achieved, or Operational Availability. An availability model is necessary to perform simple or complex calculations following a similar process of connecting series, parallel, and combinational blocks as used in the RBD. An asset's electrical and mechanical assemblies and components may be configured in a combined series/parallel configuration as shown in Figure 15.9. Figure 15.10 shows series and parallel block diagram configurations separately. Figure 15.11 provides an example of the series/parallel combined block diagram configuration. These three types of block diagram are used to create the availability model, which will be structured to match the asset configuration or a larger configuration containing multiple assets. The model will reflect the availability benefits from parallel configurations and the disadvantages of a serial configuration.

The model should be structured in one of two ways, either to assess an asset's availability as a stand-alone system, or as a member of a System-of-Systems (SoS) configuration. In some cases, the user may require that a certain number of assets out of a fleet configuration be available at any given time to fulfill the organizational objectives. Either of these model structures, stand-alone or SoS, can be created following a serial, parallel, or combined block configuration as shown in Figures 15.10 and 15.11.

Regardless of the configuration of the subsystems, systems, or assets, each block in the RBD needs to have its individual availability defined using the appropriate type of availability for the particular analysis. Once the individual block availability is determined and placed into an RBD, then the availability of the system or fleet can be determined based on the how the blocks are arranged and connected. Below are the calculations for the determining availability in the various configurations in Figure 15.10.

- Series availability
  The availability of a system with blocks in series is the simplest calculation. The total availability, A_T, is simply the product of the availabilities for the blocks: $A_T = A_1 \times A_2 \times A_3$.
  As an example, there is a system with three blocks in series, and the availability of block 1 is 0.98, block 2 is 0.96, block 3 is 0.97. The availability of the system would be:

$$\begin{aligned} A_T &= A_1 \times A_2 \times A_3 \\ &= 0.98 \times 0.96 \times 0.97 \\ &= 0.913 \\ &= 91.3\% \end{aligned}$$

  It is important to note that when calculating availability in series, the total system availability will always be less than the lowest individual block.

- Parallel availability
  Parallel systems are often used to boost the availability of a system when the individual components are below the required level of availability. This redundancy enables the system to continue to operate in

Figure 15.10 Block configurations.

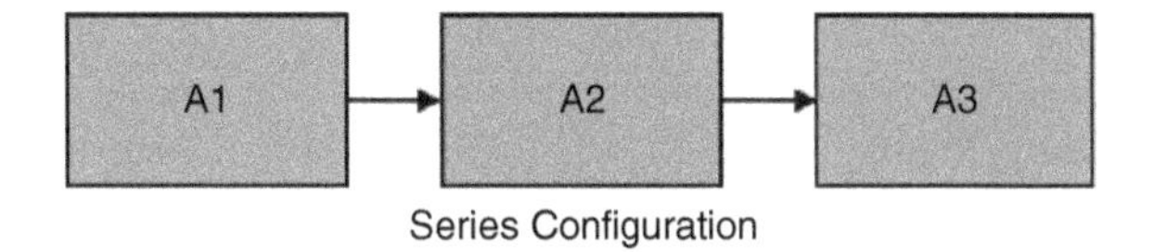

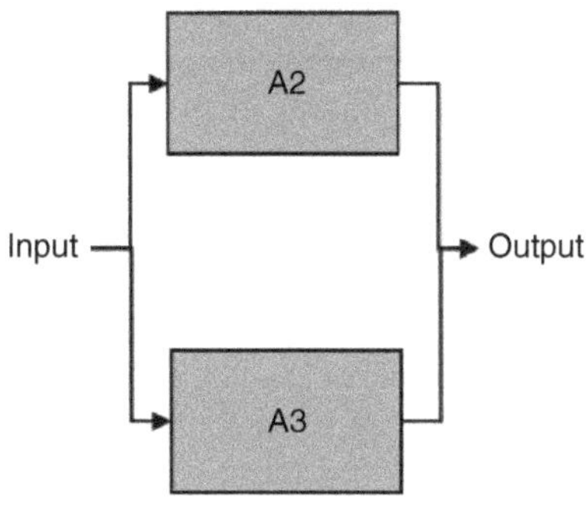

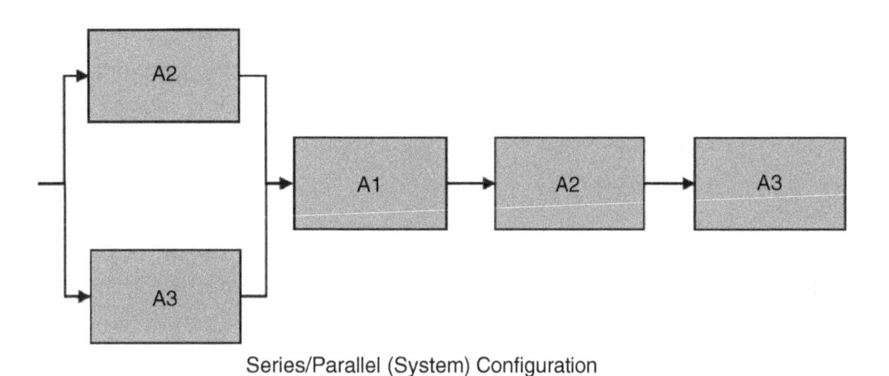

Figure 15.11 System availability calculation.

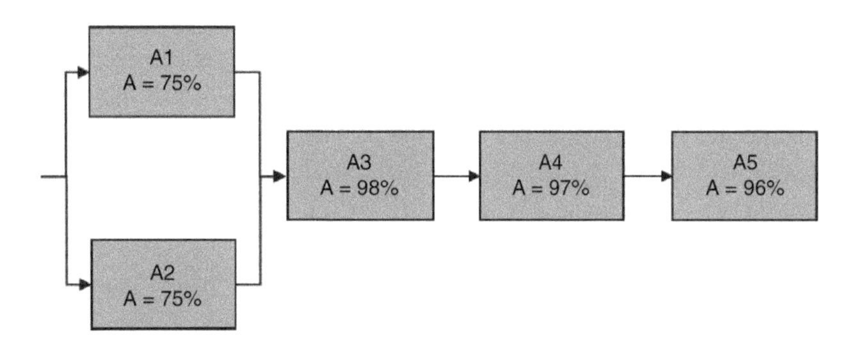

the event of a failure, minimizing the operational impact. The availability of parallel systems is calculated with the formula $A_T = 1 - (1 - A_1)(1 - A_2)(1 - A_3)$.

As an example, assuming there are three blocks in parallel, and all have an availability of 75%, the availability of the parallel system would be:

$$
\begin{aligned}
A_T &= 1 - (1 - A_1)(1 - A_2)(1 - A_3) \\
&= 1 - (1 - 0.75)(1 - 0.75)(1 - 0.75) \\
&= 1 - 0.015 \\
&= 0.984 \\
&= 98.4\%
\end{aligned}
$$

It is worth noting that when calculating availability in parallel, the total system availability will be larger than the largest individual block.

- Series/parallel (system) availability
  To calculate the total system availability for a system with both series and parallel components, the analyst would first solve for the components in parallel to reduce them to a single block, and then solve the entire system as a series configuration as shown in Figure 15.11.
  Using Figure 15.11 as an example of a system availability calculation, first the parallel components need to be solved:

$$\begin{aligned} A_{1/2} &= 1-(1-A_1)(1-A_2) \\ &= 1-(1-0.75)(1-0.75) \\ &= 0.9375 \\ &= 93.75\% \end{aligned}$$

Now with the parallel components solved, the system can be arranged into a series model to be solved:

$$\begin{aligned} A_T &= A_{1/2} \times A_3 \times A_4 \times A_5 \\ &= 0.9375 \times 0.98 \times 0.97 \times 0.96 \\ &= 0.856 \\ &= 85.6\% \end{aligned}$$

As you can see, the overall availability of the system is 85.6%, which is lower than the lowest block in the serial system model configuration, as you would expect.

Actual assets and systems are not as simple as the examples provided above. There are additional considerations that must be accounted for in the availability model, such as:

- The k-out-of-N availability model, which states how many components within the system are required to operate to ensure system operation. For example, a system may require four out of five components to operate for the system to operate. An example of this availability model feature is described above in relation to commercial aviation.
- Shared load parallel system, which represents how a parallel system shares a load between the various components all performing a similar function, such as power distribution. For example, if a system has three components in parallel, all operating at 33% of their full load capacity, the impact to availability if one fails is minimal. When one out of the three components fail, the other two components share 50% of the full load and the system will still function properly.
- Imperfect switching between parallel systems, used to represent standby redundant systems that must be manually turned on or switched over. The switchover is not instantaneous, which results in a small amount of downtime.
- Unequal failure rates between systems in parallel, used to represent differences in failure rates for similar components in parallel that are used in the same operating context.
- Active vs. standby redundancy, used to determine the timing delay to recover the system from a failure of a component and a failover sequence. A standby system requires a warm-up time before it can be operated in place of the primary unit. A redundant component is always active and takes over the function immediately. An Uninterrupted Power Supply is an example of an active redundancy system.

For further exploration of the above situations, the reader is referred to the *CRE Primer* [5]. While these calculations and situations may be complex, the more realistic the model, the more realistic the prediction will be. As the complexity of the model grows, it is recommended that the availability analyst utilizes a software tool package to model the availability of the asset, system, or SoS, to assist with quick turn analysis results, while providing a thorough understanding of how the model works and ensuring an accurate model.

15.5.3 Steps to Availability Prediction

Predicting availability requires not only a thorough understanding of the operating context, the asset or system, but also how the components within the system interact. Understanding how the system's components interact will assist in building a realistic model and replicating what is likely to occur in the real world. In order to predict the availability of the asset, a process should be followed to ensure all models are built to the same standard and contain the right data. The process to predict availability should become a Standard Operating Procedure (SOP), to ensure the result can be replicated and adjusted based on new data. The steps to predict availability are adapted from Reliasoft's *Steps in a System Reliability, Availability, and Maintainability Simulation Analysis* [6]:

15.5.3.1 Define the Problem

To build a model, it is first vital to understand the problem that is being solved. For example, is the model to evaluate proposed design changes and their impact on overall availability, evaluate the impact of various levels of support systems in the field, or to strike the right balance between reliability and maintainability? The model could also be used to establish the required number of assets to fulfill a SOS mission in a particular operating context.

By defining the problem, the model can be built to solve the specific issue at hand, while not being concerned about all of the other possible situations. However, if the problem being solved is a tradeoff between maintainability and reliability, it may be required to have multiple models that need to be evaluated against each other. A problem statement should be created, which clearly defines the problem and enables the development of useful models. A refined problem statement limits the scope and size of the analysis. The more refined the problem statement, the easier the analysis will be. The development of the problem statement may take many revisions, based upon the data available, and the customer requirements. Ideally, the problem statement is defined with a cross-functional team, to ensure it meets the needs of the end users.

15.5.3.2 Define the System

Defining the system involves building the RBD as the initial framework for the availability model. This RBD will often require inputs from various Subject Matter Experts (SMEs) on how the various component blocks will interact with each other. Ideally, the system will be kept as simple as possible, but allowing for more complexity to be added later as needed. The initial RBD should be reviewed with the various stakeholders to ensure it will meet the needs of the analysis prior to releasing it to a broad audience.

15.5.3.3 Collect the Data

With the RBD constructed, the reliability and maintainability data need to be collected. This is often the most time-consuming part of the analysis. As stated previously, the reliability and maintainability data need to be adjusted for the particular operating context for which the model is being built. Any gaps in data, flaws in the data, or data assumptions made, need to be documented at this point in time.

15.5.3.4 Build the Model

With the RBD and data collected, the next step is to build the availability model. Determining where to build the model and what modeling tool to use will depend on a variety of factors such as the anticipated model complexity, the size and magnitude of the model, the number of model changes over time, and the number of simulations to be run. Simple models can be built in spreadsheets, while larger simulations may benefit from commercially available software such as BlockSim® [7]. Where Monte Carlo analysis is required, commercially available software is the recommended approach.

15.5.3.5 Verify the Model

With the model built and data entered into the modeling tool, the model should be reviewed to ensure it is delivering the expected results. The availability analyst should utilize a cross-functional team to review the analysis results and make sure the results are accurate and meet stakeholder expectations. If the results are drastically different from what was expected, the analyst should perform a thorough review of the model calculations and data used to enter into the tool to determine the reasons for the differences.

This is one of the drawbacks when using a software package, in which the analyst enters the data without an understanding of the model or how the math works and assumes that the software is giving a correct answer. While the software may give a mathematically correct answer, the model is not representative of the real world.

15.5.3.6 Design the Simulation

With the model built and verified, the simulation needs to be finalized. The simulation will require a variety of factors to be specified, such as the mission time, or total time to be analyzed. How many different simulations will be run? How many iterations of each simulation will be run? When performing a Monte

Carlo analysis, how many histories will be performed? Should warm-up or cool-down periods be considered? Are multiple scenarios for logistical and administrative delays included? Most of the specifics should be pulled from the initial problem statement defined in step one.

15.5.3.7 Run the Simulation

Run the simulation and variations of it as required. Depending on the complexity of the system being analyzed, and the number of iterations required, this may take a few minutes, hours, or days. At the outcome of the simulation, the performance of the system and its alternatives should be understood in terms of availability.

15.5.3.8 Document and Use the Results

Using the outputs of the simulation(s), compile the results into a summary document, highlighting what was covered and what was not covered, whether the system will meet the needs of the end user, or what recommendations arose as a result of the model. This summary should be share with the cross-functional team that helped to define the problem statement. In addition to the summary document, a detailed technical report should be included, which documents the results of the simulation in detail, the assumptions made, and the validation of the results.

Once the report has been completed, the organization needs to use the findings and create a plan to drive improvements to the design of the system or the supportability program. If the report is put on the shelf never to be used, the entire analysis was a waste of time.

While not exhaustive, the steps outlined above to predict availability will allow organizations to ensure that the availability models built and analysis performed are valid, and will drive improvements to the availability of the asset or fleet.

15.6 Conclusion

Availability is a vital metric that all asset-owning/operating organizations and OEMs need to track when assurance is needed that a system or asset is ready to function and accomplish its mission when it is needed. Availability is a powerful metric that allows asset owners to understand how ready their assets are to perform intended missions when assets are deployed for continuous 24 hours by 7 days (24/7) operations. Availability allows asset owners to understand the impact to their probability of mission success as a result of poor maintainability, poor reliability, or logistical delays. Availability provides the most cost-effective solutions for asset designers as they perform a delicate balance between designing for reliability and designing for maintainability during design trade study analyses.

Availability provides insight to the organization's ability to meet their objectives. In the military or commercial aviation business, it is sorties, flights, or OEE. Availability enables asset owners to determine how many assets may be needed to ensure the achievement of the organization's objectives.

The key to success with availability within organizations is to ensure that it is measured consistently based on the goals of the organization. By leveraging the various types of availability, organizations can identify reliability, maintainability, or supportability issues. This systematic approach enables organizations to eliminate wastes within their processes or design. In addition, if there is an understanding of availability requirements, availability can be used to work backwards to establish reliability and maintainability goals while balancing the two. If design is a concern, then inherent availability may be required. If the impact to operations is the larger concern, then operational availability needs to be measured. However, measuring availability is not enough. Organizations need to act on the information gathered when measuring and calculating availability, and improve their designs for availability when the information supports it.

References

1 US Department of Defense (2005). *DOD Guide for Achieving Reliability, Availability, and Maintainability*. Washington, DC: Department of Defense.

2 US Department of Defense (1982). *Test and Evaluation of System Reliability, Availability, and Maintainability, DOD 3235.1-H*. Washington, DC: Department of Defense.

3 Society of Maintenance & Reliability Professionals (2017). *SMRP Best Practices*, 5e. Atlanta, GA: SMRP.

4 NASA (2008). *Reliability-Centered Maintenance Guide for Facilities and Collateral Equipment*. Washington, DC: National Aeronautics and Space Administration.

5 Quality Council of Indiana, Wortman, B., and Dovich, R. (2009). *CRE Primer*. Terre Haute, IN: Quality Council of Indiana.

6 Reliasoft Corporation (2009). Steps in a system reliability, availability and maintainability simulation analysis. https://www.weibull.com/hotwire/issue103/relbasics103.htm (accessed 15 September 2020).

7 Reliasoft Corporation (2020). BlockSim®. https://www.reliasoft.com/products/blocksim-system-reliability-availability-maintainability-ram-analysis-software (accessed 15 September 2020).

16

Design for Supportability

James Kovacevic

16.1 Introduction

The ability of an asset or system to function as needed is a direct result of the organization's ability to support the asset through operational and maintenance activities at an acceptable cost. The ability of the organization to support the asset is known as supportability. Supportability can be defined as the degree to which system design characteristics and planned logistics resources meet system useful life requirements and provide system operations and readiness needs throughout the system's service life at an affordable cost. It provides a means of assessing the suitability of a total system design for a set of operational goals, objectives, and needs within the intended operations and service support environment (including cost constraints) [1].

Supportability characteristics include many performance measures of the individual elements of a total system. For example: Repair Cycle Time is a support system performance characteristic independent of the hardware system. Repair Cycle Time is also called Turn Around Time (TAT). System development organizations sign agreements with customers for a specified TAT in a support agreement prior to the shipment of the first assets to customers. TAT is a supportability characteristic associated with a particular system, but is not a characteristic based on the system design. As another example, Mean-Time-Between-Failure (MTBF) and Mean-Time-to-Repair (MTTR) are reliability and maintainability characteristics, respectively. MTBF and MTTR are performance measures that are highly dependent on the system hardware and software design. These performance measures greatly impact operational support requirements of the total system, and therefore make them supportability characteristics of the system design.

Design for Supportability occurs when the characteristics of the system enable the asset to be supported cost-effectively over its life. Design for Supportability requires the design team to consider how the asset will be supported, not just in maintenance activities, but also in the supply chain, with information systems (IS). This requires that the design team consider the design and selection of the components and balance the need for specialty components, which may provide a supply chain problem, versus a Commercial Off-the-Shelf (COTS) component, which is easily attainable. It is only when these types of considerations are made in the design phase that an asset is designed for supportability.

Operational Availability (A_0) and life cycle cost are generally accepted as measures of total system supportability that may be both dependent and independent of the system hardware and software design, depending on each situation or business case. The degree to which the asset is supported will have a direct result on A_0 and life cycle cost assessments, which can be predicted and estimated through supportability analysis case studies.

Design for Maintainability, First Edition. Edited by Louis J. Gullo and Jack Dixon.

Other terms used to express similar assessments are equipment readiness and affordability [1].

The design of the asset effects supportability through standardization of components and systems across asset classes, interchangeability, accessibility, self-test, and diagnostics. While these features are inherent to the design, the decision to include these directly affects the ability to support and maintain the asset. This decision affects both Design for Supportability and Design for Maintainability (DfMn). The supportability and maintainability design features factor into the maintenance concept and design phase decisions that impact the asset's life. These design decisions must be made during the design phase, as it is virtually impossible to influence and change them after the design phase.

Supportability also includes many supply chain factors, such as the support systems required to perform and sustain normal system operations and maintenance. These supply chain factors include supplier personnel available for technical support, availability of spare parts, test equipment required, and supplier maintenance facilities required to perform the maintenance when needed.

With the vast majority of the asset's cost realized during the operational and maintenance phase (also known as the Operations and Support [O&S] phase) of the asset's life, it is vital that supportability is addressed early in the design and development phase. By addressing supportability upfront, the overall life cycle cost of the asset can be greatly reduced. Consider the need for a spare part, which is highly specialized and qualified from military applications in extremely harsh environments. Depending on where the military qualified part is built, how many are stored, and where they are stored, the asset could be down for a significant period of time when the inventory of this part is depleted, and the asset must wait for more parts to be procured. Now consider if design for supportability was emphasized earlier in the system development phase, and sole-source parts were considered and eliminated. That specialized military spare part could have been replaced with an alternative COTS part as a replacement option, with a little design work to qualify a new part supplier to minimize the risk of sole-source issues. COTS parts would likely be more available from sources of the part to fill up the inventory and be available to perform the repair much faster than equivalent functioning military parts. As a general rule, COTS parts have shorter lead times than military parts. For this reason, many military part suppliers provide parts that are considered long-lead items by the military system supply chain managers. These supply chain managers spend a lot of time worrying about these long-lead items and try hard to find alternatives from comparable COTS parts sources. This is merely one example to illustrate the need to consider supportability in the design phase. More examples are provided throughout this chapter.

To properly support an asset requires planning. This is generally accomplished through the use of an Integrated Logistics Support (ILS) plan. The ILS is a complete maintenance and supportability concept that addresses all topics related to sustainment of an asset, thus ensuring that the asset will operate as required in the field. The ILS may also be referred to as Logistics Support (LS) or Integrated Product Support (IPS). Asset sustainment encompasses all aspects of supporting an equipment or system during development, production, testing, use, and retirement from use. The ILS plan is a document that emphasizes how the asset will be supported throughout its life. More details on ILS plans will be provided later in Section 16.4.

Supportability is a vital activity that is often overlooked in the commercial sector, but has the ability to dramatically improve the operation availability of the assets and systems, upon which many depend.

16.2 Elements of Supportability

Supportability involves all related activities to achieve the totality of a logistics and service support infrastructure, from training to transportation to technical documentation, from spares and warehousing to tools and test equipment. It should be noted that the totality of ILS encompasses many topics and subtopics related to product sustainment, each of which could require an entire person's career to master all the intricate details. The supportability tasks may seem daunting at first glance, but with the right organization and structuring of the different elements of supportability, the tasks should become more understandable and manageable. This section provides that structure to help the reader to understand the basic elements of supportability. As defined by the Defense Acquisition University (DAU), there are 12 elements of Integrated Product Support (IPS) [2]. These IPS elements are:

1) Product support management
2) Design Interface
3) Sustaining Engineering
4) Supply support
5) Maintenance planning and management
6) Packaging, Handling, Storage, and Transportation (PHS&T)
7) Technical data
8) Support equipment
9) Training and training support
10) Manpower and personnel
11) Facilities and infrastructure
12) Computer resources

Supportability is crucial for the design team to consider. Consider that the vast majority of the asset's life cycle costs are determined by the design of the asset. For instance, a decision by the design team to use a specialty, custom industrial-grade or military-grade component over a readily available COTS component can add significant time and costs to repairs once the asset is deployed and used in the field. The cost of the components alone, comparing industrial-grade or military-grade to COTS grade, may range between 10 to 100 times higher or more. Part of the reason for this cost differential is the testing that must be conducted to prove an industrial-grade or military-grade component works over the environmental conditions specified in the requirements. In addition, what and how the maintenance work will be performed will have a significant impact on the A_0 and life cycle cost of the asset. Therefore, it is vital the design team consider and design for supportability as early in the design phase as possible.

The level of effort put into design for supportability will depend heavily upon the asset, the way it is used by the customers in their field applications, and its intended operating environment. Consider a military system, which must operate under the most difficult of conditions. A military system must have a rigorous Design for Supportability approach, along with extensive planning and execution of the supportability plan. On the other hand, a commercial device, like a video game console controller, does not require significant design for supportability as the consequences of the controller not being available is minimal. Regardless of the asset, the 12 elements of supportability still apply; however, the level of detail and rigor will vary.

16.2.1 Product Support Management

Product support management is the first element of supportability, which is put in place to plan and manage cost and performance across the entire life cycle (concept, design, build, operate, maintain, disposal) of the asset. This is accomplished by planning, managing, and funding supportability activities across the remaining 11 supportability elements.

Product support management is in place to provide continuous support to the asset and to monitor the support through key performance metrics, such as reliability, availability, maintainability, and total ownership costs (TOCs). Where required, the product support manager will provide targeted support activities, improve the metrics, and ensure the asset is able to perform its required functions throughout its expected life. The ultimate goal of product support management is to reduce the Total Cost of Ownership, while ensuring all performance requirements are met. In order to achieve the goals, the product support manager has 11 responsibilities [2]:

1) Develop and implement a comprehensive product support strategy for the asset.
2) Use appropriate predictive analysis and modeling tools that can improve material availability and reliability, increase operational availability, and reduce operational and sustainment costs.
3) Conduct appropriate business case cost analyses to validate the product support strategy, including life cycle cost-benefit analyses.
4) Ensure achievement of desired product support outcomes through development and implementation of appropriate Product Support Agreements (PSAs).
5) Adjust performance requirements and resource allocations across Product Support Integrators (PSIs) and Product Support Providers (PSPs) as necessary to optimize implementation of the product support strategy.

6) Periodically review Product Support Agreements (PSAs) between the PSIs and PSPs to ensure the arrangements are consistent with the overall product support strategy.
7) Prior to each change in the product support strategy, or after five years, whichever occurs first, revalidate any business-case analysis performed to assess the strategy.
8) Ensure that the product support strategy maximizes small business participation at the appropriate supply chain tiers.
9) Ensure that PSAs for the asset describe how supply chain business arrangements will ensure efficient procurement, management, and allocation of parts and inventories in order to prevent unnecessary procurements of such parts.
10) Make a determination regarding the applicability of preservation and storage of unique tooling associated with the production of program-specific components; and if relevant, include a plan for the preservation, storage, or disposal of all production tooling.
11) Work to identify future obsolete electronic parts and diminishing sources that are included in parts lists and the production Bill of Materials (BOMs). This effort is important to stay ahead of long-term commitments to use these parts to satisfy build or maintenance planning schedules, ensuring parts procured over several procurement cycles will continue to meet the specifications of assemblies used on systems for an acquisition program, and search for and approve suitable replacement parts when these parts become obsolete.

16.2.2 Design Interface

As mentioned above, the majority of life cycle costs for an asset are defined by the asset design. In some studies, the design can attribute up to 90% of the total life cycle costs [3]. As such, it is vital that the design team works to reduce the life cycle costs. This is where Design Interface comes in.

Design Interface is intended to be a set of activities and analysis to feed design changes to the design team to reduce the supportability requirements. By limiting specialty components, or test equipment, the cost for supportability can be reduced. In addition, by eliminating or reducing maintenance activities, the cost to maintain the asset can be reduced dramatically. Design Interface is ideally considered in the concept and requirements phase of the asset, as the supportability requirements should be traced directly to system-level requirements (see Figure 16.1). The Design Interface should continue past the concept phase, throughout all of the design phases. Design Interface is an iterative approach, which utilizes many types of analysis. Once the final design is completed, the supportability requirements for the asset are locked in and the requirements baseline is frozen. It is very difficult to make changes after this time.

Design Interface reflects the relationship of the asset design to support requirements. It is important to note

SUPPORTABILTY RELATED DESIGN FACTOR FOR THE F-16

Terms:	Range/Value:
Weapon system reliability	.90 – .92
Mean time between maintenance (inherent)	4.0 – 5.0 hrs.
Mean time between maintenance (total)	1.6 – 2.0 hrs.
Fix rate	60% in 2 hrs.
	75% in 4 hrs.
	85% in 8 hrs.
Total not-mission-capable rate maintenance rate	8%
Total not-mission-capable supply rate	2%
Sortie generation rate	classified (see req. doc.)
Integrated combat turn around time	15 min.
Primary authorized aircraft airlift support	6–8 C-141B equiv.
Direct maintenance personnel	7 to 12 AFSCs
Reduced number of Air Force Specialty Codes	4 to 6 AFSCs

Figure 16.1 Example of supportability requirements. Source: From US Department of Defense, *Acquisition Logistics, MIL-HDBK-502*, US Department of Defense, Washington, DC, 2013.

that the design parameters are expressed as quantitative requirements (Figure 16.1). This is done to reflect what should be achieved when the asset is in use, and forces the design team to consider supportability.

The factors that need to be considered as part of the design process include [2]:

- Reliability
- Availability
- Maintainability
- Supportability
- Suitability
- Integrated Product Support (IPS) elements
- Affordability
- Configuration management
- Safety requirements
- Environmental and HAZMAT (hazardous materials) requirements
- Human systems integration
- Calibration
- Anti-tamper
- Habitability
- Disposal
- Legal requirements

With all of the factors to consider, it is vital that the design team considers supportability as early as possible in the concept phase. This is due to the fact that the design needs to minimize logistic requirements, maximize reliability, and ensure that the asset is supported throughout its life. Additional issues that must be considered throughout the life of the asset include obsolescence management, technology refreshment, modifications and upgrades, and overall usage under all operating conditions [2].

16.2.3 Sustaining Engineering

Sustaining Engineering is the process of supporting in-service assets and systems in their operational environments. The goal of Sustaining Engineering is to provide technical expertise to identify, analyze, and mitigate risks to the continued operation and maintenance of a system in a given operating environment. Consider the challenges of operating a system in a remote desert environment. Then consider the challenges of operating the same system in a cold arctic environment. While there may be similar challenges or risk, there are definitely different challenges as well. These challenges include both asset-related challenges, such as the need for a different lubricant given the operating temperatures, to logistical challenges, such as the shelf-life of the lubricants and spare parts used or stored in the extreme temperatures of the operating or storage facility environments (e.g., a temperature-controlled facility or a tent in the middle of the desert).

Sustaining Engineering is in place to ensure that all the identified risks are addressed so the asset operates at the required capabilities, and identifies opportunities to improve the capabilities of the asset. This typically involves obtaining and analyzing all asset performance data to develop corrective and preventive actions. Each of the options is supported by a business case including a life cycle cost analysis (LCCA), and which demonstrates the value the proposed action will provide. Lastly, Sustaining Engineering also ensures that the design configuration management is followed, which ensures that any design changes to the asset are evaluated prior to implementation, and all technical drawings, parts lists, and other design documentation is updated according to the design changes.

In a typical sustaining engineering program, the sustaining team will perform the following [2]:

- Collection and triage of all service use and maintenance data
- Analysis of safety hazards, failure causes and effects, reliability and maintainability trends, and operational usage profiles changes
- Root cause analysis of in-service problems (including operational hazards, deficiency reports, parts obsolescence, corrosion effects, and reliability degradation)
- The development of required design changes to resolve operational issues
- Other activities necessary to ensure cost-effective support to achieve asset readiness and performance requirements over a system's life cycle

One of the major challenges for sustaining engineering is the rate of obsolescence in electronic equipment. Obsolescence management is a critical function to predict part supplier changes to their manufacturing lines and end production of certain parts. When this happens, suppliers send notifications to their customers of End-of-Life (EOL) decisions for parts that will require Last-Time Buys (LTBs). The customers of these parts should schedule to place LTB orders with these suppliers to build up their inventories while they prepare to redesign electronic circuit cards or find alternative replacement parts from other sources. In order to overcome this, the sustaining team must decide whether to stock enough parts to support the asset over its entire life cycle, ensuring that the supplier can and will manufacture them over the life of the asset, and/or investigate upgrade options (e.g., better-than alternative parts) and second sources to be used as the asset ages. This activity, along with all of the activities mentioned above, is vital to ensuring that the asset is able to meet the operational requirements and be cost-effective over its entire life cycle.

16.2.4 Supply Support

Supply support is a vital activity that ensures that the supplies required to repair and maintain the asset over its life are available when needed and at the lowest possible TCO. This is accomplished through a series of management actions, procedures, and techniques to determine the necessary requirements. These actions typically include a method to determine the spare part requirements, and how to catalog, receive, store, transfer, issue, and dispose of spare parts and supplies [2]. In the end, it is essentially having the right spares at the right time in the right quantity at the right place at the most cost-effective price. These actions take place during the initial acquisition phase of the asset through to the end of the system life at decommissioning and disposal of the asset.

The supply support element is the method to establish a supply chain for all of the required parts (see Figure 16.2), materials, supplies, and potential contractors to support the asset. This supply chain is not limited to the asset-owning organization and can include commercial vendors, warehouses, speciality contractors, as well as the shipping contracts. The effort to establish the supply chain is extensive, but it enables the asset to achieve operational availability at the lowest TCO.

One may think that the organization can simply stockpile parts and ship to the asset when needed. But there is often an overhead cost to stocking parts, which may range from 20–30% of the value of the parts each year. This overhead cost is known as carrying costs or inventory costs. The carrying cost is

SUPPLY SUPPORT SUMMARY

CAGE	REFERENCE NUMBER	NSN	PCCN	PLISN	ITEM NAME	UI	QP EI	SMR
97384	59822-90082-30	6130-01-279-3436	1BGLO	A003	power supply	ea	5	PAHZZ
97384	59822-90086-20		1BGLO	A004	programmer	ea	2	PAHHD
97384	59822-90086-30	5998-01-293-2774	1BGLO	A005	circuit card assembly	ea	5	PAHZZ
97384	59822-90119-21	5998-01-268-8589	1BGLO	A006	circuit card assembly	ea	8	PAHZZ
97384	59822-90119-211		1BGLO	A007	microcircuit	ea	25	PAHZZ
97384	63603-40140-20		1BGLO	A002	cabinet console	ea	1	XBHHD
97384	63603-46200-10		1BGLO	A001	Test station	ea	1	PEHHD

Figure 16.2 Example supply support summary. Source: From US Department of Defense, *Acquisition Logistics, MIL-HDBK-502*, US Department of Defense, Washington, DC, 2013.

typically made up of the overhead costs to operate the warehouse for the parts, shrinkage or losses, and the taxes that must be paid on the spare parts, along with a few other factors. Consider a spare parts inventory of $100 000 000. Using a value of 24%, that means that the cost to maintain that spare parts inventory of $100 000 000 is roughly $24 000 000. That cost may be considered extreme and not cost-effective. The supply support function is constantly looking to optimize the inventory to meet operational needs while minimizing the cost of holding the parts.

Determining where the parts are stored is another major concern of supply support. When the storage locations are distributed, the response time for a part is reduced, but additional overhead is incurred as more warehouses, staff, and support infrastructure are required. With a centralized storage location, overhead may be reduced, but shipping costs and time to respond with needed support increases. This is why the supply team must optimize the supply chain to gain the best value from the infrastructure chosen.

The supply support team must constantly work with the partners in the supply chain to ensure the availability of the parts and supplies to support the asset over its life. With the vast majority of costs occurring in the O&S phase of the asset's life, it is imperative that the supply chain be optimized to balance the costs and operational availability.

16.2.5 Maintenance Planning and Management

Maintenance planning and management is the simply the process (see Figure 16.3) of establishing the maintenance requirements to ensure operational availability is achieved at the lowest possible cost. Maintenance requirements include the repair and upkeep tasks, and the time schedules and resources required to care for and sustain the asset in its operating environment. Maintenance planning and management goes beyond identifying the tasks, but includes the identification of all the manpower and

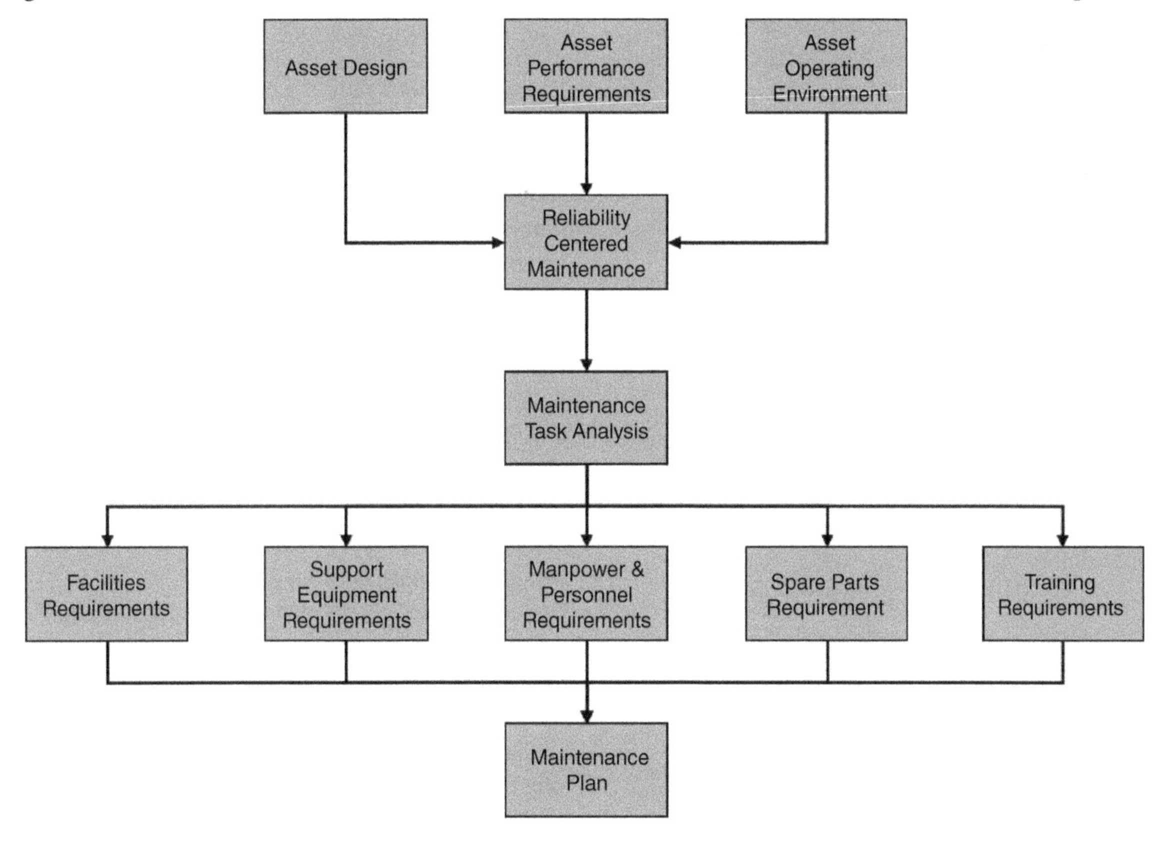

Figure 16.3 Process to establish the maintenance plan.

funding resources required to develop and implement the maintenance and modernization plan [2]. Modernization plans describe decisions made during the design and development phase of a program to provide technical insertions or technical refreshes of the technology at regular intervals during a system's life cycle to avoid the potential of part obsolescence issues and keep current on the latest technical capabilities. This can help drive sustainment costs lower and provide the users with the latest design features to improve the capabilities of the assets.

Paradigm 8: Understand the maintenance requirements

The challenge with establishing the maintenance requirements includes balancing the need for planned downtime with risk of unplanned downtime, both of which impact operational availability. To establish the balance, many analysis tools can be used, such as Level of Repair Analysis, Reliability-centered Maintenance (RCM), and Maintenance Task Analysis (MTA). These tools must also look to establish the lowest TCO. The outputs of maintenance planning and management include [2]:

- Levels of repair
- Repair times
- Testability requirements
- Support equipment needs
- Training and Training Aids Devices Simulators and Simulations (TADSS)
- Manpower skills
- Facilities
- Inter-service, internal, and contractor mix of repair responsibility
- Deployment planning/site activation
- Development of preventive maintenance programs using RCM
- Condition-based Maintenance Plus (CBM+)
- Diagnostics/Prognostics and Health Management
- Sustainment
- Performance-based Logistics (PBL) planning
- Post-production software support

Maintenance planning and management activities are heavily influenced by the design of the assets, which is why DfMn and Design for Reliability are so important. Decisions made in the design phase will have a major impact on the life cycle costs of the assets. The design team should be focused on preventing, reducing, and improving the maintenance actions required for the asset once it is in the operational environment. Once the design has been finalized, the maintenance planning and management team must decide on the maintenance strategy.

There are numerous maintenance strategies used by organizations – some better than others. These strategies are usually derived from three basic maintenance strategies. Each strategy on its own may not result in the lowest TCO possible, but when the strategies are used in combination and applied for different situations, the TCO can be optimized to be the lowest cost target level possible. The three primary maintenance strategies are:

- Preventive maintenance includes all actions performed to retain an item in a specified condition by providing systematic inspection, detection, and prevention of incipient failures. These actions include scheduled replacements, time-based overhauls, and inspections, to name but a few.
- Corrective maintenance is the ability of the asset, and the infrastructure that supports the asset, to be restored to operational readiness in the event of a failure. Corrective action often includes some level of repair or inspection to mitigate the failure. Corrective maintenance mainly consists of removal and replacement tasks, in-place repair, adjustment, and calibration, to name a few of the tasks (more details are provided in earlier chapters).
- Condition-based Maintenance is the application of the appropriate technology, and knowledge-based capabilities to identify anomalies and defects within the asset prior to the failure of the asset. This enables advanced planning to take place and prevent the asset from functionally failing. More details of Condition-based Maintenance are provided in Chapter 9.

Paradigm 4: Migrate from scheduled maintenance to Condition-based Maintenance (CBM)

Once the strategy has been developed, the maintenance planning and management team will use MTA to collect the detailed data, which are needed to establish the specific maintenance and logistical requirements for each maintenance activity. The logistical requirements include the identification of the steps, spares and materials, tools, support equipment, and personnel skill levels, as well as any facility issues that must be considered for a given repair task. The MTA measures or calculates elapsed times required for the performance of each task. MTAs cover both corrective and preventive maintenance tasks, and when completed, identify all physical resources required to support a system [2].

When the MTA is completed, MTA data are fed to the maintainability allocation and prediction processes (covered in Chapters 6 and 7). As an output from the allocation and prediction process, the entirety of the maintenance requirements may be completed and combined into a maintenance plan (see Figure 16.4). This plan prescribes the maintenance actions, maintenance intervals, repair levels and locations, personnel numbers and skills, technical data, tools, equipment, facilities, and spares and repair parts for each significant item of a system or equipment [2].

Further details on maintenance planning are contained in Chapter 3.

16.2.6 Packaging, Handling, Storage, and Transportation (PHS&T)

When the spare parts are identified and mandated as spares stockage requirements to ensure operational readiness, the processes to provide the spare parts where they are needed must be managed in a way that prevents failure of the infrastructure to provide the spares. Without managing parts while they are in storage, the organization runs the risk of installing parts that will cause a premature failure of the asset. Parts such as rubber seals, bearings, motors, and other parts have unique shelf life, packaging, and storage requirements to ensure the parts are ready to use when needed. The PHS&T element is to identify,

MAINTENANCE PLANNING SUMMARY

SECTION I: GENERAL

Inspection/ fault location to be accomplished by organizational maintenance, with follow-on inspection/ fault location and replacement of door-screen and engine assembles performed by intermediate support as well as the replacement of compressor and repair of all assemblies except the wire harness, which requires the attention of depot maintenance

SECTION II: MAINTENANCE ACTIONS

ITEM NAME	ACTION	ESTIMATED TIME	MAINT LEVEL
Engine	overhaul	4 hours	DEPOT
Pistons	remove & replace	1.13 hours	INTERMED
Plugs	remove & replace	.75 hours	INTERMED
Radio	fault locate	.25 hours	ORG

Figure 16.4 Example maintenance plan summary. Source: From US Department of Defense, *Acquisition Logistics, MIL-HDBK-502*, US Department of Defense, Washington, DC, 2013.

plan, resource, and acquire packaging/preservation, handling, storage, and transportation (PHS&T) requirements to maximize availability and usability of the materiel to include support items whenever they are needed for training or mission success. The PHS&T element is broken up into four areas [2]:

- Packaging provides for product security, transportability, and storability. The nature of an item determines the type and extent of protection needed to prevent its deterioration, physical, and mechanical damage. Shipping and handling, as well as the length and type of storage considerations, dictate cleaning processes, preservatives, and packaging materials.
- Handling involves the moving of material from one place to another within a limited range and is normally confined to a single area, such as between warehouses, storage areas, or operational locations, underway replenishment, shipboard cargo holds, aircraft, or movement from storage to the mode of transportation.
- Storage implies the short- or long-term storing of items. Storage can be accomplished in either temporary or permanent facilities of varying conditions, such as general purpose, humidity-controlled warehouses, refrigerated storage, and shipboard.
- Transportation is the movement of equipment and supplies using standard modes of transportation for shipment by land, air, and sea. Modes of transportation include vehicle, rail, ship, and aircraft.

PHS&T focuses on the unique requirements involved with packaging, handling, storing, and transporting not only for the major end items of the asset, but also the spare parts, materials, lubricants, epoxies, and supplies needed to support the asset. Figure 16.5 is an example of the PHS&T requirements for a major component, such as an engine.

As can be seen in the PSH&T summary, all of the information required to care for and transport the part is summarized on a single card. This summary enables organizations to quickly ship the parts, and ensure they arrive ready to use and undamaged.

PHS&T also has the requirement to identify and address any issues that arise with PHS&T of the materials or parts. The issues that must be monitored for and addressed are [2]:

- Transportation problems where items are delayed, or more significantly, cannot be shipped due to physical or regulatory restrictions
- Storage issues where shelf life has expired, or improper storage has caused degradation of the product
- Poor packaging or marking resulted in lost items during shipping
- Incorrect handling resulted in damage to the item being shipped

By identifying the PHS&T requirements upfront, and continually monitoring the performance of the PHS&T functions, the supply chain can be further optimized to improve operational availability and to reduce the TCO.

16.2.7 Technical Data

Technical data are required to allow the operation, maintenance, and supportability functions to make informed decisions, which reduce the Total Cost of Ownership. Often poor data lead to poor decisions, which can have a dramatic impact on the operational availability of the asset. The technical data element focuses on identifying, planning for, validating, resourcing, and implementing a system to acquire, analyze, and act on the data. The data are often used to [2]:

- Operate, maintain, and train on the equipment to maximize its effectiveness and availability.
- Effectively catalog and acquire spare/repair parts, support equipment, and all classes of supply.
- Define the configuration baseline of the asset (hardware and software) to effectively support the organization with the best capability at the time it is needed.

Paradigm 9: Support maintainability with data

Technical data come in many forms and include documents such as technical manuals, engineering drawings, engineering data, specifications, and standards. Much of these data come in both electronic and hard-copy formats, which require a well thought-out process

PACKAGING, HANDLING, STORAGE, AND TRANSPORTATION SUMMARY

SECTION I – PACKAGING, HANDLING AND STORAGE

CAGE	REFERENCE NUMBER	NAT STOCK NUMBER	ITEM NAME
10855	AA06BR200	2803-00-378-2804	engine

UI	WEIGHT	UM	LENGTH	WIDTH	HEIGHT	UM
EA	345	LB	3.0	2.0	3.5	FT

DOP	QUP	PKG-CAT	PRES MATL	WRAP MATL	CUSH MATL	UNIT CONT	SPEC MKG
8	001	8080	00	..	00	WR	99

SECTION II – TRANSPORT

MILITARY UNIT MODES OF TRANSPORT: This unit will be transported by a ground transportation company, fixed wing C-130, C-141, and C-5 units; helicopters CH-47 and CH-53 units. This unit will be used by different armored divisions.

SHIPPING WEIGHT EMPTY	SHIPPING WEIGHT LOADED	CREST ANGLE	FRONT IN	FRONT OUT	REAR IN	REAR OUT
346lbs	346lbs	N/A	N/A	105.8	N/A	N/A

LIFTING AND TIEDOWN REMARKS: The engine meets the minimum strength requirements for lifting and tie down provisions. When final configuration of the engine installed is established, all lifting and tie down provisions will have to be reevaluated.

Figure 16.5 Example PHS&T summary. Source: From US Department of Defense, *Acquisition Logistics, MIL-HDBK-502*, US Department of Defense, Washington, DC, 2013.

for organization and storage. Given that some of these data may be classified or proprietary, they also need to be securely stored with authorized access only. These data may also need to be shared with authorized users, or partners, so a secure data exchange method must be defined as well. In addition, the data must be controlled to ensure that it cannot be changed unless it follows the required configuration management process.

An often-overlooked area, at least in the commercial industry sector, is the control of software. The computer software used to operate assets is easily accessed and changed by maintenance staff. This must be addressed to ensure that only authorized changes take place and that identical assets are operated on the same version of software.

Organizing data is typically accomplished through the use of a data taxonomy. Data taxonomy is the classification of data into categories and subcategories, along with required data points. It provides a unified view of the data in an organization and introduces common terminologies and semantics across multiple systems. The data taxonomy ensures that there is a single source of truth for the data and enables organizations to monitor data quality and data completeness. According to the *IPS Elements Guidebook* by DAU, data are broken up into four primary categories [2]:

- Technical data describe product, interfaces, and decisions made; is traceable, responsive to changes, and consistent with CM requirements; is prepared and stored digitally; involves deciding what data are needed, and who controls it.

- Management data are data related to planning, organizing, and managing the project.
- Computer software documentation is a part of technical data management and is differentiated from the data category of "computer software." Computer software documentation refers to owner manuals, user manuals, installation instructions, operating instructions, and other similar documents, regardless of storage medium, that explain the capabilities of the computer software or provide instructions for using the software.
- Financial information and contract administration include contract numbers, payment due dates, contract payment terms, employee travel expenses, and contractor revenue .

The four categories are further broken down using a defined taxonomy (see Figure 16.6), which is further broken down into subcategories.

The Product Data category is further broken down into three subcategories, which include relevant information to the design for supportability of the asset. The subcategories are [2]:

- Product Definition Information: information that defines the product's requirements and documents the product's design and manufacturing. This is the authoritative source for configuration definition and control. Examples include drawings, 3D computer-aided design (CAD) models, and trade studies.
- Product Operational Information: information used to operate, maintain, and dispose of the product. Examples include records of maintenance actions, technical manuals, transportation information, and depot overhaul information.
- Product Associated Information: other product data such as test results, software or binary code embedded on memory chips, and proposed design drawing configuration changes that do not fit clearly into the other categories. This information is in the form of a living document that records the modifications, upgrades, and changes over the life cycle.

Additional requirements, such as naming, numbering, and attribute data, would be identified for each level in the data organization structure (see Figure 16.6), which is based on the economic value of the data, the cost to collect, clean, organize, and store the data. Due to the complex nature of data, it is highly recommended that the organization defines

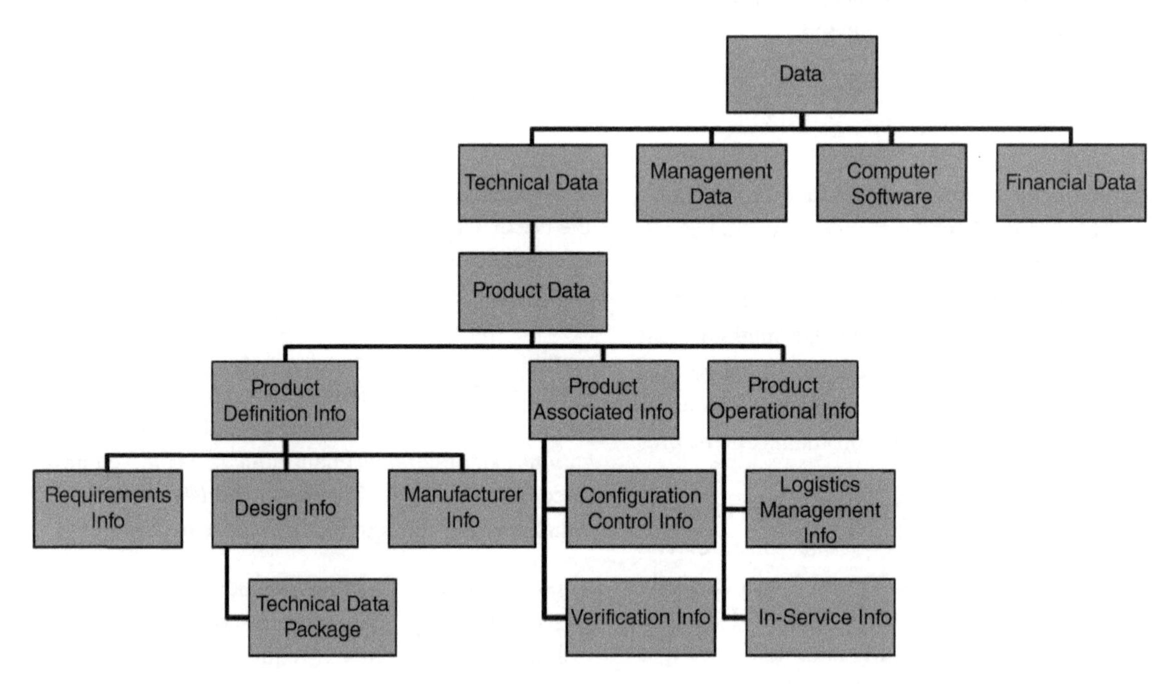

Figure 16.6 Data organization structure. Source: From *IPS Element Guidebook*, Defense Acquisition University, Fort Belvoir, VA, 2019.

what data are essential and ensures that these data are provided as part of the acquisition contract. Additional data points may be captured, but since they are not deemed essential, they are collected on an as-needed basis. Regardless of whether these data are or are not essential, the data must still be organized according to data structure and data taxonomy.

Having the right technical data is essential to enabling data-driven decision-making through the asset's life cycle. As such, technical data should be treated with the same level of care as the physical asset. The data enable the asset to be operated, maintained, and upgraded at the lowest possible TCO. These data also enable future assets or systems to be refined based on past experiences.

Being able to access the right data at the right time to make the right decisions does not happen by chance. Good data management also does not happen as a result of ordering excessive data just in case. Rather, an effective technical data strategy implementation is the product of an effective data management process [2].

16.2.8 Support Equipment

The support equipment element is in place to identify, plan, resource, and implement processes to acquire and support the asset with specialized equipment or tools. Support equipment is not limited to maintenance activities. It may include equipment used in storage facilities, which are necessary for personnel to perform their job. Support equipment categories include [2]:

- Ground support equipment
- Materials handling equipment
- Toolkits and tool-sets
- Metrology and calibration devices
- Automated test systems (general purpose electronic test equipment, special purpose electronic test equipment)
- Support equipment for on-equipment maintenance and off-equipment maintenance
- Special inspection equipment and depot maintenance plant equipment
- Industrial plant equipment

The support equipment element is critical to ensuring that assets are well maintained and properly calibrated in order to support the readiness and operational availability of the asset. Support equipment is important to understand because each piece of support equipment (see Figure 16.7) may represent its own "mini-acquisition" process within the asset's supportability program. In addition, the ability to purchase support equipment as part of the larger asset acquisition process usually leads to a lower price. Therefore, it is advantageous for the asset-owning organization to purchase the support equipment with the asset.

Another major goal of the support equipment element during the design phase is to minimize the need for new support equipment when support equipment for previous generations of assets is still suitable. By working with the Design Interface element, the need for specialized support equipment can be minimized.

16.2.9 Training and Training Support

The training and training support element is in place to plan, resource, and implement a strategy early in the asset development to train personnel to operate and maintain the asset throughout its life cycle. While identifying training can be relatively simple, there is much more to it. Training support involves identifying the TADSS to maximize the effectiveness of the manpower and personnel to operate and sustain equipment.

Training comes in many formats, which include formal and informal training activities. Formal training activities include instructor-led training, such as classroom, on-the-job training (OJT), eLearning, simulators, and refresher training. Informal training includes OJT and social learning through others. Regardless of the format, the training must be evaluated for effectiveness, and training programs updated to ensure effectiveness. The effectiveness of the training program can be related to the alignment between the asset design and the training curriculum. The more closely they resemble each other, the more effective the training will be. This is where having the right training aids comes in.

Training needs should be identified based upon who will be required to perform what specific tasks. This is often accomplished by using Bloom's *Taxonomy of Educational Objectives* [4], which is a hierarchical

SUPPORT AND TEST EQUIPMENT SUMMARY

SECTION I – TECHNICAL DESCRIPTION

SE Reference Number	CAGE	Item Name
5D43-139-A	10855	Compressor, Ring

Description And Function Of Support Equipment:
A band type sleeve with a mechanical leverage mechanism to facilitate easy reduction of ring radii.

Depth	Width	Height	UM	Weight	UM
4.0	5.0	5.0	In	3.5	Lb

Skill Specialty Code For SE	TMDE RAM Characteristics MTBF	MTTR	Cal Time	NSN and Related Data	Unit Cost
52C20	300 hrs	50 hrs	1 min.	5820-003478650	75.75

SECTION II - Unit Under Test Requirements

Supported IPC	Item Name	Characteristics I/O Parameter	Measured/ Stimulus Range From	Required Range To
005	Internal Compressor	Diameter in;	32	45
		Diameter	32	42

Figure 16.7 Example support equipment summary. Source: From US Department of Defense, *Acquisition Logistics, MIL-HDBK-502*, US Department of Defense, Washington, DC, 2013.

model used to classify educational learning objectives into levels of complexity and specificity. While a front-line maintainer may need to be trained at the application level, depot-level staff may need to be trained at the evaluation level. This is due to the fact that the depot troubleshoots and repairs equipment that the front-line maintainer will just replace. Knowing who will require what skills, the definition of training can begin. Often the training needs will be described (see Figure 16.8) in a way that calls-out who will need what skills, which may be based on rank, position, and so on. Often the training summary will also include prerequisites or education requirements as well.

Training should require that the student not only learn the theory but also the practical application of the theory. This will ensure that the operators and maintainers are well trained and able to support the asset in whichever operating environment they find themselves.

16.2.10 Manpower and Personnel

Building upon the MTA from the Maintenance Planning and Management element, the manpower and personnel needs can be identified from a maintainer standpoint. However, the manpower and personnel element must also consider the operator(s) of the asset. These requirements should be defined across all levels of maintenance (e.g., organizational, depot, etc.). Building upon the Manpower, Personnel, and Training Summary (see Figure 16.8), a metric is developed quantifying the specifics of "who will be required" and for "what duration." This metric is usually expressed in terms of hours or man-hours per year.

Manpower is the number of personnel and positions required to perform a specific task such as a maintenance activity or administrative functions. The analysts then determine the number of people required to perform that task over the course of a year. There are often multiple manpower models developed to reflect different operating environments or context.

MANPOWER, PERSONEL AND TRAINING SUMMARY

SECTION I – MANPOWER AND PERSONNEL SUMMARY

SSC	MAINTENANCE LEVEL	REQUIRED MAN-HOURS
35B20	OPER/CREW (C)	100.00
35B30	INT/DS/AVIM (F)	100.00
44E10	INT/DS/AVIM (F)	0.00
52C10	ORG/ON EQP (O)	25.00
52C20	ORG/ON EQP (O)	600.00
	INT/DS/AVIM (F)	1200.00
76J10	OPER/CREW (C)	50.00

SECTION II – NEW OR MODIFIED SKILL AND TRAINING REQUIREMENTS

ORIGINAL SSC	NEW/MOD SSC	DUTY POSITION REQUIRING NEW/MOD SKILL	RECOMMENDED RANK/RATE/GRADE MIL RANK CIVIL GRADE
52C10	52C20		

NEW OR MODIFIED SKILL REQUIREMENTS:

EDUCATIONAL QUALIFICATIONS:

ADDITIONAL TRAINING REQUIREMENTS:

Figure 16.8 Example manpower, personnel, and training summary. Source: From US Department of Defense, *Acquisition Logistics, MIL-HDBK-502*, US Department of Defense, Washington, DC, 2013.

Personnel requirements are defined by the knowledge, skills, abilities, and experience levels that are needed to properly perform the specific tasks [2]. These are often described with the use of job codes, ranks, positions, or a combination of these.

The goal of the manpower and personnel analyst is to arrive at the absolute minimum necessary personnel requirements for operation and maintenance of the asset based on the level of operational availability required. With manpower consistently the highest cost driver in many programs, typically at 67–70% of the program budget [2], it is vital that manpower costs are controlled. This may necessitate the use of contractors and internal staff to strike the right balance. As skills become more specialized, they are more expensive to build and retain. Often, very specialized skills cannot be developed internally without a large financial risk; therefore partnerships with contractors becomes a more cost-effective approach.

The manpower and personnel summary allows management to identify long-term recruitment and training of personnel to ensure that the system can be operated and maintained in the long term.

16.2.11 Facilities and Infrastructure

Facilities and infrastructure are often required to enable training, operation, maintenance, and storage of the assets. The facilities and infrastructure often consist of permanent and semi-permanent real property assets required to support a system. The facilities and infrastructure element identifies the types of facilities or facility improvements, location, space needs, environmental and security requirements, and equipment that will be needed. It includes facilities for training, equipment storage, maintenance, supply storage, and so forth [2].

The facilities should be defined by geographic location, type of facility, such as Naval Dry Dock, and the required size. To accomplish this, the supportability team must determine how many assets will be using the facility at any given point and must ensure there is ample room for storage of spare parts, training, and administrative areas. This is why it is vital to have not only the maintenance plan but also the supply requirements defined.

Once the facility is defined, this element also identifies what logistics and service support infrastructure is required in the facility – for example, what voltage plugs will be required, how many are needed, and where they should be located (e.g., on walls, in floors, on the ceiling, or suspended in the air). If compressed air is required, where should the access points be located, and what are the requirements for the compressed air system? This is often summarized in a facilities summary as shown in Figure 16.9.

Often facilities and infrastructure are reused for different types of assets, and as such, modifications may be required to accommodate the new assets. Whether the facilities and infrastructure are updated or constructed new, the lead time required is often substantial. Due to the lead time required for facilities and infrastructure, it is vital that the requirements be identified early to enable them to be ready for the arrival of the asset. Often, these requirements can be incorporated into the program cost if done early enough. If done later, there is often another project or acquisition activity required to construct the facilities and infrastructure.

16.2.12 Computer Resources

Computers and information technology touch every aspect of the asset life cycle, and as such must be planned for accordingly. Computer resources are not just the computer that the maintainer or operator will use, but includes all aspects of information technology/information support functions. The computer resources element must identify, plan, and resource all facility requirements, hardware, software, documentation, manpower, and personnel needed to operate and support mission-critical computer hardware/software systems [2]. As the primary end item, support equipment, or training device increases

FACILITIES SUMMARY

FACILITY NAME	**FACILITY CLASS**	**AREA**
Redstone Army Depot	Missile Repair Facility	15000 sq. ft.

ITEM NAME	**MAINTENANCE ACTION**
Wire Harness	test wire harness assembly repair wire harness assembly
Engine	repair engine assembly

1. **FACILITY LOCATION:**
   Redstone Arsenal, Huntsville, Alabama, Building 3441, Bay A.
2. **FACILITIES REQUIREMENTS FOR OPERATIONS:**
   Must rewire bay for forty 120 volts P/S spaced evenly along the walls.
3. **FACILITIES REQUIREMENTS FOR TRAINING:**
   2 work areas should be set aside for training
4. **FACILITY INSTALLED LEAD TIME:**
   2 years

Figure 16.9 Example facilities summary. Source: From US Department of Defense, *Acquisition Logistics, MIL-HDBK-502*, US Department of Defense, Washington, DC, 2013.

in complexity, more and more software is used. The expense associated with the design and maintenance of software programs is so high that one cannot afford not to manage this process effectively [2].

Computer resources, along with the technical data element, must also consider system security/information assurance as there is an ever-evolving cyber threat. Disaster recovery planning and execution is a requirement for mission-critical systems and will be driven by continuity of operations plans of the using organizations [2]. The need for system security must also balance the need for the transfer of data and information between systems with the use of Electronic Data Interchange (EDI). This is a constant challenge as commercial methods and standards will change many times during the operational life of an asset.

The role of information technology and computer hardware and software is becoming ever more integral to the operation and support of all complex systems. In fact, most complex systems can no longer function properly without their integrated information technology system operating correctly.

16.3 Supportability Program Planning

Supportability is about providing the 12 elements in the right balance with each other to provide the lowest TCO to the asset owner and operator. This is achieved through supportability planning; however, achieving that balance is never an easy task. While every development program is different, all supportability programs are driven by the same three major criteria: cost, equipment readiness, and manpower and personnel constraints.

Cost constraints are an inescapable economic reality. Obtaining high quality, capable, and affordable systems which meet user needs is the goal of all development programs [1]. Evaluating the TCO of a product requires consideration of all life cycle costs, as well as other acquisition costs. LCCA allows the comparison of different asset alternatives and their impact on the TCO. The LCCA should consider the support resources necessary to achieve specified levels of readiness (e.g., Operational Availability, A_0) for a range of assumptions regarding asset reliability and maintainability characteristics, usage rates, and operating scenarios. Because of the uncertainty in estimating resource costs like manpower and energy, sensitivity analyses should be performed. Sensitivity analyses help to identify and weight the various factors that drive life cycle costs. This knowledge is the key to understanding and managing program risk [1].

Equipment readiness is a measure of an organization's capability to ensure that the asset is able to achieve the required levels of A_0. Equipment readiness predictions are often the result of a simulation that takes into account all effects on A_0, such as logistical delays, usage levels of the asset, and so on. By using a Monte Carlo analysis, the supportability planning team can establish the important factors that contribute to the unavailability, as well as the level of Operational Availability that should be achieved.

Reductions in manpower and the increasing complexity of assets offer a significant challenge in acquiring affordable assets. Early consideration of manpower and personnel requirements is very important. Manpower and personnel constraints (quantities, skills, and skill levels) are major cost drivers of every total system and are as important as any other design consideration. Because of their potential impact on asset performance, readiness, and cost, all manpower and personnel requirements for new systems should be identified and evaluated early and alternatives considered. For example, use of commercial support for a low-use, highly complex product could eliminate most of the training costs associated with maintaining a qualified cadre of personnel in an environment with frequent personnel changes [1].

Estimates of manpower and personnel requirements for new systems are reported at each milestone decision point in the asset acquisition process. These requirements provide important input to resource plans, forecasts, and cost estimates, and help to formulate more cost-effective alternatives [1]. To account for these three major criteria, a supportability analysis is often used to strike the right balance, given the constraints, which will result in the lowest TOC.

16.3.1 Supportability Analysis

A supportability analysis, like many other analyses discussed in this book, is a continuously evolving process which starts in the concept phase, and continues through the design and development phase and into the O&S phase. As new information is learned about the design, the supportability analysis is updated, which in turn generates recommendations to improve the design to reduce supportability constraints. Even in the O&S phase of the asset, the supportability analysis will be updated to drive changes to any one of the 12 elements of supportability. For example, as the asset ages, there may be a need to stock more a particular type of spare part. The goals of supportability analyses are to ensure that supportability is included as a system performance requirement and to ensure that the system is concurrently developed or acquired with the optimal support system and infrastructure.

The supportability analysis is often made up of many different analyses, combined with modeling software to replicate what is likely to happen once the asset is deployed to the operating environment [5]. The integrated analyses can include any number of tools, practices, or techniques to realize the goals (see Figure 16.10). For example, Level of Repair Analysis (LORA), MTA, reliability predictions, RCM analysis, Failure Modes, Effects, and Criticality Analysis (FMECA), LCCA, and so forth, can all be categorized as supportability analyses [1].

The supportability plan must constantly be reviewed to ensure that it provides the optimized approach to ensure asset availability at the lowest TCO. One should not make the mistake of assuming that supportability is complete once the initial analysis is completed.

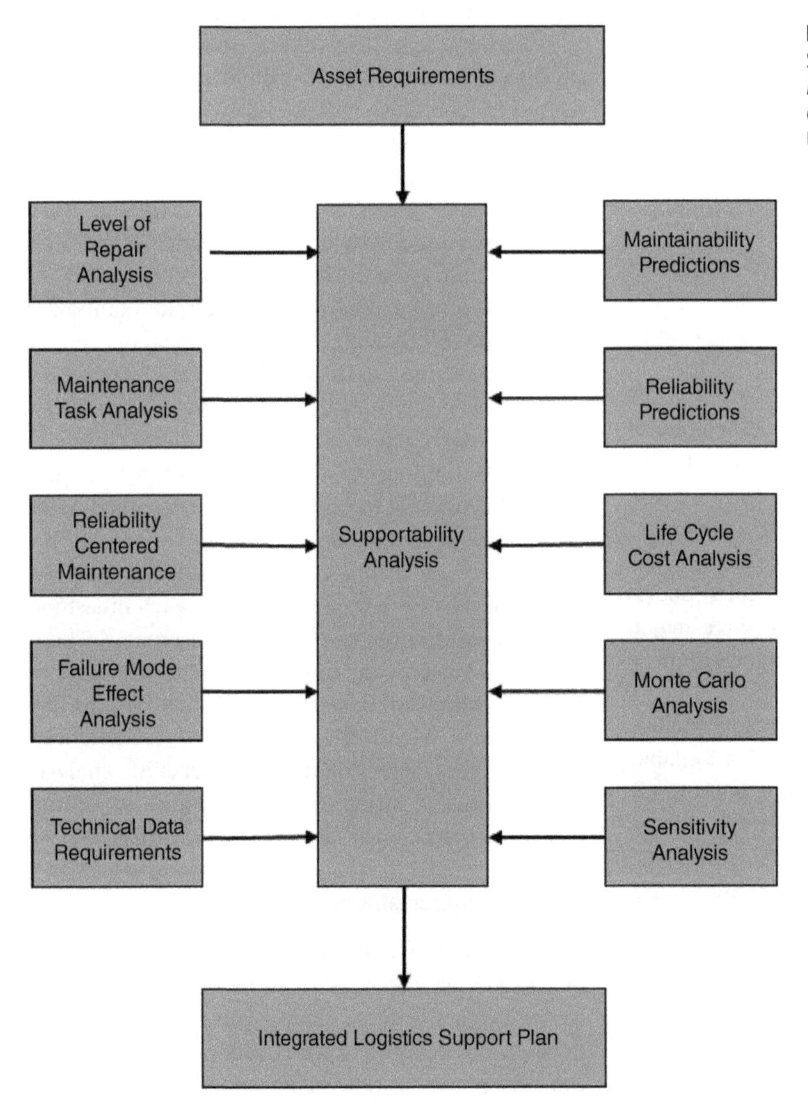

Figure 16.10 Supportability analysis. Source: Modified from Blanchard, B.S., *Logistics Engineering and Management*, 6th edition, Pearson Education Inc., Upper Saddle River, NJ, 2004.

Supportability analysis is an evolutionary process. As the system is developed and used over time through the life cycle phases, new or updated user requirements and new or revised authoritative directions or limitations may be established by the acquisition decision authority [1].

The output of the supportability analysis is the defined requirements as specified by the 12 elements. This is summarized in the ILS Plan, which the asset owning and operating organization can use to prepare for the asset.

16.4 Supportability Tasks and the ILS Plan

With the supportability analysis complete, the asset owning and operating organizations can begin to prepare for the arrival, operation, and maintenance of the asset. This is accomplished primarily through the use of the ILS Plan. The ILS Plan's objective is to ensure the asset is supported throughout its life cycle to achieve the lowest possible TCO while meeting operational requirements [5].

The ILS Plan is an integrated plan, which ensures that all 12 elements of supportability are included and that all elements are mutually supportive and non-duplicative in terms of funding, resources, and outcomes [2]. The ILS Plan is the end result of the supportability analysis, which takes into account the asset requirements, the constraints of cost, equipment readiness, and manpower and personnel needed to accomplish all required support tasks in the field applications. The ILS Plan should be a detailed plan which can be put to good use to ensure the Operational Availability of the asset in its operating environment can be achieved. The ILS will include specifics for:

- Supply support, including a list of a spare parts and materials for both the initial start-up and training of the assets as well as the operation and maintenance of the asset. Included with each spare part and material is a defined stock number, description, definition of what level of maintenance the item will be used, where the item should be procured, as well as estimated annual usage of the item. Supply support should also reference the PHS&T, technical data, and supply chain elements, where required.
- Maintenance planning and management, including a detailed maintenance plan with all preventive and corrective maintenance activities, schedules, and procedures. Where appropriate, maintenance planning and management may reference other sections such as supply support, manpower and personnel, support equipment, facilities and infrastructure, and training and training support.
- PHS&T, including a list of items requiring transportation, along with proposed methods of transportation, estimated demand rates, and the weight and size of each item. Proposed packaging methods, as well as any special handling requirements are also defined. The PHS&T may reference technical data, supply support, maintenance, and training elements.
- Technical data, including details for the classification, storage, retrieval, and security of each piece of data. Specifically, the plan should include any relevant data hierarchies, taxonomies, as well as a plan to capture, cleanse, and store each data point. Document and data control (e.g., design drawing configuration control) methodologies may also be included. The technical data may reference all other elements, especially computer resources, maintenance planning, and supply support.
- Support equipment, including a detailed list of all support equipment requirements for each level of maintenance. Also included is the plan to procure the support equipment, along with any calibration or maintenance requirements for the support equipment. This element may reference many others, specifically supply support, maintenance planning, and computer resources.
- Training and training support, including the requirements to establish a training program for both operators and maintainers. The plan should include recommended curricula, required training aids, location to acquire the training aids, as well as any recertification requirements. The initial training plan should also be included. The training and training support element may reference any of the other 11 elements.
- Manpower and personnel, including a detailed list of the required number of operators, maintainers, and proficiency levels to operate and maintain the asset.

The plan includes the use of contractors and provides recommendations for which contractors are to be used. The manpower and personnel element may reference the maintenance planning and the training elements.

- Facilities and infrastructure, including a list of property, plants, office, and warehouse requirements. Specifically, the plan should include the type, size, and construction of facilities. If an existing facility is to be used, the plan may include the capital plan to upgrade the current facility. This element may reference supply support, and maintenance planning.
- Computer resources, including a list of all required computers, software, and licensing required to support the asset throughout its life. The plan may also include procedures for updating software, and hardware. This element may reference technical data, supply support, and many others.

The ILS Plan is a fully integrated plan that must be implemented to realize the required operational availability of the asset. As such, all requirements must be identified to ensure the logistics and service support infrastructure is in place to operate and maintain the asset throughout its life.

Many ILS plans also include a verification portion, which verifies that the plan is delivering as expected, and where gaps occurs, the information is fed back to Sustaining Engineering and other parties to improve the supportability of the asset.

16.5 Summary

The ability to support the asset is both a product of the design of the asset, as well as the service support and logistics systems that are put in place to ensure the asset can be operated and maintained in its working environment. This chapter provided several examples to explain why it is important to consider design for supportability during the system design and development phase of a program. The reader will find further examples in this book, since so many applications of DfMn also benefit design for supportability.

Supportability does not happen by chance; it requires detailed analyses and a documented plan. The 12 elements of IPS are necessary to design for supportability. It is only with these 12 elements incorporated into an ILS Plan that the asset can ultimately achieve its operational objectives, while minimizing costs to the asset-owning organization.

References

1 US Department of Defense (2013). *Acquisition Logistics, MIL-HDBK-502*. Washington, DC: US Department of Defense.

2 Defense Acquisition University (2019). *IPS Element Guidebook*. Fort Belvoir, VA: Defense Acquisition University. https://www.dau.edu/tools/t/Integrated-Product-Support-(IPS)-Element-Guidebook- (accessed 18 September 2020).

3 US Department of Defense (2016). *Operating and Support Cost Management Guidebook*. Washington, DC: Department of Defense.

4 Bloom, B.S., Englehart, M.D., Furst, E.J. et al. (1956). *Taxonomy of Educational Objectives: The Classification of Educational Goals*. London: Longman, Green and Co. Ltd.

5 Blanchard, B.S. (2004). *Logistics Engineering and Management*, 6e. Upper Saddle River, NJ: Pearson Education Inc.

Suggestion for Additional Reading

US Department of Defense (1994). *Logistics Support Analysis, MIL-HDBK-1388*. Washington, DC: US Department of Defense.

17

Special Topics

Jack Dixon

17.1 Introduction

This chapter includes several topics that challenge the status quo of maintenance practices and offer insight on how to Design for Maintainability (DfMn) for the future. The topics included are:

- Reducing Active Maintenance Time with Single Minute Exchange of Dies (SMED) by James Kovacevic.
- How to Use Big Data to Enable Predictive Maintenance by Louis J. Gullo.
- Self-correcting Circuits and Self-healing Materials for Improved Maintainability, Reliability, and Safety by Louis J. Gullo.

The first topic, SMED, illustrates how using an old technique, one with origins in the 1950s, by applying it and other Lean techniques, can reduce maintenance times. It is a tribute to moving history from the past into the present by improving upon it.

While the SMED topic takes us from the old to the present, the two other topics in this chapter take us to future possibilities. The topic of Big Data opens up possibilities that machines of all kinds will know when they have a problem and will be able to relay that information to a human to help them perform the needed maintenance more quickly, efficiently, and cost-effectively. The final topic of self-correcting circuits and self-healing materials takes us to another new level of maintenance capabilities where a self-contained system contains the functionality to recognize that it has a problem when the problem first manifests, and then to cure the problem itself without external support.

So, enjoy your trip through this chapter from the historic to the futuristic.

17.2 Reducing Active Maintenance Time with Single Minute Exchange of Dies (SMED)

In the 1950s Toyota was not a very profitable organization, and as a result they had to rely on fewer pieces of equipment. This meant that in order to stamp new parts, Toyota routinely had to perform a change of the dies and tools in the presses. This changeover of dies would routinely take between two and eight hours. Obviously, two to eight hours of downtime, whether it is planned or not, is not the most efficient use of equipment, which in turn drives up costs. In addition, Toyota did not have the working capital to run large quantities of a specific part to reduce the need for these frequent die changes.

In order to address this, Toyota worked to reduce the time to change dies. This became known as Quick Die Change (QDC). QDC was based on a framework from the US World War II Training within Industry (TWI) program, called ECRS – Eliminate, Combine, Rearrange, and Simplify. Over time, the changeover time was reduced to 15 minutes in the 1960s and in

Design for Maintainability, First Edition. Edited by Louis J. Gullo and Jack Dixon.

some cases 3 minutes by the 1970s. The improvements made resulted in a significant competitive advantage for Toyota over the North American manufacturers.

SMED is a program that was built upon the Toyota QDC system. In the 1980s Shigeo Shingo brought the Toyota QDC system to North America and renamed it SMED. While the name may suggest that all changeovers should be completed in a single minute, it is not true. The goal of SMED to is reduce the planned downtime of a changeover to a single digit, that is, one to nine minutes. The SMED methodology can be applied to a variety of other organizational activities, such as planned maintenance [1].

Detecting potential failures, and correcting them before they lead to equipment downtime, is the primary purpose behind any Preventive Maintenance (PM) routine. It would stand to reason that the primary purpose of PM optimization is to detect more potential failures, earlier on the potential failure–functional failure curve (P-F curve) (see Figure 17.1). A potential failure is defined as "an identifiable condition that indicates that a functional failure is either about to occur or is in the process of occurring." A functional failure is defined as "a state in which an asset or system is unable to perform a specified function to a level of performance that is acceptable for its user" [2].

The P-F curve is used to demonstrate at which point a potential failure occurs, and when the equipment will reach a state of being functionally failed. Point P1 on the P-F curve represents the point when a potential failure occurs, while the various other P points represent the various times that the potential failure could be detected with various techniques such as vibration analysis, visual inspections, and so on. The P-F interval is the time required to move from the various P points to a complete functional failure. An understanding of the P-F curve is vital to establishing the right maintenance strategies.

Many common reliability tools, such as Reliability-centered Maintenance (RCM), Failure Modes and Effects Analysis (FMEA), and Failure Modes, Effects, and Criticality Analysis (FMECA) allow organizations to focus on identifying and detecting the potential failures. Utilizing these tools would improve the effectiveness of the PM routine, essentially improving the maintainability and availability of the equipment. These concepts are covered in depth in Chapter 15. In addition, through careful analysis using these tools, the required downtime can be predicted for both planned and unplanned downtime.

Paradigm 7: Maintainability predicts downtime during repairs

PM optimization is often thought of as an activity to improve the effectiveness of the maintenance strategy. While this is the first step, and a vitally important step to improving any maintenance strategy, it is not the only step. There are other steps. Stopping after this first step may prevent an organization from achieving their goals or best-in-class performance.

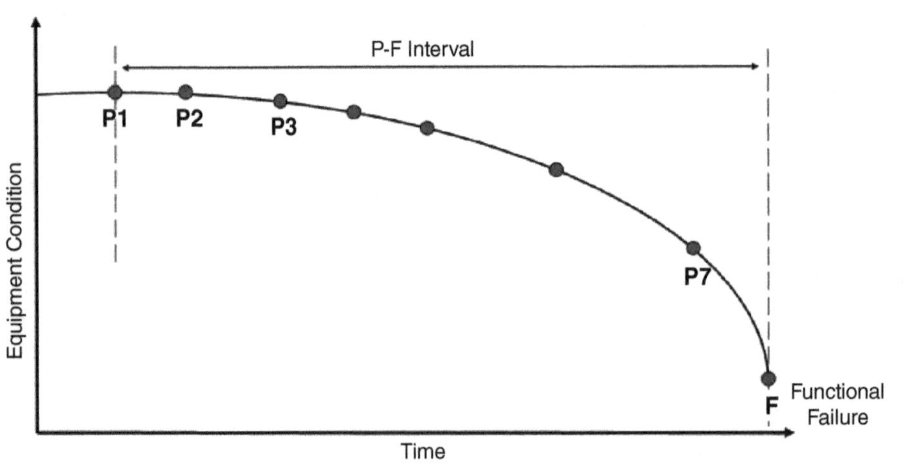

Figure 17.1 The P-F curve.

What is often overlooked is the efficiency of the PM routine or maintenance strategy. For an organization to move to world-class levels, they not only need to do the right maintenance, but they also need to do it efficiently. This is primarily where maintenance planning and scheduling comes to mind. Without planning and scheduling, the amount of inefficiency in the part of the organization responsible for maintenance is extraordinary. So, what is the second level to PM optimization?

17.2.1 Incorporating Lean Methods into PM Optimization

Once planning and scheduling have been implemented to remove waste from the maintenance process, then incorporating "Lean" into your PM routine is the next step. This is where the second level of PM optimization can unlock the hidden potential of the organization's maintenance function. When PM optimization is combined with known and accepted Lean Manufacturing (LM) [1] techniques, the efficiency of maintenance is truly unlocked.

To understand why the efficiency of PM routines is important, it is necessary to grasp some of the key metrics [3]:

- **Overall Equipment Effectiveness** (OEE) = Availability (%) × Performance Efficiency (%) × Quality Rate (%), where:
  - ○ Availability = [Uptime (hrs)]/[Total Available Time (hrs) – Idle Time (hrs)]
    - Uptime (hrs) = Total Available Time (hrs) – [Idle Time (hrs) + Total Downtime (hrs)]
    - Total Downtime = Scheduled Downtime (hrs) + Unscheduled Downtime (hrs)
  - ○ Performance Efficiency = Actual Production Rate (units per hour)/Best Production Rate (units per hour)
  - ○ Quality Rate = (Total Units Produced – Defective Units Produced)/Total Units Produced
- **Total Effective Equipment Performance** (TEEP) = Utilization Time (%) × Availability (%) × Performance Efficiency (%) × Quality Rate (%), where:
  - ○ Utilization Time = [Total Available Time (hrs) – Idle Time (hrs)]/Total Available Time (hrs)
  - ○ Availability = [Uptime (hrs) × 100]/[Total Available Time (hrs) – Idle Time (hrs)]
    - Uptime (hrs) = Total Available Time (hrs) – [Idle Time (hrs) + Total Downtime (hrs)]
    - Total Downtime = Scheduled Downtime (hrs) + Unscheduled Downtime (hrs)
  - ○ Performance Efficiency = Actual Production Rate (units per hour)/Best Production Rate (units per hour).
  - ○ Quality Rate = (Total Units Produced – Defective Units Produced)/Total Units Produced

As can be seen in both of these key metrics, when the time to perform maintenance (e.g., scheduled downtime) is reduced, the OEE or TEEP increases, which means the organization can produce more product with the existing assets. This is why there needs to be a focus on improving the efficiency of PM routines. Organizations should first utilize RCM, FMEA, and/or FMECA to ensure the maintenance strategy is effective.

17.2.1.1 Understanding Waste

The first step to introduce the second level of PM optimization with a focus on efficiency is to educate the team that will be focused on reducing the planned time to complete the maintenance. The maintenance planning team needs to understand the different types of waste, and the known Lean techniques use to eliminate waste. The eight types of waste [4] are:

- **Defects** – waste resulting from mistakes introduced during a PM activity which causes start-up failures, rework, or extended start-up time. These defects could be caused by poor instructions, a slip of the mind or the complexity of the job.
- **Overproduction** – performing excess or extra work during a PM activity. This may include moving items to get access to equipment, removing guarding to access lubrication points, or long shutdown or start-up equipment times.
- **Waiting** – waste of time due to unnecessary pauses or delays in the performance of maintenance activities. This time waste typically includes waiting for permits to be issued from a government agency, waiting for operations personnel to release equipment for maintenance purposes, and waiting for help from others either in the form

of skilled personnel assistance or detailed work instructions.

- **Not utilizing talent** – waste of talent caused by not using the right people for the job, either using over-skilled or under-skilled people. Is it value added to have skilled tradespeople perform routine cleaning tasks? Can something be visually inspected by the operator normally assigned to a piece of equipment, rather than assigning to a skilled maintenance technician?
- **Transportation** – waste caused by moving things around for accessibility purposes. Are the parts kitted and ready for the PM? Are spare parts located within close proximity to the equipment being maintained? Does the job require multiple items to be moved to gain access to the equipment that requires maintenance?
- **Inventory excess** – waste typically caused by bringing extra parts or spare items for a job in case parts are lost or damaged during the course of the maintenance task. Often these extra parts, if not used for the job, never make it back to the storeroom and become missing. These missing parts could be a source for Foreign Object Debris (FOD) that unknowingly are placed inside the equipment being maintained. To prevent FOD, operators are usually instructed to count the number of spare parts taken to the line where work is performed, and count the spare parts that return to the parts storage area for inventory control.
- **Motion waste** – waste usually caused by an inefficient sequence of work and a PM routine's order of activities. An example is a mechanic that constantly moves from one side of a machine to the other side, going repeatedly back and forth, instead of organizing all of the work to be performed on one side first, and then moving to the other side of the machine to accomplish the remainder of the work.
- **Excess processing** – this type of waste occurs when the activities are not simplified or require excess approvals, for example, having to have multiple sign-offs for completed work. Another example is not utilizing visual factory methods to simplify the inspection process.

Once the maintenance planning team is aware of the waste, there needs to be an unrelenting focus on eliminating the waste and minimizing planned downtime. This is where Lean techniques really come in.

17.2.1.2 Apply Lean Techniques to Eliminate Waste

In order to eliminate the waste from a PM routine, a Lean technique called SMED could be utilized [5]. SMED is a system for dramatically reducing the time it takes to complete equipment changeovers, which are the various tasks that will take place during the planned downtime. The essence of the SMED system is to convert as many changeover steps as possible to "external" (e.g., performed while the equipment is running) and to simplify and streamline the remaining steps. This very same approach can be used on PM routines to improve efficiency.

The steps to an SMED are quite simple and can be applied to a PM optimization as described and shown:

1. Measure scheduled downtime – Measure and set a goal to reduce the scheduled downtime. To measure the time, list all of the tasks and record the time it takes to perform each one (see example in Figure 17.2). Validate the time estimates by video-recording the actual PM routine as it is being performed. Following the PM routine, debrief the video with the team performing the PM routine, to learn what they did right or wrong, and record the specific start and end time of each PM routine activity.
2. Separate internal and external downtime – With the current time established, the tasks need to be separated by internal and external tasks. Internal tasks can only be done when the equipment is locked out and not operating. Next, identify what tasks can be done while the equipment is operating – these are known as external tasks. This step is vital to determine what activities required planned downtime.
3. Convert internal tasks to external tasks – Carefully review each internal task and determine whether there is a way that the task can be done with the equipment running. Start with the simple tasks; do not jump to re-design right away. Ensure all preparation work (such as gathering materials, specifications, tools, etc.) is done before the equipment is shut down. There may be times that re-design is an option. Consider a lubrication task

PM ACTIVITY OBSERVATION SHEET Date: Machine: Residual Slop Management				PM Routine: Weekly Operator Round
No Step by Step action element	**Can be Split**		**Timing**	**Observations**
	Int	Ext	Unit:	
1 Go to Centrifuge			4	
2 Go to Tool Crib			3	
3 Get Tool Kit			3	Operator waiting for turn at tool kit
4 Go to Centrifuge			3	
5 Perform visual inspect on gauges			5	Operator had no reference of specification for the readings
6. Shutdown Centrifuge			2	
7 Perform lubrication			10	Had to remove guarding to access lubrication points
8 Startup centrifuge			2	
9 Go to Pump			3	
10 Shutdown pump			2	
11 Clean out strainer			10	Toolkit did not have the right wrench to access the strainer
12 Restart pump			2	
13 Monitor flow rates of pump			2	
14 Go to Silo			3	
15 Perform visual inspection			6	
Total:			60	

Figure 17.2 Example SMED task sheet.

that required the machine to be shut down. By adding some lubrication lines, the lubrication task can now be done while the equipment is running, which is ideal from a lubrication best-practice standpoint as well. Re-designing guarding and access points are also common approaches to changing internal to external tasks. Due to the cost of re-design, a cost–benefit analysis must be performed to determine whether the time saving (in terms of lost production) is worth the cost of the re-design. An example is shown in Figure 17.3.

Other methods to convert internal to external tasks can include the use of jigs to calibrate or adjust specific elements offline, before being installed. Modularization is another approach, which enables routine adjustments, repairs, and inspections to be performed offline and quickly swapped during the planned downtime. The goal in this step is to reduce the total downtime so that it is at, or close to, the goal established previously. The various methods are used to reduce or eliminate the wastes in both internal and external tasks.

Paradigm 6: Modularity speeds repairs

4. Eliminate internal waste – During this phase, identify how each task can be simplified or eliminated. One way to eliminate internal waste is to sequence the tasks in the PM routine to prevent the technician from moving around from side to side of the equipment unnecessarily. Using the video-recording taken during the PM routine in step 1, a map can be created of the actual path (also called a spaghetti diagram) taken by the staff during the PM routine. This diagram will allow the team to visualize the amount of time and steps spent going around the equipment (see Figure 17.4). Once the team is aware of the lost time, they can begin to

PM ACTIVITY OBSERVATION SHEET

Date:

Machine: Residual Slop Management — PM Routine: Weekly Operator Round

No Step by Step action element	Can be Split Int	Ext	Timing Unit:	Observations
1 Go to Centrifuge			4	
2 Go to Tool Crib				
3 Get Tool Kit				g for turn at tool kit
4 Go to Centrifuge				
5 Perform visual inspect on gauges			5	Operator had no reference of specification for the readings
6 Shutdown Centrifuge	X		[illegible]	
7 Perform lubrication	X			brication points
8 Startup centrifuge	X		2	
9 Go to Pump		X	[illegible]	
10 Shutdown pump	X		2	
11 Clean out strainer	X		10	Toolkit did not have the right wrench to access the strainer
12 Restart pump	X		2	
13 Monitor flow rates of pump		X	2	
14 Go to Silo		X	3	
15 Perform visual inspection		X	6	
Total:	28	32	60	

Install Remote Lube Lines to eliminate the need to shutdown

Install second strainer in parallel to enable cleaning while running

Figure 17.3 Example SMED task sheet with internal to external conversion.

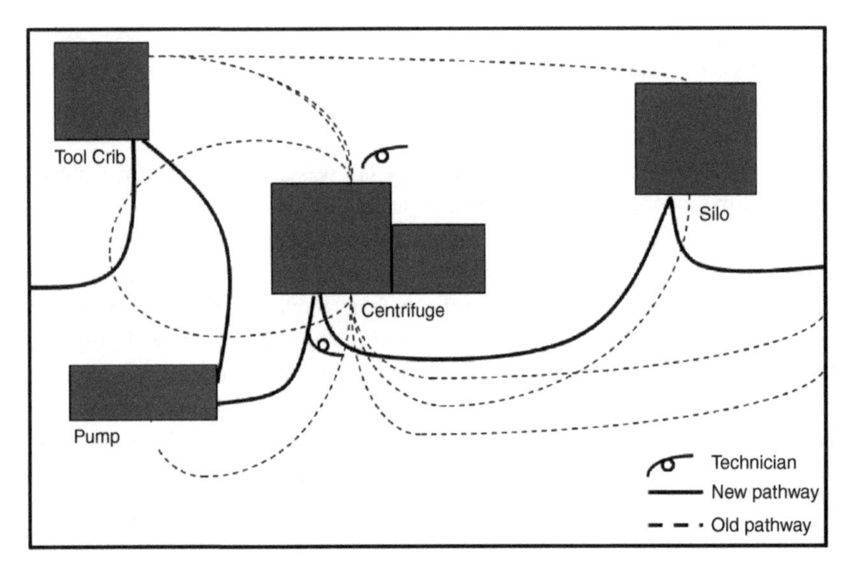

Figure 17.4 Example spaghetti diagram.

sequence the task in a way that minimizes the travel performed by the technician.

Once the new maintenance sequence has been established, there needs to be a focus on how to shorten the time spent on the tasks. For example, use quick-turn fasteners in place of bolts, use stop blocks to reduce the time to position a part, or have all settings/specifications clearly defined. This is also another opportunity to use a jig, if swapping out components for calibration or adjustment is not an option. If the PM routine requires extended downtime, consider adding technicians to perform the tasks in parallel to reduce the overtime scheduled downtime. If possible, eliminate adjustments or potential adjusts and replace with fixed points. Lastly, try to standardize hardware to reduce the number of tools required to perform the PM.

5. Eliminate external waste – While this step will not reduce the scheduled downtime, it will increase the overall efficiency in the maintenance department. Look at decentralized tool and part storage to reduce travel times, establish a standard pick list (i.e., a list of materials and supplies required to complete the task), simplify paperwork, and lastly, have a good procedure to ensure that the technicians have all of the parts/supplies/tools ready (see example in Figure 17.5).
6. Standardize and maintain best practice – Once the best practice for the PM routine has been developed, document it in a standard job plan. Take this new job plan and try it out. Record the process, as in step 1, and look for further opportunities to work out any issues in the new routine. Once the team is happy with the new PM routine, continue to track and compare the actual scheduled downtime with the planned downtime to identify opportunities for further improvement.

17.2.1.3 Continually Improve the PM Routine

The focus on improving efficiency extends past the initial PM optimization and should become part of the PM routine itself. At the conclusion of the PM routine, a quick analysis should be performed to identify any sources of waste and what can be done to reduce or eliminate it.

PM ACTIVITY OBSERVATION SHEET

Date:

Machine: Residual Slop Management — PM Routine: Weekly Operator Round

No Step by Step action element	**Can be Split** Int	Ext	**Timing** Unit:	**Observations**
~~1 Go to Centrifuge~~		~~X~~	~~4~~	
2 Go to Tool Crib		X	3	
3 Get Tool Kit		X	3	Operator waiting for turn at tool kit
4 Go to Centrifuge		X	3	
5 Perform visual inspect on gauges		X	5	Operator had no reference of specification for the readings
~~6. Shutdown~~ Centrifuge	~~X~~		~~2~~	
7 Perform lubrication		X	5	Had to remove guarding to access lubrication points
~~8 Startup centrifuge~~	~~X~~		~~2~~	
9 Go to Pump		X	3	
~~10 Shutdown pump~~	~~X~~		~~2~~	
11 Clean out strainer		X	7	Toolkit did not have the right wrenchto access the strainer
~~12 Restart pump~~	~~X~~		~~2~~	
13 Monitor flow rates of pump		X	2	
14 Go to Silo		X	3	
15 Perform visual inspection		X	6	
Total:	8	44	52	

Implement Visual Factory for gauge readings

Figure 17.5 Example SMED task sheet with external waste reduction.

The human elements are typically much quicker and less expensive to implement than design changes. Changing the sequence, separating internal and external activities, standardizing roles, and developing procedures and specifications are examples of the human elements. Usually, when working with maintenance staff, they will look to implement re-designs and other technical elements, so it is imperative that the facilitator focuses the team on the human element first.

Paradigm 5: Consider the human as the maintainer

The optimization process may also be completed over multiple iterations. Often a full optimization requires extensive time and resources and is met with some skepticism. If these constraints are encountered, perform the first and second steps and try the first revision to demonstrate some results and build buy-in by the management and maintenance teams. Once the benefits and results are observed, take it to the next step. After the first one or two optimizations have taken place, the organization will be more likely to support full optimizations moving forward.

Another word of caution when deciding which equipment to perform these analyses on; it is important to focus only on the critical assets or bottlenecks when utilizing SMED for PM routines. Not every piece of equipment needs to have PM optimization performed to this level. The equipment that the organization needs to squeeze every ounce of uptime out of should be focused upon first.

17.2.2 Summary

In order to drive up the maintainability and availability of the assets, organizations need to focus not only on doing the right maintenance, but ensuring that the maintenance is effective and efficient. This improvement in efficiency is used to drive a reduction in planned downtime, ensuring an increase in availability. Without this focus, organizations are at risk of developing extensive maintenance routines that require lengthy planned downtime, which results in lower availability and higher operating costs.

Conducting the SMED analysis after the asset is built is not ideal. While it can drive a reduction in planned downtime, organizations would benefit greatly from including this type of review during the design of the asset. It is much more cost-effective to make changes to the asset in the design phase than after the asset is built.

The SMED technique is covered in extensive detail in various forms of published literature describing Design for Manufacturing (DfMa). DfMa is concerned with manufacturability and producibility as engineering disciplines conducted during the development phase of a program. DfMa focuses on engineering a design for ease of assembly during the process of initial build of a system or product. DfMa describes the methods to ensure a product design is economically produced when it is transitioned to the manufacturing environment or factory. DfMa integrates the design of the system or product with the successful manufacturing concepts of Just-In-Time (JIT) and Lean Manufacturing to assemble and test new designs in a high-volume manufacturing flow. SMED offers value to DfMn as well as DfMa.

In conclusion, when the eight types of waste are targeted and reduced using Lean techniques, the amount of planned downtime is reduced. This allows the organization to achieve even higher levels of availability, OEE or TEEP, allowing an increase in the performance of their assets.

17.3 How to use Big Data to Enable Predictive Maintenance

Imagine a world where your electronics, such as your phone or tablet, knows that it has a problem, a health pain, and can predict impending debilitation. In this world, your electronics warns you that it has a problem (pain), and needs a remedy, just like the human body warns the brain that it feels a pain, and the person visits a medical doctor. The electronics might be in the form of a robot, and the robot has sensors to assist in its motions as well as diagnose components that have exceeded their stress limits. In this world, the electronics would need hundreds of sensors with millions of bits of data accumulated every hour to have the self-awareness of its pain conditions and be able to adequately inform an electronics repair person. This type of electronics requires lots of data processing power with lots of data memory and storage. The solution to handling this data processing is Artificial Intelligence (AI) and Machine Learning (ML), (see details on this topic already covered in Chapter 8). For the handling of large amounts of data, data processing capabilities with data buffering, memory, and storage, discussion on the topic of Big Data [6] is necessary.

17.3.1 Industry Use

Big Data is a topic that was briefly mentioned in Chapter 8, but has not been explained elsewhere in this book, so it is covered here. Until now, Artificial Intelligence and Machine Learning were discussed for their abilities to process data autonomously without regard for the size of the data-set being processed. If the data-set is relatively small in size, small enough for a personal computer (PC) to handle, then the time to process is relatively short. On the other hand, if the data-set is relatively large, such that a PC could not process the data in a few hours and because the time to process the data goes up considerably as the data size increases, then new complexities are introduced which require different technologies to work with the systems that would benefit by the incorporation of Artificial Intelligence and Machine Learning. These complexities only occur with extremely large data-sets. These complexities must be dealt with in the design of the electronics, otherwise the system will not be able to effectively and efficiently process the data. The design of the system can vary widely, depending on the type of industry that wants to use it and the benefits that industry is looking to obtain. Here are some examples of systems in industries that use Big Data and the general benefits that Big Data offer with regards to maintainability.

Industries Using Big Data

- Aviation
- Energy (e.g., power generation, wind farms, smart grid)
- Manufacturing
- Railroads
- Automotive (e.g., smart cars)

General Benefits

- Enables predictive maintenance
- Enables optimization
- Use in spare parts management
- Asset management
- Internet of Things (IoT)

Big Data reduces costs through predictive maintenance in the commercial aviation industry, as one example of an industry-specific benefit that is revolutionizing the way in which businesses are able to make decisions [7]. "As new aircraft generate more in-flight data compared to older ones, innovative analysis methods summarized by Big Data Analytics enable the processing of large amounts of data in short amount of time. Recent studies show a reduction of maintenance budgets by 30 to 40% if a proper implementation is undertaken. As a result, predictive analytics of flight recorded data is an exciting and promising field of aviation that airliners are starting to develop. Unlocking the valuable information within this data is referred to as Data Mining" [7].

Businesses rely on the IoT as a technology with connectivity from the Edge to the Cloud, using a robust telecommunications system able to handle large-scale data transfers. The role of Big Data in IoT is to process and store large amounts of data on a real-time or near-real-time basis. IoT processes involving Big Data may follow these four steps:

1. Enterprise-level data storage system leverages wired or wireless telecommunications networks to collect raw data with sizes ranging from gigabytes (GB) to trillions of GB, which are generated by numerous IoT devices for any particular business customer over a specific period of time and frequency of occurrence.
2. A widely distributed database system extracts and manages the large amounts of time-stamped data in files and folders using tight configuration controls and software fault-tolerant architecture for handling data errors.
3. Application tools reduce down and analyze the data to provide value-added solutions for customers. Big Data is analyzed using any one of a plethora of tools, such as Hadoop, or tools used specifically for Machine Learning or predictive analytics. "Hadoop

is an open-source software framework for storing data and running applications on clusters of commodity hardware. It provides massive storage for any kind of data, enormous processing power and the ability to handle virtually limitless concurrent tasks or jobs" [8].

4. The final step is report creation, where manual report-writing instructions are performed, or automated routines generate reports of analyzed data or digital dashboards for continuous parameter metrics tracking.

Big Data almost always includes some level of automation or autonomics from the initial collection and storage of the raw data to the generation of reports that contain actionable information or knowledge for which customers pay high prices. These reports summarize the results of enormous volumes of data that exist, which can be used to figure out why and how things happen, and then use the results to predict what will happen.

17.3.2 Predicting the Future

In the case of a catastrophic event prediction, you can never have enough data. The bigger your data-set, the more accurate are your predictions of a future event. The success of predictive analytics depends on the confidence in the data, which relates to the accuracy and the size of the data-set. Maybe it is not obvious how this claim of success in using Big Data to predict the future is possible. Consider this example:

You are a manager of a fleet of delivery trucks that is dispatched every day to pick up and deliver consumer products from distribution centers to retail stores where shelves are stocked every night with these products. Whether you know it or not, your job depends on the availability of Big Data to ensure you are successful. Big Data is accumulated and accessible to those with a will to use it for their benefit and the benefit of their company. If you take your trucks to a service center under a maintenance contract to repair your trucks when needed, and keep your trucks in good working condition with regular periodic scheduled maintenance, then it can be reasonably assured that the service center collects lots of data on your truck repair and maintenance histories. If the data-set is large, and it is merged with even larger data-sets for other customers to create a pool of data for like-type vehicle models, then this could be your Big Data opportunity to mine for gold to enable your success as the Fleet Manager. Now the question is where to start your mining for gold using Big Data. Let's start with the basic type of Big Data for service centers that you can immediately use to help your situation – maintenance records.

By applying data science and predictive analytics to the Big Data in the service center for all trucks with the same vehicle model and configuration, maintenance records and work order history gives you knowledge of future failure occurrence well beyond the data you could collect for your limited fleet, providing you with a holistic view of the health of your trucks. This holistic view allows you to score each truck against each other to flag the ones with the highest probability of a failure occurrence for the most severe repair issues. Furthermore, this holistic view provides intelligent equipment maintenance through the development of a critical maintenance item hit-list of the top priority service issues for each truck. The purpose of this hit-list is to ensure preventive maintenance can be performed to thwart a costly repair action out on the road in between deliveries, and avoid delivery delays due to poor availability of your trucks to meet your customers' needs. The bottom line is that Big Data is the key to improving the customer-to-fleet relationship.

Assuming you have a mobile phone with a long-term wireless Internet Service Provider (ISP) contract for data uploads and downloads, you can use your phone to alert you to when a critical component in one of your trucks in the fleet of trucks that you manage is about to fail. To generate the alert on your phone, a Condition-based Maintenance (CBM) system (see Chapter 9 of this book for more details of a CBM system) assesses your trucks' health and provides a status of each truck's probability of failure on a daily basis. The CBM data are combined with the results of oil sample viscosity testing, the last date of scheduled periodic preventive maintenance inspection, the results of the last inspection, and any outstanding actions from the last three inspections. Now you can decide when it is economically feasible to bring a truck into the maintenance facility for the critical component part replacement, while completing the outstanding actions that were previously recommended for servicing at the next available opportunity.

"The system then recommends when and how you should replace the part, incorporating weather data to suggest the optimal day to perform the necessary maintenance – identifying a window of time when rain is forecasted [9]."

Unbeknownst to you, the critical part inventory is low in the maintenance shop where your vehicle must be repaired. The needed part is a high failure rate item that gets replaced often, so the demand for this part is high and the supply of this part needs replenishment often. At the same time that the alert was sent to you, the alert also went to the maintenance facility's parts stock room to generate an order to purchase more of the same parts to increase the inventory. This order could be filled in a few days without jeopardizing the downtime exposure for your truck when it arrives at the maintenance facility for the preventive repair work, and ensure it will be back on the road making deliveries to satisfy your delivery requirements and timelines.

"The value of data in equipment maintenance cannot be understated. Without data-driven insights, maintenance processes are manual, time-consuming, and difficult to perform, in a best-case scenario. As a worst-case scenario, maintenance processes are inefficient and costly in terms of time and money lost."

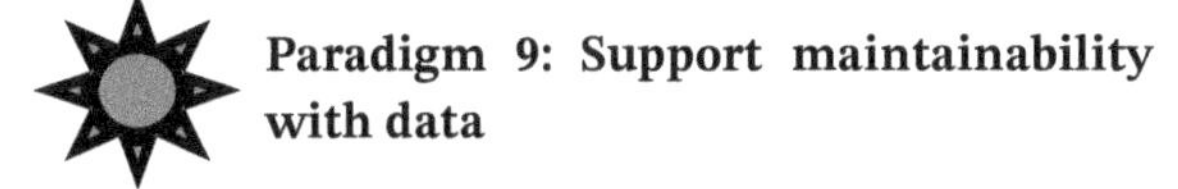

Paradigm 9: Support maintainability with data

Knowing in advance of a part failure by collecting symptomatic data measuring critical part performance parameters to identify early in the degradation process, and to estimate the best time to fix the part, is infinitely more valuable than knowing when the part has already failed.

Not all equipment data are created equal. Visual inspection data might be valuable to the user if an indication of wear-out or signs of overstress are apparent without performing a test. An example of a sign of overstress using an unintentional visual indication is discoloration on a printed circuit board (PCB). This visual indication is due to the PCB getting hotter than normal at the location on the PCB where a component has failed. The hot spot is seen as a slightly darker colored area on the PCB from the rest of the normal PCB color that is seen over the rest of the board surface. When these hot spot visual indications are found during inspections, they should be documented in the failure data as symptoms of the failure.

If no obvious visual inspection indication anomalies are present in the data, then more specific data to assess the health of the circuits must be collected through health checks, status monitoring of critical performance parameters, and diagnostic testing. The data-sets that may be collected at any one time may amount to 1 Mb (1 million bits or 1 megabit) in data size. Then, if this data-set is collected every minute, the size of the data-set for one day will be 1.44 Gb (1.44 gigabits). Referring back to the fleet of delivery trucks example, this is the data-set for one truck in the fleet for one day. Now multiply this data size for 100 trucks in the fleet and one year of operation per truck. The data-set size grows to a tremendous 52 560 Gb or 52.56 Tb (terabits). Handling this amount of data requires a unique type of data science. It requires a specific mindset to build a data-handling system, and an economic method to understand what the data say, determine how to interpret the data, and maximize its value over a period of time to realize a return on investment.

17.3.3 Summary

This section addresses the facts that Big Data is here, and there needs to be an automated way to deal with it. For the benefits to be achievable for maintenance cost reductions using IoT technology and Machine Learning, the four steps described to process and store large amounts of data on a real-time or near-real-time basis must be established. By following these steps to handle Big Data, a business will truly be able to unlock the benefits of IoT and Machine Learning in designing for maintainability.

17.4 Self-correcting Circuits and Self-healing Materials for Improved Maintainability, Reliability, and Safety

In designing for maintainability, the designer wants to deliver a system or product to the customer with less maintenance than the customer had to previously perform for predecessor systems or products that did a similar function. For maintainability success, the fewer repairs you do, the better the system or product is. More embedded intelligence, automated processes, and robust design features that reduce maintenance are good from the customer perspective.

There are many real-world situations, some of which have already been discussed in this book, where maintenance reduction improves reliability and reduces the safety risks of a design. The same is also true where improvements in reliability and safety design characteristics minimize maintenance. If a product or system never fails and never experiences an accident or mishap, the corrective maintenance time to repair the product or system is zero. PM or CBM tasks might be partly the reason for this outcome, so some type of maintenance may be necessary. Even though the design may be architected for some level of PM or CBM, the ultimate preferred situation is no maintenance.

Paradigm 4: Migrate from scheduled maintenance to Condition-based Maintenance (CBM)

The challenge in maintainability is how to design a system or product where maintenance is minimized or avoided. Self-correction techniques may be the solution to this challenge. Self-correction means that a self-contained system or product has embedded functionality to recognize an internal problem when design characteristics coded in data parameters begin to show a problem has surfaced, and to remedy this problem without external support.

Zero maintenance is possible with self-correcting techniques such as autonomous design capabilities, intelligent materials for self-healing, machine learning, error detection and correction algorithms, remote health monitoring, predictive analytics, and fault-tolerant architecture. Some of these topics have already been discussed in the book, such as the use of fault-tolerant architecture and Error-correcting Code (ECC) memory in Chapter 11.

17.4.1 Self-correcting Circuits

In this section, ECC memory is explained in more detail with an emphasis on the overall strategy for self-correcting circuits. Maintenance can be reduced or potentially eliminated with advanced technology and self-correcting electronic circuits. Along with self-correcting electronic circuits, we also explore self-healing materials, leveraging physics and special intelligent material properties.

The ultimate goal in designing for zero maintenance is to self-correct problems discovered in the field. Some of the solutions to develop self-correcting problems have existed for some time, but there is far more capability needed, which is still in early research. Some of this research is in the enhancements of integrated circuit (IC) chip design. "The first part is to analyze the design using machine learning algorithms and embed Agents that continuously create new data on the chip's profiling, health, and performance," Brousard said (quoted in [10]). This initial stage in the research for self-correcting circuits in IC chip design is imperative for the success of autonomous vehicles, like self-driving cars. "The algorithms that run the autonomy platform in a car are generally made by machine learning. They're not human-written, and they're not easily dissectible or readable for a software team to go in and code check" (Phillips, quoted in [10]). For more information on Machine Learning, see Chapter 8.

A big part of self-correcting circuits in IC chip design is the use of ECC, for the data transmitted between components in a computer, such as between the central processing unit (CPU) and its memory or data storage devices. ECC memory is a type of recording media for digital data storage that can detect and correct different types of data corruption symptoms, such as the random flipping of bits as seen with Multiple Bit Errors (MBEs), or Single Bit Errors (SBEs) where one bit is stuck low or stuck high. SBEs where a single bit is randomly flipped is also possible [11].

As an example of a SBE with a random bit flip error that could be ignored by a system with no error-checking, or would halt a machine with parity checking, or would be invisibly corrected by ECC, consider a numerical spreadsheet storing numbers in ASCII format. The ASCII data are loaded into the CPU. The spreadsheet data contain the character "8" (decimal value 56 in the ASCII encoding), which is stored in the byte at a specific memory location that contains a stuck bit at its lowest bit position (0011 1000 binary). Then, new data is written to the spreadsheet in memory by the CPU, and the new data is saved. The new data was not intending to change the value of the "8" stored in memory, but as a result of the spreadsheet data update, the "8" (0011 1000 binary) has unintentionally changed to a "9" (0011 1001).

Electrical or magnetic interference inside a computer system can cause an SBE or MBE within Dynamic Random-Access Memory (DRAM) to spontaneously flip bits from one state to the opposite state, such as a bit change from a logic "0" to a logic "1." This type of SME or MBE, which is sometimes referred to as a "soft error," was initially discovered to be caused by alpha particle emissions from contaminants in IC chip packaging material. The latest research found that the majority of these SBE/MBE soft errors in DRAMs occur as a result of background cosmic radiation, causing free neutron scatter, which randomly changes the data in memory locations, or interferes with the CPU circuitry that reads or writes the data to memory. The error rates may rapidly increase as the altitude at which the system is operating increases. For example, the rate of neutron flux is 3.5 times higher at 1.5 km altitude compared with the normal levels of neutron flux at sea-level. Furthermore, the neutron flux is 300 times higher at 10–12 km altitude compared with the normal levels of neutron flux at sea-level. As a result, systems operating at high altitudes, such as commercial airline systems, require special provision for enhanced reliability and maintainability, and therefore are big users and proponents of self-correcting circuits [11].

ECC memory is usually immune to SBEs, so the data read by a CPU from each memory location contain the exact stream of bits, forming bytes or words, as the data stream that was written to that memory location. ECC also reduces the number of CPU hang-ups or processing crashes, or potentially eliminates the possibility of these CPU hang-ups or crashes. ECC memory is used with many types of CPUs where these types of MBE and SBE data corruption cannot be tolerated.

"The consequence of a memory error is system-dependent. In systems without ECC, an error can lead either to a crash or to corruption of data; in large-scale production sites, memory errors are one of the most common hardware causes of machine crashes. Memory errors can cause security vulnerabilities. A memory error can have no consequences if it changes a bit which neither causes observable malfunctioning nor affects data used in calculations or saved. A 2010 simulation study showed that, for a web browser, only a small fraction of memory errors caused data corruption, although, as many memory errors are intermittent and correlated, the effects of memory errors were greater than would be expected for independent soft errors [11]."

As stated earlier in this book, fault tolerance can be achieved by designing a system that anticipates every type of exceptional condition and aims for self-stabilization so that the system converges towards an error-free state. If a fault-tolerant system anticipates every fault condition and executes a process to mitigate the effects of the fault without alerting an operator or maintainer to perform a maintenance action, it is said that the system performs with zero maintenance. Another approach for fault-tolerant architecture is redundancy. Redundant hardware and software components allow for failures to occur in a particular system operation by leveraging spare capabilities or backup features. For more information on fault-tolerant architecture, refer to Chapter 11.

17.4.2 Self-healing Materials

Self-healing materials are artificial or synthetically-created substances, which are able to repair damage to themselves without any external diagnosis or human support. Generally, materials will degrade over time due to fatigue or damage incurred during operation. Cracks and other types of damage on a microscopic level have been shown to change the properties of materials. The propagation of cracks eventually leads to physical failure of the material. "Self-healing materials counter degradation through the initiation of a repair mechanism that responds to the micro-damage. Some self-healing materials are classed as smart structures, and can adapt to various environmental conditions according to their sensing and actuation properties" [12].

The most common types of self-healing materials are polymers or elastomers, but self-healing covers all classes of materials, including metals, ceramics, and cementitious materials. Self-healing mechanisms may vary from an intrinsic repair of the material to the addition of a repair agent contained in a microscopic vessel. Self-healing polymers may be activated in response to an external stimulus, such as visible or ultraviolet light, or temperature, to initiate the healing processes [12].

A study on the damage sensing and self-healing of carbon fiber polypropylene (CFPP) and carbon nanotube (CNT) nanocomposite was performed and written about in a paper in 2018 [13]. This work, titled: "Damage detection and self-healing of carbon fiber polypropylene (CFPP)/carbon nanotube (CNT) nano-composite via addressable conducting network," was based on an Addressable Conducting Network (ACN). To increase damage sensing resolution of CFPP/CNT nanocomposite, electrical conductivity was improved by adjusting press conditions and by spraying CNTs between prepregs. Prepregs are composite materials of a high-strength reinforcement fiber, which is pre-impregnated with a thermoset or a thermoplastic resin [13].

From the results, electrical resistivity in thickness direction was reduced to 19.44 Ω-mm under 1.0 MPa and 1.0 wt% of CNT condition. Also, self-healing efficiency was examined with respect to the temperature and time via resistive heating of the CFPP/CNT nanocomposite. As a result, the optimized fabrication and self-healing condition exhibited high resolution of damage sensing with outstanding self-healing efficiency (96.83%) under the fourth cycle of a repeated three-point bending test [13].

Arizona State University's (ASU) Adaptive Intelligent Materials and Systems (AIMS) Center is a leader in innovation research in the United States [14]. The ASU AIMS Center has a wide spectrum of innovation that includes advanced self-healing and self-sensing materials. The Center is researching and making advances into intelligent materials that hold promise for self-healing properties. Some of these intelligent materials are [15]:

- Novel hybrid materials such as light-responsive mechanophores for damage sensing and self-healing, self-assembled emulsions, colloids, and gels for their sensitivity and responses to environmental changes, and ion-containing block copolymers for stimuli-responsive solution assemblies and hierarchical separation membranes.
- Soft matter and composites for organic opto-electronics, electrically conductive polymer sensors/actuators, and biologically inspired material.
- Carbon nanotube (CNT) membranes such as buckypaper (BP) with self-sensing properties that activate on the formation of cracks.

In one particular study at ASU, BP membranes were found to provide self-sensing capability in glass fiber reinforced polymer matrix laminates. This self-sensing capability was achieved by embedding BP membranes in the interlaminar region of the laminates. Piezo-resistive stress–strain characterization studies were conducted by subjecting the self-sensing specimens to cyclic mechanical fatigue loading. The fatigue crack growth rates and crack propagation observations in baseline and experimental specimens were compared. A measurement model for real-time quantification of fatigue cracking was developed using resistance measurements. This empirical model was used to quantify and simulate fatigue crack length and crack growth rates in real time. The results from ASU show that the introduction of buckypaper reduced the average crack growth rate by an order of magnitude as a result of crack tip blunting during fatigue, while facilitating real-time strain sensing and damage quantification [15].

17.4.3 Summary

We are living in exciting times, considering all the advancements that are being made with self-correcting electrical circuits and self-healing materials. The future looks bright for these technologies. There will soon be components placed on systems and equipment in harsh environmental applications that will be able to self-correct and self-heal to automatically recover from failure modes when they occur.

17.5 Conclusion and Challenge

Several special topics and their applications have been covered in this final chapter that hopefully will stimulate the reader to consider these topics and dream about what the future of maintainability engineering may bring and what they can bring to it. As the SMED article demonstrated that an old technique can lead to new improvements, so too can other applications offer similar opportunities that could reap similar, or even larger, benefits. Likewise, Big Data and self-correcting systems open up many possibilities for innovation. Innovation can come from transitioning technology from one application, where the technology is mature and proven to work, to other applications that no one has thought of doing before. Such innovation may help to solve large problems facing the world today. It is up to the imagination of the reader to discover those opportunities. We encourage you to explore the possibilities, use the techniques described in our book, and become that innovator.

References

1 Roser, C. (2014). *The History of Quick Changeover (SMED)*. All About Lean. https://www.allaboutlean.com/smed-history/ (accessed 18 September 2020).

2 Nowlan, F.S. (1978). *Reliability Centered Maintenance*. San Francisco: United Airlines.

3 Society of Maintenance & Reliability Professionals (2017). *SMRP Best Practices*, 5e. Atlanta, GA: SMRP.

4 McGee-Abe, J. (2015). The 8 Deadly Lean Wastes. http://www.processexcellencenetwork.com/business-transformation/articles/the-8-deadly-lean-wastes-downtime.

5 Lean Production. (n.d.). *SMED (Single-Minute Exchange of Dies),* http://www.leanproduction.com/smed.html (accessed 6 October 2020).

6 Wikipedia Big data. https://en.wikipedia.org/wiki/Big_data (accessed 18 September 2020).

7 Exsyn Aviation Solutions Big Data in Aviation – Reduce Costs thru Predictive Maintenance. https://www.exsyn.com/blog/big-data-in-aviation-predictive-maintenance (accessed 18 September 2020).

8 SAS Hadoop: What is it and why it matters. *SAS Insights*. https://www.sas.com/en_us/insights/big-data/hadoop.html (accessed 18 September 2020).

9 Novak, Z. (2018). How Big Data Makes Predictive Equipment Maintenance Possible. https://www.uptake.com/blog/how-big-data-makes-predictive-equipment-maintenance-possible-1 (accessed 18 September 2020).

10 Sperling, E. (2019). Different Ways to Improve Chip Reliability. *Semiconductor Engineering*. https://semiengineering.com/different-ways-to-improving-chip-reliability/ (accessed 18 September 2020).

11 Wikipedia ECC memory. https://en.wikipedia.org/wiki/ECC_memory (accessed 18 September 2020).

12 Wikipedia Self-healing material. https://en.wikipedia.org/wiki/Self-healing_material (accessed 18 September 2020).

13 Joo, S.-J., Yu, M.-H., Kim, W.S., and Kim, H.-S. (2018). Damage detection and self-healing of carbon fiber polypropylene (CFPP)/carbon nanotube (CNT) nano-composite via addressable conducting network. *Composite Science and Technology* **167**: 62–70. https://doi.org/10.1016/j.compscitech.2018.07.035.

14 Arizona State University (ASU) Adaptive Intelligent Materials and Systems (AIMS) Center brochure. https://aims.asu.edu/ (accessed 18 September 2020).

15 Datta, S., Neerukatti, R.K., and Chattopadhyay, A. (2018). Buckypaper embedded self-sensing composite for real-time fatigue damage diagnosis and prognosis. *Carbon* **139**: 353–360. https://doi.org/10.1016/j.carbon.2018.06.059.

Suggestions for Additional Reading

Gullo, L.J. and Dixon, J. (2018). *Design for Safety*. Hoboken, NJ: Wiley.

Raheja, D. and Allocco, M. (2006). *Assurance Technologies Principles and Practices*. Hoboken, NJ: Wiley.

Raheja, D. and Gullo, L.J. (2012). *Design for Reliability*. Hoboken, NJ: Wiley.

Appendix A

System Maintainability Design Verification Checklist

A.1 Introduction

This checklist provides a wide-ranging collection of suggested guidelines, design criteria, practices, and check points that draws from numerous knowledgeable sources and references. Among these sources are a plethora of Department of Defense (DoD) MIL-STDs and MIL-HDBKs, various military guidance documents, industry training and guidance manuals, white papers and reports, and many years of experience of the authors.

This checklist is intended to provide these guidelines, design criteria, practices, and check points for thought-provoking discussions, design trade studies, and considerations by a design team. It can instigate design requirements generation for specification of maintainability design aspects of the product, its use, and its maintenance, and to verify the acceptance of the implementation of these design requirements in the product design. This checklist is not intended to be a complete, all-encompassing list of all possible maintainability design considerations for any and all products for any customer. The reader may wish to tailor this checklist for the items that are applicable to the particular product, customer, use application, and service support situation.

A.2 Checklist Structure

This checklist is divided into multiple sections to categorize similar types of questions and design considerations into common groupings. The sections are listed:

Section 1: Requirements Management
Section 2: Accessibility
Section 3: Tools
Section 4: Maintenance
Section 5: Software
Section 6: Troubleshooting
Section 7: Safety
Section 8: Interchangeability
Section 9: Miscellaneous Topics

Design for Maintainability, First Edition. Edited by Louis J. Gullo and Jack Dixon.

Checklist Section 1: Requirements Management

Section 1: Requirements Management	VERIFY	REMARKS
1.1 Prior to beginning the design process, did the design team identify the customer and define who the users are?		
1.2 Prior to beginning the design process, did the design team use the customer's Concept of Operations (CONOPS), or was the CONOPS for the product defined by the design team?		
1.3 Prior to beginning the design process, did the design team define where and how the product would be used, transported, maintained, and stored?		
1.4 Did the design team create maintainability design requirements using appropriate requirements management processes and tools?		
1.5 Did the design team manage maintainability requirements using appropriate requirements management and change control processes/tools?		
1.6 Did the design team decompose the product concept of operations and maintenance into a set of design maintainability requirements for the system?		
1.7 Did the design team institute maintainability best practices, leveraging existing maintainability requirements derived from historical maintainability definitions, analyses, and maintenance management processes?		
1.8 Did the design team create maintainability requirements that are specific, singular, unambiguous, quantifiable, and measurable? Ensure each requirement is allocated to and traceable to specific product features or parts.		
1.9 Did the design team ensure that system-level maintainability requirements are complete and written at the appropriate level? Good requirements creation and management dictates that all requirements are traceable to a higher level requirement all the way to the top system-level requirements.		
1.10 Did the design team allocate lower-level requirements (i.e., subsystem, function, group, unit, LRU, component, etc.) to the appropriate product part and ensure that it is linked to the higher-level requirement?		
1.11 Did the design team ensure that each maintainability requirement is allocated to and traceable to specific product features or parts?		
1.12 Are the maintainability requirements decomposed to the lowest level of maintenance for the lowest repairable component or assembly that is realistic for the product?		
1.13 Do all derived maintainability requirements completely and accurately define the interpretation and understanding of the original requirement?		
1.14 Are all derived maintainability requirements treated the same as any other maintainability requirement?		
1.15 Did the design team separate out compound, multi-part requirements into singular requirements and ensure that each new requirement is traceable to the higher multi-part requirement and allocated to the product as any other requirement?		
1.16 Prior to system or unit testing, did the design team audit the requirements to verify and document (via the checklists or a requirements management tool) that these requirements have been satisfied?		

Checklist Section 2: Accessibility

Section 2: Accessibility	VERIFY	REMARKS
2.1 Does the product ensure accommodation, compatibility, operability, and maintainability by the intended maintainers and users?		
2.2 Does the product ensure adequate clearance for movement for ingress and egress of the work area?		
2.3 Does the product ensure adequate visibility to perform all required maintenance operations?		
2.4 Are the components laid out so they are accessible, with the emphasis for easy access placed on items that require frequent inspection and maintenance?		
2.5 Are components that require frequent visual inspection, check points, adjustment points, and cable-end connectors located in positions that can be easily viewed?		
2.6 Are component labels located in positions that can be easily viewed?		
2.7 Are small hinge-mounted units which require access to the back to open their full distance and be capable of remaining open without being held open?		
2.8 Are fuses located so they can be seen and replaced without tools or removing other parts or subassemblies?		
2.9 Do structural members or permanently installed equipment visually or physically obstruct adjustment, servicing, removal of replaceable equipment, or other required maintenance tasks?		
2.10 Are check points, adjustment points, test points, cables, connectors, and labels accessible and visible during maintenance?		
2.11 Are items requiring visual inspection (hydraulic reservoirs, gauges, etc.) located so personnel can see them without removing panels or other components?		
2.12 Are access openings provided for all equipment or components that require testing, calibrating, adjusting, removing, replacing, or repairing large enough to accommodate hands, arms, tools, and provide full visual access to the task area?		
2.13 Are access covers equipped with grasp areas or other means for opening them that accommodates gloves or special clothing that may be worn by the maintainer?		
2.14 Is direct tool access provided to allow for torqueing without the use of irregular extensions?		

Checklist Section 3: Tools

Section 3: Tools	VERIFY	REMARKS
3.1 Is the equipment designed to minimize the numbers, types, and complexity of tools required for maintenance?		
3.2 Does the product design minimize the need for maintainers to acquire and use special test equipment?		
3.3 Does the design minimize the number and type of different tools required, focused on utilizing the existing tool-set of the intended maintainers?		
3.4 Early in the design process, did the team determine and define what tools the intended maintainers have or will use?		

Checklist Section 4: Maintenance

Section 4: Maintenance	VERIFY	REMARKS
4.1 Does the design minimize permanent connections (such as soldering) for parts, connectors, and wiring?		
4.2 Does the design ensure that the analyses for part repair vs. replace are documented, accurate and consistently applied?		
4.3 Does the design ensure that the repair levels for the product match the repair capabilities of the maintainers for the product?		
4.4 Does the design provide remote access to the product for operational checks, performance, and status reporting, and maintenance activity, such as remote BIT operation and remote updating of software?		
4.5 Are replaceable items capable of being removed and replaced without having to remove or replace other items or components?		
4.6 Does the design provide for equipment removal and replacement by one person where permitted by structural, functional, and weight limitations?		
4.7 Does the product design allow for ease of stowage and removal for maintenance purposes under any and all extreme environmental conditions, especially harsh cold weather where bulky clothing items are worn?		
4.8 Does the design incorporate error-proofing in equipment mounting, installing, interchanging, and connecting?		
4.9 Does the design provide for the use of speed-enhancing tools (e.g., ratchets, powered screwdrivers, or power wrenches) to accommodate torque requirements considering space limitations?		
4.10 Does the design provide quick-connect and quick-disconnect devices that fasten and release easily, without requiring tools?		
4.11 Does the design provide for square-head or hexagonal-head screws which provide better tool contact, have sturdier slots, and can be removed with wrenches compared to round or flat screws?		
4.12 Are valves or petcocks used in preference to drain plugs? Where drain plugs are used, do they require only common hand tools for operation?		
4.13 Are lifting handles provided to make lifting and carrying easier?		
4.14 Where there are calibration or adjustment controls which are intended to have a limited degree of motion, are there mechanical stops sufficiently strong to prevent damage by a force or torque 100 times greater than the resistance to movement within the range of adjustment?		
4.15 Where the operator or maintainer is subjected to disturbing vibrations or acceleration during the adjustment operation, are suitable hand or arm supports provided near the control to facilitate making the adjustment?		
4.16 Are captive fasteners used where dropping or losing fasteners could cause damage to equipment or create a difficult or hazardous removal problem?		
4.17 Are captive fasteners provided for access covers requiring frequent removal?		
4.18 Does the design provide for, whenever possible, identical screw and bolt heads to allow panels and components to be removed with one tool?		
4.19 Does the design provide for captive washers and lock washers when loss would otherwise present a hazard to equipment or personnel?		
4.20 Are replaceable items removable along a straight or slightly curved line, rather than through an angle?		

Section 4: Maintenance	VERIFY	REMARKS
4.21 For items which must be precisely located, does the design provide guide pins, or their equivalent, to assist in alignment of the item during mounting?		
4.22 For items that incorporate rack and panel connectors, does the design provide guide pins, or their equivalent, to assist in alignment during mounting?		
4.23 Does the design ensure that rollout racks pulled to the fully extended position do not shift the center of gravity to the point where the rack or console becomes unstable?		
4.24 Does the design provide for rollout racks that automatically lock in both servicing and operating positions?		
4.25 Does the design provide limit stops on racks and drawers that are required to be pulled out of their installed positions?		
4.26 Does the design provide for cables to be labeled to indicate the equipment to which they belong and the connectors with which they mate.		
4.27 Does the design prevent fluid and gas lines from spraying or draining fluid on personnel or equipment?		
4.28 Does the design provide fluid and gas cutoff valves at appropriate locations to permit isolation or drainage of the system for maintenance or emergency purposes?		
4.29 Does the design provide plugs requiring no more than one turn, or other quick-disconnect plugs, whenever feasible?		
4.30 Does the design provide connector designs in which it will be impossible to insert the wrong plug into a receptacle or to insert a plug into the correct receptacle the wrong way?		
4.31 Where high torque is required to tighten or loosen the connector, does the design provide space for use of a connector wrench?		
4.32 Does the design provide test points that are permanently labeled to identify them in the maintenance instructions?		
4.33 Does the design provide test points that are located on surfaces or behind accesses which may be easily reached or readily operated when the equipment is fully assembled and installed?		
4.34 Do all stored energy devices have a hazard warning sign attached to the device?		
4.35 In establishing the available workspace, does the design provide considerations as to the number of personnel required to perform the work and the body positions required to do the work?		
4.36 Does the design include considerations for how components should be arranged and located to provide rapid access to those components with lower reliability, that will probably require maintenance most frequently, or whose failure would critically degrade the end item's performance?		
4.37 Did the design team identify tools and test equipment already in the operational support environment that can be made available for maintenance of the product?		
4.38 Does the design include modular or unit packaging or throw-away components and techniques?		
4.39 Does the design utilize self-lubricating principles?		
4.40 Does the design utilize sealed and lubricated components and assemblies?		
4.41 Does the design provide Built-In-Test (BIT) with embedded diagnostics and calibration features for major components and functions?		

Section 4: Maintenance	VERIFY	REMARKS
4.42 Does the design provide self-adjusting mechanisms?		
4.43 Does the design minimize the number and complexity of maintenance tasks?		
4.44 Does the design provide for quick recognition of malfunctions or marginal performance?		
4.45 Does the design provide for quick identification of the replaceable defective components, assemblies, and parts?		
4.46 Does the design minimize skills and training requirements of maintenance personnel?		
4.47 Does the design provide for maximum safety and protection for personnel and equipment?		
4.48 Does the design facilitate manual handling required during maintenance, and comply with established manual force criteria?		
4.49 Does the design permit maintenance from above and outside, in contrast to requiring access for maintenance from underneath?		
4.50 Does the design result in equipment with no more than two-levels-deep mechanical packaging to expedient assembly and disassembly?		
4.51 Does the design minimize the need for torque wrenches and incorporate self-locking nuts and bolts, as appropriate?		
4.52 Does the design, wherever possible, utilize self-aligning bearings instead of ball caps and sockets in worm gear mechanisms?		
4.53 Are related subassemblies grouped together as much as possible?		
4.54 Does the design ensure that maintenance required on a given unit or component can be performed with the unit or component in place, where possible, and without disconnection, disassembly, or removal of other items?		
4.55 When it is necessary to place one unit behind or under another, does the design ensure the unit requiring the most frequent maintenance is most accessible?		
4.56 Does the design ensure irregular, fragile, or awkward extensions, such as cables and hoses, are easily removable before the unit is handled?		
4.57 Does the design include tapered alignment pins, quick-disconnect fasteners, and other similar devices to facilitate removal and replacement of components?		
4.58 Does the design include quick-connect and quick-disconnect devices which fasten or unfasten with a single motion of the hand?		
4.59 Does the design provide an obvious indication when quick-connect and quick-disconnect devices are not correctly engaged?		
4.60 Does the design incorporate, whenever possible, fasteners that require a minimum of repetitive motion for removal and installation, such as quarter-turn connect and release fasteners?		
4.61 Does the design include captive fasteners where "lost" screws, bolts, or nuts might cause a malfunction or excessive maintenance time?		
4.62 Does the design incorporate cables or lines which must be routed through walls or bulkheads for easy installation and removal without the necessity for cutting or compromising the integrity of the system?		
4.63 Does the design protect electrical wiring from contact with fluids such as grease, oil, fuel, hydraulic fluid, water, or cleaning solvents?		
4.64 Where cable connections are maintained between stationary equipment and sliding chassis or hinged doors, does the design provide service loops to permit movement, such as pulling out a drawer for maintenance without breaking the electrical connection?		

Section 4: Maintenance	VERIFY	REMARKS
4.65 Does the design provide for service loops that have a return feature to prevent interference when removable chassis are replaced in the cabinet?		
4.66 Does the design provide test and service points that offer positive indication, by calibration, labeling, or other features, of the direction, degree, and effect of the adjustment?		
4.67 Does the design provide test and service points that offer lead tubes, wires, or extended fittings to bring hard-to-reach test and service points to an accessible area?		
4.68 Does the design provide components that require little or no preventive or scheduled maintenance actions?		
4.69 Does the design minimize mechanical adjustments?		
4.70 Have the design choices that affect maintenance processes been reviewed, considering any simplified alternatives?		
4.71 Can the product technical manuals be understood by an average person with a junior high school (seventh or eighth grade maximum) education?		
4.72 Can the product technical manuals be understood by an average person where English is not their first language?		
4.73 Did the design team conduct trade studies to consider minimizing components to design a function (where more components decrease reliability), while considering functional requirements for redundancy (where more components improve reliability)?		
4.74 Does the design feature oil sight gages that are positioned to be directly visible to the service crew without the use of special stands or equipment?		
4.75 Does the design feature wrenching functions designed for the same size wrench? Same torque values?		
4.76 Does the design provide quick disconnects for hydraulic, fuel, oil, and pneumatic line couplings for all components subject to time replacement or minimum service life and for all modular components?		
4.77 Does the design avoid the use of parts or materials that are known to have caused reliability and maintainability (R&M) problems in earlier designs?		
4.78 Does the design feature diagnostic techniques that are simplified?		
4.79 Does the design include access openings without covers where they are not likely to degrade performance?		
4.80 Does the design provide units placed so that structural members do not prevent access to them?		
4.81 Does the design include components placed so that all throwaway assemblies or parts are accessible without removal of other components?		
4.82 Does the design feature units laid out so that maintenance technicians are not required to retrace their movements during equipment checking?		
4.83 Does the design feature provisions for support of the units while they are being removed or installed?		
4.84 If the design has delicate fragile parts, does the design include rests or stands on which units can be set to prevent damage to delicate parts during maintenance?		
4.85 Does the design consider human strength limitations in designing all devices that must be carried, lifted, pulled, pushed, and turned?		
4.86 Does the design include illumination for internal parts where necessary?		
4.87 If fuses are clustered, is each one identified?		
4.88 Are fuses located so that they can be seen and replaced without removal of any other item?		

Section 4: Maintenance	VERIFY	REMARKS
4.89 Does the design provide handles on smaller units that may be difficult to grasp, remove, or hold without using components or controls as handholds?		
4.90 Does the design include handles on transit cases to facilitate the handling and carrying of the unit?		
4.91 Does the design provide that all handles are placed above the center of gravity and positioned for balanced loads?		
4.92 Does the design include recessed handles located near the back of heavy equipment to facilitate handling?		
4.93 Does the design include handles located to prevent accidental activation of controls?		
4.94 Does the design provide for cables to be routed so they need not be sharply bent or unbent when being connected or disconnected?		
4.95 Does the design incorporate cable and wire bundles that are routed so they cannot be pinched by doors or lids, or so they will not be stepped on or used as handholds by maintenance personnel?		
4.96 Does the design include electrical wiring that is routed away from all lines that carry flammable fluids or oxygen?		
4.97 Does the design include direct routing of wires and cabling away from congested areas wherever possible?		
4.98 Does the design feature modules with as much self-fault testing and isolation capability as possible?		
4.99 Does the design feature modules designed to maximize the potential for discard-at-failure, rather than repair, for all modules planned for on-equipment replacement?		
4.100 Does the design include long-lived parts that are not scrapped because of failure of short-lived parts?		
4.101 Does the design include situations where the failure of inexpensive parts results in the disposal of an expensive module?		
4.102 Does the design include low-cost and noncritical items that are made disposable?		
4.103 Does the design feature diagnostics strategies which minimize "cannot duplicate," "bench checked serviceable," "retest OK," and false alarm conditions?		
4.104 Does the design feature diagnostics techniques which maximize vertical testability (i.e., from system, to subsystem, to subassembly, to part level)?		
4.105 For test points that require test probe retention, does the design include fixtures to hold test probes so that the technician will not have to hold the probe while working?		
4.106 Are test points located so the technician operating the associated control can read the signal on display?		
4.107 Are test points coded or cross-referenced with the associated units to indicate the location of faulty circuits?		
4.108 In the design process, did test point selection, design, and testing partitioning play a major role in the layout and packaging of the system?		
4.109 In the design, is each test point labeled with a name or symbol appropriate to that point?		
4.110 In the design, is each test point documented in maintenance instructions with the in-tolerance signal or limits that should be measured, along with their tolerance ranges for the measured values?		

Section 4: Maintenance	VERIFY	REMARKS
4.111 In the design, are test lead connectors used that require no more than a fraction of a turn to connect?		
4.112 In the design, are test points located close to the controls and displays with which they are associated?		
4.113 Are test points planned for compatibility with the maintenance skill levels involved?		
4.114 Does the design test architecture and implementation provide fault isolation to the desired replacement level?		
4.115 In the design, are filler areas for combustible materials located away from sources of heat or sparking, and are spark-resistant filler caps and nozzles used on such equipment?		
4.116 In the design, are fluid-replenishing points located so there is little chance of spillage during servicing, especially on easily damaged equipment?		
4.117 In the design, are filler openings located where they are readily accessible and do not require special filling adapters or work stands?		
4.118 In the design, are drain points located to permit fluid drainage directly into a waste container without use of adapters or piping?		
4.119 If the design uses lubrication points with different or incompatible lubricants, are fittings easily distinguished with different mechanical fitting shapes or types?		
4.120 In the design, are components which are subject to steam or solvent cleaning (or random contact during equipment cleaning) properly sealed to prevent interior damage?		
4.121 Does the design locate wiring and cabling and support structures where they do not interfere with maintainer ingress, egress, or access to the product and related equipment?		
4.122 Early in the design process, did the design team define the user maintenance concept and capabilities for the product?		
4.123 During the design process, did the design team define where the product is to be maintained, along with the expectations of who the maintainers are, considering their knowledge, skills, and experience with the product in that maintenance location?		
4.124 During the design process, did the design team define the maintenance resources available and how long maintenance is expected to take, realizing that the time to perform maintenance and the amount of equipment to be maintained must be balanced with the amount of resources available?		
4.125 Early in the design process, did the design team define the maintenance facility work conditions at specified or potential locations where maintenance is expected to be performed?		
4.126 Early in the design process, did the design team identify and define the operational environment the product will experience, to include all operational environments, especially scenarios where operational environments change, such as aircraft and submarines?		
4.127 Did the design team revisit and update the operational maintenance environment throughout the development process?		
4.128 During the design process, did the design team socialize the expected operational maintenance environment with the customer, user, and support communities?		
4.129 Does the design minimize the removal of other equipment to access and repair/replace a faulty part?		
4.130 Does the design provide for remote maintainer connectivity with the product via an electronic communications link to allow external product maintenance support either from close proximity or from long distance, while local maintenance personnel have both hands free to do other things?		

Section 4: Maintenance	VERIFY	REMARKS
4.131 Does the design provide reach back capability to the maintainer for ready electronic access to technical documentation, troubleshooting procedures, checklists, and so on?		
4.132 During the design process, did the design team identify the expected storage, transportation, packaging, and handling environments for the product and create appropriate maintainability requirements to reflect them?		
4.133 During the design process, did the design team identify all lifting restrictions or space restrictions and create requirements to reflect them?		
4.134 Does the design provide easy and simple ways to assess whether the product has been damaged during shipping or handling?		

Checklist Section 5: Software

Section 5: Software	VERIFY	REMARKS
5.1 Does the product software user interface allow for error correction and changes of entered data such that the user is able to change previous entries or selections, by delete and insert actions?		
5.2 Does the product software user interface present only context-appropriate menu selections for actions that are current to the maintainer available and for the user access authority?		
5.3 Does the product software user interface present only menu selections for actions that are applicable to the maintainer access authority?		
5.4 Is the product software user interface consistent in implementation across all functional parts accessed by the maintainer?		
5.5 Whenever the maintainer interacts with the product software, is there clear feedback of what has been done? This feedback may include sensory outputs of any type (such as tactile feedback when pressing a key, auditory feedback when moving a pointer, or visual feedback when selecting an option on a display screen)?		
5.6 Does the product software user interface show visual and/or other feedback indications to the maintainer at the location on the display screen where the action was taken?		
5.7 Does the product software user interface show visual and/or other feedback indications to the maintainer immediately following the action (less than a 3 second delay)?		
5.8 Does the product software include error messages that focus on the procedure for correcting the error, not on the action that caused the error?		
5.9 Does the product software include security protocols that are compatible with the maintainer security environment?		
5.10 Does the product software require personnel training and skills that are compatible with the human resources planned for maintainer support at all planned maintenance levels/locations?		
5.11 Does the product software require maintenance support equipment and support software that are compatible with the equipment resources planned for maintainer support at all planned maintenance levels/locations?		
5.12 Does the product software provide for easy and error-free software updating by the maintainer?		
5.13 Does the product software provide for easy and error-free database updating by the maintainer?		
5.14 Is the product software properly documented in appropriate "best-in-class" technical documentation?		
5.15 Are the product software support and reprogramming environment, and needed resources well documented?		

Checklist Section 6: Troubleshooting

Section 6: Troubleshooting	VERIFY	REMARKS
6.1 Are the product Built-In-Test (BIT) and self-check diagnostic capabilities embedded in the product to reduce the maintenance time and the need to transport and connect to external support equipment?		
6.2 Is the product BIT designed so that it identifies that a true fault exists as close to 100% of the time as possible, to minimize false fault reports and mission aborts?		
6.3 Is the product BIT designed so that it correctly identifies the true faulty component as close to 100% of the time as possible, to minimize maintenance time, providing a high probability of fault detection and isolation?		
6.4 Is the product BIT designed so that it minimizes replacement of parts that are not actually faulty, minimizing the ambiguity group during troubleshooting?		
6.5 Is the product BIT designed to provide fault detection and reporting capabilities that operate in the background to monitor and report operational and equipment problems?		
6.6 Is the product BIT designed to provide the operator and maintainer with a way to run this capability on command?		
6.7 Do the product BIT error messages explicitly provide as much diagnostic information and remedial direction as can be inferred reliably from the error condition?		
6.8 Do the product BIT error messages describe a way to remedy, recover, or escape from the error?		
6.9 Does the product BIT provide a way for the maintainer to request more detailed diagnostic information about the error?		
6.10 Does the product BIT design facilitate rapid fault detection and isolation of defective items?		
6.11 Does the product design provide test points used for adjustment that are located sufficiently close to the controls and displays used in the adjustment so that the maintainer can remain in place during the adjustment process?		
6.12 Does the product design provide test points for adjustment that are physically and visually accessible in the installed condition by the maintainer without removing other items?		
6.13 Does the product design provide for troubleshooting that does not require removal of subassemblies from assemblies?		
6.14 Are the product test points accessible without removing modules and/or components?		
6.15 Are the product BIT and troubleshooting connection points located within the equipment so that maintainers have easy access to important test points without having to struggle or probe into equipment, possibly causing unnecessary damage?		
6.16 Does the product design minimize the number and type of different test equipment required?		
6.17 Does the product design focus on utilizing the existing test equipment of the intended maintainers?		

Checklist Section 7: Safety

Section 7: Safety	VERIFY	REMARKS
7.1 Does the product design include emergency shutoff controls that are readily accessible, positioned within easy reach of the maintainer?		
7.2 Does the product design include emergency shutoff controls that are located to prevent accidental activation?		
7.3 Are the product components located to minimize the possibility of equipment damage or personnel injury?		
7.4 If the product design includes any hazardous areas or conditions, such as exposed high-voltage conductors behind access panels, are there physical barriers over the access area equipped with an interlock that will de-energize the hazardous equipment when the barrier is opened or removed?		
7.5 Appropriate warning labels should also be provided.		
7.6 If the product design includes any hazardous conditions, are both the presence of the hazard and the fact that an interlock exists noted on the equipment case or access cover, such that it remains visible when the access is open?		
7.7 In the product design, are any hazard warnings removed or visually obstructed when opening or removing an access cover?		
7.8 Does the system/product design include features to prevent hazards, such as auxiliary hooks, holders, lights, outlets, nonskid treads, expanded metal flooring, or abrasive coating on surfaces for walking, climbing, or footholds, as appropriate?		
7.9 Where bending is required to facilitate maintenance, does the product design consider the frequency and time in the bent position such as not to cause fatigue or injury?		
7.10 In the product design, are electrical plugs and receptacles identified by color, shape, size, key, or equivalent means to facilitate identification when multiple, similar connectors are used in proximity to each other?		
7.11 In the product design, are all colors used for coding readily discriminable from each other under operational lighting conditions?		
7.12 In the product design, are similar-shaped connectors keyed or coded by using various shapes of matching plugs and receptacles to differentiate connectors and to prevent incorrect placement?		
7.13 In the product design, are test points designed with guards and shields to protect test and service equipment, particularly if the equipment must be serviced while operating?		
7.14 In the product design, are test points designed with guards and shields to protect personnel, particularly if the equipment must be serviced while operating?		
7.15 In the product design, are test points designed with tool guides and other design features to facilitate operation of test or service points which require blind operation?		
7.16 In the product design, are test voltages below acceptable hazardous levels? If not, are appropriate warning labels provided?		
7.17 In the product design, are components positioned such that the maintainers do not have to reach too far for lifting or moving heavy units beyond their strength capability?		
7.18 In the product design, if test stimuli are introduced into a system or component to determine its status, are safeguards in place to ensure that the stimuli will not cause damage to the system or injury to personnel?		

Section 7: Safety	VERIFY	REMARKS
7.19 In the product design, if test stimuli are introduced into a system or component to determine its status, are safeguards in place to ensure that the stimuli will not cause inadvertent operation of critical equipment?		
7.20 Does the product design feature fail-safe precautions in all areas where failure can cause catastrophe or injury through damage to equipment?		
7.21 Does the product design feature fail-safe precautions in all areas where failure can cause injury to personnel?		
7.22 Does the product design feature fail-safe precautions in all areas where failure can cause inadvertent operation of critical equipment?		
7.23 Are the product weights and reach characteristics within allowable human design criteria for the intended maintenance population?		
7.24 Does the product design have clear indications for power-off or power-on?		

Checklist Section 8: Interchangeability

Section 8: Interchangeability	VERIFY	REMARKS
8.1 Does the product design make maintenance-significant parts as interchangeable as possible?		
8.2 Does the product design avoid developing custom sizes, shapes, connections, and fittings for a part or component, and minimize special use parts, such as left side vs. right side connectors?		
8.3 Does the product design have physical size and shape features that preclude connecting parts any way but the right way, and preventing parts from being connected that do not belong together?		
8.4 Does the product design have hole patterns such that it is not possible to mate two or more parts incorrectly?		
8.5 If the product design includes more than one size or type fastener on the same equipment or cover, does the fastener/equipment/cover interface permit the maintainer to readily distinguish the intended location of each fastener?		
8.6 Does the product design ensure that identical fasteners are not used where removal of the wrong fastener can result in equipment damage or change to calibration settings?		
8.7 Does the product design ensure that fasteners with left-hand threads, where required, are identified so they are distinguishable from right-hand threaded fasteners?		

Checklist Section 9: Miscellaneous Topics

Section 9: Miscellaneous Topics	VERIFY	REMARKS
9.1 Has the design maintainability model been maintained in sync with the design of the product, and does it accurately reflect the actual product design?		
9.2 Was a Failure Modes and Effects Analysis available and used as a part of the testability analysis?		
9.3 Was a Level of Repair Analysis used as part of the testability analysis?		
9.4 Was the product designed so maintainers can readily perform all assigned maintenance tasks wearing Personal Protective Equipment (PPE), Mission Oriented Protective Posture (MOPP) gear, cold weather gear, rain gear, hazardous material gear, biohazard gear, or any other protective items that might be needed?		
9.5 Was the product designed based on the worst-case maintenance in various environmental stress conditions and scenarios?		
9.6 During the preparation for design reviews, were checklists for the maintainability requirements used to audit the design status and readiness for the reviews?		
9.7 Does the product or system design maximize use of standardized common usage piece parts and components, such as screws, washers, nuts, and bolts, to minimize spare parts inventory for customer repair service in the various field applications?		
9.8 Does the product or system design maximize use of a standard design template composed of standard dimensions, shapes, sizes, power, interfaces (mechanical and electrical), signal connections, and modular elements?		
9.9 Does the product or system design plan to collocate and package all piece parts and components required for a particular maintenance task into a single kit?		

Index

Design for Maintainability, First Edition. Edited by Louis J. Gullo and Jack Dixon.

d

g

h

i

j

k

l

n

o

p

q

r

s

t